19th Coherent Laser Radar Conference (CLRC 2018)

Okinawa, Japan
18 - 21 June 2018

ISBN: 978-1-5108-7033-8

Printed from e-media with permission by:

Curran Associates, Inc.
57 Morehouse Lane
Red Hook, NY 12571

Some format issues inherent in the e-media version may also appear in this print version.

Printed by Curran Associates, Inc. (2018)

For permission requests, please contact Cooperative Institute for Research in Environmental Sciences (CIRES) at the address below.

Cooperative Institute for Research in Environmental Sciences (CIRES)
University of Colorado Boulder
216 UCB
Boulder, CO 80309
USA

Phone: 303-492-1143

cires.colorado.edu

Additional copies of this publication are available from:

Curran Associates, Inc.
57 Morehouse Lane
Red Hook, NY 12571 USA
Phone: 845-758-0400
Fax: 845-758-2633
Email: curran@proceedings.com
Web: www.proceedings.com

TABLE OF CONTENTS

Coherent Laser Radar - from global wind sensing to the detection of Gravitational Waves

Robert L. Byer
Stanford University

Abstract: Coherent Laser Radar based on solid state lasers was motivated by the need for efficient, highly coherent lasers on orbit for observation of global winds. The challenge led to the invention of the monolithic non-planar ring oscillator (NPRO) in Nd:YAG. The Laser Interferometer GravitationalWave Observatory (LIGO) selected Nd:YAG lasers to power the 4km-scale interferometers. Today, advanced LIGO uses the NPRO and a 180W Nd:YAG power amplifier. The detection of gravitational waves opened a new era of multi-messenger astronomy with observations in gravitational waves (audio frequency region) and electromagnetic waves. The principles of coherent laser radar are now powering not only engineering applications in transportation and communications but also fundamental science and astrophysics applications at unprecedented resolutions and sensitivities.

ESA Wind Lidar mission AEOLUS

A. Dabas (a), T. Flament (a), F. de Bruin (b), A. Straume (b), T. Kanitz (b), O. Reitebuch (c), U. Marksteiner(c), D. Huber (c), L. Isaksen (d), M. Rennie (d), A. Stoffelen (e), G.-J. Marseille (e), J. de Kloe (e), O. Lecrenier (f), J. Lochard (f), P.McGoldrick (g)

(a) Météo-France
Toulouse, France
(b) European Space Agency
Noordwijk, The Netherlands.
(c) Deutsches Zentrum für Luft und Raumfahrt
Oberpfaffehoffen, Germany.
(d) European Center for Medium Range Weather Forecast
Reading, United Kingdom.
(e) Koninklijk Nederlands Meteorologisch Instituut
De Bilt, The Netherlands.
(f) Airbus Defence and Space
Toulouse, France
(g) Airbus Defence and Space
United Kingdom.

Lead Author e-mail address: alain.dabas@meteo.fr

Abstract: AEOLUS is a space-based wind lidar. Developed in the frame of the Earth Explorer program of the European Space Agency, it will be launched in late August 2018. Build by Airbus Defence and Space, it will be the first Doppler lidar in space. It will measure winds all over the globe from the surface to the low stratosphere, as well as the optical properties of clouds and aerosols. Its data shall be used for weather and air quality prediction and help better understand the climate of the Earth.

Keywords: lidar, space, wind, clouds, aerosols

1. AEOLUS mission

AEOLUS should be launched on the 21st of August from Kourou, French Guyana. The satellite will bear the Doppler wind lidar ALADIN as single instrument. The lidar ALADIN operates in the UV at $\lambda = 355nm$. It will implement two detection channels, one designed for the detection of the molecular return from the atmosphere, and the other one for the return of aerosol or cloud particles. ALADIN is thus a Mie-Rayleigh lidar.

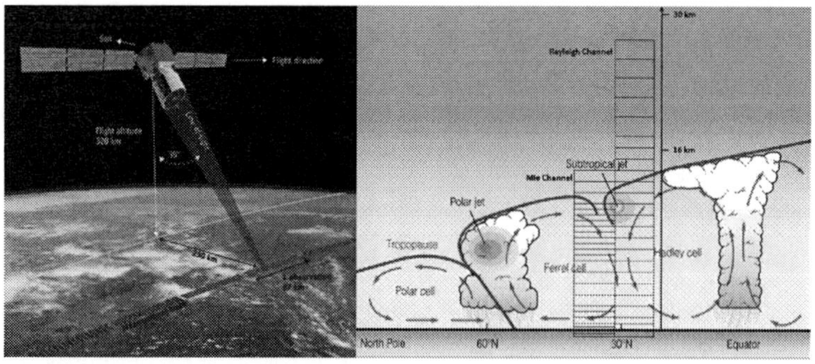

Figure 1: Mesurement geometry of AEOLUS. The satellite altitude is 320km. The lidar beam is 35 degrees off-nadir in the East-West direction. The standard data granularity is 87km. The return signals are vertically integrated on board in height-bins. Height-bins are 250m to 2km thick.

Laser	Wavelength	$\lambda = 354.8\ nm$
	Energy/pulse	$80mJ$
	Pulse repetition frequency	$50Hz$
Telescope	Type	Cassegrain
	Diameter	$\emptyset = 1.5m$
Rayleigh receiver	Type	Double Fabry-Perot
	Free-Spectral Range	$FSR = 10938\ MHz$
	Full-Width-Half-Max	$FWHM \approx 1550\ MHz$
Mie receiver	Type	Fizeau
	Free-Spectral Range	$FSR = 2190\ MHz$
	Full-Width-Half-Max	$FWHM \approx 160\ MHz$

Table 1: Main characteristics of ALADIN

AEOLUS will fly on a sun-synchronous orbit, crossing the equator at about 6 o'clock local solar time. Originally at 400km, the orbit altitude has been recently lowered to 320km. The beam of the lidar is directed 35° off-nadir and is mostly east-west (see Figure 1). The signals returned by the atmosphere are integrated in height-bins of settable thicknesses from 250m up to 2km. The number of height-bins is Mie and Rayleigh channels is 24. The standard horizontal granularity of data is 87km (called observation), but raw data are downlinked with a finer granularity of ~3km (measurement).

The major characteristics of the lidar are listed in Table 1 and the overall optical arrangement of the receiver is depicted in Figure 2. The light captured by the telescope is directed first to the Mie receiver (Fizeau). The reflection on the Fizeau then illuminates the Reyleigh receiver (double Fabry-Pérot).

The raw data are downlinked once per orbit. Level 2 data shall be made available to users in less than 3 hours worldwide (near real-time), and even less than 1.5 hours in Europe (quasi real-time) so they can be used for weather or air-quality prediction. Data will be distributed via the Archive and Data Dissemination Facility (ADDF) set up by ESA.

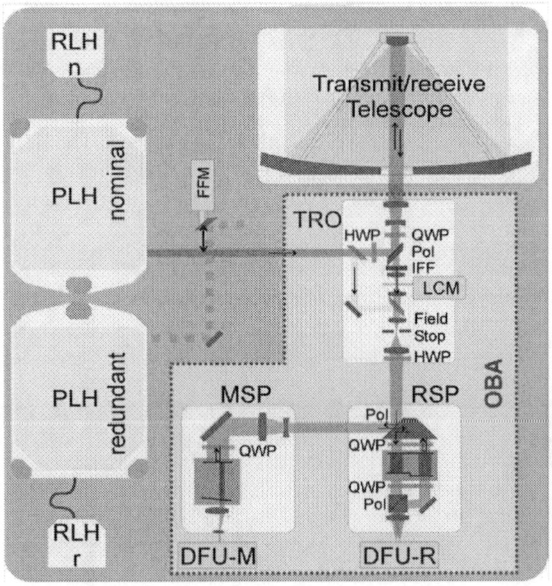

Figure 2: Optical arrangement of ALADIN (PLH: Power Laser Head, RLH: Reference Laser Head, FFM: Flip-Flop Mirror, LCM: Laser Chopper Mechanism, TRO: Transmit/Receive Optics, RSP: Rayleigh Spectrometer, DFU: Detection Front-end Unit, QWP: Quarter-Wave Plate, HWP: Half-Wave Plate, Pol: Polariser, IFF: Interference Filter.

2. Wind measurements

Winds measurements are derived from both the Rayleigh and the Mie channel. In the Rayleigh channel, the wind Doppler shift is obtained as explained in Figure 3 [1]. The number of photons N_A and N_B out the two Fabry-Perot interferometers FPA and FPB are counted. The Rayleigh response $R = (N_A - N_B)/(N_A + N_B)$ is formed and inverted using a calibration curve. In practice, the response of the Rayleigh receiver also depends on the pressure and temperature via the shape and width of the Rayleigh spectrum [2], and the presence of aerosols. The Level 2B processor of the mission is in charge of the inversion taking into account all these factors [3]. It uses pressures and temperatures from numerical weather prediction models. The L2B processor of the mission will be run at the European Center for Medium Range Weather Forecast, but a portable version can be installed at any meteorological service that desires to implement its own version within its weather numerical prediction system.

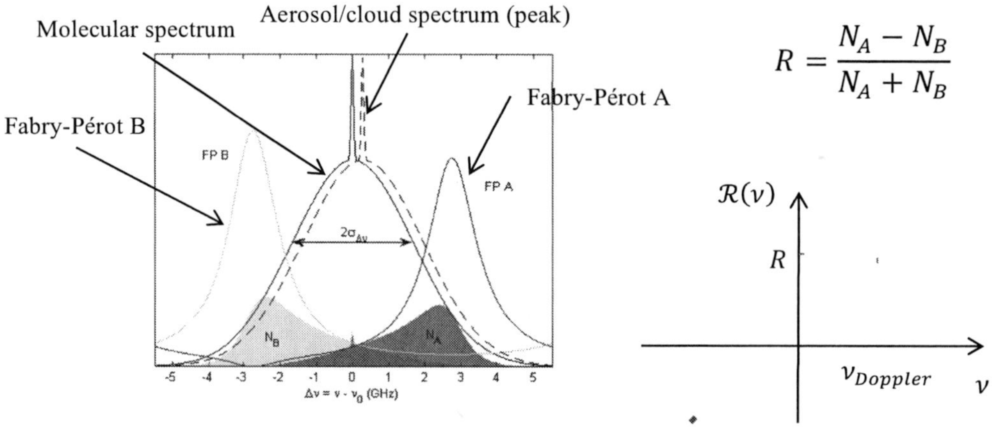

Figure 3: Diagram showing how the frequency Doppler shift is estimated from the double Fabry-Pérot of the Rayleigh channel. The green and blue lines on the left are the transmission curves of the two Fabry-Perot interferometers FP A and FPB. In the x-axis, the frequencies are counted relative to the laser emission. The number of photons out of FPA and FPB are N_A and N_B.

The derivation of winds from the Mie channel is depicted in Figure 4. The light captured by the telescope illuminates the Fizeau. The Mie (particulate) peak forms a fringe inside the Fizeau at a position that depends on its frequency. The fringe is imaged on a CCD and its position is retrieved from the CCD counts.

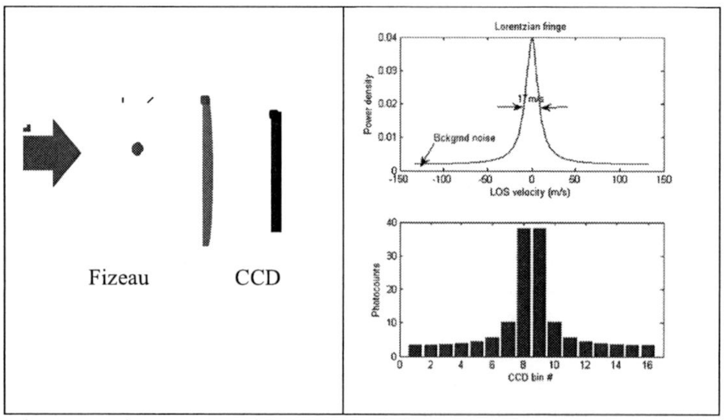

Figure 4: Diagram showing how the Mie receiver measures winds.

3. Cloud and aerosols products

The Fizeau and double Fabry-Perot interferometers in the Mie and Rayleigh channels act respectively as a narrow and a wide-band filter. The narrow filter is mostly sensitive to the aerosol/cloud backscatter as it forms a narrow peak in the returned spectrum. The wide-band filter is sensitive to both molecular and particle backscatter. Denoting β_p, α_p, β_{mol}, and α_{mol} the particulate and molecular backscatter and extinction, the signals S_{Mie} and S_{Ray} detected in the Mie and Rayleigh channels can be written

$$S_{Mie} = \frac{K_{Mie}}{R^2}\left[C_1\beta_{mol} + C_2\beta_p\right]\exp\left(-2\int \alpha_{mol}(x) + \alpha_p(x)dx\right)$$

$$S_{Ray} = \frac{K_{Mie}}{R^2}\left[C_1\beta_{mol} + C_2\beta_p\right]\exp\left(-2\int \alpha_{mol}(x) + \alpha_p(x)dx\right)$$

with

$$C_1 > C_2 \quad and \quad C_3 > C_4$$

The system can be easily inverted

$$X_{mol} = \beta_{mol}\exp\left(-2\int \alpha_{mol}(x) + \alpha_p(x)dx\right) \qquad X_p = \beta_p\exp\left(-2\int \alpha_{mol}(x) + \alpha_p(x)dx\right)$$

From which the particulate backscatter and extinction can be obtained

$$\beta_p = \frac{X_p}{X_{mol}}\beta_{mol} \qquad \alpha_p = \frac{1}{2}\frac{d}{dz}\ln\left(\frac{X_{mol}}{\beta_{mol}}\right) - \alpha_{mol}$$

The retrieval of cloud/aerosol products is done by the Level 2A processor of the mission. Due to the insensitivity of the receiver to cross-polarized light, the backscatter coefficient measured by ALADIN applies to the parallel-polarization.

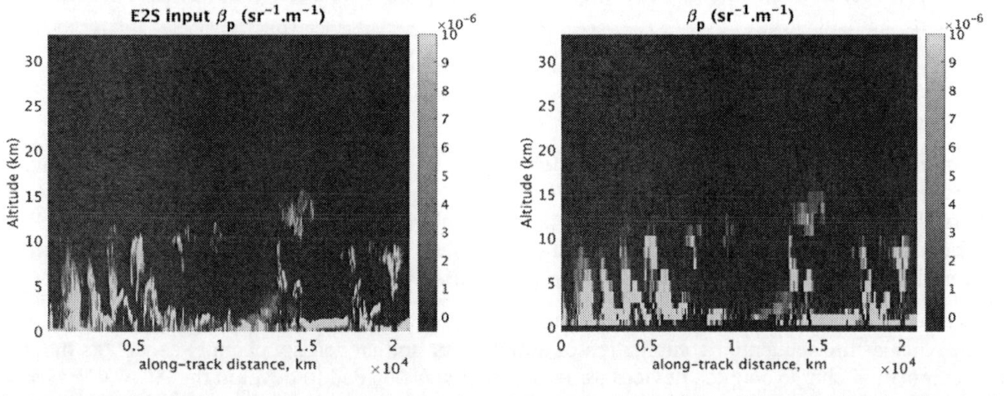

Figure 5: Particulate backscatter retrieved by the L2A processor (right) applied to data simulated with the End-to-End Simulator (E2S) of AEOLUS on the basis of particulate backscatter measured by the space-borne lidar CALIPSO in the visible and converted in the UV.

4. References

[1] Souprayen, C., A.Garnier, A. Hertzog, A. Hauchecorne, J. Porteneuve, « Rayleigh-Mie Doppler wind lidar for atmospheric measurements. I. Instrumental setup, validation, and first climatological results. Applied Optics, **38**, 2410-2421 (1999).

[2] Witschas B., "Analytical model for Rayleigh–Brillouin line shapes in air," Appl. Opt. 50, 267–270 (2011).

[3] A. Dabas, M. Denneulin, P. Flamant, C. Loth, A. Garnier, and A. Dolfi-Bouteyre, "Correcting winds measured with aRayleigh Doppler lidar from pressure and temperature effects," Tellus A 60, 206–215 (2008).

Results from recent airborne campaigns aiming at the preparation of ESA's Aeolus Wind Lidar Mission

Benjamin Witschas, Christian Lemmerz, Oliver Lux, Uwe Marksteiner, Stephan Rahm, Oliver Reitebuch, Andreas Schäfler, Fabian Weiler

Deutsches Zentrum für Luft- und Raumfahrt (DLR), Institute of Atmospheric Physics, 82234 Oberpfaffenhofen, Germany

Benjamin.Witschas@dlr.de

Abstract: Almost 20 years after its selection as an Earth Explorer Core Mission of ESA's Living Planet Program, Aeolus is ready to be launched in August 2018. Aeolus will carry the first wind lidar instrument in space, ALADIN (Atmospheric Laser Doppler Instrument), aiming at the measurement of wind profiles globally from ground up to 25 km.
To validate ALADIN's measurement principle and calibration routines, DLR has developed an airborne Calibration and Validation (Cal/Val) payload based on a direct detection wind lidar (355 nm), and a coherent detection wind lidar (2 µm) which was deployed in several field campaigns over the last years.
From September to October 2016, as part of the international campaign NAWDEX (North Atlantic Waveguide and Downstream Impact Experiment), 14 flights were performed from the airport of Keflavik, Iceland. In this paper, an overview of results obtained during NAWDEX and their conclusion on the Aeolus mission will be discussed.

Keywords: Aeolus, Wind Lidar, Coherent Wind Lidar, Direct Detection Wind Lidar, Airborne Wind Lidar Measurements

1. Introduction

Accurate knowledge of wind profiles on a global scale is of highest priority in order to improve the quality of numerical weather prediction [1]. For that reason, the European Space Agency ESA initiated the Aeolus mission in 1999 as an Earth Explorer Core Mission of ESA's Living Planet Program. Aeolus will carry the direct-detection Doppler Wind Lidar (DWL) ALADIN which will deliver global wind profiles from ground up to 25 km after being launched in August 2018.

As the ALADIN instrument combines various techniques that were not implemented in wind lidars before, such as the sequential combination of a molecular and an aerosol channel as well as the usage of accumulation charge coupled devices as detectors, it was decided to develop the ALADIN Airborne demonstrator (A2D) in 2003 [2], which has since then been deployed in several field campaigns aiming at the validation of the satellite instrument as well as the optimization of its retrieval algorithms and calibration schemes.

To complete the airborne Cal/Val payload for ALADIN, DLR additionally implemented a coherent DWL onboard the DLR-Falcon aircraft which serves as a reference system for the A2D and ALADIN and which has been operated at DLR already since 1999. The 2-µm DWL has contributed to a variety of atmospheric-science related airborne campaigns e.g. [3,4] as well as to Aeolus pre-launch Cal/Val campaigns [5-7].

As a rehearsal for the upcoming Cal/Val activities foreseen to be performed after the satellite launch, the Aeolus Cal/Val payload was recently deployed in several field campaigns. In May 2015, the joint ESA-DLR-NASA airborne campaign WindVal I was conducted based in Keflavík, Iceland. For the first time, four DWLs were operated together onboard the NASA DC-8 and the DLR-Falcon aircraft.

One and a half years later, in autumn of 2016, the Aeolus Cal/Val payload was employed in the framework of the NAWDEX campaign, again based in Keflavík. This international field campaign focused on investigating the influence of diabatic processes on the evolution of disturbances of the North Atlantic jet stream. For this purpose, four research aircraft equipped with diverse payloads were employed which allowed for the observation of a large set of atmospheric parameters using a multitude of state-of-the art remote sensing instruments [5,6].

With a view to the forthcoming Aeolus mission, the NAWDEX campaign was an ideal platform for extending the wind dataset obtained with the Aeolus Cal/Val payload, as it offered the opportunity to perform wind measurements in dynamically complex scenes, including strong wind shear, varying cloud conditions and high wind speeds in the vicinity of the jet stream. In addition, the unique setting in Iceland with its large-scale cooperation of atmospheric research groups from around the world was beneficial for the preparation of the upcoming launch of Aeolus.

In this paper, an overview of the wind measurements of the A2D and the 2-µm DWL performed during NAWDEX is given. A statistical comparison of the A2D and 2-µm DWL winds is done based on a wind scene measured on 4 October 2016 illustrating the good agreement of the wind data acquired with both DWLs.

2. DLR's Aeolus Cal/Val payload

DLR's Aeolus Cal/Val payload consists of the direct detection DWL A2D and the coherent detection 2-µm DWL, both integrated onboard the DLR-Falcon research aircraft. A picture of the Falcon itself and of both wind lidars integrated within the aircraft is shown in Figure 1.

Details about the A2D instrumental setup and wind retrieval procedures are given in [2,6,7], and more specifically, the laser frequency stability is discussed in [8]. Details about the 2-µm DWL instrumental setup and wind retrieval routines can be found in [3,9].

Figure 1. The left picture shows the three aircraft operated during the NAWDEX campaign from the airport in Keflavik, Iceland, namely DLR's Falcon (left), the HALO aircraft (middle) and the French SAFIRE Falcon (right). The right picture illustrates the Aeolus Cal/Val payload integrated in DLR's Falcon aircraft with the 2-µm and A2D DWL including data acquisition unit (DAQ).

3. Wind measurements during the NAWDEX campaign

On 4 October 2016 a flight was performed in order to measure the high wind speeds and wind speed gradients in the vicinity of the jet stream which was located south-east of Iceland. For this purpose, the Falcon crossed the jet stream twice, as it flew two legs back and forth between the way points located at 66.0°N, 17.5°W and 64.0°N, 7.0°W (see Figure 2, left). The corresponding wind speed measured with the 2-µm DWL during the entire flight is shown in Figure 2, right. It can be seen that the 2-µm

DWL could already measure during ascent from about 8 km altitude on. Furthermore, the pronounced jet-stream structure with wind speeds up to 72 m/s becomes obvious from the wind measurements (~800 km to 1600 km). The white areas indicate regions with low aerosol content and thus, no valid wind data that could be retrieved with the coherent detection system here. Furthermore, the vertical data gaps at around 600 km and 1900 km are caused by a failure of the GPS module that led to a stop of the data acquisition and thus to missing data in this region.

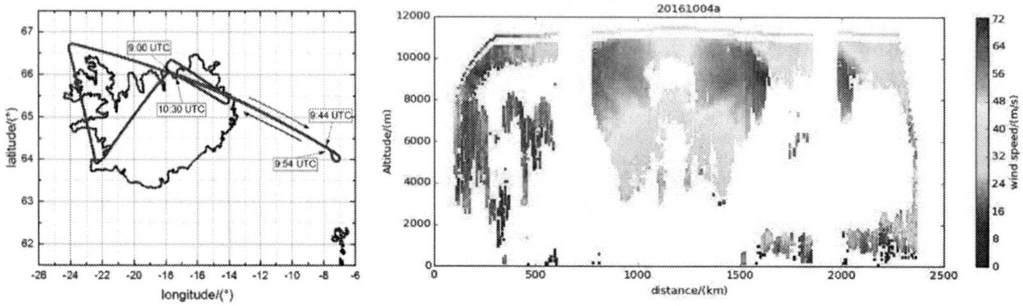

Figure 2. The flight track flown on 4 October 2016 (left) and the corresponding horizontal wind speed measured with the 2-μm DWL (right).

At the same flight, the A2D measured wind profiles during the periods from 09:00 to 09:44 UTC (~750 km to 1100 km) and from 09:54 to 10:30 UTC (~1300 km to 1650 km) that can be used for comparison with 2-μm DWL data. However, as the A2D only measures the line-of-sight (LOS) wind speed at an off-nadir angle of 20°, the 2-μm data first has to be projected on the A2D viewing direction. Furthermore, as both systems have a different temporal resolution (A2D ~ 14 s for one LOS, 2-μm ~ 42 s for the wind vector), a weighted interpolation has to be performed before the data can be compared. Details about the projection and weighted interpolation can be found in [6,7].

The A2D wind speed obtained from the Mie and Rayleigh channel (combined), as well as the 2-μm DWL winds projected onto the A2D LOS are shown in Figure 3 top and bottom, respectively.

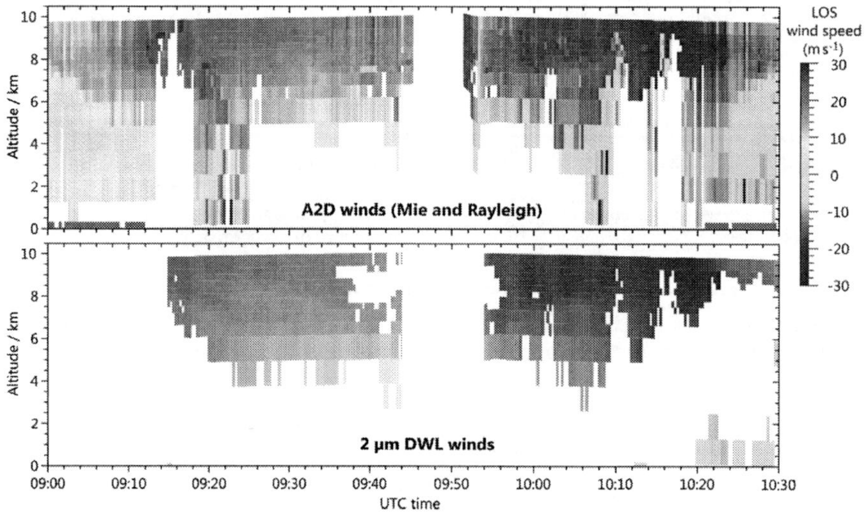

Figure 3. LOS wind profiles during the flight on 4 October 2016 between 09:00 UTC and 10:30 UTC from A2D (top panel – combined Mie and Rayleigh winds) and the 2-μm DWL (bottom).

Since the wind was blowing towards the A2D LOS on the first leg, positive LOS wind speeds of up to 25 m/s (projection to horizontal LOS: 73 m/s) were measured, whereas negative winds of the same magnitude were detected on the flight leg back to Iceland. Furthermore, it can be seen that the coverage for A2D and 2-µm DWL winds is different. 2-µm DWL winds are available in the vicinity of the jet stream where cirrus clouds are prominent and lead to particulate backscatter. Thus, this is also the region where winds from the A2D Mie channel are available. In addition to that, the A2D also provides wind measurements in the aerosol-poor regions before and after the jet stream from the Rayleigh channel based on molecular scattering. Hence, this nicely demonstrates the complementarity of the Mie and the Rayleigh channel and at the same time shows the limits of a coherent detection DWL system.

The wind measurements shown in Figure 3 are now used for a statistical comparison as indicated by Figure 4. Here, the A2D winds are plotted versus the corresponding interpolated 2-µm DWL winds, resulting in a cloud of data points that ideally lie on the dashed line representing $v_{A2D} = v_{2\mu m}$. The non-weighted linear fit $v_{A2D} = A \cdot v_{2\mu m} + B$ through the real data provides values for the slope A and intercept B. Before comparison, eleven gross outlier with $v_{A2D} - v_{2\mu m} > \pm 10$ m/s were excluded from the Rayleigh wind data.

As the 2-µm relies on particulate backscatter, a more sophisticated comparison can be performed for Mie winds (Figure 4, blue points). In particular, 1246 bins are available for comparison. Still, due to the very high sensitivity of the coherent detection of the 2-µm DWL, even 168 valid bins are available for comparing wind from the A2D Rayleigh channel (Figure 4, green points).

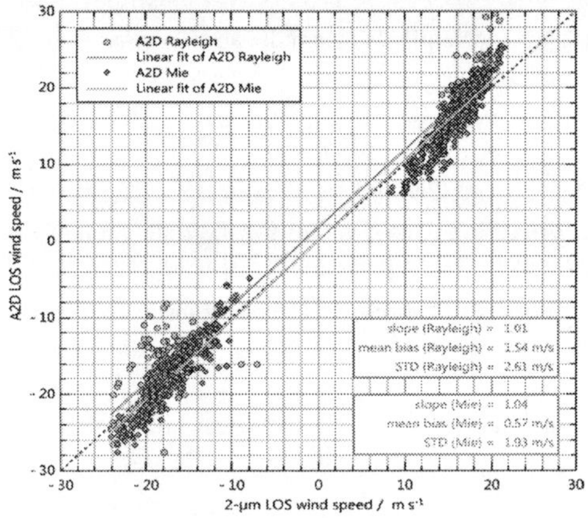

Figure 4. Statistical comparison between LOS winds measured by A2D and the 2-µm DWL during the flight on 4 October 2016.

It can be seen that the comparison of A2D and 2-µm data shows good agreement. For the Mie winds, the mean bias is 0.57 m/s and for the Rayleigh winds it is 1.54 m/s. The corresponding standard deviations are 1.93 m/s and 2.61 m/s. As already mentioned, the comparison to Rayleigh winds is more difficult as only few 2-µm DWL data is available in the clear atmosphere. Hence, the overlap of valid data from both systems is restricted to regions where weak particulate backscattering occurs which is still sufficient to be analyzed by the sensitive 2.µm DWL, while significant Mie contamination of the A2D Rayleigh signal is negligible according to the current wind retrieval settings.

The shown result demonstrates that DLR has a well-established Aeolus Cal/Val payload based on two different DWLs that is perfectly suited for the upcoming Cal/Val activities after the satellite has been launched in August 2018. In addition to the comparison shown here, further comprehensive comparisons to measurements from dropsondes launched form the HALO aircraft and to wind measurements from radar and lidar flown onboard the French SAFIRE Falcon were performed and confirmed the good performance of the 2-µm DWL and the A2D based on the NAWDEX data set.

Acknowledgements

The development of the ALADIN Airborne Demonstrator and the work carried out during the WindVal I and II were supported by the German Aerospace Center (Deutsches Zentrum für Luft- und Raumfahrt e.V., DLR) and the European Space Agency (ESA), providing funds related to the preparation of Aeolus (WindVal, contract No. 4000114053/15/NL/FF/gp).

4. References

[1] Andersson E., Statement of Guidance for Global Numerical Weather Prediction (NWP), World Meteorological Organisation, 2016, https://www.wmo.int/pages/prog/www/OSY/SOG/SoG-Global-NWP.pdf (accessed on 20 February 2018)

[2] Reitebuch, O., Lemmerz, C., Nagel, E., Paffrath, U., Durand, Y., Endemann, M., Fabre F., and Chaloupy, M., "The airborne demonstrator for the direct-detection Doppler wind lidar ALADIN on ADM-Aeolus. Part I: Instrument design and comparison to satellite instrument", J. Atmos. Oceanic Technol., **26**, 2501-2515, 2009.

[3] Witschas, B., Rahm, S., Dörnbrack, A., Wagner, J., and Rapp, M., "Airborne Wind Lidar Measurements of Vertical and Horizontal Winds for the Investigation of Orographically Induced Gravity Waves", J. Atmos. Oceanic Technol., **34**, 1371–1386, 2017.

[4] Weissmann, M., Busen R., Dörnbrack A., Rahm S. and Reitebuch O., "Targeted observations with an airborne wind lidar", J. Atmos. Oceanic Technol., **22**, 1706–1719, 2005.

[5] Schäfler A., et al., "The North Atlantic Waveguide and Downstream Impact Experiment", Bull. Am. Meteorol. Soc., in press, 2018.

[6] Lux, O., Lemmerz, C., Weiler, F., Marksteiner, U., Witschas, B., Rahm, S., Schäfler, A., and Reitebuch, O., "Airborne wind lidar observations over the North Atlantic in 2016 for the pre-launch validation of the satellite mission Aeolus", Atmos. Meas. Tech. Discuss., https://doi.org/10.5194/amt-2018-19, in review, 2018.

[7] Marksteiner, U., "Airborne Wind-Lidar Observations for the Validation of the ADM-Aeolus Mission", Ph. D Thesis at Technical University Munich, DLR FB 2013-25, 2013.

[8] Lemmerz, C., Lux, O., Reitebuch, O., Witschas, B., and Wührer, C., „Frequency and timing stability of an airborne injection-seeded Nd:YAG laser system for direct-detection wind lidar", Appl. Opt., **56**(32), 9057-9068, 2017.

[9] Chouza, F., Witschas, B., and Reitebuch, O., "Heterodyne high-spectral-resolution lidar", Appl. Opt. **56**, 8121-8134, 2017.

Evaluation of potential impacts of future Japan's space-based Doppler Wind Lidar (DWL) on polar- and tropical-orbiting satellites

Kozo Okamoto(a,b), Toshiyuki Ishibashi(a), Shoken Ishii(b), Philippe Baron(b), Kyoka Gamo(c), Taichu Tanaka(a), Koji Yamashita(d), Takuji Kubota(e)

(a) Meteorological Research Institute of Japan Meteorological Agency, Tsukuba, Japan
(b) National Institute of Information and Communications Technology, Kobanei, Japan
(c) FUJITSU FIP Corporation, Tokyo, Japan
(d) Meteorological Satellite Center of JMA, Tokyo, Japan
(e) Japan Aerospace Exploration Agency, Tsukuba, Japan
kokamoto@mri-jma.go.jp

Abstract: The feasibility of coherent Doppler wind lidars (DWLs) has been investigated based on a sensitivity observation system simulation experiment (OSSE). The pseudo-truth atmospheric field is generated from the Sensitivity Observing System Experiment (SOSE) approach. Hourly aerosols are produced to simulate DWL by a global aerosol chemical transport model in which wind field is nudged with pseudo-truth. Simulated measurement of horizontal line-of-sight wind speeds is assimilated by using the four-dimensional variational (4D-Var) scheme of the operational global data assimilation system at Japan Meteorological Agency. We evaluated potential impacts of DWLs onboard polar and low-inclination orbiting satellites in January and August in 2010. We found positive impacts of DWLs on either satellites and greater impacts in the January experiment. The results also showed seasonal dependency of impacts and importance of quality control and observation error setting.

Keywords: Data assimilation, OSSE, satellite

1. Introduction

The 3D global wind observations are essential for numerical weather prediction (NWP) but current global observation network does not satisfy the NWP needs. A space-based Doppler wind lidar (DWL) is one of good candidates and its feasibility study and development have been underway in many studies. The Japanese scientific community is also investigating the impact of DWLs on NWP by using an observing system simulation experiment (OSSE). Compared with many previous studies on the OSSE for DWLs, this study simulated hour-by-hour aerosol fields and then realistic DWL data distributions and quality information, which allowed us to discuss the importance of quality control and observation error setting in a data assimilation system. The paper gives a brief description on our OSSE approach and results of DWL assimilation. The detail is found in [1].

2. OSSE configuration

We adopted the OSSE based on a sensitivity observing system experiment (SOSE) approach [2] because it could use all existing observations as opposed to a traditional nature-run OSSE that requires simulating them. Our SOSE-OSSE configuration is shown in Fig.1. First, we simulated DWL winds and their quality information such as the signal-to-noise ratio from a pseudo-truth (PT) atmospheric field, and clouds and aerosols. The PT field was constructed by correcting an original initial (analysis) field using the adjoint sensitivity structure and by merging it with existing observations. Clouds were generated from the global forecast model through data assimilation cycle of the PT generation process. Aerosols were calculated using a global chemical transport model [3] nudged with PT winds. PT atmospheric field, clouds and aerosols were generated hourly and consistent each other.

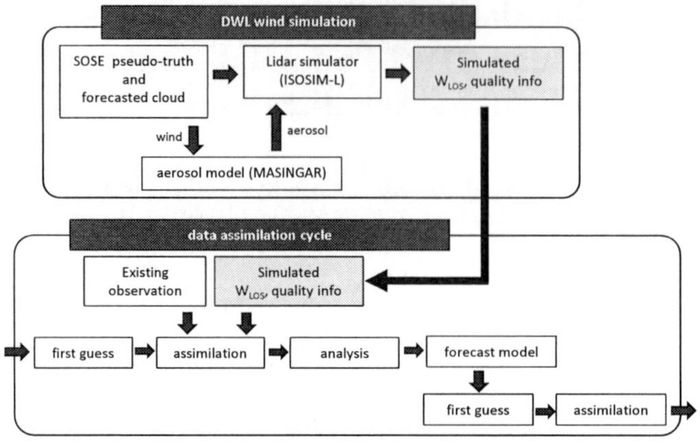

Figure 1. SOSE-OSSE scheme. The upper block is an offline simulation process that produces PT atmospheric fields, clouds, and aerosols and then simulates DWL winds by using ISOSIM-L. The lower block is a regular assimilation process, except for adding DWL simulations.

The end-to-end lidar simulator, named the Integrated Satellite Observation Simulator for a Space-Borne Coherent Doppler Lidar (ISOSIM-L) calculates backscattered power, background noise power, signal-to-noise ratio (SNR), noisy Doppler-shifted signal, retrieved winds in the direction of line-of-sight, and retrieval quality information. With respect to DWL, we assumed two coherent receivers pointing toward the ground at 35° off-nadir with azimuthal angles of 45° and 135° along the satellite track. Laser wavelength, pulse energy and repetition frequency are 2.0 μm, 125 mJ, and 30 Hz, respectively. Target vertical resolution and wind speed accuracy are set to 0.5 km and 1 m s^{-1} at an altitude between 0 and 3 km, 1 km and 3 m s^{-1} between 3 and 8 km, and 2 km and 3 m s^{-1} between 8 and 20 km. A candidate platform for the DWL is a super-low-altitude satellite flying at 220 km or lower. The detail of ISOSIM-L and DWL instruments are described in [4,5], respectively.

3. Data assimilation experiments

We assimilated simulated observations of horizontal line-of-sight (HLOS) wind speed. In the preprocessing procedures of data assimilation, we removed the anomalous data flagged by ISOSIM-L, low SNR data and data substantially departed from first-guess (short-range forecast). The quality control (QC) procedures with rigid thresholds successfully removed DWL winds that disagreed from PT: 85 % of all data were excluded (mainly data above 10 km and near the surface), of which 60 % by SNR-based QC. Observation error R was defined with $R = C\sqrt{E_m^{\,2} + 1.0^2}$: E is measurement error estimated from ISOSIM-L according to altitude and season. Other error components such as observation operator and representativeness error were assumed to be 1.0 m s^{-1}. Finally, we inflated the observation error by multiplying an experimental factor of C (now set to 4.0) to avoid degradation of forecast skills through data assimilation experiment. This degradation was probably attributed to inappropriate treatment of oversampling of DWLs and PT error.

We conducted analysis–forecast cycle experiments to assess the impacts of the DWL data simulated by ISOSIM-L in two one-month assimilation experiments in January and August 2010. The data assimilation system was the low-resolution version of the operational global data assimilation system of JMA as of 2010 [6]. The analysis system was an incremental 4D-Var method with an inner loop of horizontal resolution of 120 km and outer loop of 60 km. We performed 6 h data assimilation cycles from December 20, 2009, to February 7, 2010, (January experiment) and from July 20, 2010, to September 9, 2010 (August experiment). We also ran 120 h forecasts at 12:00 UTC from January 1 to 31, 2010, for the January experiment and from August 1 to 31, 2010, for the August experiment. The results are shown from three experiments with different observation configurations. One is a reference experiment, denoted as CNTL, where all of the observations used in the operational system were assimilated. The second experiment, (TESTP) assimilated DWL HLOS winds from the polar-orbiting satellite in addition to CNTL observations. The third experiment (TESTT) assimilated DWL HLOS winds from tropical satellite with low-inclination of 35.1° in addition to CNTL observations.

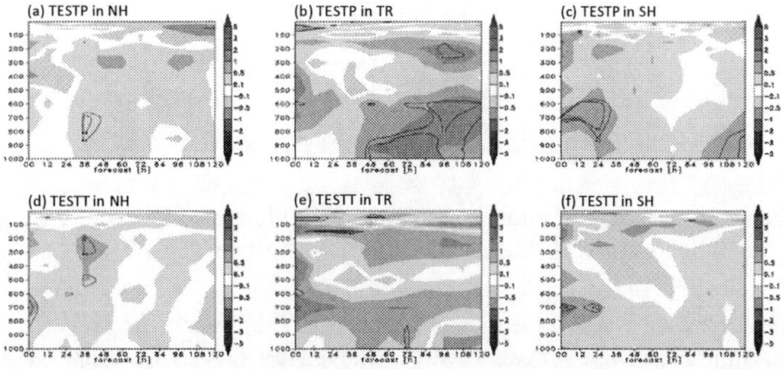

Figure 2. Relative forecast error reduction (%) of zonal wind as a function of forecasts up to 120 h for (upper) TESTP and (lower) TESTT for the January experiments. It is verified against PT in (a, d) the northern hemisphere (20°N–90°N), (b, e) tropics (20°N–20°S), (c, f), and southern hemisphere (20°S–90°S). Positive values correspond to forecast improvement in TESTs. The contour lines indicate the statistical significance at the 90% and 95% confidence levels.

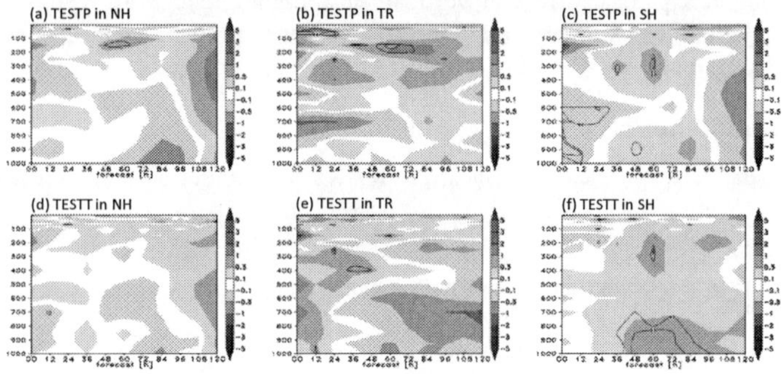

Figure 3. Similar to Fig. 2 but for the August experiment.

Figure 2 shows the relative forecast error reduction of zonal wind for TESTP and TESTT in the January experiment. The relative forecast error is defined with root mean square (RMS) difference from PT field and normalized by CNTL RMS forecast error. Positive impact (error reduction) is evident in both hemispheres and in the tropics through wide tropospheric layers over a broad forecast range up to day 5 for TESTP and TESTT experiments. Improvement is clear, particularly in the upper and lower troposphere in the tropics, with statistical significance. Figure 3 shows that impacts for the August experiment are also generally positive with some exceptions of negative impact in short-range forecasts in the southern hemisphere for TESTP. Because we found simulated DWL quality was a little lower in August (not shown), those data probably contaminated analyses and then forecasts even after the rigid QC. This suggests that more appropriate QCs, such as having more situation dependency, should be developed. We found clear positive impacts on temperature forecasts for TESTP and TESTT in the January and August experiments (not shown). The track forecasts of tropical cyclones (TCs) was slightly improved for both TESTP and TEST when their center positions of 18 cases for 5 TCs were verified against best track dataset. Finally, we directly compared the impacts between TESTP and TESTT. The direct comparison of TESTT and TESTP in Fig. 4 shows that relative impacts are mixed and, interestingly, that TESTT is superior to TESTP up to Day 2 but is inferior beyond Day 2 in the tropics for both the January and August experiments. We need more investigation to make a firm conclusion on whether TESTT or TESTP are superior.

4. Summary and final comments

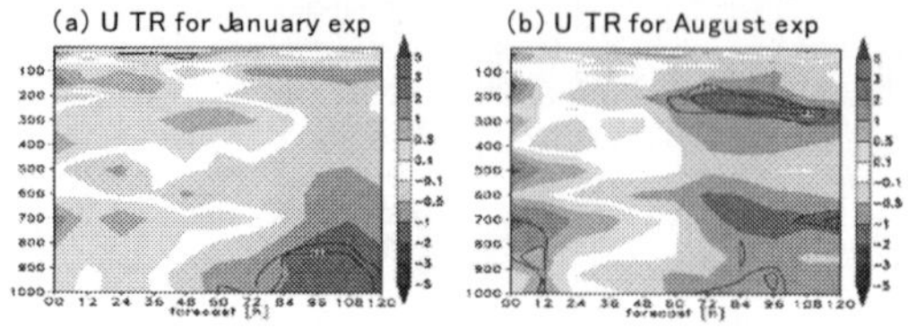

Figure 4. Relative forecast error reduction (%) of zonal wind speed for TESTT against TESTP. A positive value indicates that TEST has a smaller RMSE than TESTP.

We found that a space-based DWL gave us significant positive forecast impacts from the SOSE-based OSSE study. The similar positive impacts were achieved by DWLs on polar- and tropical-orbiting satellites and decision of the superiority of DWL on either satellite could not be made at this stage of the study. We also found positive impacts of DWL in both August and January and greater in January. Also, the results demonstrated the importance of appropriate QC and observation error assignment and that suggested more positive impacts would be obtained by refining them. For example, the simple observation error inflation setting in this study could be replaced with an adaptive inflation according to the density of existing observations. Furthermore, some observation errors were not taken into account such as neglecting local effect of vertical wind component. Future study will include these developments. We will investigate the complementarity between DWL and atmospheric motion vectors (AMVs) derived from tracking successive imageries of passive infrared and visible imagers [7].

5. Acknowledgements

A part of this research was supported by JSPS KAKENHI under Grant Numbers 15K05293 and 15K06129.

6. References

[1] Okamoto, K., T. Ishibashi, S. Ishii, P. Baron, K. Gamo, T. Y. Tanaka, K. Yamashita, and T. Kubota: "Feasibility study for future space-borne coherent Doppler wind lidar. part 3: Impact assessment using sensitivity observing system simulation experiments", *J. Meteor. Soc. Japan*, **96**, 179-199, (2018).

[2] Marseille, G. J., A. Stoffelen, and J. Barkmeijer: "Sensitivity observing system experiment (SOSE): A new effective NWP - based tool in designing the global observing system," *Tellus A*, ***60***, 216-233 (2008).

[3] Tanaka, T. Y., and M. Chiba, "Global simulation of dust aerosol with a chemical transport model, MASINGAR", *J. Meteor. Soc. Japan*, ***83A***, 255-278 (2005).

[4] Baron, P., S. Ishii, K. Okamoto, K. Gamo, K. Mizutani, C. Takahashi, T. Itabe, T. Iwasaki, T. Kubota, T. Maki, R. Oki, S. Ochiai, D. Sakaizawa, M. Satoh, Y. Satoh, T. Y. Tanaka, and M. Yasui: "Feasibility study for future spaceborne coherent Doppler Wind Lidar. Part 2: Measurement simulation algorithms and retrieval error characterization," *J. Meteor. Soc. Japan*, ***95***, 319–342 (2017).

[5] Ishii, S., K. Mizutani, P. Baron, M. Aoki, K. Mizutani, M. Yasui, S. Ochiai, A. Sato, Y. Satoh, T. Kubota, D. Sakaizawa, R. Oki, K. Okamoto, T. Ishibashi, T. Y. Tanaka, T. T. Sekiyama, T. Maki, K. Yamashita, T. Nishizawa, M. Satoh, and T. Iwasaki: "Feasibility study for future spaceborne coherent doppler wind lidar, Part 1: Global Wind Profile Observing System," *J. Meteor. Soc. Japan*, **95**, 301–317 (2017) .

[6] Japan Meteorological Agency, *Outline of the operational numerical weather prediction at the Japan Meteorological Agency* [Available at http://www.jma.go.jp/jma/jma-eng/jma-center/nwp/outline-nwp/index.htm] (2007).

[7] Ishii, S., K. Okamoto, P. Baron, T. Kubota, Y. Satoh, D. Sakaizawa, T. Ishibashi, T. Y. Tanaka, K. Yamashita, S. Ochiai, K. Gamo, M. Yasui, R. Oki, M. Satoh, and T. Iwasaki, "Measurement performance assessment of future space-borne Doppler wind lidar," *SOLA*, **12**, 55–59 (2016).

DAWN in CPEX campaign and Space pathfinder coherent wind lidar development at LaRC

Jirong Yu (a), Michael J. Kavaya (a), Larry B. Petway (a), Songsheng Chen (a), John E. Marketon (a), Sammy Henderson, and David Emmitt
(a) NASA Langley Resear Center

Abstract: Recently, the NASA Langley high power Doppler AerosolWiNd (DAWN) instrument participated in NASA 2017 CPEX campaign, a field experiment to study Convective Processes in the Tropics. It provides previously unavailable high resolution and high sensitivity 3D wind profiles with a vertical resolution of 50m as well as LOS wind measurements every 25-50 m horizontally over the open ocean. Currently, NASA Langley is developing a space pathfinder coherent wind lidar instrument with a lidar performance meeting the space wind lidar figure of merit requirements. It operates at 200 pulse repletion range, 55mJ pulse energy, 150-200ns pulse width and with beam quality M2 value better than 1.1. This instrument is suitable for International space station platform to measure global 3D winds. In this paper, we will describe the DAWNinstrument sensitivity improvements and its performance during the CPEX campaign, and the space pathfinder coherent wind lidar instrument development progresses.

High-average-power, conductively cooled Tm,Ho:YLF laser for Doppler wind lidar

Atsushi Sato (a, b), Makoto Aoki (b), Shoken Ishii (b), Kohei Mizutani (b), Satoshi Ochiai (b), and Katsuhiro Nakagawa (b)
(a) Tohoku Institute of Technology
35-1 Yagiyama-Kasumi, Taihaku-ku, Sendai, 982-8577, Japan
(b) National Institute of Information and Communications Technology
4-2-1 Nukui-kitamachi Koganei, Tokyo, 184-8795, Japan
E-mail address: atsushi@tohtech.ac.jp

Abstract: A conductively cooled Tm,Ho:YLF laser is one of the most promising laser sources for a spaceborne Doppler wind lidar. In this study, high-average-power operation of the conductively cooled Tm,Ho:YLF laser was demonstrated at a crystal temperature of −80 °C. At a pulse repetition frequency of 70 Hz, a Q-switched pulse energy of 104 mJ was obtained. This corresponds to an average output power of 7.28 W, which is the highest average power for a 100-mJ-class Q-switched Tm,Ho:YLF laser oscillator with conductive cooling. The maximum output energy reached 125 mJ at 50 Hz. Based on these results, an optical design of a master oscillator and power amplifier system operating at −40 °C was investigated to reduce the power consumption of the laser transmitter.

Keywords: Lidar transmitters, 2-micron lasers, conductive cooling, high-power lasers.

1. Introduction

A conductively cooled, Q-switched Tm,Ho:YLF laser is suitable for use as a transmitter in a spaceborne Doppler wind lidar (DWL). Recently, a super low altitude test satellite was successfully launched by the Japan Aerospace Exploration Agency in December, 2017 [1]. Although it moderates the requirements for the DWL transmitter, a 100-mJ-class laser source is still required [2]. For DWL transmitters, not only single-longitudinal-mode operation but also high-average-power operation is needed because the figure of merit (FOM) of the DWL system is defined as a product of the pulse energy, the square root of the pulse repetition frequency (PRF), and the telescope diameter. Under the same telescope condition, the FOM depends only on the product of the pulse energy and the square root of the PRF. Conductively cooled Tm,Ho:YLF lasers for spaceborne DWL applications have been developed at the National Institute of Information and Communications Technology. In our previous work, an average output power up to 2.4 W has been achieved at a crystal temperature of −80 °C [3, 4]. However, further power scaling of the laser is required for the DWL measurements from a super low altitude satellite (~220 km). Our goal is to develop the Tm,Ho:YLF laser transmitter capable of producing a pulse energy of 125 mJ at a PRF of 30 Hz, corresponding to an average output power of 3.75 W. In addition, the higher operating temperature (~ −40 °C) is preferable to reduce the power consumption of the cooling system. In this paper, we report on high-average-power operation of the Q-switched Tm,Ho:YLF laser at −80 °C. Based on the results of laser experiments and numerical simulations, a DWL transmitter design for high-power operation at −40 °C is also investigated.

2. Experimental setup

The YLF laser rod codoped with 4% Tm and 0.4% Ho was 4 mm in diameter and 33 mm in length was mounted into three copper heat-sink blocks using 50-μm-thick indium foils for better thermal contact between the laser rod and the heat sink. The pump source was a three-stacked laser diode (LD) with a pitch of 400 μm and was temperature-tuned to the 792 nm absorption peak of Tm,Ho:YLF. The pump module consisted of a set of three LDs arranging side by side and a quartz light guide plate with an end-face dimension of 33.7 mm × 1.2 mm. Three pump modules were arranged 120° apart around the laser rod. The Cu heat sinks for the laser rod were liquid-cooled by circulating Fluorinert coolant. The laser head was placed inside a vacuum chamber to prevent condensation and frost. The resonator configuration of the Q-switched Tm,Ho:YLF laser is shown in Fig. 1. A 3.86-m-long ring resonator was used in our experiment. The flat output coupler (OC) had a reflectivity of 61% or 74% at 2050 nm. The resonator consisted of the OC and three flat mirrors having a high-reflection coating for a wavelength of 2050 nm. A crystal-quartz acousto-optic (AO) Q-switch was used as a Q-switch element. Two convex lenses with a focal length of 3 m (L1, L2) were inserted into the resonator for the compensation of the negative thermal lens in the Tm,Ho:YLF rod. When the focal length of thermal lensing is 1 m or longer, the resonator is always stable, and the TEM_{00} mode radius at the laser rod remains almost constant around 1 mm. This leads to a good overlap between the cavity mode and the central pumped region with a diameter of approximately 2 mm.

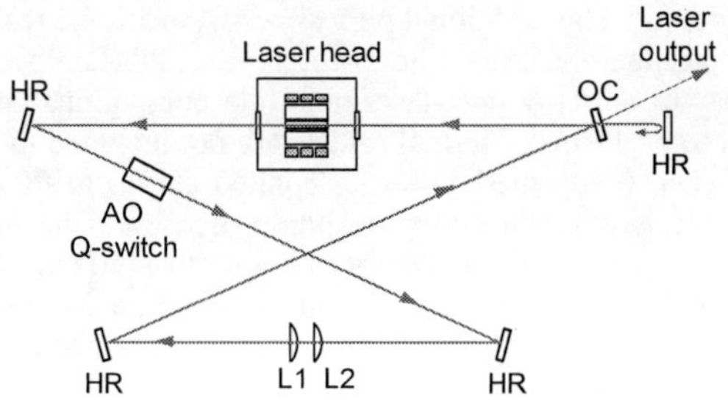

Figure 1. Resonator configuration of the Q-switched Tm,Ho:YLF laser.

3. Experimental results

The lasing characteristics in both normal- and Q-switched-mode operations were measured at 50 and 70 Hz. The results are shown in Fig. 2. When the laser was operated in normal mode at 50 Hz with a 0.8 ms pump pulse, the threshold pump energy and the slope efficiency were 0.63 J and 16.7%, respectively. The maximum normal-mode pulse energy of 200 mJ was obtained for a pump energy of 1.85 J. The same performance as in normal-mode operation was obtained in Q-switched-mode operation for a pump energy lower than 1 J. However, the slope efficiency in Q-switched-mode operation decreased as the pump power

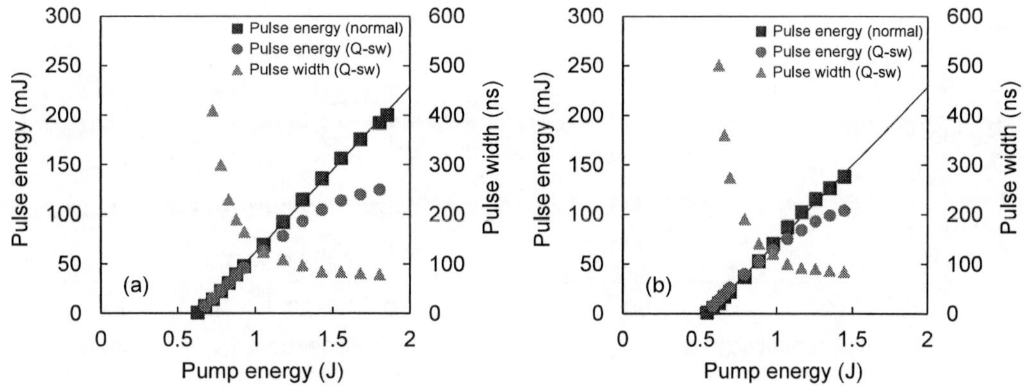

Figure 2. Pulse energy and pulse width of the Tm,Ho:YLF laser as a function of pump energy. The laser was operated (a) at 50 Hz with a 0.8 ms pump pulse and (b) at 70 Hz with a 0.6 ms pump pulse.

increased under intense pumping. This can be explained by the decrease in the fractional population of the Ho upper-laser manifold in the coupled Tm-Ho system resulting from the depletion of the Ho lower-laser manifold due to the high pumping density. Nevertheless, a Q-switched pulse energy of 125 mJ was achieved at 50 Hz, corresponding to an average output power of 6.25 W. The pulse width was then 80 ns. As can be seen in Figs. 2(a) and 2(b), the threshold pump energy at 70 Hz was lower than that at 50 Hz. The reason for the lower threshold at 70 Hz is that the short pumping interval leads to a large residual population in the Ho upper-laser manifold for the next pump pulse. The maximum average output power at 70 Hz reached 7.28 W [5]. While the M^2 value increased with increasing average pump power, it was not higher than 1.5 even at the maximum average pump power.

In order to investigate the contribution of the residual population in the Ho upper-laser manifold to the next pump pulse, pulse energies and pulse widths were measured for several different PRFs. Fig. 3 shows the experimental and calculated results for a pump energy of 1.45 J. Numerical simulations were performed by using a rate-equation model [6]. The additional pump energy due to the residual population after lasing was estimated by taking into account the second quasi-thermal equilibrium between the Tm and Ho upper-laser manifolds and the decay of the Ho upper-laser-manifold population. The lifetime of the Ho upper-laser manifold was assumed to be 15 ms. Thermal effects are not included in the calculation. An obvious enhancement of the output energy appeared in the calculated curves at PRFs higher than 20 Hz. This indicates that there is a trade-off between the additional pump energy and the temperature rise at high PRFs. However, the measured output energy decreased at PRFs higher than 50 Hz so that the PRF range of 30 to 50 Hz might be optimal choice in our system. The Q-switched-mode operation at −40 °C was also demonstrated at a PRF of 30 Hz, as shown in Fig. 3. The output energy decreased from 103 to 45 mJ as the temperature of the laser rod was increased from −80 to −40 °C. Although the high output energy was obtained at −80 °C, the pulse width was one-half of our target. On the other hand, the pulse width observed at −40 °C fulfilled the requirement of the DWL system. Therefore, a master oscillator and power amplifier (MOPA) is a promising candidate for the laser transmitter operating at −40 °C.

4. Comparison of oscillator and MOPA

Since the experimental results shown in Fig. 3 were in reasonable agreement with the calculated results, the MOPA system operating at −40 °C was investigated with numerical simulations. Figs. 4(a) and 4(b) show the results of simulations for the laser oscillator with two laser heads and the MOPA, respectively. It is

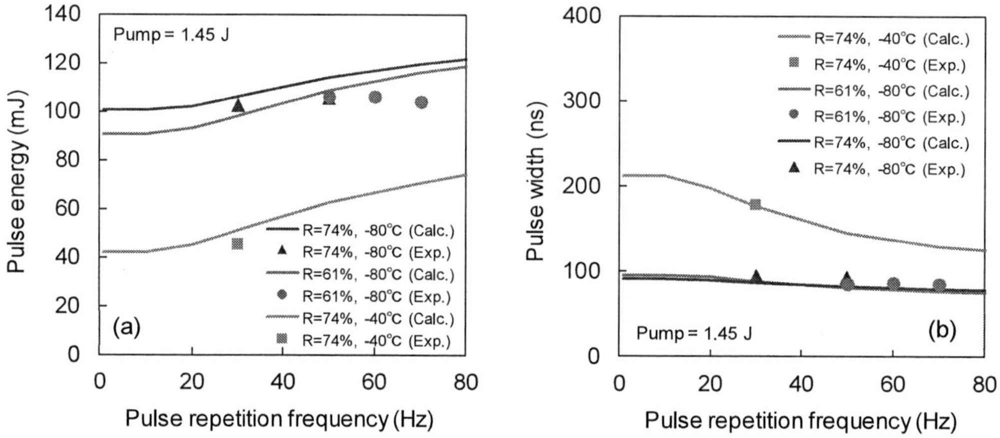

Figure 3. (a) Pulse energy and (b) pulse width of the Q-switched Tm,Ho:YLF laser as a function of pulse repetition frequency. The pump energy was 1.45 J. The solid curves were calculated with a rate-equation model.

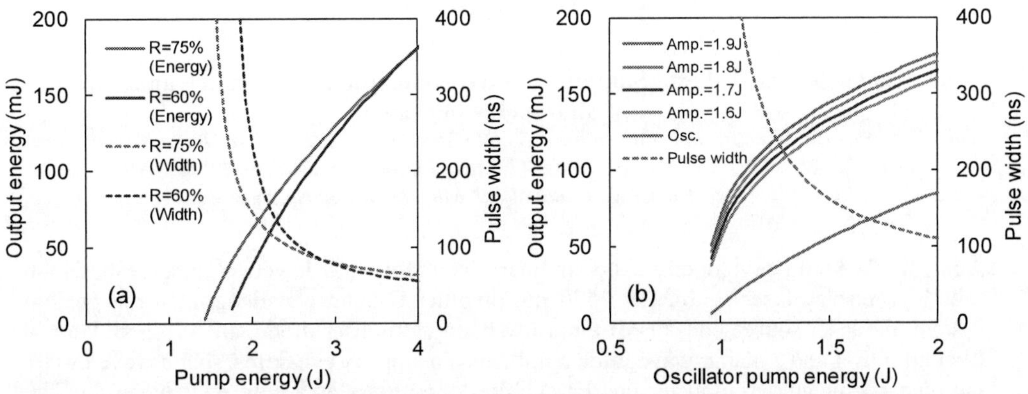

Figure 4. Results of numerical simulations of output energies and pulse widths for (a) the oscillator with two laser heads and (b) the MOPA with the single laser-head oscillator and the single laser-head 2-pass amplifier.

assumed that the MOPA system consists of the single laser-head oscillator and the single laser-head 2-pass amplifier. Pulse energies and pulse widths were calculated for different amplifier pump energies as a function of the oscillator pump energy. As shown in Fig. 4(a), the pump energy required for 125-mJ output was calculated to be 3.1 J, and the pulse width of the oscillator is shorter than 80 ns at this output level. On the other hand, an output energy of 125 mJ can be obtained in the MOPA with total pump energies of 3.1−3.2 J. In this condition, the pulse width is longer than 160 ns. This result supports the superiority of the MOPA system.

5. Conclusion

We have demonstrated a conductively cooled Q-switched Tm,Ho:YLF laser operating at −80 °C with an average output power higher than 7 W. To reduce the power consumption of the cooling system, a design of the laser transmitter operating at −40 °C was investigated. As a result, we found that a MOPA system is a promising candidate for a 3.75-W DWL transmitter operating at −40 °C.

6. References

[1] http://global.jaxa.jp/projects/sat/slats/.

[2] S. Ishii, P. Baron, M. Aoki, A. Sato, K. Okamoto, T. Ishibasi, T. Y. Tanaka, T. T. Sekiyama, T. Maki, K. Yamashita, T. Nishizawa, M. Satoh, T. Iwasaki, M. Yasui, K. Mizutani, S. Ochiai, Y. Satoh, D. Sakaizawa, R. Oki, and T. Kubota, "Feasibility study for future space-borne coherent Doppler wind Lidar. Part 1: Instrumental overview for global wind profile observation," J. Meteor. Soc. Japan, vol. 95, no. 5, pp. 301–317 (2017).

[3] S. Ishii, K. Mizutani, H. Fukuoka, T. Ishikawa, B. Philippe, H. Iwai, T. Aoki, T. Itabe, A. Sato, and K. Asai, "Coherent 2-μm differential absorption and wind lidar with conductively-cooled laser and two-axis scanning device," Appl. Opt., vol. 49, no. 10, pp. 1809–1817 (2010).

[4] K. Mizutani, T. Itabe, S. Ishii, M. Aoki, K. Asai, A. Sato, H. Fukuoka, T. Ishikawa, and K. Noda, "Diode-pumped 2-μm pulse laser with noncomposite Tm,Ho:YLF rod conduction-cooled down to −80°C," Appl. Opt., vol. 54, no. 26, pp. 7865–7869 (2015).

[5] A. Sato, M. Aoki, S. Ishii, R. Otsuka, K. Mizutani, and S. Ochiai, "7.28-W, high-energy, conductively cooled, Q-switched Tm,Ho:YLF laser," IEEE Photon. Technol. Lett., vol. 29, no. 1, pp. 134–137 (2017).

[6] A. Sato, M. Aoki, S. Ishii, R. Otsuka, K. Mizutani, and S. Ochiai, "High-average-power operation of a 100-mJ-class, conductively cooled, Q-switched Tm,Ho:YLF laser," 12th Conference on Lasers and Electro-Optics Pacific Rim (CLEO-PR 2017), (Singapore, July 31 - August 4, 2017), paper Oral 1-4H-4.

9-mJ laser amplifier at 1540 nm for a coherent lidar system with 250 Hz repetition

Kenichi Hirosawa, Takeshi Sakimura, Narito Samejima, Takayuki Yanagisawa, and Shumpei Kameyama
Mitsubishi Electric Corporation, Information Technology R&D Center, 5-1-1 Ofuna Kamakura Kanagawa 247-8501, Japan
Hirosawa.Kenichi@db.MitsubishiElectric.co.jp

Abstract: We have developed a coherent lidar system for wind detection since 1998, which uses high energy laser sources at 1550 nm or other C-band wavelength. In the present system, the laser source consists of a narrow-bandwidth laser diode, an AO modulator, a fiber amplifier and a planar waveguide amplifier. To amplify eye-safe C-band wavelength, the planar waveguide amplifier has Er, Yb co-doped glass as a core of waveguide. The planar waveguide suppress nonlinear effects such as stimulated Brillouin scattering because it accepts larger beam width than single mode optical fibers. Therefore, higher output energy can be achieved than conventional fiber amplifiers. In this paper, we modified the structure of the planar waveguide, and achieved 9 mJ with 250 Hz repetition rate. The power performance with various repetition rates will be presented.

Keywords: Coherent Laser Radar, Erbium Lasers, Planar Waveguide Amplifier

1. Introduction

We have developed a coherent lidar system for wind detection since 1998 [1], which uses high energy laser sources at 1550 nm or other C-band (1530 nm - 1565 nm) wavelength. C-band is eye-safety wavelength so that it is allowed to emit quite high power to the atmosphere, and many commercial fiber-based optical components are available in the wavelength range. Fiber lasers and fiber amplifiers are also available, however stimulated Brillouin scattering (SBS) limits transmitting peak power below approx. 50 W in conventional optical fibers. For long range detection, lidar systems need higher power than conventional optical fibers can transmit, and therefore long-range lidar systems need to adopt a specialty fiber or a free-space-based light source. Free-space-based solid state lasers for lidar systems have been reported and work quite well [2,3], however long term reliability is still concern. Because of such a background, we have developed a planar-waveguide amplifier using Er, Yb:glass [4]. The planar waveguide structure has more long term reliability than free-space-based solid state lasers and accepts large-area beam maintaining beam quality unlike conventional optical fibers. As a result, it has achieved output power of 3.3 kW with the pulse width of 580 ns (i.e. pulse energy of 1.9 mJ) and the repetition rate of 4 kHz in addition to the good beam quality of M2: 1.3. In 2011, we developed the lidar system using this amplifier and demonstrated the >30 km long-range wind sensing [5].

In our system, the output power nearly reached to eye-safety limit. Therefore, improving pulse energy and decreasing repetition frequency (to keep average output power under the eye-safety limit) is effective to improve lidar performance. For this purpose, we modified the form of the planar waveguide amplifier. To enlarge beam area and tolerate higher pulse energy, core thickness and incident angle were changed. In this paper, we report on the evaluation result of new planar waveguide, especially focusing on the output pulse energy with low repetition rate.

2. Experiments

Experimental setup is shown in Fig. 1. It consists of a narrow-bandwidth laser diode, an AO modulator, a fiber amplifier and a planar waveguide amplifier. The seed laser is FBG (Fiber Bragg Grating) wavelength-stabilized laser diode at 1540 nm, which has < 1 MHz bandwidth. The seed laser is modulated by an acousto-optic (AO) modulator into approx. 1-µs-duration pulses. The optical pulse is amplified by the fiber amplifier to approx. 37 µJ of pulse energy.

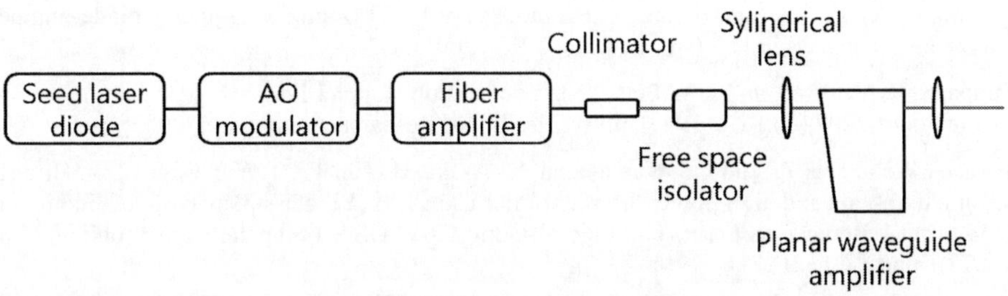

Figure 1. Experimental setup.

Output power of the planar waveguide amplifier is shown in Fig. 2. Pulse energy increased as repetition frequency decreased. Highest pulse energy was approx. 9 mJ with 250 Hz repetition. With 500 Hz repetition, pulse energy of 8 mJ was also obtained. The narrowest output pulse width was approx. 450 ns.

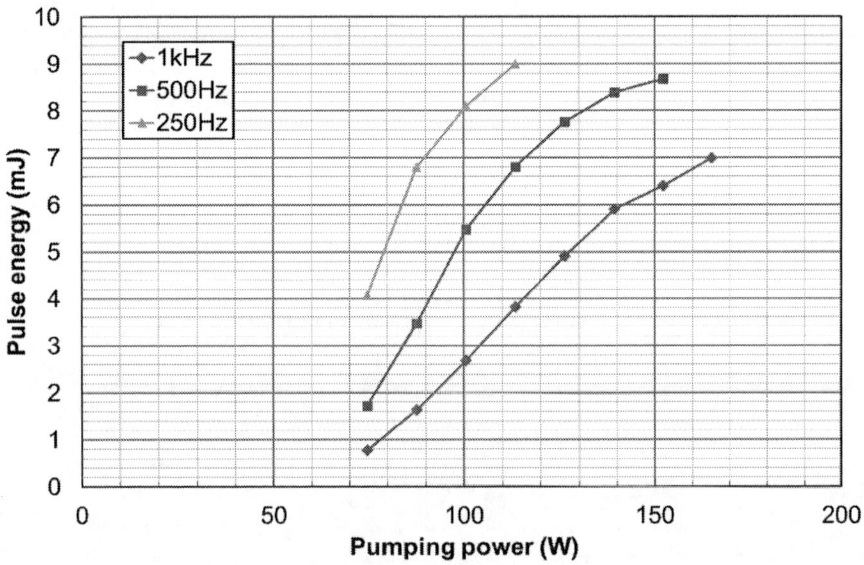

Figure 2. Output pulse energy with various repetition rate.

3. Conclusion

To improve lidar performance, we modified the structure of the planar waveguide to tolerate higher pulse energy. Consequently, 9 mJ with 250 Hz repetition rate was achieved. With 500 Hz repetition, pulse energy of 8 mJ was also obtained. The improvement of pulse energy would contribute extension of detection range and signal to noise ratio of our coherent Doppler lidar.

[1] S. Kameyama, T. Yanagisawa, T. Ando, T. Sakimura, H. Tanaka, M. Furuta, and Y. Hirano, "Development of wind sensing coherent Doppler LIDAR at Mitsubishi Electric Corporation ~from late 1990s to 2013," in 17th Coherent Laser Radar Conference (CLRC 2013) (Universities Space Research Association, 2013), p. 12

[2] T. Yanagisawa, K. Asaka, K. Hamazu, and Y. Hirano, "11-mJ, 15-Hz single-frequency diode-pumped Q-switched Er, Yb:phosphate glass laser," Opt. Lett. **26**, 1262 (2001).

[3] S. M. Hannon, S. R. Vetorino, and J. V. Pelk, "Next generation Doppler lidar sensor at 1.6μm," in 14th Coherent Laser Radar Conference 2007 (CLRC XIV) (Universities Space Research Association, 2007), p. 59

[4] T. Sakimura, Y. Watanabe, T. Ando, S. Kameyama, K. Asaka, H. Tanaka, T. Yanagisawa, Y. Hirano, and H. Inokuchi, "1.5-μm high-gain and high-power laser amplifier using a Er,Yb:glass planar waveguide for coherent Doppler LIDAR," in Conference on Lasers and Electro-Optics 2012 OSA Technical Digest (online) (Optical Society of America, 2012), paper CTu2D.7

[5] S. Kameyama, T. Sakimura, Y. Watanabe, T. Ando, K. Asaka, H. Tanaka, T. Yanagisawa, Y. Hirano, H. Inokuchi, "Wind sensing demonstration of more than 30km measurable range with a 1.5μm coherent Doppler lidar which has the laser amplifier using Er,Yb:glass planar waveguide," Proc. of SPIE, 85260E-1 (2012).

Photonic Integrated Circuit FMCW Lidar On A Chip

Paul J.M. Suni (a), James R. Colosimo (a) S.J. Ben Yoo (b), John Bowers (c), Larry Coldren (c), Jonathan Klamkin (c)

(a) Lockheed Martin Advanced Technology Center, Louisville, CO, USA
(b) University of California Davis, Davis, CA, USA
(c) University of California Santa Barbara, Goleta, CA, USA
paul.suni@lmco.com

Abstract: Photonic integrated circuits (PICs) enable construction of complex optical systems at a miniaturized scale not possible with bulk elements. All active and passive components required to fabricate chip-scale coherent lidar systems have been demonstrated and system demonstrations are currently underway. In this paper we describe our efforts to demonstrate a chip-scale frequency modulated continuous-wave (FMCW) lidar. This system incorporates wavelength tuning and phased array operation to enable 2D non-mechanical beam steering. We further discuss approaches aimed at scaling aperture dimensions from mm scales to 1 cm and 10 cm scales. These efforts include development of micron-sized mirror arrays to enable coupling between optical layers and demonstration of wafer-scale routing of light at high efficiency.

Keywords: FMCW lidar, PIC, Photonic Integrated Circuits

1. Introduction

Chip-scale lidar systems have been sought for many years due to the potential for low size, weight, and power (SWaP) remote sensing systems. The possibility of fabricating parts at the wafer-level is also an attractive means to reduce cost for high volume applications. The recent explosion in photonic integrated circuit (PIC) development has now reached a state where coherent complex active optical systems can be constructed, including coherent lidar.

Power handling of sub-micron waveguides and other functional elements remains a fundamental limitation. Depending on the materials used typical upper limits before non-linear optical effects set in are tens to hundreds of mW in silicon (Si), limited by two-photon absorption (TPA), while silicon nitride (SiN) can handle up to ~1 W power levels. This means that very high peak power lidar systems are unlikely to be completely developed in PIC form. However, coherent systems, such as frequency modulated continuous-wave (FMCW) [1], are entirely feasible. We also note that the power limits apply to single waveguides. By constructing systems with large numbers of parallel channels with embedded semiconductor optical amplifiers (SOA), the total power can be scaled up by orders of magnitude.

The DARPA Modular Optical Aperture Building Blocks (MOABB) program was initiated in 2016 with the goal of developing aperture scalable coherent optical systems with lidar as the demonstrator application. In Phase 1, which is ending as of this writing, the goal was to construct a 1 mm^2 coherent aperture. The program goals increase to 1 cm^2 in Phase 2 and 10 cm x 10 cm in Phase 3. Total power output goals for the three phases are 5 mW, 0.5 W, and 100 W, respectively.

A critical aspect of the MOABB program is incorporation of non-mechanical beam steering (NMBS). Previous efforts, including the DARPA SWEEPER program, have developed techniques to do so. To date the best way [2] to implement NMBS in two dimensions creates diffraction gratings in waveguides and tunes the laser wavelength to steer the beam in one dimension at a rate of ~0.14 degrees/nm of laser tuning. In the second (lateral) dimension a large number of parallel waveguides incorporating phase shifters are used to create transverse linear phase gradients on the beam and steer laterally as an optical phased array (OPA).

This document was cleared by DARPA on May 14, 2018.
Distribution Statement "A" (Approved for Public Release, Distribution Unlimited).

2. Technology Challenges

One challenge in fully integrated PIC-based coherent lidar incorporating NMBS is heterogeneous integration of a laser and detectors. Silicon is an indirect bandgap semiconductor and hence cannot be used to make lasers and detectors (in the infrared). To incorporate these elements indium phosphide (InP) chips need to be bonded to silicon-based structures that contain the remaining functional lidar elements [2].

A greater challenge is to produce an OPA with low or no sidelobes. This requires ultra-dense waveguide pitches. Small pitches lead to cross-talk for long parallel waveguides unless measures are taken to reduce the effect. Figure 1 illustrates the relationship between coupling length and sidelobe-free steering angle for conventional waveguides. Color coding in the right plot indicates etch depth between waveguides. Note that to get sidelobe-free steering over ±90° requires a waveguide pitch of $\lambda/2$. We have to date demonstrated 1.3 μm waveguide pitches which produce a ±10° sidelobe-free steering range [3]. The program goals require much greater steering and we are exploring novel techniques aimed at producing crosstalk-free waveguide pitches of ~0.85 μm to enable ±55° sidelobe-free steering [4].

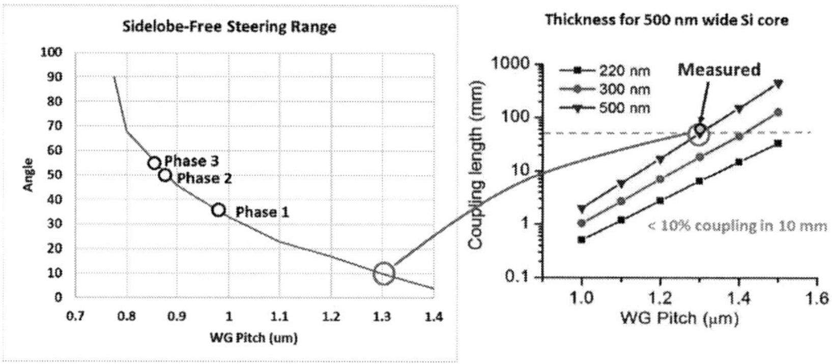

Figure 1. Relationship between sidelobe-free OPA steering, cross-talk coupling lengths, and waveguide pitch

3. Demonstrator System

We have developed a system design that incorporates all elements to demonstrate coherent lidar operation. Figure 2 illustrates the layout of the PIC. The transceiver section incorporates a widely tunable single-frequency sampled-grating distributed Bragg reflector (SG-DBR) laser [5], LO splitter, semiconductor optical amplifiers (SOA), and a balanced detector pair. Light is transmitted into a star coupler [6] which splits the light into *N* secondary waveguides. Light is transported into an equal number of phase shifters which can be independently controlled. From there light propagates into a grating section where weak gratings diffract light into free space along the length of the 5-10 mm long grating structure. In the lidar case light is received in the same waveguide structure and propagates back through the system, is picked off at a transmit/receive 50/50 coupler, and propagates to the detectors.

Figure 3 left illustrates the physical layout of the transceiver chip. The right figure shows a photo of a completed chip mounted to a "supercarrier" and connected to two small boards containing laser control electronics and the receiver front end. As of this writing the transceiver chip is undergoing final testing and characterization.

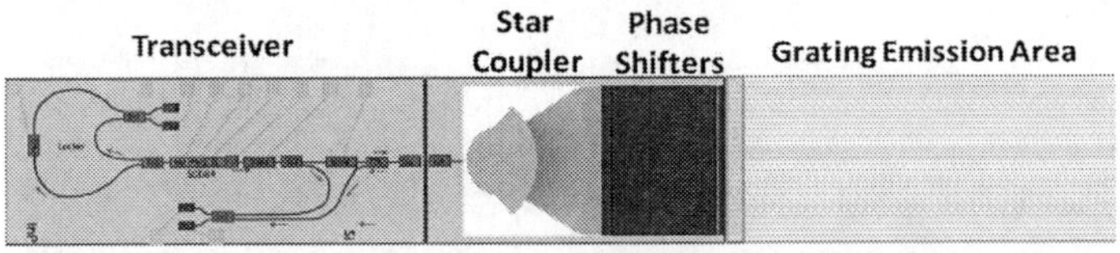

Figure 2. PIC architecture implementing a coherent FMCW lidar on a chip

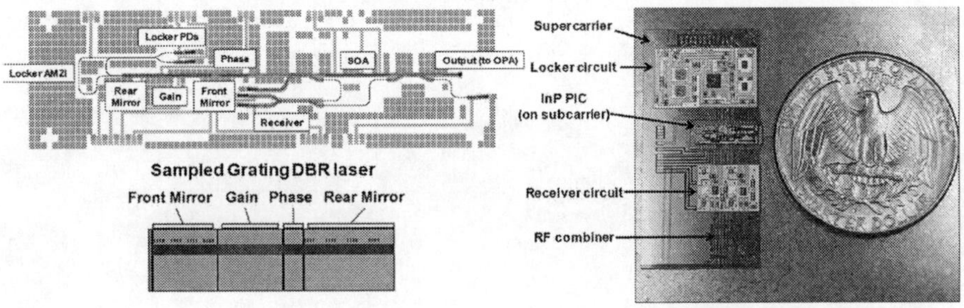

Figure 3. Transceiver front end as fabricated.

To demonstrate NMBS we used a 120-channel chip and an external tunable SG-DBR laser from Freedom Photonics. Figure 4 shows the hardware used for this demonstration. Left – system with control electronics NMBS demonstration. Center – PIC chip on silicon interposer with fiber input from external laser. Right – pattern written on wall using wavelength and OPA steering.

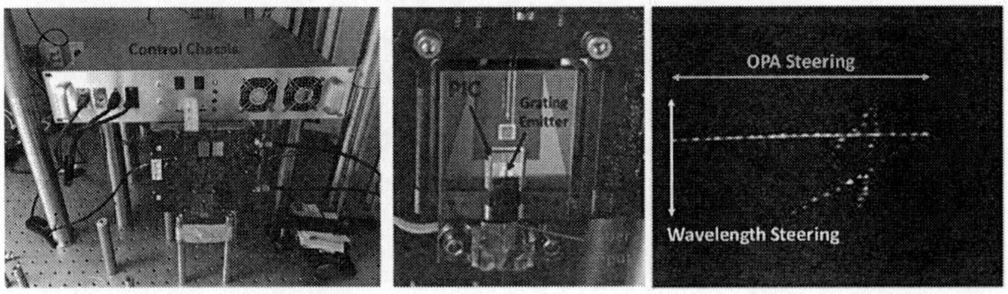

Figure 4. Demonstration of 2D NMBS

4. Next Generation PIC Lidar

The emission area of the first chip is approximately 1 mm^2, which is useful for demonstrations, but not for practical applications. We have developed concepts to scale emission apertures to much larger dimension. Doing so while maintaining high emission area fill factor requires 3D integration of optical layers as well as electronics for SOA drivers and phase shifters. Extensive work has been carried out to demonstrate 3D integration aspects and is described in greater detail by Yu et al. [7]. Figure 5 illustrates some of these efforts. From left to right: two-layer optical chip with splitters and phase shifters in one layer and emitter grating in second layer; sub-micron 45° mirror pairs used to couple light between layers; ultrafast laser inscription (ULI) used to write 3D optical waveguides of arbitrary 3D shape for stitching parts together optically.

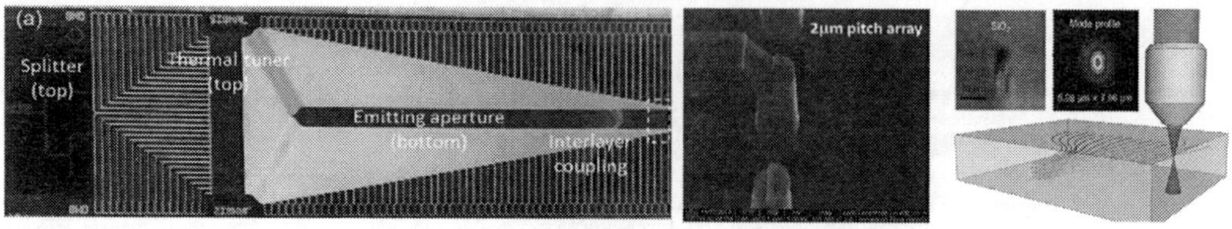

Figure 5. 3D structures developed for future large aperture scaling.

Figure 6 illustrates concepts for scaling these systems to larger aperture dimensions. Left – planar structure with 1 cm^2 grating emission area fed by external laser. Center – 3D structure with parts folded under grating layer. Right – concept for construction of 10 x 10 cm flat system outputting 100 W optical power.

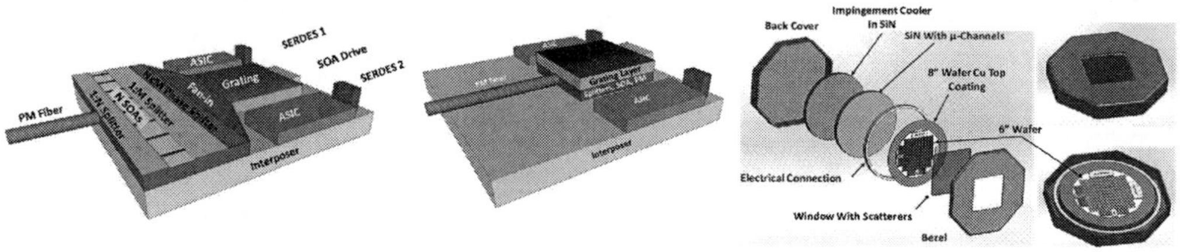

Figure 6. Concepts for future aperture scaling to larger aperture dimensions.

5. References

[1] B. W. Krause, B. G. Tiemann, and P. Gatt, "Motion compensated frequency modulated continuous wave 3D coherent imaging ladar with scannerless architecture", *Applied Optics,* vol. 51, pp. 8745-8761, 2012.

[2] J. C. Hulme, J. K. Doylend, M. J. R. Heck, J. D. Peters, M. L. Davenport, J. T. Bovington, L. A. Coldren, and J. E. Bowers, "Fully integrated hybrid silicon two dimensional beam scanner", *Optics Express*, vol. 23, pp. 5861-5874, 2015/03/09 2015.

[3] Yu Zhang, Yi-Chun Ling, Yichi Zhang, Kuanping Shang, and S. J. Ben Yoo, "Sub-wavelength-pitch Silicon-Photonic Optical Phased Array for Large Field-of-Regard Coherent Optical Beam," submitted for presentation at European Conference on Optical Communications, 2018.

[4] T. Komljenovic, R. Helkey, L. Coldren, and J. E. Bowers, "Sparse aperiodic arrays for optical beam forming and LIDAR", Optics Express, (25)3, 2511-2528, February 6, 2017

[5] A. Sivanathan, Hyun-chul Park, Mingzhi Lu, John S. Parker, Eli Bloch, Leif Johansson, Mark Rodwell, and Larry Coldren, "Integrated Linewidth Reduction of a Tunable SG-DBR Laser", Conf. on Lasers and Electro-Optics, OSA Technical Digest (Optical Society of America, 2013), paper CTu1L.2.

[6] E. J. Stanton, N. Volet, T. Komljenovic, and J. E. Bowers, "Star coupler for high-etendue LIDAR," in Conf. on Lasers and Electro-Optics, OSA Technical Digest (online) (Optical Society of America, 2017), paper STh1M.4.

[7] Yu Zhang, Yi-Chun Ling, Yichi Zhang, Kuanping Shang, and S. J. Ben Yoo, "High-Density Wafer-Scale 3D silicon-photonic integrated circuits," to be published IEEE Journal of Special Topics in Quantum Electronics, Special Issue on Emerging Areas in Integrated Optics, 2018.

6. Acknowledgement

This research was developed with funding from the Defense Advanced Research Projects Agency (DARPA) under contract HR0011-16-C-0106. The views, opinions and/or findings expressed are those of the authors and should not be interpreted as representing the official views or policies of the Department of Defense or the U.S. Government.

Advantages of 3D Imaging Coherent Lidar for Autonomous Driving Applications

Stephen Crouch
Blackmore Sensors and Analytics, Inc.
136 Enterprise Blvd., Bozeman, MT, 59718
crouch@blackmoreinc.com

Abstract: Blackmore has developed the first real-time, 3D imaging lidar product based on frequency modulation and coherent detection for autonomous driving applications. Units began shipping to automotive customers in late 2017. The technology provides long range performance (>300m), impressive point throughput (300k to 1.2M points per second per scanned beam), total interference immunity, and a unique ability to measure radial velocity on each 3D data point. In addition, continuous wave transmission is an ideal fit for integrated photonics and solid-state scanning solutions such as optical phased arrays. These advantages highlight an incredible market opportunity for coherent lidar systems in automotive and other consumer applications where direct detection lidar fails to meet system requirements. Technology building blocks in high-throughput processing, integrated photonics and optical phased array technology, all under development at Blackmore, will be presented in addition to exciting mobile, geo-registered 3D point cloud data results.

Keywords: Coherent Laser Radar, 3D Imaging, Autonomous Driving, Range-Doppler Imaging

1. Introduction

Autonomous transportation systems have seen a spike in interest in recent years. Successful research and pilot programs by industry leaders such as Waymo have demonstrated the possibility of autonomous systems to make transportation safer, more efficient, and easier to access [1]. A variety of sensors including cameras, radars, ultra-sonics, and lidar work together through complicated artificial intelligence (AI) stacks to complete the necessary navigation and avoidance tasks. Of these sensors, lidar is the most nascent and highest cost. Yet experts agree that lidar technology is critical to the success of the "self-driving" revolution.

Automotive lidar market leader Velodyne distributes a lidar sensor based on pulsed, time-of-flight ranging operated at 905nm wavelength. While this sensor has proven to be indispensable for this task, the limited range performance (<100m typical), high cost, and issues with sunlight interference have triggered a market for better lidar technology. According to released specifications, the pulse energy per measurement of a Velodyne system is approximately 300nJ [2].

The most capable automotive lidar systems to-date have transitioned to 1550nm wavelength sources. At this wavelength, the systems can emit significantly higher optical power [3]. This allows manufacturers to make more measurements at longer ranges (~200m). While systems such as Luminar have succeeded in hitting most automotive OEM specifications, the inefficient use of optical power via fiber laser amplifiers underlines a questionable pathway to scalability. In some cases, manufacturers have claimed to use in excess of 10uJ per pulse leading to time averaged optical powers of >10 watts per system [4].

In the search for optimal lidar solutions, many have dismissed coherent detection as "impossible" from both technical and cost perspectives. However, Blackmore's products prove that coherent systems have an important role in autonomous vehicle sensor suites. The presented technology represents a step for coherent laser radar (lidar) into mass market applications.

2. Coherent Lidar Operational Advantages

Coherent lidar has several operational advantages that have been demonstrated in over 18 months of on-vehicle, mobile lidar operation. These advantages include range performance, interference rejection, and direct velocity measurement. The long-range performance of Blackmore's lidar system is shown in Figure 1 where the data was collected with a real-time processing and geo-registration.

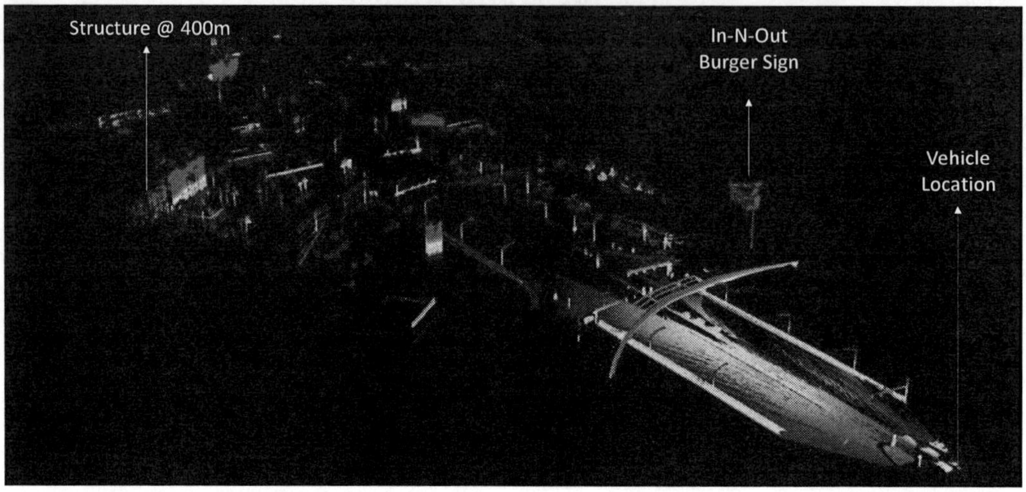

Figure 1. A few seconds of lidar data is accumulated to demonstrate the fill factor and range of a coherent lidar imaging system.

The range demonstrated in Figure 1corresponds to a system operating with a 100mW continuous wave output power and 305kHz repetition frequency. The observable range extends beyond 450m for brighter diffuse targets. System parameters are included in a proprietary link budget model for a 10% diffuse reflector which shows good agreement with traditional models and supports good SNR at a 200m range [5-7].

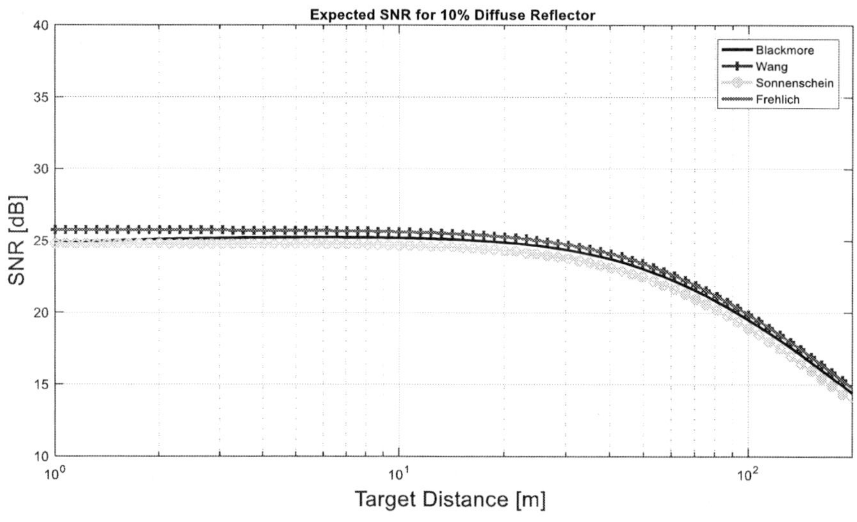

Figure 2. Blackmore's link budget model is compared to classic coherent detection papers and shows good agreement.

Coherent systems are very sensitive to the doppler effect. While operating with this effect can be technically challenging, signal processing solutions enable excellent velocity precision (<0.25 m/s) over very short measurement time intervals (3.3 us). The rapid measurement of range and velocity sets this approach apart from automotive radar systems that need to operate on timescales of tens of millisecond to effectively resolve similar Doppler frequencies. Example Doppler imagery is shown in Figure 3.

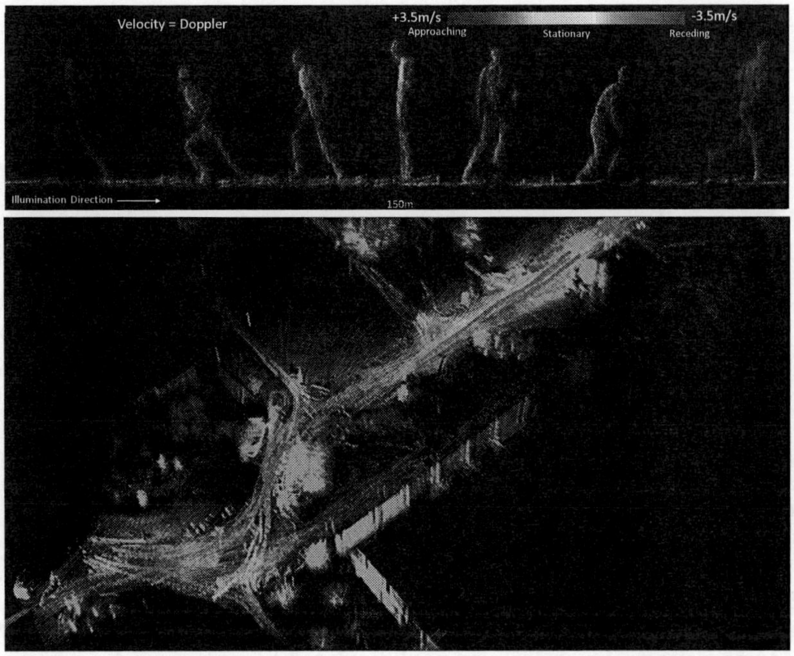

Figure 3. (TOP) A series of frames is from Blackmore's automotive lidar system is shown. Points are colored by velocity showing clear resolution of motion components on the moving target. (BOTTOM) Mobile lidar data showing pedestrian tracks when colored by the Doppler data field.

The Doppler field allows more reliable measurement of scene activity with lower latency. This translates into more efficient artificial intelligence for vehicle navigation by using velocity-based scene segmentation, tracking, and object identification. Ultimately this increases the measurement certainty and overall "value" of each 3D data point.

3. The Pathway to Scalability

The discussion in sections 1 and 2 highlights a stark contrast between the optical power requirements for pulsed/direct-detect and CW/coherent scanned lidar approached. The >10x difference in optical power efficiency of coherent detection on a per measurement basis is an important consideration in discussing the scalability of lidar designs. High optical powers will not translate to small, mass manufacturable chip designs.

Optical fiber communications points to a solution. This space has long targeted coherent detection as the preferred implementation for high bandwidth, long-haul, power efficient systems [8]. Many of the same technology building blocks such as narrow linewidth lasers and balanced receivers, along with broader supply chain capacity, can be leveraged in the construction of scalable coherent lidar systems. An example of InP chips from on of Blackmore's recent wafer runs is shown in Figure 4. These chips represent a major subsystem for future automotive lidar design at Blackmore and a step towards "lidar on a chip".

Non-mechanical beamsteering is often cited as a long-term requirement for automotive lidar. The low optical power requirements offer a match with integrated photonics for non-mechanical beamsteering where high peak powers are not acceptable. Blackmore is working with Sandia National Labs to develop silicon photonics based non-mechanical beamsteering devices for use in future systems. While these systems are indeed challenging to realize and require special attention to insertion loss mechanisms, the opportunity to realize complete "lidar chipsets" is clear. Ultimately, lidar sensors need to be chip-scale to ever realize the low cost and volume potential of the automotive market.

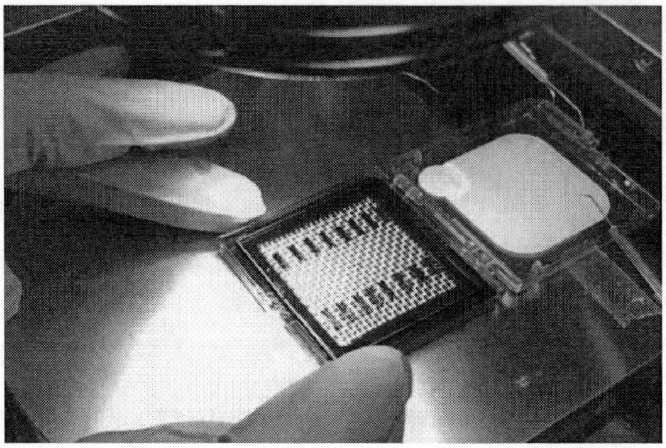

Figure 4. A set of InP chips with several lidar subsystem building blocks is shown in testing. The chips were developed in conjunction with Fraunhofer HHI and JePPIX.

As a final technical point, the author notes that coherent detection has significantly less susceptibility to interference from sunlight or other lidar system due to spatial, spectral, temporal, and other filter options that make interference an impossibility. As automotive lidar scales to the broader market, this advantage will prove to be invaluable as interference rejection in coherent lidar is inherent and does not require expensive optical bandpass filters which may degrade across temperature and field of view. In a safety critical application, performance must be guaranteed across the broadest possible operational set. This includes immunity to interference.

4. Conclusions

The requirements for automotive lidar are challenging for any optical technology to address. However, coherent detection offers a unique set of capability and scalability advantages that will support significant future investment as other technologies fail. The low optical power requirements relative to pulsed, direct-detect systems will be the major differentiator as the automotive industry pivots to coherent lidar. Through prototype systems and technology investments, Blackmore has demonstrated a roadmap for driving coherent laser radar towards mainstream applications.

5. References

[1] T. Higgins and C. Dawson, "Waymo Orders Up to 20,000 Jaguar SUVs for Driverless Fleet," Wall Street Journal, 27-Mar-2018.

[2] Velodyne. "HDL-64E, Resource Manual, Laser Safety Parameters," November 7, 2007.

[3] "Encyclopedia of Laser Physics and Technology - eye-safe lasers, retina, corneal injuries, erbium, thulium." [Online]. Available: https://www.rp-photonics.com/eye_safe_lasers.html.

[4] Jeff Hecht. "Under the Hood of Luminar's Long-Reach Lidar." IEEE Spectrum. July 5, 2017.

[5] R. G. Frehlich and M. J. Kavaya, "Coherent laser radar performance for general atmospheric refractive turbulence," Appl. Opt., AO, vol. 30, no. 36, pp. 5325–5352, Dec. 1991.

[6] C. M. Sonnenschein and F. A. Horrigan, "Signal-to-Noise Relationships for Coaxial Systems that Heterodyne Backscatter from the Atmosphere," Appl. Opt., AO, vol. 10, no. 7, pp. 1600–1604, Jul. 1971.

[7] J. Y. Wang, "Detection efficiency of coherent optical radar," Appl. Opt., AO, vol. 23, no. 19, pp. 3421–3427, Oct. 1984.

[8] F. Corporation, "Finisar Introduces Industry's Smallest Coherent Optical Assembly for High-Density Line Card and Transceiver Designs at OFC 2018," 13-Mar-2018. [Online]. Available: http://globenewswire.com/news-release/2018/03/13/1421255/0/en/Finisar-Introduces-Industry-s-Smallest-Coherent-Optical-Assembly-for-High-Density-Line-Card-and-Transceiver-Designs-at-OFC-2018.html.

Coherent Lidar for 3D-imaging through obscurants

Aude Martin (a), Jérôme Bourderionnet (a), Luc Leviander (a),
John F. Parsons (b), Mark Silver (b), Patrick Feneyrou (a)
(a) Thales Research & Technology France, 1, Avenue Augustin Fresnel, 91767, Palaiseau, Cedex, France
(b) Thales Ltd, 1 Linthouse Road, Glasgow, G51 4BZ, UK.,
Email: aude.martin@thalesgroup.com

Abstract: In the context of unmanned vehicles, eye-safe LiDARs able to measure simultaneously range and speed in degraded visual environments, are required. The ability of frequency modulated continuous wave (FMCW) coherent LiDAR at 1.55 μm to differentiate hard target from diffuse ones like clouds is used here to demonstrate 3D-imaging through obscurants. Detection experiments of a moving target up to 50 m using a compact FMCW LiDAR based on silicon photonics are also highlighted. The photonic integrated circuit, without any moving parts, allows emission and detection in 8 different collimated directions spread over the desired angle.

Keywords: Coherent Laser Radar, Frequency Modulated Continuous Wave Lidar, Photonic Integrated Circuit.

1. Introduction

Multidimensional imaging systems either use an array of detectors such as gated viewing cameras [1]; single photons array [2], or use a single pixel technology such as laser ranging with a beam scanning technology [3] or a single pixel camera [4]. In this work, we have used the FMCW LiDAR technology with a scanning head for imaging through obscurants and used a non-mechanical scanning system with a LiDAR on-chip to detect a hard target.

The main idea for FMCW lidar is to obtain spatial resolution using a frequency modulation in a coherent detection scheme. Both local oscillator and emitted signal are frequency modulated. The backscattered signal is time-delayed due to the light propagation time from the lidar to the target and frequency shifted due to Doppler shift. Hence, the interference between local oscillator and backscattered light shows frequency plateaus, the average of which corresponds to the Doppler shift (speed measurement) and the difference of which corresponds to the range multiplied by the frequency modulation slope.

As illustrated in Figure 1, this type of detection also enables FMCW LiDAR to distinguish a (relatively weak) static target signal from a much larger signal backscattered by a cloud. Since the cloud presents a significant thickness, the power spectral density of the backscattered light is spread over a large number of frequency bins. In contrast, the signature of a hard target consists of two very sharp peaks. By only selecting narrow peaks, the system can reduce false alarms and, hence, increase the signal processing speed. The signal processing used in the following experiments [5] also enables detection of multiple targets within the collimated beam.

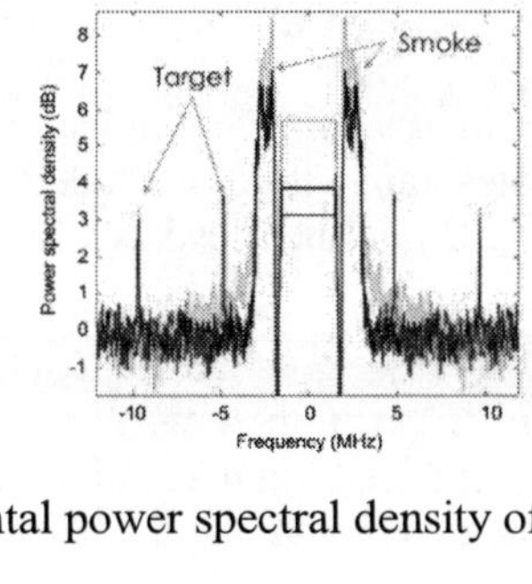

Figure 1. Experimental power spectral density of a target behind a cloud

2. Differentiating objects based on their speed

In these 3D-detection experiments, a RIO ORION laser with a linewidth of 1 kHz was modulated with slopes of 5 and 6 MHz/µs and at a period of 3.8 kHz. In a single pixel configuration, with the same LiDAR and a different optical head, targets up to 10 km were detected [6] with a mean optical power of 200 mW. A 2-axis piezo tip/tilt mirror (PI), with maximum angular range of 10 mrad, was used to scan the beam to create 2-D images. The signal processing is performed in two steps:

- Windowing and Fourier transforms are computed in real time in the FPGA using a FFT (Fast Fourier Transform) algorithm. The FFTs are averaged in power (incoherent integration) and the result is transferred to the laptop.

- Peak detection and computation of range and speed from the detected frequencies achieved using a program coded in C.

In Figure 2, a scene was mapped using the system on a clear day with 2 mW of output power and an integration time of 1 ms. The red square in Figure 2a, corresponds to the field of view of the optical head with a 10 mrad angle. A tilted sign, vegetation and a building are present at ranges between 70 and 120 m. The color coded detected ranges are plotted along the (x,y) coordinates in Figure 2b. These data with a focus on the 67 to 68 m ranges are then plotted in a 2D plot where the tilted sign and the mast are visible with a resolution of a few tens of centimeters that matches the expected resolution of $\Delta R = 37$ cm. Figure 2d shows the speed of each pixel. Green areas correspond to non-moving objects like the sign and the building while the vegetation that is moving at up to 5 cm/s can be easily identified. The ability to simultaneously measure the range and speed of an object helps identifying different objects in a complex scene.

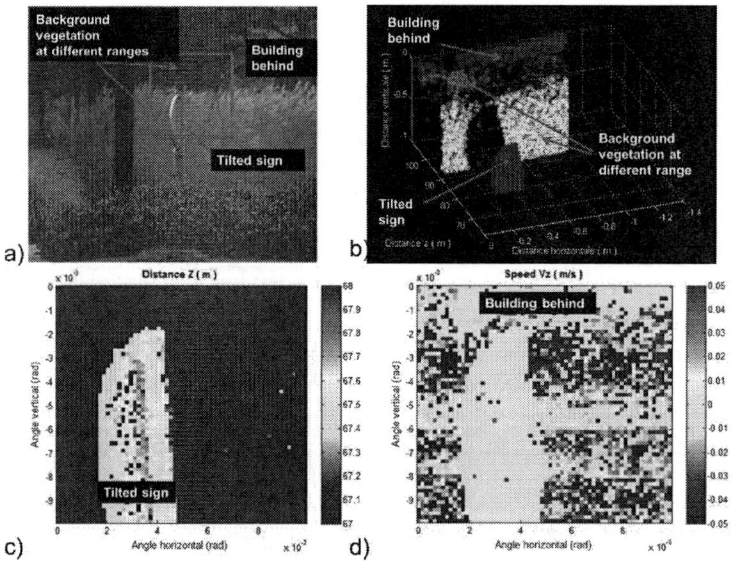

Figure 2. a) Picture of the scene, the field of view is highlighted in red; b) 3D plot of the range of the pixels as a function of the (x,y) coordinates, range is colour coded; 2D plots of the range with a focus on the 67 to 68 m range (c), and the speed as a function of the angle.

3. Large dynamic range of the LiDAR in the presence of smoke obscurants.

Range and speed measurements of a hard target in the presence of obscurants were carried out at DSTL Battery Hill facility (Porton Down, UK). Artificial fog made of a mixture of oil and water was used to fill a 10 m tent that obscured the view between the LiDAR and a hard target (a mannequin as shown in inset of Figure 3b). A picture of the scene (Figure 3a) shows the optical head and the open door of the 10 m long tent filled with artificial fog which obscures the mannequin. . Figure 3b presents detected ranges between 72 and 74.5 m as a function of the 2D scan angle. Note that the fence behind the mannequin can still be observed. Figure 3c presents the SNR for each detected pixel. Acquisition starts at the top right of the image and the scanning pattern describes a series of switchbacks. Smoke generation in the tent starts at the beginning of the acquisition and the target (head of the mannequin) is detected clearly with 20 to 30 dB of SNR. The neck and shoulders of the mannequin are not detected, as the smoke builds-up to a level that fully blocks the return signal. The smoke generation is then stopped and, as the cloud dissipates, the signal from the hard target reappears when the signal to noise ratio goes above 2 dB. In this configuration, the FMCW lidar can detect and evaluate range and speed from targets with a SNR ranging from 2 dB up to over 30 dB without recalibrating any system parameters.

During the smoke generation, the transmission at 1.55 μm was measured using a corner cube and transmissometer in order to have an independent measure of the smoke attenuation through which the FMCW could detect the target. We conclude that the Lidar was able to detect a target through smoke down to 0.04% of transmission (double pass), which corresponds to 3.9 attenuation lengths (single pass).

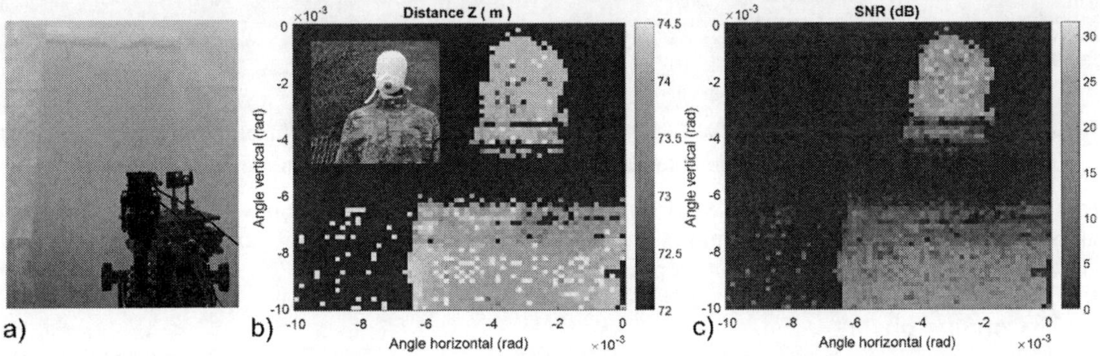

Figure 3. a) Picture of the scene with the optical head (front), the tent filled with smoke and the mannequin (not visible but indicated by an arrow), 2D plots of the range b) and the SNR c) as a function of the angle.

4. Integrating an FMCW Lidar on-chip

In order to develop 3-D mapping functionalities for unmanned vehicles, compact eye-safe LiDARs, which are able to measure simultaneously range and speed with no or few moving parts, are required. Photonic integrated circuits (PICs) are considered a viable option as they allow monolithic integration of both electronic and optical devices on the same chip and offer good thermal and mechanical stability. In addition, silicon PIC technology is becoming mature thanks to foundries, which supply parallel manufacturing paving the way for mass production. As high peak powers need to be handled with care with PICs, pulsed LiDARs are challenging to implement on-chip. Besides since the transparency window of silicon imposes to use wavelength above 1100 nm, the implementation of the architecture of an FMCW LiDAR at 1.55 μm on-chip is particularly straightforward [7].

Except from the DFB and output circulators, the optical part of the LiDAR is fully implemented on a 3x3 mm silicon chip, as shown on Figure 4a. It consists of 8 emission channels and 8 collection channels addressed using phase modulators inserted in Mach Zehnder interferometers and a waveform calibration channel. In Figure 4a, the switch network for the reception ports (SN1) and emission ports (SN2) are colored green and red respectively. Emission channels are successively addressed and connected via optical fibered circulators (FC) to the corresponding on-chip Balanced PhotoDiodes (BPD) and the output collimator. Thus we avoid the need for mechanical scanning as the PIC is addressed to select the channel and collimator of the desired output beam. The waveform calibration is achieved with the on-chip delay line interferometer (DLI). The silicon chip is then bonded to a Printed Circuit Board to read the balanced photodiodes signals.

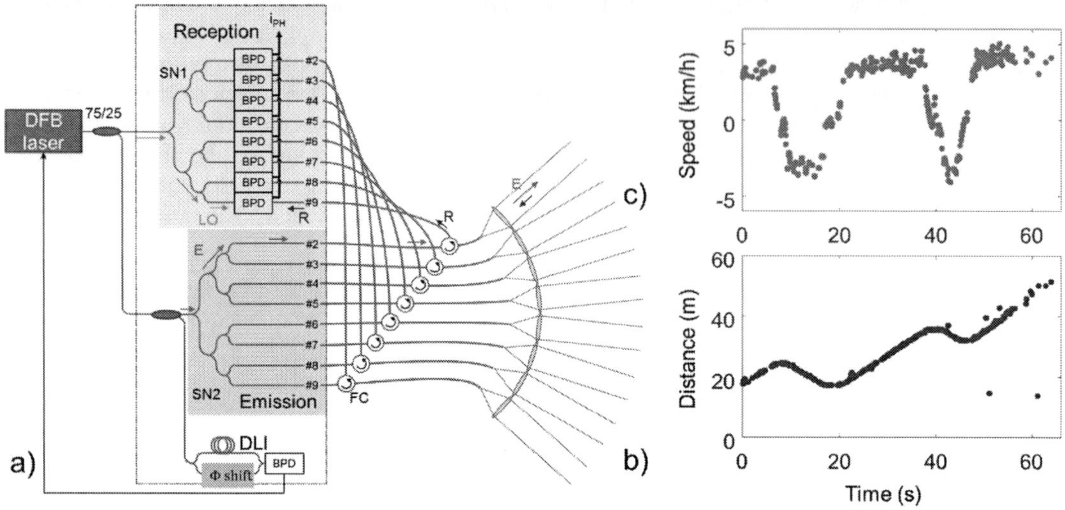

Figure 4. a) Architecture of the FMCW Lidar system (the rectangle delimits the chip perimeter), Range b) and speed c) of a moving target detected by the PIC LiDAR with less than 5 mW of output power.

We have demonstrated up to 30 mW of output power for 200 mW input power and a homogeneous routing of the 8 channels. Figure 4b and c) present the detection experiment of a moving target at up to 50 m of range with less than 5 mW of output power.

5. Conclusion

In this article, two different scanning systems were demonstrated; first, a 2-axis scanning head was combined with an FMCW Lidar allowing 3D measurements and second, switching networks implemented within a PIC LiDAR steered non-mechanically an optical beam. In the first case, clear air measurements allowed demonstrating a resolution limited by the frequency modulation of the laser and measurements in a dense fog highlighted the large detection dynamics of the LiDAR. This allows imaging of complex scenes with multiple targets with different albedos and at (very) different ranges. The performances of the system were experimentally measured to 3.9 attenuation length detection capabilities (single pass) at 70 m. In the second system, this architecture was integrated on-chip to demonstrate range and speed measurements up to 50 m with less than 5 mW of output power. This last system was not tested for 3D measurements but the scanning method potentially increases the speed and ruggedness of the system.

6. References and Acknowledgements

We acknowledge the funding of the DSTL under DSTL1000114607 “Seeing Through the Clouds”. Funding is also acknowledged by the European Union Seventh Programme for research, technological development and demonstration under grant agreement No 318178 - PLAT4M. We thank IMEC foundry for the fabrication of the chips and Tyndall University for the packaging.

[1] Busck, J., and Heiselberg, H “Gated viewing and high-accuracy three-dimensional laser radar”, Applied optics, 43(24), 4705-4710 (2004)

[2] Pawlikowska, A., Halimi, A., Lamb, R., and Buller, G. "Single-photon three-dimensional imaging at up to 10 kilometers range," Opt. Express 25, 11919-11931 (2017)

[3] Aflatouni, F., Abiri, B., Rekhi, A., and Hajimiri, A. "Nanophotonic coherent imager", Opt. Express 23, 5117-5125 (2015)

[4] Hardy, N., and Shapiro, J. "Computational ghost imaging versus imaging laser radar for 3D imaging", Physical Review A, 87, 023820 (2013)

[5] Feneyrou, P., Leviandier, L., Minet, J., Pillet, G., Martin, A., Dolfi, D., ... & Midavaine, T. "Frequency-modulated multifunction lidar for anemometry, range finding, and velocimetry–1. Theory and signal processing.", Applied Optics, 56(35), 9663-9675 (2017).

[6] Feneyrou, P., Leviandier, L., Minet, J., Pillet, G., Martin, A., Dolfi, D., ... & Midavaine, T. " Frequency-modulated multifunction lidar for anemometry, range finding, and velocimetry—2. Experimental results.", Applied optics, 56(35), 9676-9685 (2017).

[7] Poulton C. V. Yaacobi, A., Cole D., Byrd M., Raval M.,Vermeulen D. & Watts, M. "Coherent solid-state LIDAR with silicon photonic optical phased arrays", Optics Letters, 42(20), 4091-4094 (2017)

An optical scanner on silicon photonics integrated circuit for LIDAR application

Daisuke Inoue, Tadashi Ichikawa, Akari Kawasaki, Tatsuya Yamashita
Toyota Central Research and Development Laboratory incorporated
and 41-1, Yokomichi, Nagakute, Aichi, Japan
Author e-mail address:daisuke-i@mosk.tytlabs.co.jp

Abstract: The reliability and small size of solid state scanners makes them ideal for LIDAR. We fabricated an optical scanner using silicon photonics integrated circuit technology. The scanner comprises a ring resonator multiplexer and a number of grating arrays, and employs a beam switching method. The beam switching method means that the scanner is movement-free. We demonstrated the successful operation of this scanner. The multiplexer determines the optical path and one of the gratings is selected for the emitter. The scanning angle obtained was 6 degrees. LIDAR sensors can be used for automated cruising in automotive applications.

Keywords: Coherent Laser Radar, etc.

1. Introduction

Recently, the advent of automated cruising has intensified the competition in developing the sensor technologies required for this. LIDAR sensors are regarded as being important perimeter monitoring sensors. For automotive applications the sensors need to be compact because automotive components are densely packed. We consider that LIDAR based on silicon photonics integrated circuit technology is suitable for automotive applications.

We fabricated and demonstrated an optical scanner fabricated using silicon photonics integrated circuit technology. The beneficial features of this device are that it is movement-free and of small size.

2. Principle

The device is based on a beam switching method. As shown in Figure 1, it consists of an imaging lens and a photonics integrated circuit. The integrated circuit comprises an optical switch and a number of grating arrays. The optical switch selects the destination of the light path, and light is emitted from the selected grating. The gratings are arranged in front of the lens, which provides an image of the emitted light.

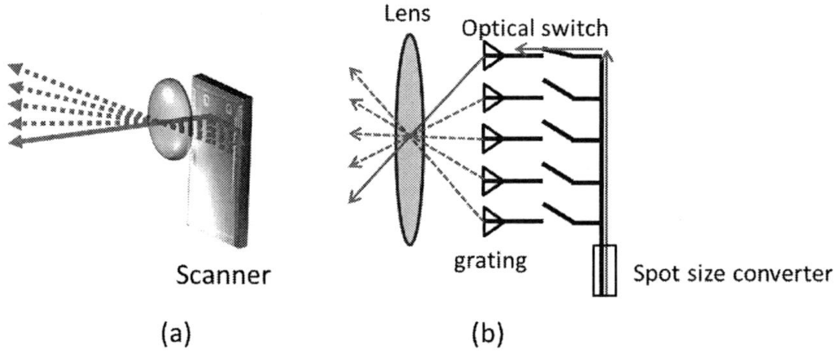

Figure 1. (a) Conceptual diagram of the movement free scanner and (b) schematic diagram

3. Device Structure

Figure 2 shows the structure of the integrated photonics chip. The devices are fabricated on silicon on insulator (SOI) substrates with a 0.22 μm thick silicon layer on top of a 3 μm thick buried oxide (BOX) layer. The waveguides are 0.4 μm wide and formed in the top silicon layer of the SOI substrate. These are covered with a 3 μm thick layer of silicon oxide, which in turn is covered with a 3 μm thick silicon dioxide layer. The silicon oxide layer is rich in silicon and has a higher refractive index than that of the silicon dioxide layer. So, light can be confined and can propagate in the silicon oxide waveguides.

Thin film heaters made of 0.4 μm thick tantalum are deposited on the silicon dioxide layer. The width of the thin film heaters is 10 μm. Current injected into the heater heats the waveguide, changing its refractive index.

The optical switch consists of a ring resonator, waveguides, and thin film heaters. Switching is controlled by injecting current into the thin film heaters installed on the waveguides.

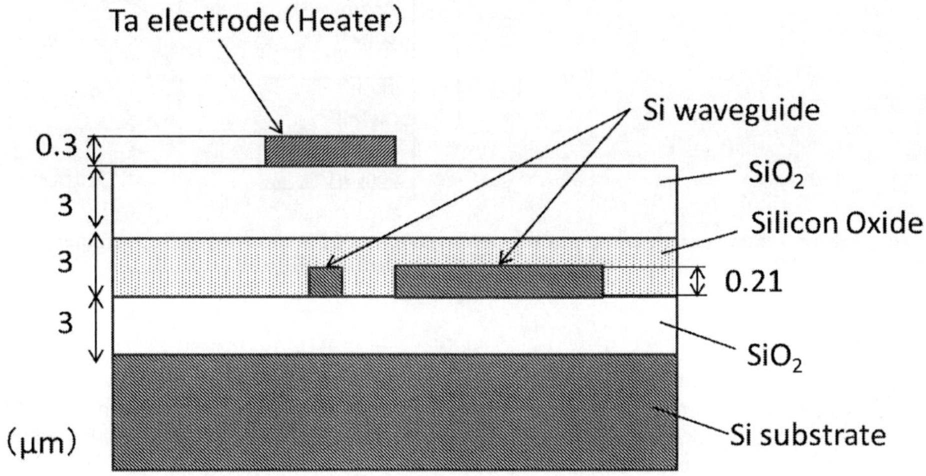

Figure 2. Cross sectional view of the photonics integrated circuit for the movement free scanner

Figure.3 shows the control circuit of the movement free scanner. This comprises a USB-SPI converter, an analog-digital converter (ADC) with 20 channels, and 20 buffer amplifiers. The USB-SPI converter decodes the command from a personal computer and transfers the output voltage to the ADC. The resolution of the ADC is 16-bit. The ADC cannot drive the heaters directly because the current consumed by the heaters exceeds 20 mA. The buffer amplifiers are connected to the ADC to drive the thin film heaters.

The heat density limits the size of the circuit and the maximum output current of buffer amplifiers.

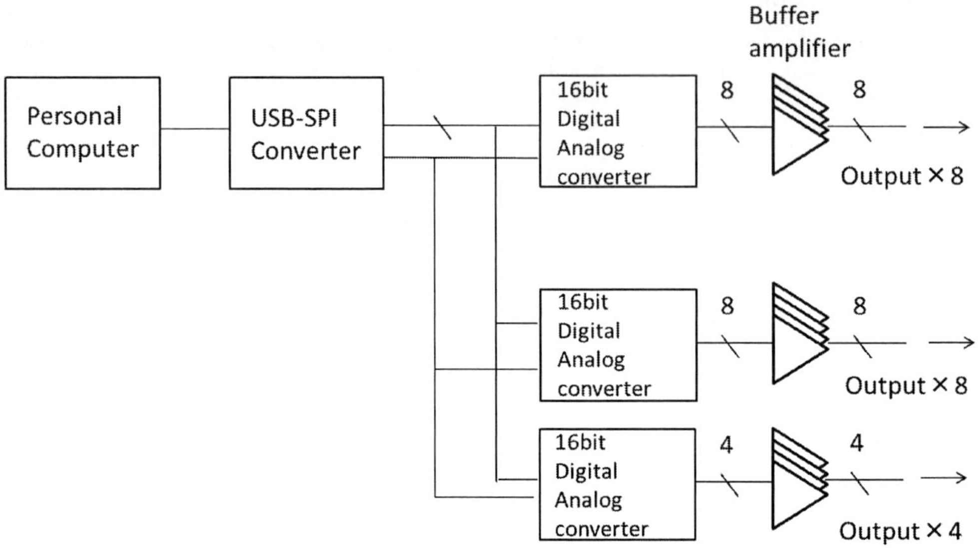

Figure 3. Schematic diagram of the control circuit of the movement free scanner

4. Result

We observed the far field pattern of this scanner using an infrared camera. Figure 4 shows the far field pattern of the solid state scanner. The scanning angle and resolution were measured and found to be 6 degrees and 0.3 degree, respectively, and these measurements agree with the values obtained from the pitch of the gratings and the focal length of the lens.

Measurements of the response time of the optical switch for this scanner at room temperature gave a value of 0.3 μs.

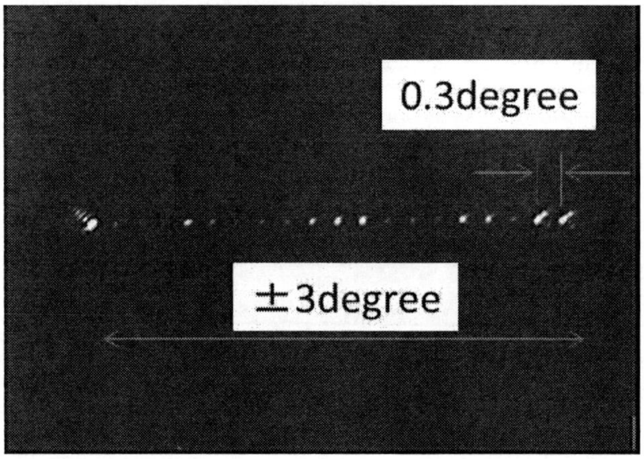

Figure 4. The far field pattern emitted from movement free scanner

5. Discussion

The scanning angle of our scanner is 6°. We can increase this by increasing the number of gratings or shrinking the focal length of the lens. Increasing the number of gratings by shrinking the pitch of the gratings increases the resolution.

Other kinds of scanners based on photonics integrated circuits have recently been reported. [4]-[6] The use of phased array antennas based on silicon photonics is one promising approach. The resolution of photonic phased array antennas is considered to be high. A comparison between optical phased array antennas and beam switching methods shows that both methods have merits and demerits. Phased array antennas require higher precision semiconductor processes. Beam switching methods require a lot of electrodes. However, integration of driver circuits on photonics integrated circuits can solve the issue of the number of electrodes. 0.13 μm technology is sufficient for beam switching methods.

6. Summary

We fabricated an optical scanner using silicon photonics integrated circuit technology. The scanner comprises a ring resonator multiplexer and a number of grating arrays, and employs a beam switching method. The beam switching method means that the scanner is movement-free. We demonstrated the successful operation of this scanner. The multiplexer determines the optical path and one of the gratings is selected for the emitter. The scanning angle obtained was 6 degrees. LIDAR sensors using such scanners can be used for automated cruising in automotive applications.

[1] Daisuke Inoue, Tadashi Ichikawa, Hiroyuki Matsubara, Manabu Kagami, "Improvement of Highly sensitive LIDAR with a thumb-sized sensor-head built using an optical fiber preamplifier(3)", 8731, SPIE defense, security and sensing 2013(USA)

[2] Daisuke Inoue, Tadashi Ichikawa, Hiroyuki Matsubara, Xueon Mao, Mitsutoshi Maeda, Chie Nagashima, Manabu Kagami, "Improvement of Highly sensitive LIDAR with a thumb-sized sensor-head built using an optical fiber preamplifier", 8379-33, SPIE defense, security and sensing 2012(FL, USA)

[3] Daisuke Inoue, Tadashi Ichikawa, Hiroyuki Matsubara, Xueon Mao, Mitsutoshi Maeda, Chie Nagashima, Manabu Kagami, "Improvement of Highly sensitive LIDAR with a thumb-sized sensor-head built using an optical fiber preamplifier", 8037-09, SPIE defense, security and sensing 2011(FL, USA)

[4] Firooz Aflatouni, Behrooz Abiri, Angad Rekhi, and Ali Hajimiri, "Nanophotonic coherent imager," OPTICS EXPRESS 23, 5117-5125 (2015).

[5] J. K. Doylend, M. J. R. Heck, J. T. Bovington, J. D. Peters, L. A. Coldren, and J. E. Bowers, "Two-dimensional free-space beam steering with an optical phased array on silicon-on-insulator r," OPTICS EXPRESS 19, 21595-21604 (2011)

[6] Jie Sun, Erman Timurdogan, Ami Yaacobi1, Ehsan Shah Hosseini & Michael R.Watts, "Large-scale nanophotonic phased array,",NATURE 493, 193-199(2013)

[7] Isamu Takai, Hiroyuki Matsubara, Mineki Soga, Mitsuhiko Ohta, Masaru Ogawa and Tatsuya Yamashita, " Single-Photon Avalanche Diode with Enhanced NIR-Sensitivity for Automotive LIDAR," Sensors MDPI, 16, 459 (2016).

Synergetic observations of downbursts using a Doppler lidar and radars in Okinawa, Japan

Hironori Iwai(a), Makoto Aoki(a), Shinsuke Satoh(a), Takeharu Koketsu(a), Hiroyuki Yamada(b), Tadayasu Ohigashi(c), Taro Shinoda(d), Kazuhisa Tsuboki(d)
(a) National Institute of Information and Communications Technology
Koganei, Tokyo, Japan
(b) University of the Ryukyus
Nishihara, Okinawa, Japan
(c) National Research Institute for Earth Science and Disaster Resilience
Tsukuba, Ibaraki, Japan
(d) Nagoya University
Nagoya, Aichi, Japan
Lead Author e-mail address: iwai@nict.go.jp

Abstract: On 28 July 2016, an isolated convective storm produced two downbursts within about 10 km north from a Doppler lidar at NICT's Okinawa Electromagnetic Technology Center in Okinawa, Japan. Concentric circular patterns of the Doppler velocity, spectral width, and carrier-to-noise ratio were detected near the sea surface in the 0°-elevation sector plane position indicator scans of the Doppler lidar. An X-band phased array weather radar (PAWR) detected the first echo approximately 13 min prior to the near-surface divergent signature of the downbursts. The PAWR provided 30-sec volumetric data of the storm and observed rapid evolution of reflectivity cores with >65-dBZ radar reflectivity. The cores descended rapidly toward the sea surface. The associated differential reflectivity minimum was observed by an X-band polarimetric radar. In this paper, we show that the synergetic use of different remote sensing instruments is important for better understanding the physical processes of downburst evolution.

Keywords: Downburst, Doppler lidar, Phased array weather radar, Polarimetric radar.

1. Introduction

Downbursts are near-surface small-scale diverging flow that can generate damaging winds (Fujita 1981) [1]. Wet downbursts are typically produced by strong downdrafts from convective storms. Ohno et al. (1996) analyzed 75 downbursts occurred in Japan and reported all of these events were wet downbursts [2].

Although sensitive Doppler radars can detect downbursts, the Doppler radars often suffer from significant ground clutters as well as an inability to detect clear-air echoes due to low radar reflectivity. Several previous studies demonstrated that Doppler lidars were a relevant remote sensing instrument for detecting the downbursts. Weber et al. (1988) have firstly demonstrated that the Doppler lidar has the capability of observing downbursts [3]. Menzies and Hardesty (1989) and Thobois and Soderholm (2015) showed the descending signature from cloud base and divergent outflow at the surface by range-height indicator (RHI) scans of Doppler lidars [4,5]. Hannon et al. (2005) showed the concentric circular pattern of a downburst and an outward propagating flow by plan position indicator (PPI) scans of a Doppler lidar [6].

This study is based on data collected from a Doppler lidar and radars at Okinawa, Japan. The Doppler lidar detected horizontally divergent flows near the sea surface. The data observed by the radars can be used to investigate rapid evolution of an isolated convective storm that produced two downbursts.

2. Instruments

NICT installed a Doppler lidar (Windcube 400S, Leosphere) and an X-band (9.32 GHz) phased array weather radar (PAWR) in NICT's Okinawa Electromagnetic Technology Center (26.50ºN, 127.84ºE; Fig. 1) to detect hazardous weather phenomena such as typhoons, tornadoes, and waterspouts in the subtropical region. The Doppler lidar and PAWR and were collocated on a steel tower. The Doppler lidar was placed at 25 m above mean sea level (MSL). Its range gate is a long and thin cylindrical shape with a length of 150 m and a diameter of 0.08 m. The Doppler lidar performed horizontal (0°-elevation) sector plane position indicator (PPI) scans from 225° to 15° in azimuth in clockwise direction. The scan rate and azimuthal resolution of sector PPI scans were 2° s^{-1} and 1°, respectively. The sector PPI scans provided information about the horizontal variability of the air flow near the sea surface. The PAWR has a slotted waveguide array antenna (horizontal polarization), which consists of 128 elements. The PAWR transmits broad beam using 12 to 24 elements and receives scattered signals with all 128 elements. Digital beam forming is performed for received signals with 109 elevation angles. The array antenna rotates in azimuth for 30 s. The PAWR has a spatial resolution of 100 m in the radial direction, 1.2° in the azimuthal direction, and about 1° in the elevation direction. An X-band (9.415 GHz) polarimetric radar (hereafter KIN radar) was installed on the roof of a building at the Senbaru Campus of the University of the Ryukyus (26.25°N, 127.77°E; Fig. 1). The KIN radar was operated in the volume scan mode at 14 elevation angles within 6 min. A Ka-band (34.87 GHz) polarimetric radar was installed on the roof of a building at the Sesoko Station of the University of the Ryukyus (26.64°N, 127.87°E; Fig. 1). The Ka-band radar was operated in the volume scan mode at 12 elevation angles within 10 min.

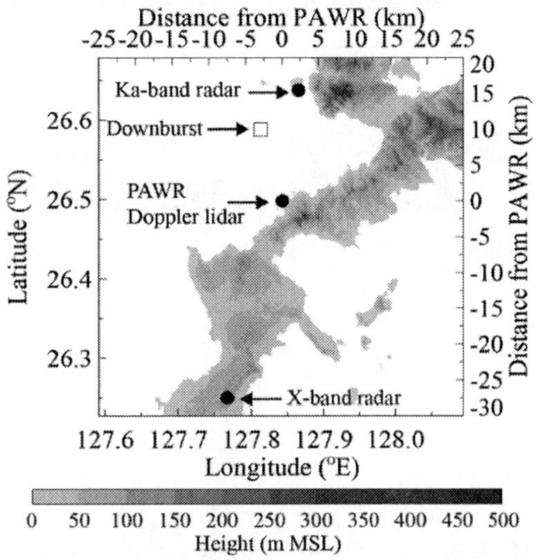

Figure 1. Map and topography of Okinawa Main Island. The grayscale bar shows the height of the topography in m MSL. The closed circles indicate the locations of the Doppler lidar, PAWR, X-band radar, and Ka-band radar. The square shows the position of the downburst detected by the Doppler lidar.

3. Results

Figure 2 shows the horizontal variability of the Doppler velocity of the near-sea-surface flow observed by the Doppler lidar with the 0°-elevation sector PPI scan. The Doppler lidar detected a signature of a downburst A but no significant horizontally divergent outflow at 1618:00 Japan Standard Time (JST) (not shown). Then, a concentric circular pattern of the Doppler velocity was detected and the peak Doppler velocity of –10.8 m s^{-1} was observed at 1625:00 JST (Fig. 2a). The downburst A expanded, becoming less intense, and it was advected south with a background flow. At 1627:00 JST, the

Doppler lidar detected another signature of a downburst B within the downburst A (not shown). Then, two concentric circular patterns of the Doppler velocity were detected by the Doppler lidar (Fig. 2b). The downburst B was less intense (smaller peak velocity) than the downburst A.

On 28 July 2016, convective rainfall had already begun in the morning around Okinawa Main Island. At 1605:00 JST, the PAWR detected the first echo about 11 km north-northwest of the PAWR at 3 km MSL (not shown). The radar reflectivity of the core A reached 50 dBZ at 4 km MSL at 1612:00 JST (Fig. 3a) and had values exceeding 65 dBZ at 1613:30 JST. The core A descended rapidly and the core A with >50-dBZ radar reflectivity reached sea surface at 16:18 JST (Fig. 3b). The core A produced the downburst A shown in Fig. 2a. At 1618:00 JST, a new precipitation core B developed at 2 km MSL about 1 km northeast of the core A. Although the 20-dBZ echo-top height was low (about 3-4 km MSL), the radar reflectivity of the core B reached 60 dBZ (Fig. 3c). The core B with 50-dBZ radar reflectivity reached sea surface at 1626:30 JST (Fig. 3d). However, the near-surface divergent signatures of the downburst A and B were not detected by the PAWR due to low radar reflectivity.

Figure 4 show vertical cross sections of differential reflectivity of the precipitation core A. The vertical cross sections shown in Fig. 4 are reconstructed from the volume scans of KIN radar. Based on Fig. 4a, the differential reflectivity values reveal that the core A was composed of raindrops at 1612:00 JST. In Fig. 4b, there was a local minimum of differential reflectivity (about 1.5 dB) in the descending reflectivity core A at 1618:00 JST. The minimum of differential reflectivity in the descending reflectivity core suggests melting hails were presented [7]. However, there was no local minimum of differential reflectivity in the descending reflectivity core B (not shown).

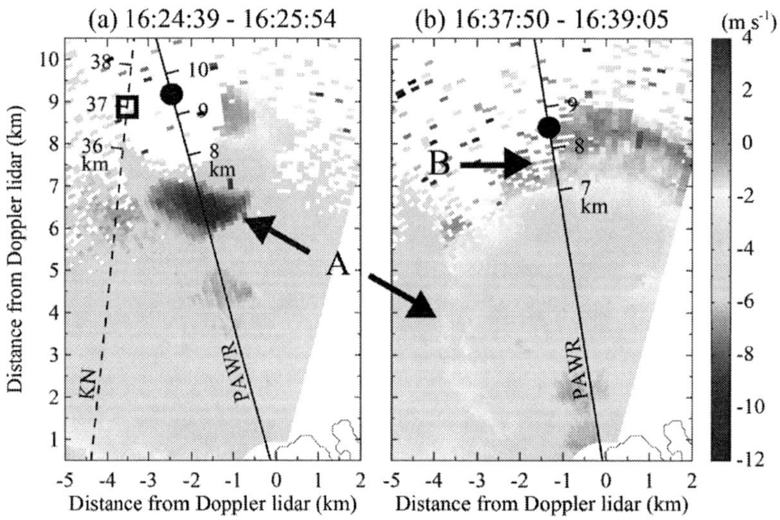

Figure 2. Doppler lidar 0°-elevation sector PPI scans of Doppler velocity (color bar in m s^{-1}) at (a) 1625 and (b) 1638 JST on 28 July 2016. Positive and negative velocities indicate flows away from and toward the Doppler lidar, respectively.

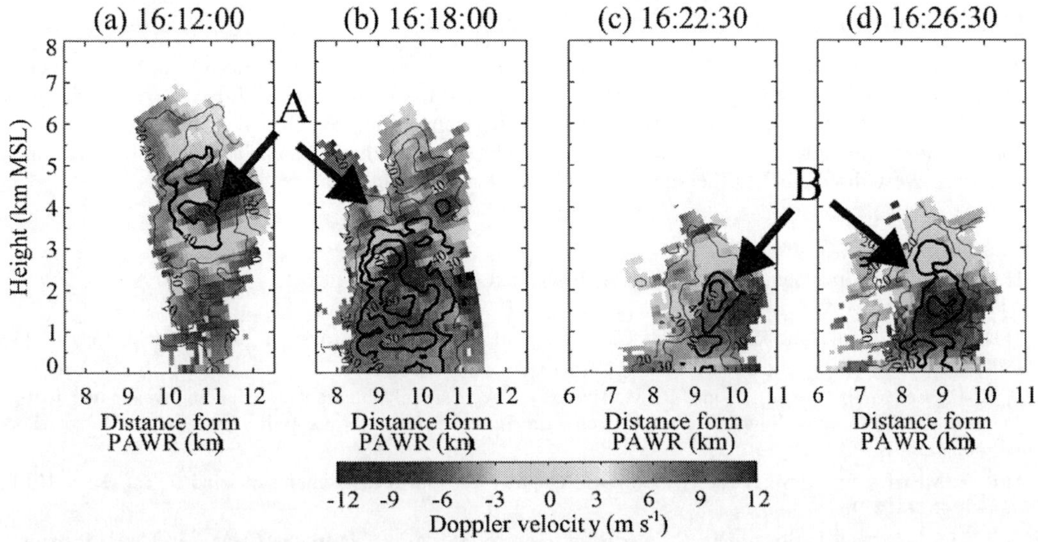

Figure 3. Vertical cross sections of Doppler velocity (color bar in m s^{-1}) observed by the PAWR at (a) 1612:00, (b) 1618:00, (c) 1622:30, and (d) 1626:30 JST on 28 July 2016. The verical cross sections of radar reflectivity are superimposed on the Doppler velocity (black lines). Thick black lines are radar reflectivity values with echoes greater than 40 dBZ. The horizontal axes are along the black line labeled PAWR in Fig. 2a (a,b) and Fig. 2b (c,d), respectively.

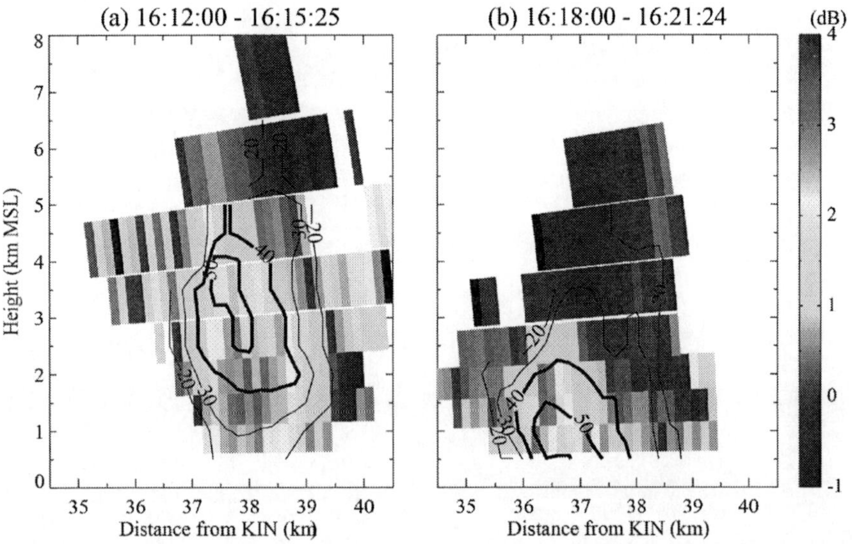

Figure 4. Vertical cross sections of differential reflectivity (color bar in dB) observed by the KIN radar at (a) 1612:00 and (b) 1618:00 JST on 28 July 2016. The vertical cross sections of radar reflectivity are superimposed on the Doppler velocity (black lines). Thick black lines are radar reflectivity values with echoes greater than 40 dBZ. The horizontal axes are along the black dashed line labeled KIN in Fig. 2a.

4. Summary

In this study, we described the dynamics of two downbursts and the downburst-producing convective storm using Doppler lidar and radar observations collected in Okinawa on 28 July 2016. Doppler lidar and radars captured the life cycle of these atmospheric phenomena. We show that the synergetic use of Doppler lidar and radar is important for better understanding the physical processes of downburst evolution. We will conduct further analysis using the Ka-band radar data.

5. References

[1] T. T. Fujita, "Tornadoes and downbursts in the context of generalized planetary scales," J. Atmos. Sci., **38**, 1511−1534 (1981).

[2] H. Ohno, O. Suzuki, and K. Kusunoki, "Climatology of downburst occurrence in Japan," Tenki, **43**, 101−112 (1996) (in Japanese).

[3] B. F. Weber, F. F. Hall, Jr, and R. M. Hardesty, "A positively buoyant downburst investigated in two dimensions with Doppler lidar," Proc. 15th Conf. on Severe Local Storms, Baltimore, Amer. Meteor. Soc. 403−406 (1988).

[4] R. T. Menzies and R. M. Hardesty, "Coherent Doppler lidar for measurements of wind fields," Proc. IEEE, **77**, 449−462 (1989).

[5] L. Thobois and J. Soderholm, "Nowcasting Severe Storms - Observing Clear Air Close Proximity Environment of Severe Storms," Meteorol. Technol. Int., 132–135 (2015).

[6] S. M. Hannon, K. S. Barr, D. K. Jacob, M. W. Phillips, "Application of pulsed Doppler lidar in the airport terminal area," Proc. SPIE, **5653**, Lidar Remote Sensing for Industry and Environmental Monitoring V (2005).

[7] R. M. Wakimoto and V. N. Bringi, "Dual-polarization observations of microbursts associated with intense convection: The 20 July Storm during the MIST Project." Mon. Wea. Rev., 116, 1521−1539 (1988).

Estimating Atmospheric Boundary Layer Height using Large-scale Coherent Doppler Lidar

Kohin Hirano(a), Shin-ichi Suzuki(a), Takeshi Maesaka(a), S. B. Grimmond(b)
(a) National Research Institute for Earth Science and Disaster Resilience
3-1 Tennodai Tsukuba Ibaraki 305-0006, JAPAN
(b) University of Reading
Gate PO Box 243 Reading RG6 6BB, UK
Lead Author e-mail address: hirano@bosai.go.jp

Abstract: The atmospheric boundary layer is important because it affects exchanges of momentum, heat, moisture, and aerosol between the surface and the free atmosphere above. Several methods have been proposed to estimate boundary layer height (BLH) using various remote sensing instruments. As part of the National Institute for Earth Science and Disaster Resilience (NIED, Japan) efforts to improve detection of storm clouds forming and weather forecasting through data assimilation, it has established a ground base remote sensing network in great Tokyo area since 2014 FY. The network includes 3 Large-scale Coherent Doppler Lidars (LDL) manufactured by Mitsubishi Electric, which have high power laser amplifier using Er, Yb:glass planer waveguide and output Doppler velocity, Doppler velocity width and signal-to-noise rate (SNR). This paper reports the estimation of BLH using these lidars and compares the results with subjective methods.

Keywords: Coherent Doppler Lidar, Boundary layer height, Doppler velocity, Velocity width, Signal-noise ratio.

1. Introduction

The atmospheric boundary layer (ABL) is defined as the part of the troposphere directly influenced by the earth's surface [1]. Determining the atmospheric boundary layer height (BLH) is important for weather, climate and air quality because it is where any moment, heat, moisture and aerosols from the surface exchange with the free atmosphere above. Traditionally, the BLH has been determined from temperature, humidity or wind profiles measured by radiosondes [2-3]. However, radiosondes cannot provide an adequate resolution of BLH since it is quite variable in both time and space, and, usually, it is difficult to set operational radiosonde stations in urban areas. Remote sensing technologies allow us to retrieve BLH using non-traditional ways. Several approaches have been carried out by using Ceilometer, Raman Lidar, and vertical-wind Doppler Lidar [4-6], however, all these approaches have their own restrictions on continuous operational use. The objective of this paper is to analyze Coherent Doppler Lidar data obtained by a preliminary observation over 48 hours in 2016 FY to assess the characteristic of BLH in Tokyo and discuss the availability of LDL in continuous operational BLH observations.

2. Instruments and Observations

The preliminary observations were made at the Ookayama campus of Tokyo Technology University, in the southwest of Tokyo (Figure 1) from 18:00 Japanese Standard Time (JST, GMT+9) 9 July 2016 to 20:00 JST 11 July. This was just after rainfall caused by a front 200 km to the south passed after Tokyo metropolitan area. This period was chosen to switch the sampling to the IOP scan patterns as no intense precipitation was expected for 48 hours but as it was the middle of summer intensive heating was expected to generate intensive boundary layer growth during the daytime.

The utilized Large-scale Coherent Doppler Lidar (DIABREZZA, LDL hereafter), manufactured by Mitsubishi Electric [7-8], was installed on the top of Ishikawadai Building 4 (lat. 35° 36' 19.5" N, lon. 139° 40' 56.1" E) at 79.8 m above sea level. This L_{TEC} was approximately 500 m to the northeast of the

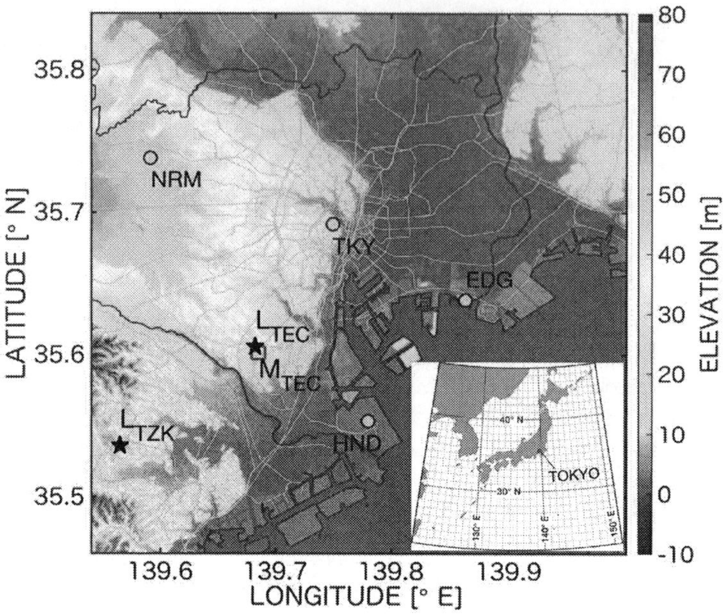

Figure 1. Location of instruments and the surrounding topography (5 m Mesh DEM Data, Geospatial Information Authority of Japan). Black stars (L_{TEC}, L_{TZK}) denote the position of Doppler Lidar, red square (M_{TEC}) is the MWR, and green circles (NRM, TKY, EDG, and HND) show the position of AMeDAS stations. Light gray lines denote the railway lines to central Tokyo.

Radiometer Physics GmbH Humidity And Temperature PROfilers (RPG-HATPRO) MWR (MTEC, lat. 35° 36' 4.3" N, lon. 139° 41' 6" E).

The wind profile was observed using the Doppler Beam Swinging (DBS) method, with the beam pointed vertically to take 3 observations, then tilted 15° to the north, and then to the east with 3 observations at each direction. Since the pulse repetition frequency and pulse hits are 4000 Hz and 8000 respectively, the LDL point in one direction for 6 s (3 × 8000 / 4000) and it takes 28 s to complete the entire scan cycle. The scan cycle repeats every 120 s.

The Radiometer Physics GmbH Humidity And Temperature PROfilers (RPG-HATPRO) MWR allow vertical profiles of atmospheric temperature and humidity with high temporal and spatial resolution to be derived. The RPG-HATPRO works in two scanning modes: zenith point scanning mode and boundary layer scanning mode that producing 30 m resolution temperature profile for 0-1200 m from Plan Position Indicator (PPI) scans at 6 elevations. The temperature data applied in this study is the component of the boundary layer scanning (lower layers) and the zenith scanning (upper layers). The vertical resolution of zenith scanning mode is 200 m for range 1,200-5,000 m, and 400 m from 5,000 to 10,000 m. For liquid water profile, the vertical resolution is 250 m to 2,000 m, 300 m from 2,000 to 5,000 m and 400 m from 5,000 to 10,000 m. MTEC sensor has a boundary layer scan every 5 minutes, which takes approximately 3 minutes. The sensors point towards the zenith direction for the remaining time. Thus, the MWR provides a temporal and spatial average for each 5 minutes.

3. Results and Discussion

The Lidar observations from zenith-direction scan with threshold exceeded SNR (10 dB) are shown in Figure 2. The results are intermittent at approximate 300 m but appear consistent above this. As the manufactures specified minimum detectable distance is 400 m, the data below 300 m are ignored. For the first 12 hours, the available data are restricted to the lower levels, but after 06:00 this limitation is overcome. This is likely caused by the low aerosol concentration after rainfall and then the increasing urban activities. The striped pattern around 9:00-16:00 on both 10th and 11th are associated with stronger vertical motions, whereas weaker upward motions occur at the other times.

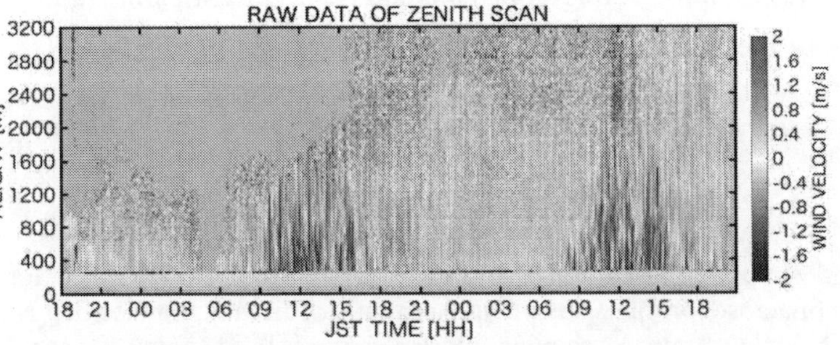

Figure 2. LDL zenith scan raw data

The BLH is estimated from all three output parameters of LDL, Doppler velocity, velocity width, and signal-noise ratio (SNR), respectively, using a combination of gradient method and variance method.

All methods captured the diurnal cycle of BLH, but differences are found among gradient method, which based on the behavior of SNR and velocity width, and velocity variance methods. This is attributable to the definition of BLH since SNR responding to the aerosol concentration density, but velocity variance represents turbulence scale. Hence, a hybrid method is necessary for estimating BLH continuously.

4. References

[1] Stull, R. B., *An Introduction to Boundary Layer Meteorology* (Kluwer Academic Publishers, Norwell, MA, 1988), Chap. 1.

[2] Grossman, R. L., and N. Gamage, "*Moisture flux and mixing processes in the daytime continental convective boundary layer*", J.Geophys. Res., 100D, 25 665–25 674 (1995).

[3] Guo, J. P., and Coauthors, "*The climatology of planetary boundary layer height in China derived from radiosonde and reanalysis data*", Atmos. Chem. Phys., 16, 13 309–13 319 (2016).

[4] Seibert, Petra, Frank Beyrich, Sven-Erik Gryning, Sylvain Joffre, Alix Rasmussen, and Philippe Tercier. "Review and intercomparison of operational methods for the determination of the mixing height." Atmospheric environment 34, no. 7, 1001-1027 (2000).

[5] Peng, J., Grimmond, C. S. B., Fu, X., Chang, Y., Zhang, G., Guo, J., Tang, C., Gao, J., Xu, X., and Tan, J., "*Ceilometer-based analysis of Shanghai's boundary layer height (under rain- and fog-free conditions)*", Journal of Atmospheric and Oceanic Technology, 34(4), 749–764 (2017).

[6] Baars, H., Ansmann, A., Engelmann, R., & Althausen, D., "*Continuous monitoring of the boundary-layer top with lidar*", Atmospheric Chemistry and Physics, 8(23), 7281–7296 (2008).

[7] Sakimura, T., Y. Watanabe, T. Ando, S. Kameyama, K. Asaka, H. Tanaka, T. Yanagisawa, Y. Hirano, and H. Inokuchi, "*3.2 mJ, 1.5 μm laser power amplifier using an Er, Yb: glass planar waveguide for a coherent Doppler LIDAR*", Proc. 17th Coherent Laser Radar Conf., (2013).

[8] Kameyama, Shumpei, Takayuki Yanagisawa, Toshiyuki Ando, Takeshi Sakimura, Hisamichi Tanaka, Masashi Furuta, and Yoshihito Hirano, "*Development of wind sensing coherent Doppler lidar at Mitsubishi Electric Corporation from late 1990s to 2013.*" In Abstracts Int. Coherent Laser Radar Conf., (2013)

Aerosol profile measurement by a high-spectral-resolution lidar with a scanning interferometer

Yoshitaka Jin(a), Nobuo Sugimoto(a), Tomoaki Nishizawa(a), and Hajime Okamoto(b)

(a) National Institute for Environmental Studies
16-2 Onogawa, Tsukuba, Ibaraki 305-8506, Japan
(b) Research Institute for Applied Mechanics, Kyushu University
6-1 Kasuga Park, Kasuga, Fukuoka 816-8580, Japan
Lead Author e-mail address: jin.yoshitaka@nies.go.jp

Abstract: A simple high-spectral-resolution lidar (HSRL) is developed for continuous observation of aerosol profile. The second harmonic of an injection-seeding Nd:YAG laser is used for the light source. A scanning interferometer is used to separate the Rayleigh and Mie backscattering components. The interferometer is periodically scanned in the span of one fringe. Lidar signals at the minimum and maximum Mie scattering transmission are obtained. Rayleigh scattering signals are derived by subtracting the bias (cross-talk) component from the lidar signals at the minimum Mie scattering transmission, and the aerosol extinction coefficient is retrieved. The interferometer transmittance and the fringe position are calibrated with the reference signals taken from a portion of the transmitted laser beam. The method is originally developed for an HSRL with a multimode laser, but the measurement sensitivity is higher when using a single-mode laser.

Keywords: High-spectral-resolution lidar, aerosols, interferometry.

1. Introduction

Aerosols play important roles in the Earth's climate through scattering and absorbing solar radiation. Observation of aerosol optical properties and spatial distribution is needed to understand effects of aerosols on Earth's radiation budget. Aerosol backscattering and extinction coefficients can be obtained from lidar observation. For elastic lidars, extinction-to-backscattering ratio (lidar ratio) is assumed to retrieve extinction coefficient. To investigate aerosol optical properties quantitatively, extinction coefficient should be measured independently from the backscattering coefficients. The independent measurement can be realized using High-spectral-resolution lidars (HSRL).

HSRL measures Rayleigh scattering separately from Mie scattering. The wavelength of Rayleigh and Mie scattering is the same as that of the transmitted laser because of the elastic scattering, but there is a big difference between two scattering in terms of the Doppler broadening since the motion velocities for atmospheric molecules and aerosols are different. The full width at half maximum (FWHM) of Rayleigh spectrum is ~2.5 GHz at 532 nm while that of Mie spectrum is ~100 MHz that is almost the same as the transmitted narrow band laser. High resolution spectral filters such as Interferometers or vapor filters are used to separate the Rayleigh scattering from the Mie scattering [1,2]. To match the laser wavelength with the minimum transmission of the spectral filters to block the Mie scattering, the laser should be controlled properly. The control is not easy, and the lidar system is to be complicated.

In this study, we develop a simple HSRL using a scanning interferometer. The interferometer is periodically scanned in the span of one fringe, and the lidar signals at the minimum and maximum Mie scattering transmission are obtained. The interferometer transmittance and the fringe position are calibrated with the reference signals taken from a portion of the transmitted laser beam. The scanning method does not need any complicated controls and may be suitable for continuous observations of aerosol profiles. The method was originally developed for an HSRL with a multimode laser [3].

The multimode HSRL system uses a long-arm interferometer to have a free spectral range (FSR) of ~255 MHz to be matched with the mode spacing of the multimode laser. Therefore, the rejection rate for the Mie scattering is poor. Using single-mode lasers, the interferometer length can be optimized to obtain better rejection rate for the Mie scattering. In this study, we investigate how the measurement errors are improved for the single-mode HSRL compared to the multimode HSRL.

2. System and analysis method

A block diagram of the HSRL is shown in Figure 1. The second harmonic of an injection-seeding Nd:YAG laser is used as the transmitter. The pulse energy is 200 mJ and pulse repetition rate is 10 Hz. The spectral width of the laser is approximately 120 MHz. An 8-inch telescope is used as the receiver. Field-of-view (FOV) of the receiver is 0.1 mrad. A 2-inch telescope with 2 mrad FOV is also used to correct the overlap function of the lidar signals for the 8-inch telescope. A portion of transmitted laser is used for the reference light. The reference and lidar signals are recorded every shot.

A scanning Mach-Zehnder interferometer is used as the high resolution spectral filter. Scanning frequency of the interferometer is 1 Hz. The FSR of the interferometer is set to 2 GHz. The rejection rate for the Mie scattering increases as the FSR, but the FSR should be smaller than the spectral width of the Rayleigh scattering to reduce variation of the interferometer transmittance for the Rayleigh scattering. The interferometer transmittance for the Rayleigh scattering can be approximated as 0.5 for 2 GHz FSR.

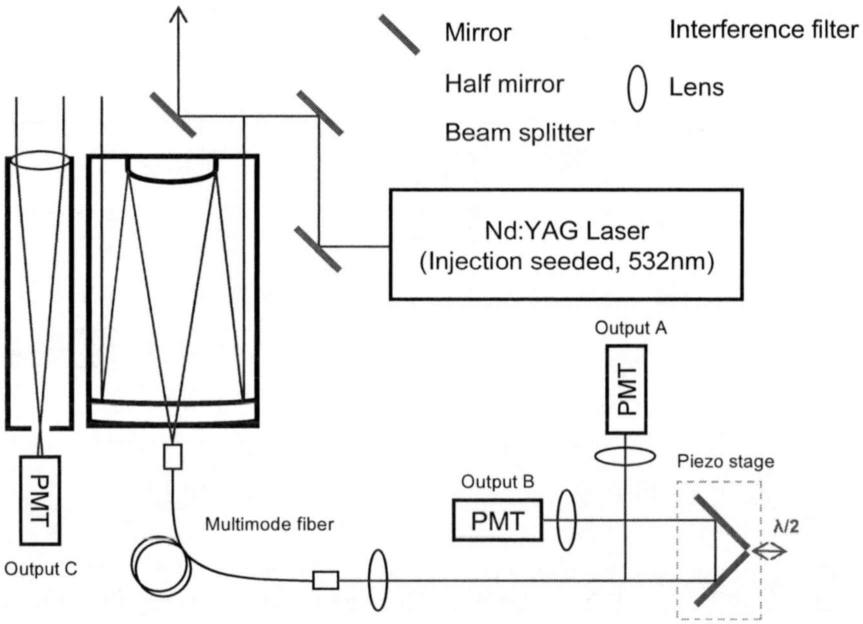

Figure 1. Block diagram of HSRL using a single-mode laser and a scanning interferometer.

Lidar signal intensity (output A and B in Figure 1) changes depending on the relative position of the laser and the interferometer spectra. When the transmittance minimum matches the laser, the Mie scattering component is rejected most efficiently, resulting in the lowest signal strength. Part of the Mie scattering component remains as a bias (cross-talk) due to the laser spectral width and the interferometer imperfection. The bias component must be removed from the lowest signals to obtain Rayleigh scattering signals. The bias component is estimated by measuring reference light with the same interferometer.

The lowest lidar signals are estimated by fitting a curve for the scan data. The minimum of the fitting curve will be the lowest lidar signal. To improve signal-to-noise ratio (SNR), the fitting is performed after integrating the scan data for a certain period of time. In this study, we integrate the scan data for 5 minutes (i.e., 3000 shots). Since the system does not control the laser, the laser wavelength can shift depending on the air temperature. Therefore, the phase between the laser and the interferometer can change with time. The phase shift is corrected every scan using the reference signals.

3. Results

Figure 2 shows range-corrected signals and scan data at three heights that were measured during the daytime on February 20, 2018 at Tsukuba, Japan. The scan data are normalized by the total intensity (i.e., $P_A / (P_A + P_B)$). For reference signals (Figure 2(b)), the scan data represents the interferometer transmittance for the Mie scattering. The minimum transmittance is used to estimate the bias component. For atmospheric scattering signals (Figure 2(c) and 2(d)), the amplitude of the scan data at a height of 1.5 km is larger than that at 3 km since Mie scattering is stronger.

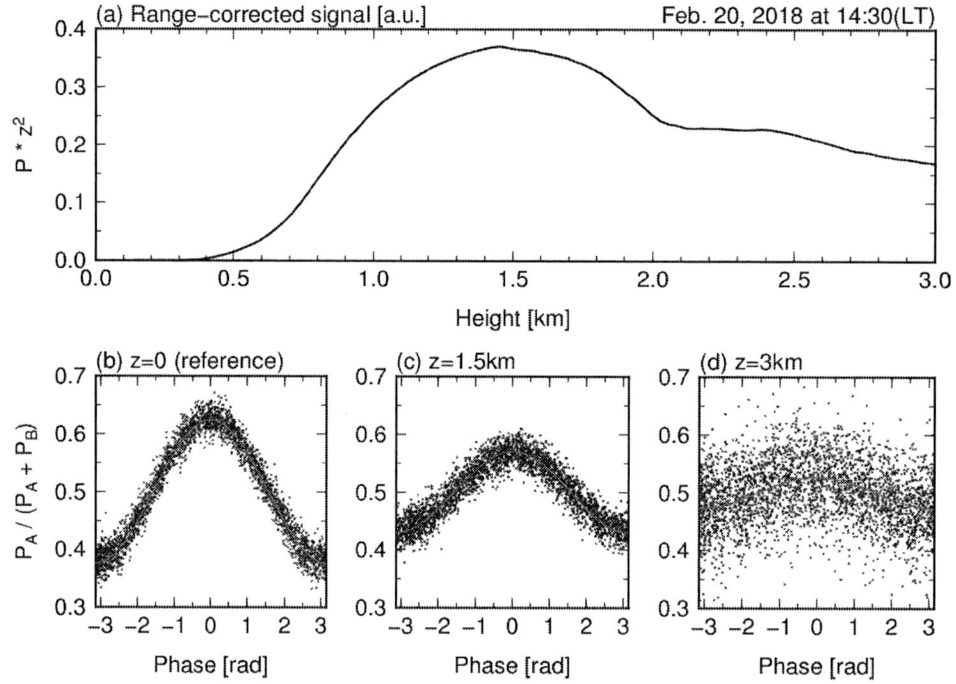

Figure 2. (a) Range-corrected signal measured at 14:30 (local time) on February 20, 2018 at Tsukuba, Japan, and scan data (b) for the reference signals and for atmospheric scattering signals (c) at 1.5 km and (d) at 3 km. Red lines in (b)–(d) denote fitting curves for the scan data.

The backscattering coefficient is retrieved from the amplitude of the scan data for reference and atmospheric scattering signals. Figure 3(a) shows retrieved backscattering coefficient. Well mixed aerosols in the boundary layer are observed. The extinction coefficient is retrieved from the Rayleigh scattering signals obtained by subtracting the bias component from the lidar signals at the minimum Mie scattering transmission. Retrieved extinction coefficient is shown in Figure 3(b). The random errors are calculated from the SNR of the lidar signals. The errors in the extinction coefficient are approximately 10 times smaller than the one measured by the multimode HSRL.

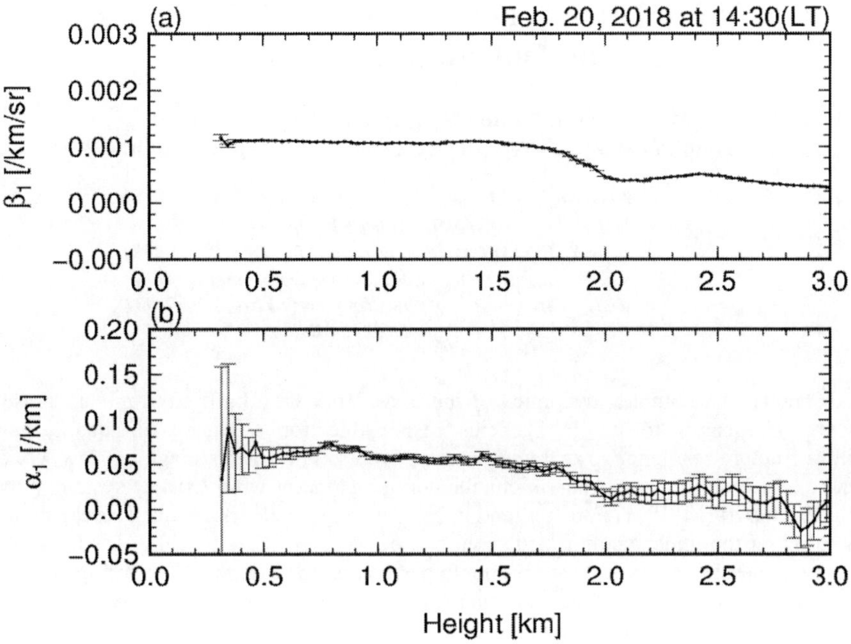

Figure 3. (a) Backscattering coefficient and (b) extinction coefficient measured at 14:30 (local time) on February 20, 2018 at Tsukuba, Japan. The error bars denote random errors estimated from the signal-to-noise ratio.

4. Conclusions and future works

Using a single-mode laser and a scanning interferometer, a simple HSRL is developed. Since the SNR and the rejection rate for the Mie scattering are improved compared to the multimode HSRL, the measurement errors considerably decrease (approximately one-tenth). However, the errors below 500 m are still large. This is because the narrow FOV is employed to reduce the beam divergence in the interferometer. To make the FOV wider, we plan to insert a glass block at the interferometer arm to compensate the angle dependence of the interferometer. Moreover, we will compare the interferometer based HSRL with an iodine filter based HSRL for a performance evaluation.

5. Acknowledgement

This work is supported by JSPS kakenhi JP17H06139.

6. References

[1] S. T. Shipley et al., "High spectral resolution lidar to measure optical scattering properties of atmospheric aerosols. 1: theory and instrumentation," Appl. Opt. **22**, 3716 - 3724 (1983).

[2] Z. Liu, I. Matsui, and N. Sugimoto, "High-spectral-resolution lidar using an iodine absorption filter for atmospheric measurements," Opt. Eng. **38**, 1661 - 1670 (1999).

[3] Y. Jin et al., "Measurement method of high spectral resolution lidar with a multimode laser and a scanning Mach-Zehnder interferometer," Appl Opt. **56**(21), 5990 - 5995 (2017).

Vertical Wind Patterns in the NYC Built Environment: Wind LIDAR Observations of Dispersion and Downburst Behavior

Mark Joseph Campmier (a), Aris Fernandez (b, c), Yonghua Wu (b), Fred Moshary (a, b, c), Mark Arend (b, c)
(a) Environmental Engineering Program,
CUNY City College of New York, New York, NY, 10031
(b) NOAA-CREST Center
CUNY City College of New York, New York, NY, 10031
(c) Electrical Engineering Department
CUNY City College of New York, New York, NY, 10031
mcampmier@gmail.com

Abstract: The complex dynamics of the New York City built environment present many challenges to public agencies responsible for assessing air quality, and infrastructure resilience to extreme weather events. The City College of New York's suite of remote sensing observations including a Doppler wind LIDAR system, micro pulse LIDAR (MPL), ceilometer, and microwave radiometer have aided researchers in improving the understanding and modeling of weather and air quality for hazardous events. Over the course of a seven month period incorporating the late spring, summer, and early autumn, the ORSL implemented a number of measurement strategies including PBL height measurements, VAD, and vertical stares to increase understanding of wind shears, and their effects on air pollution transport and infrastructure. Initial results indicate that an integrated approach to urban observations is necessary for robust air quality monitoring during extreme weather events.

Keywords: Doppler Laser Radar, Air Quality Monitoring, New York City

1. Introduction

New York City (NYC) is a uniquely challenged urban environment. Its dense population and costal land-ocean-atmospheric interactions leave it vulnerable to many environmental risks. The United States Environmental Protection Agency (EPA) has designated NYC a nonattainment area for violating ozone standards [1]. Tropospheric ozone formation is driven by sun driven photochemical reactions between NO_x and VOCs, leaving the city especially vulnerable during clear sky heat waves

Figure 1: Satellite imagery of New York City with Manhattan Island featured in the center and CCNY's Hamilton Heights neighborhood in the yellow highlighted region.

[2]. During heat waves the planetary boundary layer (PBL), the lowest layer of the atmosphere in which there is the most turbulent mixing, reaches heights not typically seen in cooler seasons [3]. The PBL acts as a limit to the transport and mixing of pollutants from the surface [3]. As a result, heat waves have been associated with ozone events since the intense insolation, and relatively high PBL height increases the likelihood of ozone formation from vertical mixing in the lower atmosphere [2].

Laser Radar (LIDAR) can offer insights into the vertical mixing, and horizontal transport of the pollutants via wind. Specifically, Coherent Doppler LIDAR systems are capable of taking minimally supervised eye-safe measurements 24-hour each day. At the City College of New York (CCNY), the research team at the NOAA-CREST Optical Remote Sensing Lab (ORSL) operates a suite of remote sensing instruments dedicated to measuring the thermal and momentum fluxes, and structures in the urban atmosphere. Over the course of the past year, the team has been calibrating and testing various scan strategies of a recently acquired Coherent Doppler LIDAR system. All of these initial tests have relied on the operation of a single Doppler LIDAR system as opposed to cooperating systems used to attain wind vector information from two or more systems.

2. Methodology

The system used is a Leosphere Windcube 200s with a 1543 nm laser. As expected from a Doppler LIDAR system, the relative velocity of the wind is measured by calculating the Doppler shift from the laser backscatter from aerosols. Using an internal algorithm, the system returns relative wind speeds along the line of sight (LOS) at given range gates. In the urban environment, instrument placement can be difficult since surrounding buildings can easily block the signal. The team at the ORSL decided to place the LIDAR on the roof of CCNY's Steinman Hall, to minimize obstructions and maintain connections to power and ethernet.

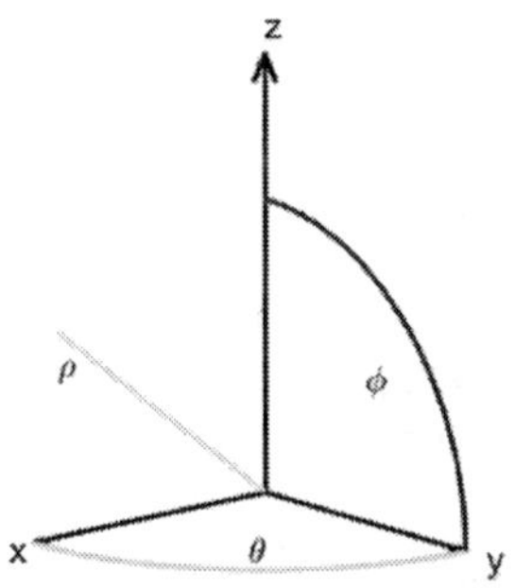

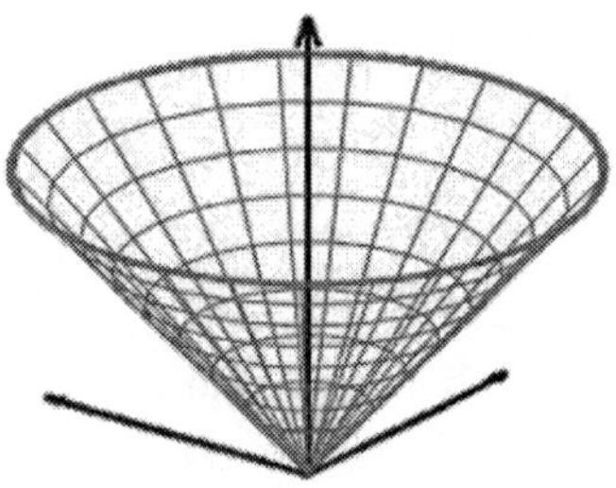

Figure 2-3: (L) Figure 2 depicts the variables used to represent the Cartesian and spherical coordinate systems used in the analysis. (R) Figure 4 is a photo of the LIDAR arrangement on the roof.

For the arrangement used at CCNY, the axes are defined as follows: x represents east-west, y represents north-south, z represents altitude above roof level (approximately 30 m above ground level), θ represents the azimuth angle (degrees increasing clockwise from north which is 0°), φ represents the elevation angle with 0° parallel to the surface, and ρ representing the LOS of the laser. For measurements of the vertical component of the wind (w), the ρ is simply aimed at φ equal to 90°. Positive values are wind vectors moving toward the LIDAR, and negative values are moving away from it. Since the x and y axis components of the wind (u and v respectively) are orthogonal to the beam, the measurement requires no additional processing past the radial wind speed (V_R) output from the internal algorithm.

Measuring u and v, collectively referred to as the horizontal wind velocity (V_H), is much more difficult to measure especially with a single Doppler LIDAR. To measure V_H, the Velocity-Azimuth Display (VAD) scan strategy was implemented. VAD utilizes a conical scan strategy from the LIDAR to measure V_R around the entire 360° of θ at a given z and a static φ [4]. Provided the LIDAR can attain more than 6 measurements at an elevation z, the signal can be represented with the following equation (1), and fit to a curve using Fourier analysis to find V_H [4].

$$(1)\ V_R = -V_H(\theta * (\theta - \beta)) * cos(\varphi) + V_f(\theta) * sin(\varphi)$$

In equation (1) β represents the direction of the V_H, and V_f represents the fall velocity or the speed of falling particles such as rain [4]. Following the curve fit the following equations can be used to calculate V_H, V_f, and ∇V_H[4]. The Fourier coefficients are represented by a_n and b_n.

$$(2)\ V_H = \frac{-\sqrt{a_1^2+b_1^2}}{\cos(\varphi)}$$

$$(3)\ V_f = \frac{\frac{1}{2}*(a_0-a_2)}{\varphi}$$

$$(4)\ \nabla V_H = -\frac{a_0}{z*\cos^2(\varphi)} + 2 * \frac{V_f}{z*\cos(\varphi)} + \tan(\varphi)$$

$$(5)\ \beta = \frac{3\pi}{2} - \arctan\left(\frac{a_1}{b_1}\right)$$

$$(6)\ u = V_H * \sin(\beta), v = V_H * \cos(\beta)$$

It is vital to note that the VAD scan strategy must assume a homogenous wind field and as such these calculations assume w is approximately equal to 0, so during episodes with intense w, VAD will yield unreliable results [4]. During the heat wave episode discussed in the results section, horizontal wind speeds were slow, so the VAD will be more useful for detecting nighttime dispersion of pollution generated during days with intense insolation, or measuring intense storms. In addition to the Doppler LIDAR system mentioned previously, a 1054 nm Backscatter LIDAR, ceilometer, and microwave radiometer were utilized to provide supporting evidence, however they will not be the focus of the analysis.

3. Results

On July 20, 2017 NYC experienced a heat wave event experiencing temperatures as high as 34°C and low horizontal wind surface wind speeds averaging approximately 1.8 m/s. The ceilometer scans from the day with only scattered clouds, so the LIDAR should be able to scan high without obstruction. Figure 5 shows the atmospheric potential temperature profile from the microwave radiometer. There

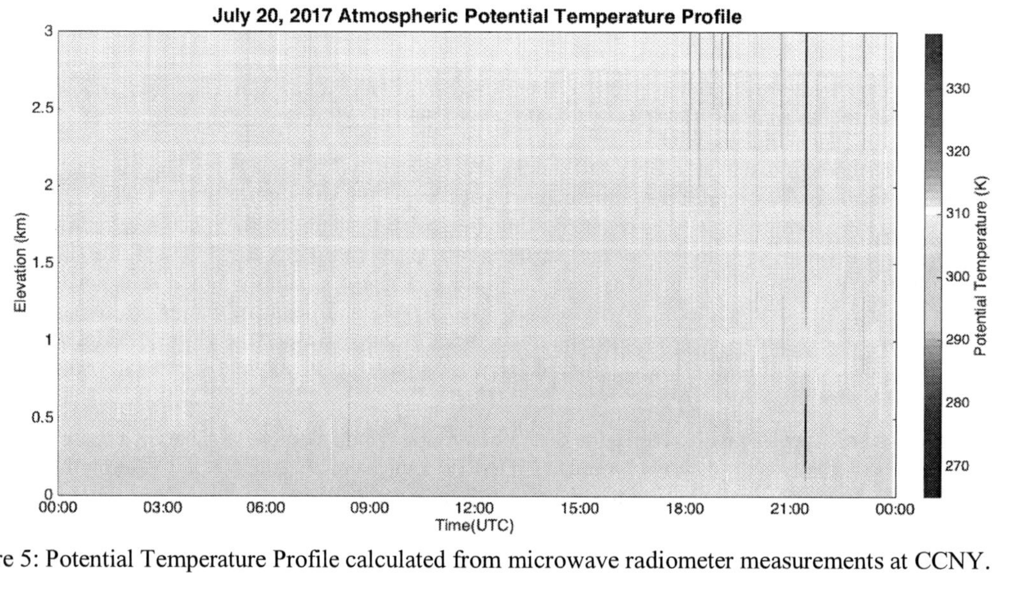

Figure 5: Potential Temperature Profile calculated from microwave radiometer measurements at CCNY.

are no obvious capping inversions, instead the temperature profile of the lowest 3 km of the atmosphere are almost entirely homogenous pointing to the possibility of intense mixing conditions.

During the day, the LIDAR operated 2 minute vertical scans alternated with 6 minute VAD scans at φ equal to 60°. The data was processed following World Meteorological Organization 20 minute standard for wind measurements. The figure below summarizes results from July 20th and feature wind w wind measurements from the lowest 3 km of the atmosphere relative to the LIDAR. The figure on beneath it depicts the results of the VAD scan for a five-day time frame (July 19-22) reflecting conditions before and during the heat wave episode.

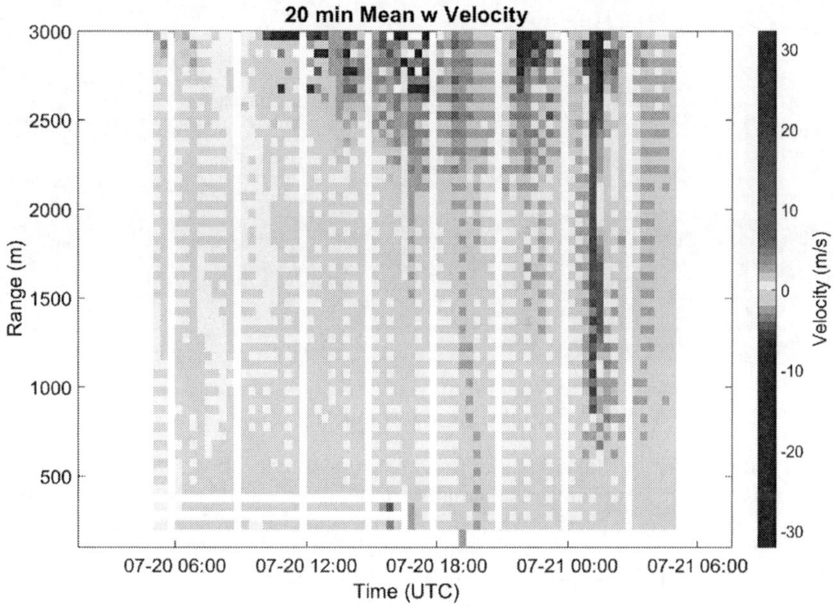

Figure 5: The w profile of the atmosphere for July 20, 2017 (EDT) or 6:00 July 20, 2017 to 6:00 July 21st (UTC)

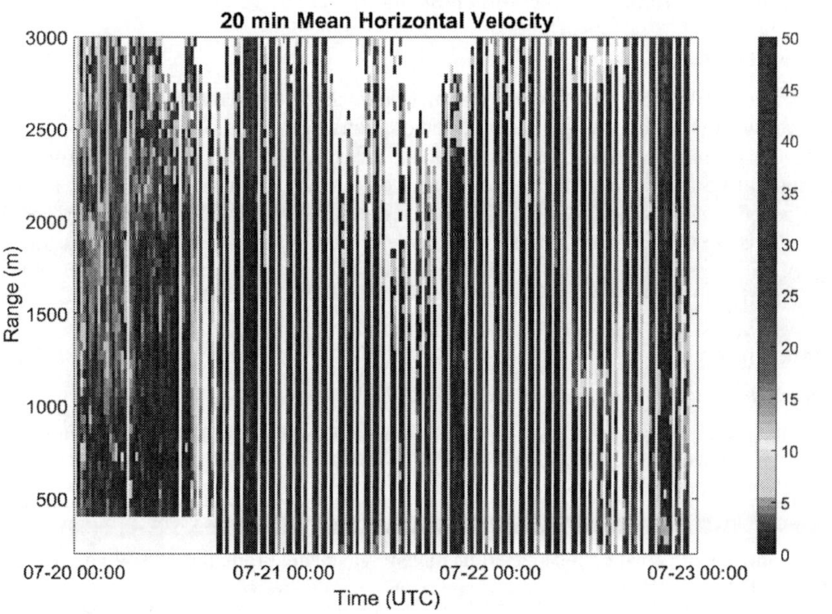

Figure 6: The V_H profile of the atmosphere for July 19-22, 2017 (EDT) or 6:00 July 23, 2017 to 6:00 July 21st (UTC)

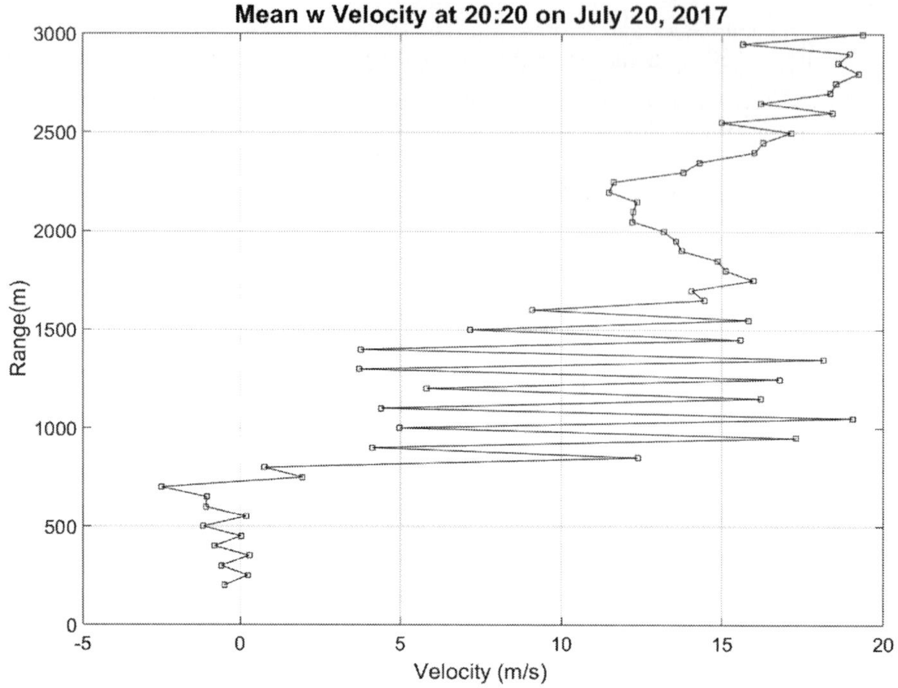

Figure 7: Profile of the observed downburst at 20:20 (EDT) on July 20, 2017

The w profile shows clearly illustrates strong convective activity throughout the day and night with very few periods of calm confirming the supposition that vertical mixing would take place based on the temperature profile. As expected, when the w started to rise, the V_H VAD calculations ceased to represent reality, as values approached > 50 m/s, which was not corroborated with other data sources. However, before the heat wave, the VAD calculated values behaved similar to expected values based on other sources of upper air. One particular event of importance was the downburst, which occurred at 20:20 on July 20th. Figure 7 shows the intense downward motion of the wind indicative of the possible mixing which promotes ozone formation during heat wave events.

4. Conclusions

After processing wind measurements and calculations using a Coherent Doppler Laser Radar system, in cooperation with other active and passive remote sensing technology, the team was able to detect intense convective mixing, and identify a strong downburst. Additional post-processing statistics including qualitative and quantitative assessments of turbulence and mixing offer more evidence to support the usage of the LIDAR system as a robust meteorological tool. Air pollution will continue to be a challenge to NYC and other dense urban centers, however, as researchers gain more experience with this tool and capture more data, it can be used to inform policy makers and the general public on weather and air quality patterns and predictions.

5. References

[1] J.J. Snyder, *Designation Recommendations for the 2015 Ozone NAAQS*, (NYS DEC, 2016)

[2] J.K. Rosenthal, P.L. Kinney, K.B. Metzger, "Intra-urban vulnerability to heat-related mortality in New York City, 1997–2006," Health & Place **30**, 45-60 (2014).

[3] J. Liu, J. Huang, B. Chen, T. Zhou, H. Yan, H. Jin, Z. Huang, B. Zhang, "Comparisons of PBL heights derived from CALIPSO and ECMWF reanalysis data over China," JQSRT **153**, 102-112 (2015).

[4] K.A. Browning and R. Wexler, "The Determination of Kinematic Properties of a Wind Field Using Doppler Radar, " JAM **7**, 105–113 (1968).

Antenna Efficiency Optimization in Coherent Lidar Systems

Sammy Henderson, Pat Kratovil, and Charley Hale
Beyond Photonics, Boulder, CO, 80301, sammy@beyondphotonics.com

Abstract: Maintaining very high antenna efficiency is crucial in achieving a high level of performance from coherent lidar systems. This is especially important for space-based operation where available spacecraft prime power limits the transmitter power that can be utilized. We review the impact of wavefront aberrations and misalignment, including lag angle misalignment, on the performance of coherent lidar systems and describe requirements for achieving high antenna efficiency. In space-based large aperture coherent lidar systems, spacecraft rotations result in unacceptable lag angle misalignment loss. A lag-angle compensation system is described allowing for maintenance of high efficiency in large-aperture coherent lidar systems.

Keywords: Coherent Lidar, Wind Measurement

1. Introduction

Global measurements of altitude-resolved vector winds from an Earth-orbiting platform are of interest for both Earth science and weather forecasting applications. The potential for global wind measurements using lidar systems has been recognized for many years [1-3], and the technology has now advanced such that the first global wind measurement mission using direct detection Doppler lidar technology is scheduled to launch in late 2018 [4]. In recent years coherent lidar systems have primarily been considered for space-based wind measurements only in regions of the atmosphere where the aerosol backscatter is relatively high, e.g., the boundary layer and elevated aerosol layers higher in the troposphere. In a companion paper [5], we describe the potential for a coherent detection wind lidar that has sufficient backscatter sensitivity to allow for high percentage coverage throughout the troposphere and lower stratosphere. To achieve high performance the coherent lidar signal detection efficiency must be maintained at a very high level. Below we summarize coherent lidar signal detection efficiency loss mechanisms and describe requirements and techniques to ensure the efficiency is maintained at a high level.

2. Signal Detection Efficiency for Coherent Lidar

As described in the literature [2, 5] the total coherent detection signal is proportional to the total lidar efficiency. We represent the total lidar efficiency as $\eta_T = \eta_{eo}\eta_a = \eta_{to}\eta_{ro}\eta_q\eta_{sn}\eta_a$. In this equation η_{eo} is defined as the lidar electrooptic efficiency, which contains all the transmit and receive optical path losses, η_{to} and η_{ro}, the detector quantum efficiency, η_q, and the coherent detection shot noise efficiency, η_{sn}. The antenna efficiency is the product of the transmit beam truncation efficiency and the heterodyne efficiency. The heterodyne efficiency contains losses resulting from wavefront mismatch between the received signal and local oscillator fields. Reduction of antenna efficiency occurs due to non-optimal sizing of the transmit and local oscillator beams, misalignments, wavefront aberrations in the lasers or optics, and atmospheric refractive turbulence.

The antenna efficiency, which can most conveniently be computed from the overlap integral of the target plane normalized irradiance of the transmit and BPLO fields [2], is given by

$$\eta_a(z) = \eta_{Tt}\eta_{Tb}\frac{\lambda^2 z^2}{A_r}\iint I_{nt}(x,y,z)\, I_{nb}(x,y,z)\,dxdy\,, \quad (1)$$

where η_{Tt} and η_{Tb} are the truncation efficiencies of the transmit and BPLO beams, respectively.

3. Electro-optic Efficiency

The electrooptic efficiency in well-designed coherent lidar systems can exceed 70%. An example configuration having >66.5% electrooptic efficiency is shown in Figure 1. We emphasize that the electro-optic efficiency does not include antenna efficiency. For example, for designs using single mode optical fibers to collect the signal light (like shown in the Figure), the loss due to mismatch of the received signal field with the fiber single mode field is part of the antenna efficiency. The 99% efficiency shown in the figure is only due to imperfect AR coatings on the fiber input face.

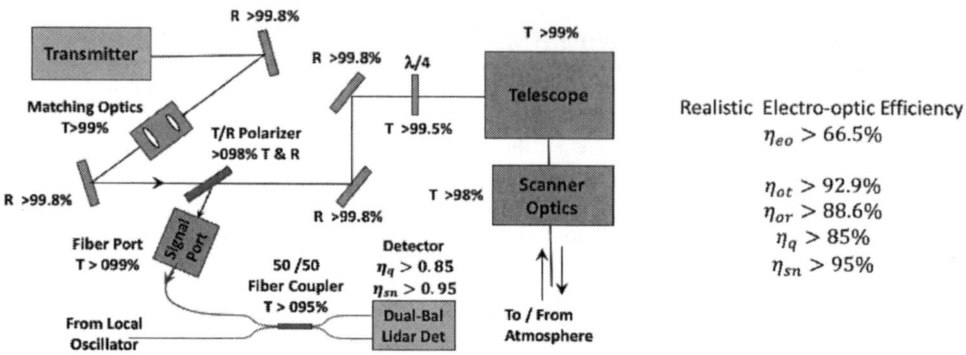

Figure 1. Simplified coherent lidar layout showing electro-optic efficiency greater than 66.5%.

4. Antenna Efficiency

The antenna efficiency is generally defined in Equation 1. Building on our previous analytic antenna efficiency approximation work [6], we have generalized the analytic equations allowing for unmatched transmit and back-propagated local oscillator (BPLO) beam sizes, focus conditions, truncation ratios, and beam qualities (aberrations). The transmit and BPLO beams are assumed centered in the transmit aperture. Our latest improved analytic approximation of the antenna efficiency [7] is

$$\eta_a(z) \approx 2\eta_{clt}\eta_{Tt}\eta_{clb}\eta_{Tb}\rho_{Tb}^2\psi(z)\cdot exp\left[-2\psi(z)\left(\frac{\delta\theta}{\theta_{db}}\right)^2\right], \tag{2}$$

where

$$\eta_{cli} \approx 1-0.162\ \exp\left(-\frac{0.25}{\rho_{Ti}}-\frac{0.164}{\rho_{Ti}^2}-\frac{0.873}{\rho_{Ti}^3}\right), \tag{3}$$

is the central lobe energy of the far field diffraction pattern for either the transmit or BPLO beam, $\theta_{db} = \lambda/\pi\omega_{ob}$ is the diffraction limited divergence of the untruncated BPLO beam, $\delta\theta$ is the misalignment angle between the BPLO and transmit beams, and

$$\eta_{Ti} = 1 - exp\left(-\frac{2}{\rho_{Ti}^2}\right) \tag{4}$$

is the truncation efficiency, with $\rho_{Ti} = \omega_{oi}/a$ being the aperture truncation ratio of the transmit or BPLO beam, ω_{oi} the e^{-2} intensity radius of the beam at the aperture location (prior to truncation), and a the radius of the circular truncating aperture (usually the same as the primary mirror diameter).

The effective Gaussian beam antenna efficiency parameter is given by

$$\psi(z) = \frac{\omega_d^2(z)}{\omega_{et}^2(z)+\omega_{eb}^2(z)} = \left(\frac{\omega_{et}^2(z)}{\omega_d^2(z)}+\frac{\omega_{eb}^2(z)}{\omega_d^2(z)}\right)^{-1}, \tag{5}$$

Again, we note that these equations assume that both beams are centered in the aperture, and therefore at that location they are centered with respect to each other.

Although we have developed the analytic equations for the effective Gaussian beam sizes, $\omega_{ei}^2(z)$, at arbitrary range (e.g., for the near, intermediate, and far field), due to space limitations we only describe the far field expressions here, which are sufficient for space based lidar performance consideration. In the far field, the effective beam size in Equation 5 is given by

$$\omega_{ei,F} = M_i^2\frac{\lambda z}{\pi\omega_{oi}},\ \text{with}\ M_i^2 \approx M_{iL}^2 M_{io}^2 M_{iT}^2\ . \tag{6}$$

In this expression, M_{iL}^2, is the beam spreading (or beam quality) factor of the lasers themselves, M_{iO}^2, is the beam spreading due to aberrations in the transmit or BPLO optics, and $M_{iT}^2 \approx \sqrt{\eta_{cli}\eta_{Ti}/f_{c2i}S_{i,F,T}}$ is the beam spreading factor due to the truncation by the aperture. In words, for small to moderate aberration levels, the total far field beam spreading factor is well approximated as the product of individual beam spreading factors as long as the various contributors are statistically independent. The far field Strehl ratio associated with the telescope truncation is $S_{i,F,T} = [1 - \exp(-1/\rho_T^2)]^2$, and f_{c2}, is a correction factor (see Figure 2) needed to match the analytic equations to numerical calculations. This correction factor is related to the correction factor from earlier work [6] by $f_{c2} = f_c/\eta_{cl}$.

Approximations for the optical path related aberration losses were described in the earlier paper [6], and similar more detailed expressions should be used the get higher accuracy, but an even simpler approximate analytic expression has now been developed, which is a good approximation for weak to moderate aberration levels in most efficient coherent lidar designs which have $0.75 < \rho_{Ti} < 1.2$. The simpler expression is $M_{iO}^2 \approx e^{0.5(\gamma\pi\sigma_{iO})^2}$, with the coefficient γ varying between 1.8 and 2.8 depending on the type of aberration that dominates. The aberration type resulting in the highest antenna efficiency reduction (worst case) is the spherical aberration, which is well approximated using $\gamma = 2.8$ for any truncation ratio value, even those outside the 0.75 to 1.2 range.

Using the above equations, the far field antenna efficiency parameter becomes

$$\psi_F \approx \left[{M_{bL}^2}^2 {M_{bO}^2}^2 {M_{bT}^2}^2 + \left(\frac{\rho_{Tb}}{\rho_{Tt}}\right)^2 {M_{tL}^2}^2 {M_{tO}^2}^2 {M_{tT}^2}^2\right]^{-1}. \quad (7)$$

If the transmit and BPLO beam sizes are matched at the aperture, the far field antenna efficiency is

$$\eta_{a,F} \approx 2\eta_{amax,F}\left[{M_{bL}^2}^2 {M_{bO}^2}^2 + {M_{tL}^2}^2 {M_{tO}^2}^2\right]^{-1} exp\left[-2\left[{M_T^2}^2\left({M_{bL}^2}^2 {M_{bO}^2}^2 + {M_{tL}^2}^2 {M_{tO}^2}^2\right)\right]^{-1}\left(\frac{\delta\theta}{\theta_{db}}\right)^2\right], \quad (8)$$

where $\eta_{amax,F}$ is the far field antenna efficiency for the lidar design in the absence of laser or optical wavefront aberrations (e.g., 0.4 for the Wang lidar design [2]).

The impact of overall wavefront aberrations on the antenna efficiency for a coherent lidar with matched transmit and BPLO beam sizes at the aperture and with truncation ratio of 0.8 (Wang design) is shown in Figure 3. In the figure, the curves represent: solid - numerically calculated using primary aberrations, light dots – numerically calculated using random aberrations, and dashed - aberration analytic model [6] prediction. For the curves and simulation points plotted in the figure, the rms aberrations are assumed to be the same for both the BPLO and transmit beams. If the aberrations are dominated by the telescope, this would be the case, but if the aberrations were dominated by the lasers themselves or by other optics in the system the impact of aberrations would differ from that shown in the figure.

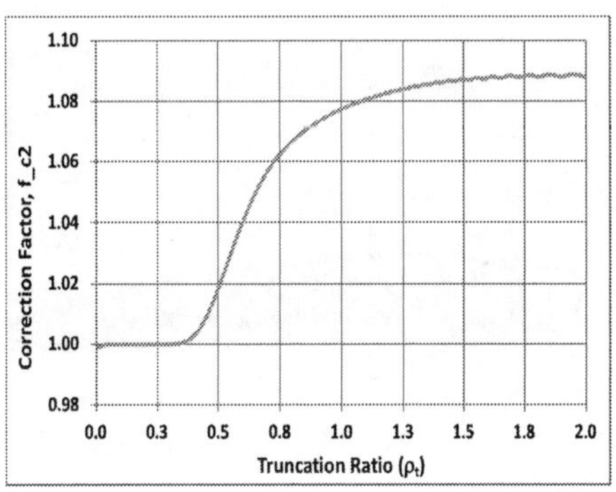

Figure 2. Correction factor f_{c2} used in analytic antenna efficiency equations.

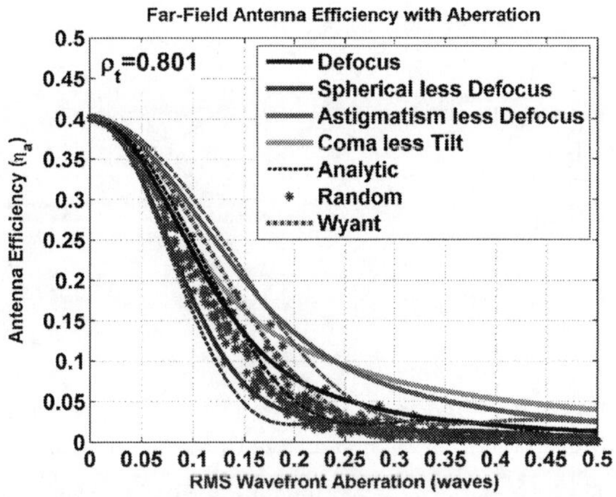

Figure 3. Antenna efficiency reduction due to wavefront aberrations for Wang design lidar.

Figure 3 illustrates the importance of having near diffraction limited optics in the coherent lidar system. For example, if the aberrations are dominated by spherical aberrations the rms wavefront variation must be held to about $\lambda/20$ in order to maintain >80% of peak (unaberrated) performance. This is easier accomplished at longer wavelengths – e.g., $\lambda/20$ for 2 μm is equivalent to ~$\lambda/6.3$ at 633 nm.

Equation 9 is convenient to estimate the far field antenna efficiency of coherent lidar systems with non-perfect lasers and optics. For example, for a Wang design lidar with: no misalignment, transmit and BPLO beam qualities of $M_{tL}^2 = 1.1$ and $M_{bL}^2 = 1.03$, and transmit and BPLO optics rms aberrations dominated by spherical with $\sigma_{to} = 0.05$ ($M_{to}^2 = M_{to}^2 \approx 1.1$) we find $\eta_{a,F} \approx 0.73 \cdot \eta_{amax,F} = 0.29$. So, the efficiency of this example coherent lidar is only about 73% of a perfect lidar of the same design.

Maintaining antenna efficiency loss due to aberrations at levels within 80% of the peak unaberrated performance is possible at longer wavelengths but aberrations must be carefully minimized.

5. Misalignment and Lag Angle Losses

For coherent lidar systems the only signal content that is mode matched the local oscillator (LO) field will be coherently detected. Misalignment of the received field with respect to the LO results in signal loss. Combining Equations 2 and 8 the alignment related efficiency is

$$\eta_{a,\theta}(z) \approx exp\left[-2\psi(z)\left(\frac{\delta\theta}{\theta_{db}}\right)^2\right] \approx exp\left[-2\left[M_T^{2^2}\left(M_{bL}^{2\,2}M_{bo}^{2\,2}+M_{tL}^{2\,2}M_{to}^{2\,2}\right)\right]^{-1}\left(\frac{\delta\theta}{\theta_{db}}\right)^2\right], \quad (9)$$

The antenna efficiency for the Wang and Rye unmatched coherent lidar designs [2] assuming no aberrations vs. misalignment angle between the transmit and BPLO is shown in Figure 4. Note the Rye design has higher overall efficiency and is less sensitive to misalignment, so should it be considered for space-based and other lidar designs, but the Wang design is easily manufactured and can use single mode fiber to collect the signal light and mix it with the LO, so it is frequently utilized. For the Wang design to maintain an alignment efficiency above 95% requires that $\delta\theta < 0.27\theta_{db}$. For example, for a space-based lidar having a 1.5 meter diameter aperture, and no beam aberrations this alignment requirement is $\delta\theta_{lb} < 0.29\ \mu rad$ in the large beam space, which in the small beam space prior to the telescope expansion, where alignment is actually maintained, is $\delta\theta_{sb} < 0.29 \cdot M\ \mu rad$, where M is the magnification of the telescope. For example, if the net magnification, from where the transmit and BPLO beams are aligned to the final expanded output, is $M = 200$, possibly with a two-stage expansion, then $\delta\theta_{sb} < 58\ \mu rad$, which sounds less ominous.

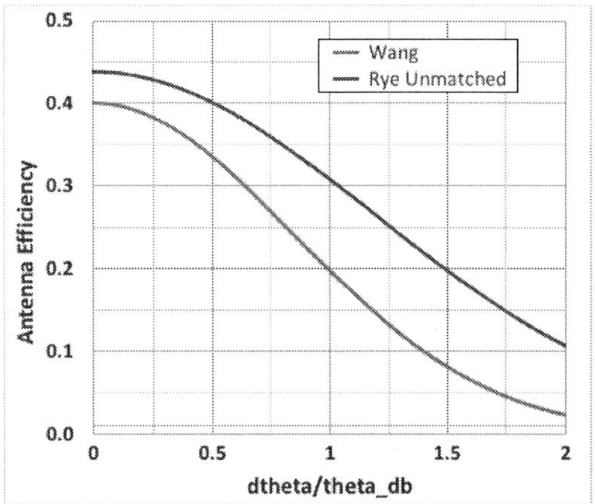

Figure 4. Reduction in antenna efficiency with misalignment for Wang and Rye Unmatched designs.

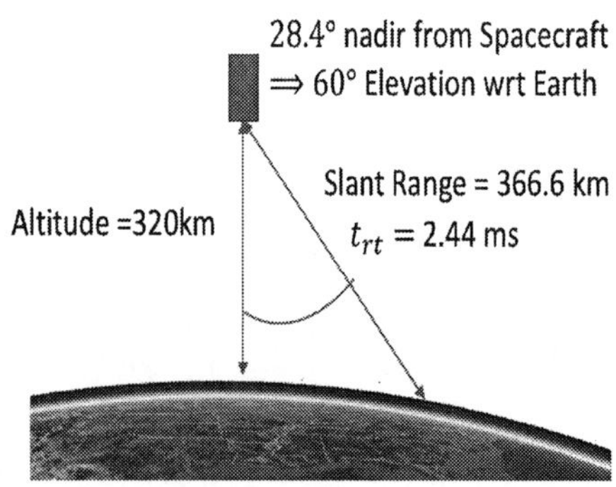

Figure 5. Example measurement geometry for a space-based lidar.

If the lidar platform is rotating, or if a rotating scanner is used to direct the beam, the induced lag angle between the signal return and the LO is given by $\theta_L = \Omega t_{rt}$, where Ω is the rotation rate and $t_{rt} = 2R/c$ is the round-trip time to the measurement volume, with R the range and c the speed of light. An example space-based measurement geometry is illustrated in Figure 5. At the orbit altitude of 320 km the spacecraft circles the Earth every 90.78 minutes, so the spacecraft rotates at a rate of rotates at $\Omega_{orb} = 1.15\ mrad/s$. Given the measurement geometry shown in the figure, the resulting lag and from the orbital rotation alone is $\theta_L = 2.8\ \mu rad$. If uncorrected this lag angle would result in a 20.5 dB loss in antenna efficiency. Clearly this loss is unacceptable and must be removed. For this large 1.5 m diameter aperture example, rotation rates $> 119\ \mu rad/s$ must be sensed and compensated in order maintain the antenna efficiency to within 95% of its peak value.

6. Auto-alignment and Lag Angle Compensation

We are developing a lidar auto-alignment and lag and compensation capability based on using a precision angle measurement sensor to sense instantaneous platform rotation rates $< 100\ \mu rad/s$ and feed this forward to an alignment system in the lidar shown in concept in Figure 6. A laser that is back propagated from the signal fiber is sensed and compared to the transmitted beam. With the rotation rate knowledge, the angle can be offset by the proper amount to allow the lag angle loss to be removed. In addition, any misalignments within the lidar can also be removed automatically. This will be described in more detail in the presentation.

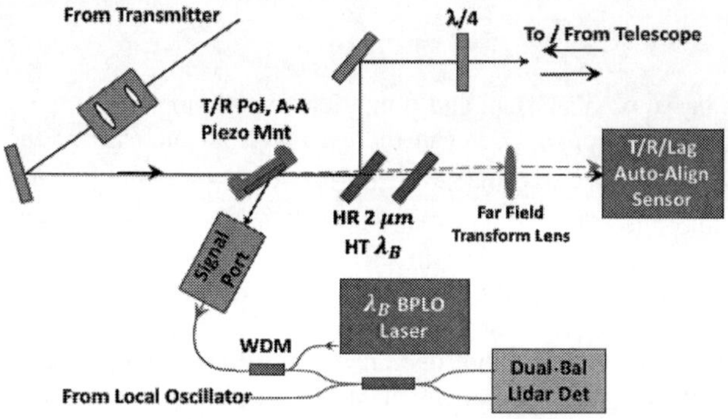

Figure 6. Auto-alignment and lag angle compensation concept.

7. Acknowledgments

Portions of this work were funded under a NASA Phase I SBIR. We are grateful for this support and the many useful technical discussion with Michael Kavaya the NASA technical monitor

8. References

[1] R.M. Huffaker, et.al., "Feasibility studies for a global wind measuring satellite system (Windsat): Analysis of simulated performance," Appl. Opt., 23, 2523-2536, 1984

[2] S.W. Henderson, et.al, "Wind Lidar" in *Laser Remote Sensing,* T. Fujii and T. Fukuchi, Eds., CRC Taylor and Francis, 469-722 (2005)

[3] W. E. Baker, eta.al., "Lidar-Measured Wind Profiles: The Missing Link in the Global Observing System," Bulletin of the American Meteorological Society, Vol 95, No 4, (2014)

[4] A.M Dabas, et,al., "ESA wind lidar mission Aeolus", 19th Coherent Laser Radar Conference, paper Mo2, Okinawa, Japan (2018)

[5] S.W. Henderson, "Space-based Coherent Lidar for Wind Measurements with High-Percentage Tropospheric Coverage," 19th Coherent Laser Radar Conference, paper We4, Okinawa, Japan (2018)

[6] S.W Henderson and D Jacob, "Improved Analytic Equations for Estimating the Performance of Truncated and Aberrated Gaussian Beam Coherent Lidar," 18th CLRC, paper W8, Boulder CO (2016)

[7] S. W. Henderson, "Improved Analytic Equations for Coherent Lidar Antenna Efficiency," Beyond Photonics Internal document, available on request, (May 2018)

Amplification of the Mean Echo Signal Power of a Coherent Lidar in a Turbulent Atmosphere

V.A. Banakh (a), A.V. Falits (b)
(a, b) V.E. Zuev Institute of Atmospheric Optics SB RAS 1 Akademik Zuev Square, Tomsk, 634050, Russia
banakh@iao.ru

Abstract: The mean power of a coherent lidar echo signal from a topographic target under different turbulent conditions in the atmosphere is studied experimentally. It is shown that with intensification of refractive turbulence the mean power of echo signal of a coherent lidar increases due to the effect of backscatter amplification in random media.

Keywords: Coherent lidar, backscatter amplification, refractive turbulence

1. Introduction

In [1,2] it is shown based on analytical and numerical calculations that the mean power of the echo signal of a pulsed coherent Doppler lidar can increase with an increase of the structure constant of turbulent fluctuations of the refractive index of air C_n^2. According to [1,2], such behavior of the mean power is a consequence of the effect of backscatter amplification in random media, which arises at paths with reflection due to correlation of waves in the forward (probing beam) and backward (scattered radiation) directions, which pass through the same turbulent inhomogeneities of the refractive index of air in the atmosphere. However, the effect of amplification of the echo signal of a lidar with the coherent detection of the scattered radiation was not demonstrated experimentally yet. In this paper, we report the results of an experiment on measurement of the mean power of echo signal of a coherent lidar from a topographic target under various turbulent conditions in the atmosphere.

2. Experimental Overview

The field experiments were carried out with the HALO Photonics pulsed coherent Doppler lidar Stream Line [3] at the Basic Experimental Station of the Institute of Atmospheric Optics, Siberian Branch of the Russian Academy of Sciences, in suburbs of Tomsk in July 20-25 of 2017. To avoid the influence of fluctuations of the atmospheric aerosol concentration, in the experiment with a coherent lidar the signal was backscattered from a diffuse screen. The block diagram of the experiment is shown in Figure 1.

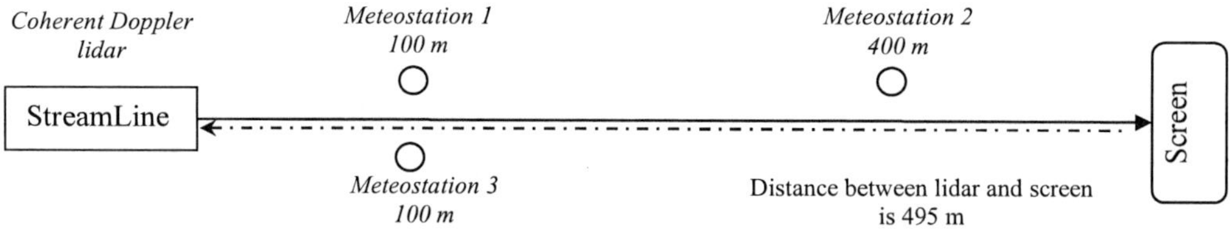

Figure 1. Block diagram of the experiment.

The lidar was set at a distance of 495 m from a screen. The propagation path of the lidar probing beam was nearly horizontal. Three AMK-03 sonic meteostations ensuring measurements of temperature and wind velocity fluctuations were installed just near the path at distances of 100 and 400 m from the lidar. In the experiment, the mean power of the scattered radiation $< P_S > =$ SNR was determined as a ratio of the mean power of echo signal to the noise power in the pass band of the lidar detector [2]. The Stream Line lidar combines the transmitting and receiving system, and therefore it records the radiation scattered strictly in the backward direction, and the amplification of the lidar echo signal due to turbulence should occur.

The measurements were conducted continuously for several days. SNR values were recorded at the following parameters of the lidar: wavelength of 1.5 μm, laser pulse repetition frequency of 15000 s^{-1}, pulse duration (FWHM power) of 172 ns, beam collimated, diameter (exp(-2) level) of 56 mm. The mean value of SNR was estimated for the time interval of 2 min for the data obtained on July 20 and 21 (range gate length of 18 m, pulses/ray of 5000) and 4 min for the data obtained in the period of July 22-25 (range gate length of 48 m, pulses/ray of 15000). Simultaneously, the sonic meteostations measured the air temperature and wind velocity with a frequency of 80 Hz. The data obtained were used to calculate the mean values of the structure constant of turbulent fluctuations of the refractive index of air C_n^2 averaged over 4 min time interval.

3. Calculations

For the experimental conditions, the normalized mean power of the echo signal of coherent lidar $< P_S > \equiv$ SNR was calculated by the equation for the echo signal power averaged over microphysical parameters of the scattering medium $\bar{P}_S(t)$ [2, Eq. (1.94)]:

$$\bar{P}_S(t) = \frac{\eta E_P c}{2h\nu B_F} \beta_\pi(R) T_A^2(R) \left(\frac{\lambda}{\pi a_0^2} \right)^2 \int_{-\infty}^{+\infty} d^2\rho I_{PN}^2(R, \mathbf{\rho}, t) . \qquad (1)$$

In Eq. (1), I_{PN} is the normalized (dimensionless) intensity of the lidar probing beam [2], t is time, R is path length, $\beta_\pi(R)$ is the aerosol backscattering coefficient, a_0 is the initial radius of the probing beam, $T_A(R)$ is the atmospheric transmittance, E_P is the probing pulse energy, B_F is the receiver pass band, η is the quantum efficiency, $h\nu$ is the photon energy, h is the Planck's constant, c is the speed of light, λ is the wave length. The averaging $\bar{P}_S(t)$ over an ensemble of random realizations $< \bar{P}_S(t) > \equiv < P_S > \equiv$ SNR gives the mean power of the echo signal, and the angular brackets denote the averaging over an ensemble.

It follows from Eq. (1) that to estimate $< P_S >$, it is required to simulate the propagation of the probing beam along the measurement path in the atmosphere, obtain random realizations of two-dimensional distributions of I_{PN} and then average I_{PN}^2 over an ensemble. The propagation of the probing beam in the turbulent atmosphere was simulated based on solving the parabolic wave equation by the split step method [1, 2 (Section 1.4.2)]. The same model random realizations of I_{PN} were used to calculate the mean power of the echo signal in the absence of amplification $< P_S >_M$ for the bistatic scheme, when the ensemble-average $< I_{PN}^2 >$ in the calculations of $< P_S >$ is replaced with $< I_{PN} >^2$.

Figure 2 shows the mean power of the lidar signal $< P_S >$ calculated with Eq. (1) upon normalization to the power $P_0 = \frac{\eta E_P c}{4h\nu B_F \pi} \beta_\pi \left(\frac{\lambda}{a_0^2} \right)^2$ at the beginning of the path as a function of the distance. Figure 3 shows the calculated ratios of the mean power of the lidar signal in the turbulent atmosphere to the mean power of the echo signal in the absence of turbulence at the monostatic $\frac{< P_S >}{P_S(C_n^2 = 0)}$ and bistatic $\frac{< P_S >_M}{P_S(C_n^2 = 0)}$ schemes of detection as functions of C_n^2. From Figure 2 it follows that at strong turbulence, when $C_n^2 = 10^{-13}\ m^{-2/3}$, the mean power $< P_S >$ first increases as compared to P_0 with increasing the path length to 500 m and only then it begins to decrease and becomes less than in the absence of turbulence at distances exceeding 1 km.

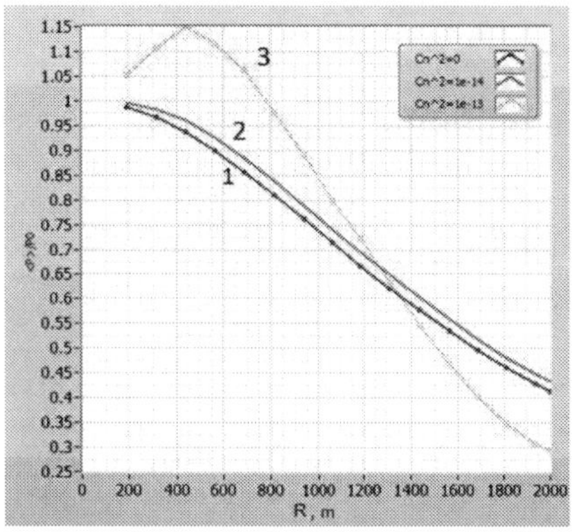

Figure 2. Normalized mean power of lidar echo signal $< P_S > / P_0$ as a function of the distance at (1) $C_n^2 = 0$; (2) 10^{-14} ; (3) $10^{-13}\ m^{-2/3}$.

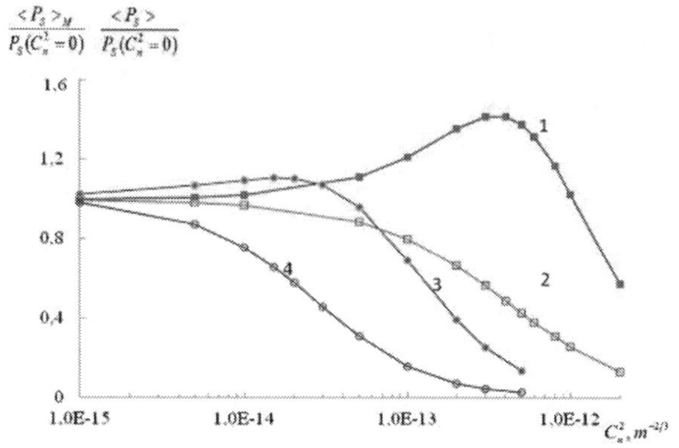

Figure 3. Ratio of the mean power of the lidar echo signal in the turbulent atmosphere to the mean power of echo signal in the absence of turbulence as a function of C_n^2 at distances of 500 m (1, 2) and 2000 m (3, 4): (1, 3) – monostatic $< P_S >$ and (2, 4) bistatic $< P_S >_M$ schemes.

It is seen from Figure 3 that at the monostatic scheme the mean power of the lidar signal $< P_S >$ first increases with respect to that in the absence of turbulence due to the effect of backscatter amplification as turbulent fluctuations of the refractive index of air in the atmosphere become more intense. It achieves the maximum at $C_n^2 = 2 \cdot 10^{-14}\ m^{-2/3}$ on 2 km path and at $C_n^2 = 4 \cdot 10^{-13}\ m^{-2/3}$ on 500 m path. Then it decreases and becomes smaller than the mean power of the lidar signal in the absence of turbulence. At the bistatic scheme the mean power of the lidar echo signal decreases monotonically with intensification of refractive turbulence. It is always less than the mean power of the signal at the monostatic scheme. The more C_n^2, the less $< P_S >_M$ as compared to $< P_S >$.

4. Experimental Results

Figure 4 shows the mean power of the signal of the Stream Line lidar recorded upon scattering of the probing beam at a screen remote at a distance 495 m from the lidar (Fig.1). It is seen that the dynamics of variation of the recorded mean power of the lidar echo signal coincides with the dynamics of strength of refractive turbulence in the atmosphere. The mean power of the echo signal increases simultaneously with the increase of the structure constant, while the decrease of the structure constant leads to a decrease of the lidar signal-to-noise ratio. The experimental data are in a good agreement with the results of numerical calculations. In particular, it follows from the results of measurements on July 20 of 2017. At C_n^2 approximately equal to $5 \cdot 10^{-13}\ m^{-2/3}$ the mean power of the lidar signal (SNR) is maximal, in line with the calculation for 500 m path shown in Fig.3, and SNR decreases drastically approximately 1.6 times from 24 to 15 at a decrease of C_n^2 down to $5 \cdot 10^{-15}\ m^{-2/3}$, as in Figs. 2 and 3.

Figure 4 shows the data of measurements only for the periods, when variations of the recorded mean power of the lidar echo signal and the strength of refractive turbulence had the same dynamics. As a rule, this was observed in daytime, when the conditions of developed turbulence occur. In nighttime, turbulence was weak in the period of measurements, and no correlated variations of the mean power of the lidar echo signal and the structure constant of fluctuations of the refractive index were observed in the experiments. It is agreed with the calculation results for weak ($C_n^2 < 10^{-14}\ m^{-2/3}$) turbulence in Figs 2, 3.

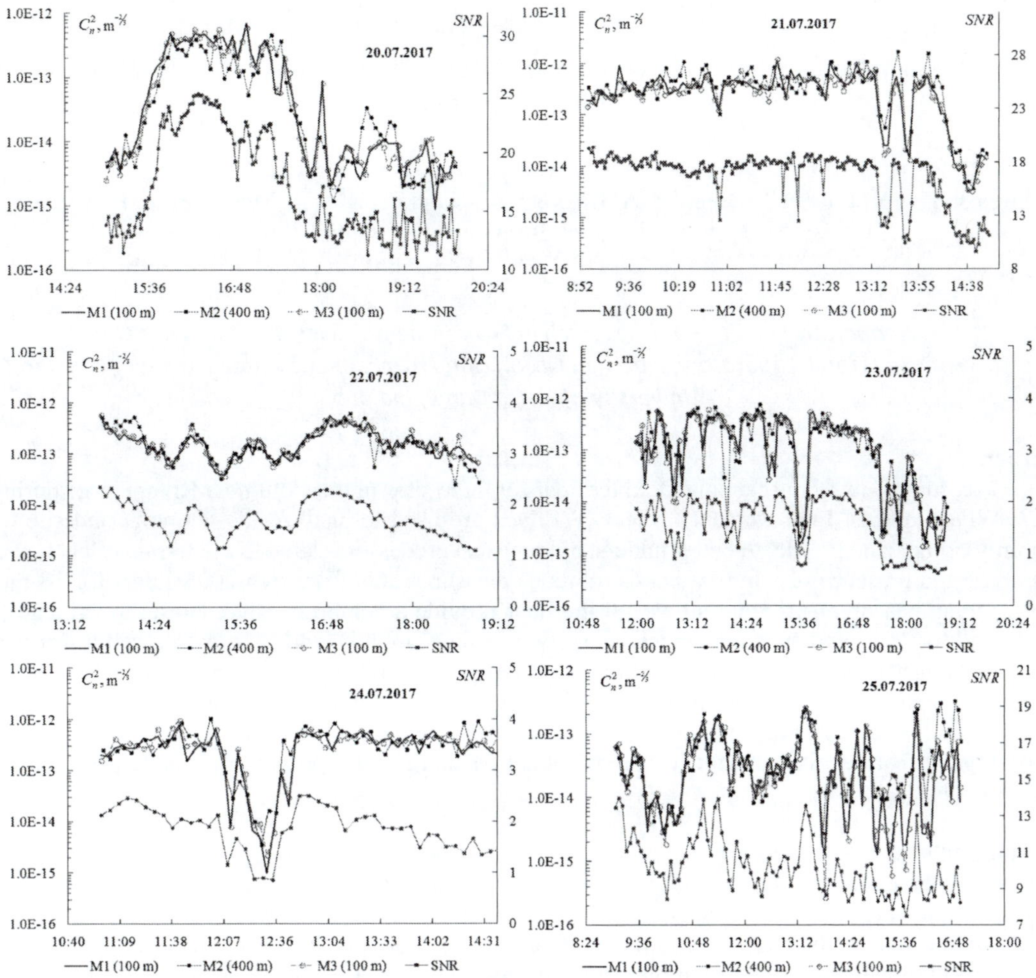

Figure 4. Temporal variations of the structure constant of turbulent fluctuations of the refractive index of air and the signal-to-noise ratio of the Stream Line lidar. M1, M2, M3 correspond to C_n^2 calculated from the data of meteostations 1, 2, 3 shown in Fig. 1. Local time.

It is commonly known that as the refractive turbulence becomes more intense, the spatial coherence of laser radiation deteriorates and the efficiency of heterodyning, that is SNR of a coherent lidar, decreases. The results depicted in Fig. 4 demonstrate that this does not always occur. As the turbulence strength increases, the effect of backscatter amplification in random media can, to the contrary, lead to an increase of the mean power of the echo signal of a coherent lidar.

ACKNOWLEDGMENTS: This study was supported, in part, by the Russian Foundation for Basic Research, Project No. 18-05-00220.

5. References

[1] V.A. Banakh, I.N. Smalikho, and Ch.Werner, "Numerical simulation of effect of refractive turbulence on the statistics of a coherent lidar return in the atmosphere", Applied Optics, **39**(30), 5403-5414 (2000).

[2] V.A. Banakh, and I.N. Smalikho, *Coherent Doppler wind lidars in a turbulent atmosphere* (Artech House. Boston-London. 2013. 250 p.)

[3] http://halo-photonics.com

Tu4

Scanning Doppler lidars for Wind Energy research and validation of NWP model wind forecast in complex terrain.

Pichugina, Y[1, 2], R. Banta[2], T. Bonin[1, 2], A. Brewer[2], A. Choukulkar[1, 2], K. McCafrey[1, 2], B. McCarty[1, 2], J. Kenyon[1, 2], J. Olson[1, 2], and C. Bonfanti[1, 2], S. Baidar[1, 2], S. Sandberg[2], A. Weickmann[1, 2], M. Hardesty[1, 2], H. J. S. Fernando[3], and R. Krishnamurthy[3]

[1]Cooperative Institute for Research in Environmental Sciences, Boulder, CO
[2]Earth System Research Laboratory, National Oceanic and Atmospheric Administration, Boulder, CO
[3]University of Notre Dame, Indiana

Abstract.

Measurements from scanning Doppler lidars at three sites in the Columbia River Gorge during the second Wind Forecast Improvement Project (WFIP-2) are used to analyze the temporal and site-to-site variability of the wind profile, to better understand physical processes related to the terrain effects, weather conditions, and presence of wind farms. Continuous measurements from two 200S lidars for 18-months and year-long measurements from a StreamLine lidar provide a comprehensive dataset to validate NWP model wind forecasts and evaluate model accuracy over days, months and seasons, as well as for periods of interesting meteorological events including wind ramps, cold fronts, or marine push in the study area. Verification metrics such as bias, RMSE, MAE, and the correlation coefficient between observed and modeled wind variables are analyzed as a function of height, time, and forecast hour, revealing meteorological features that are not well represented by models and provide insight to potential model improvements.

Introduction.

The basin of the Columbia River in Oregon and Washington is the site of many large wind farms and a major source of wind energy to the Western U.S. The imposing Cascade Mountain Range provides a natural barrier between the Pacific marine air to the west and the arid conditions over the wind-farm study area to the east. Wind flow through the Columbia River Gorge, the only sea-level gap through this range. Typical weather patterns can be identified for summer and winter conditions over the Basin. During summer a persistent offshore subtropical ridge to the west of the range and very strong diurnal heating of the dry interior to the east, including the wind-farm region, drive a predominant westerly wind flow having a diurnal cycle that peaks at midnight local time. In winter, frequent storm and frontal systems, and resulting low surface pressure offshore combine with high pressure inland, often associated with multiday persistent cold-air pools and their strong inversions in the basin regions. This configuration produces a pressure gradient favoring easterly flow through the Columbia Gorge and cold air pool conditions over the Basin, including a very weak, mostly easterly but highly variable drift to the winds there.

Properties of winds at the hub-height and within the turbine rotor layer, their temporal variability for monthly, seasonal, and annual time scales as well as their spatial variability over the complex terrain of this large basin are important for wind industry operations.

In this study wind properties are measured using scanning, pulsed Doppler lidars sited along the Columbia River wind corridor at three sites along a distance of 70 km (Fig.1).

Lidar measurements provide wind speed and direction profiles every 15 min at high precision (<10 cm s^{-1}) and resolution (vertical spacing of ~10 m or less through the rotor layer). Consecutive measurements from all 3 lidar during a 12-month period in 2016 are used to quantify spatial and temporal wind variability ay each site and between sites, and to validate numerical weather prediction (NWP) model winds in the lowest 1 km above ground level (AGL) and to obtain model-error metrics for any given height, time of day, season or over particular layers of the atmosphere, focusing on the hub-height and the rotor-layer that are important to wind energy.

The model used for comparison was the NOAA/NCEP HRRR NWP operational forecast model as described by Benjamin et al. (2016).

The lidar deployments were part of a larger field campaign, the Second Wind Forecast Improvement Program, or WFIP-2, from September 2015 through March 2017. Overviews of the WFIP-2 experiment and scientific objectives, detailed description of instruments involved in the experiment as well as the logistics of their deployments, and the improvements of NOAA/ESRL and NOAA/NCEP operational models as well as better-resolution experimental models are in preparation as well as many other papers coming out of the analysis of this reach data set.

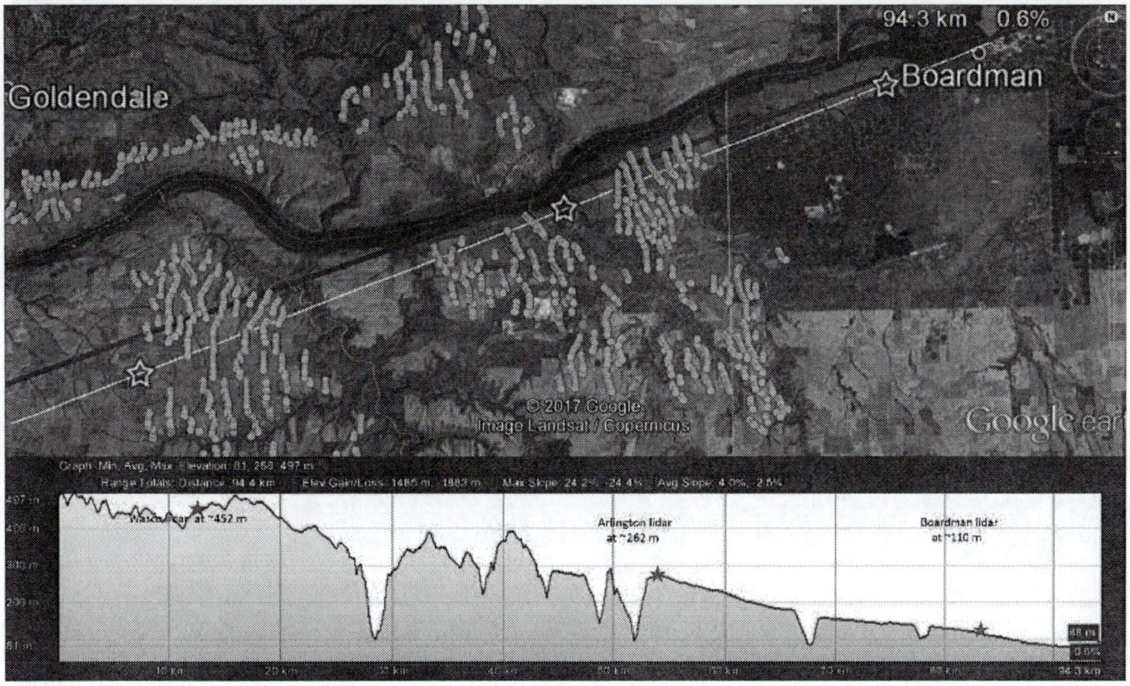

Figure 2. The top part shows Google map of the Eastern Gorge part of the study area shows the location of 2 NOAA Doppler lidars (200S) at Wasco site at 469 m above Ground level (AGL) and Arlington site at 179 m AGL and the Notre Dame University HALO lidar at Boardman site at 111 m AGL. The distance between two NOAA lidars along the white line is about 40 km, and between lidars at the Arlington and Boardman sites is about 31 km. The surrounding wind turbines are indicated by the clusters of dark yellow circles. The blue line indicates the East-West Transect of the study region along the prevalent wind directions observed from surface measurements and models during previous studies in this area. The bottom part shows terrain elevation along this line and lidar locations shown by blue stars.

Results.

To illustrate the nature of the data and model information being used, Fig. 2 shows lidar time-height cross sections for each of the three sites for two individual days: 26 February, a wintertime day, and 28 June, a summer day. Also shown are the corresponding HRRR model cross sections for the 1- and 3-h forecasts. The bottom rows show time-height cross sections of the model-minus-lidar wind-speed differences (model bias errors). The wind patterns on 26 February consisted of diminishing easterly gap flow due to a synoptic slackening of the pressure gradient across the Cascades, and the buildup of a weak-wind, cold-air-pool layer. The model predicted the weak-wind layer at Arlington and Boardman, but at Wasco the stronger winds persisted, indicating that the top of the weak-wind layer was predicted to be below the elevation of Wasco during this period. The timing of the wind-down ramp and the vertical structure of the departing westerlies are major sources of model error through the lowest km. The wind-speed patterns on 28 June reflect the summertime diurnal cycle of winds due to daytime heating and nighttime cooling of the Basin.

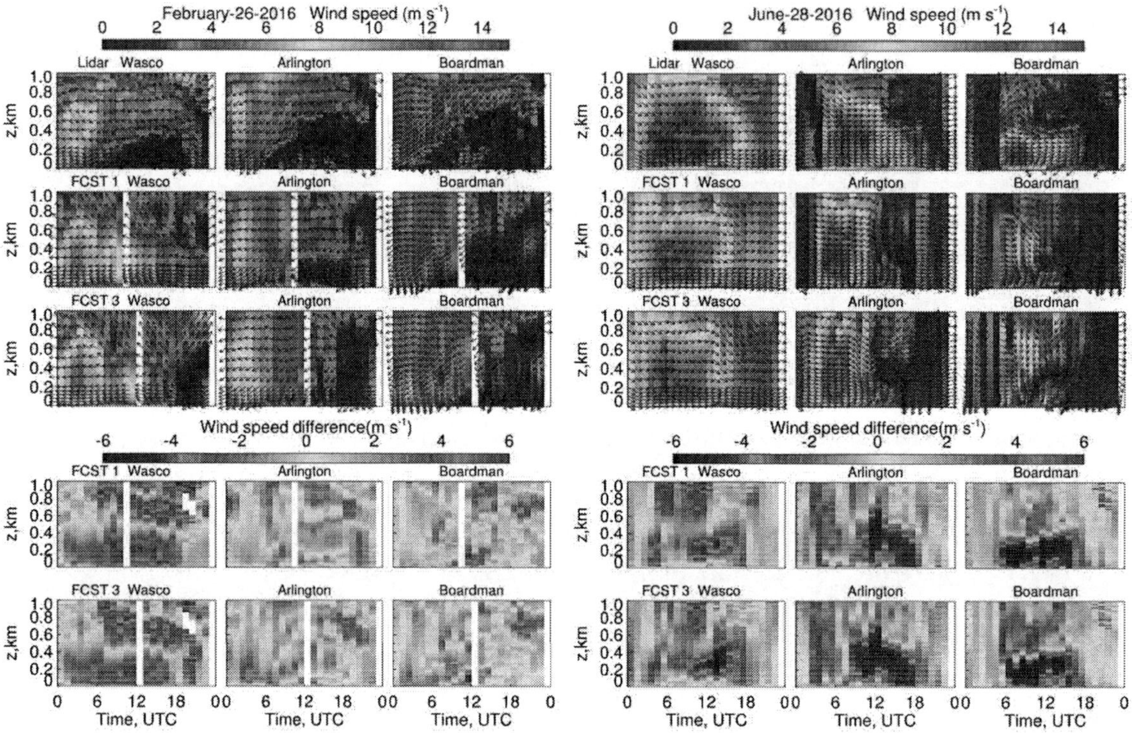

Fig. 2. Daily time-heights of mean wind from (a) lidar, (b) HRRR NCEP forecast hour 1, and wind speed difference for (c) forecast hour 1, (d) forecast hour 3 at 3 sites are shown for (26 February) cold pool and (28 June) marine intrusion events.

Figure 3 shows the annually averaged model-verification statistics for wind speed within the rotor layer as a function of (a) forecast lead time and (b) diurnal cycle. RMSE values (Fig. 3a) were ~3 m s^{-1} for the first 2 h at this height, modestly increasing with lead time to ~3.5, 3.1, and 3.1 m s^{-1} at Wasco, Boardman, and Arlington, respectively. The winds at the lower two sites showed low biases of less than 0.5 m s^{-1}, the largest magnitudes occurring at 5 and 6 h lead time, but at Wasco, the speeds were biased high by approximately 1 m s^{-1}. Thus, annually averaged model forecast skill did not significantly degrade with lead time, out to 15 forecasts hours (Fig 3a), but show some diurnal variability (Fig. 3b). Similar analysis for monthly and seasonal averages show that larger positive errors of a hub-height winds at the Wasco site are mostly attributed by errors from the cold period months. The time-of-day variations of model error were not strong but did occur. For example, the summertime negative biases at Arlington and Boardman had the largest magnitudes at night, when the marine-intrusion/ sea-breeze flows would be occurring at those sites. Another exception was that the R2 correlations were especially bad during the day at Wasco in fall, and to a lesser extent in winter. This behavior was consistent with daytime mix out in the model, reinforcing the premature mix out hypothesis at Wasco.

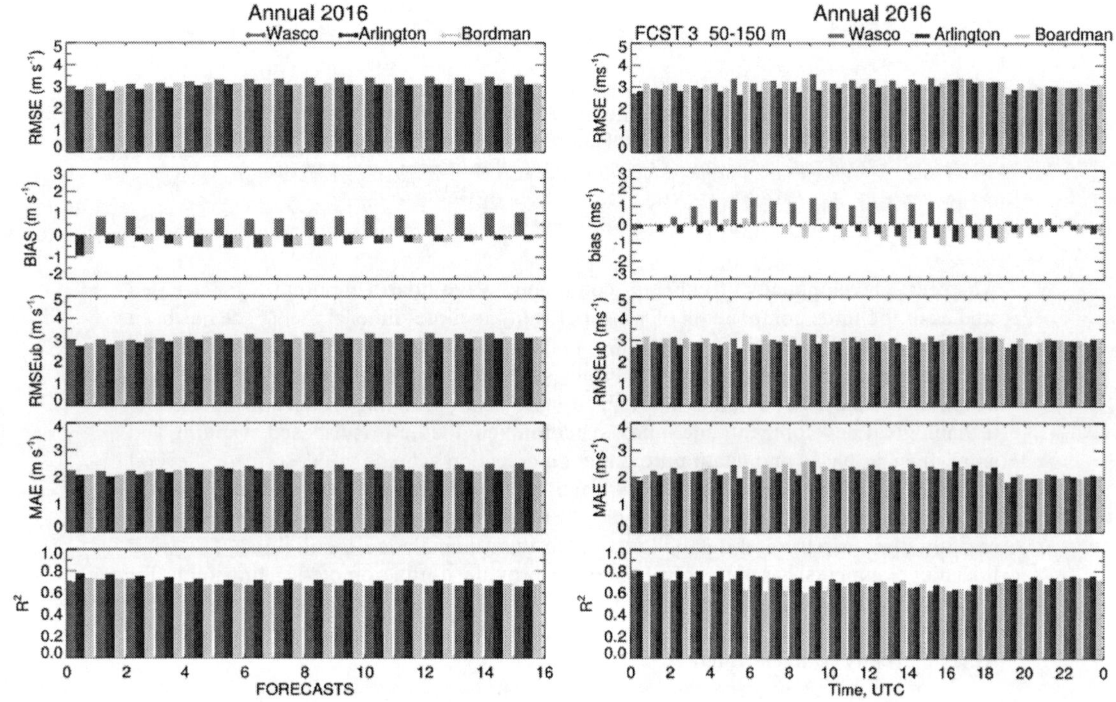

Figure 3. Validation metrics between measured and modeled wind speed at three sites for the annual period January-December 2016 are shown as: (a) function of forecast hours, (b) function of time (UTC).

Conclusions:

Mean values of hub-height wind speed and direction can vary significantly over a wind plant, especially in terrain as complex as central Oregon and Washington. Measurements from three scanning lidars separated from each other by 30-40 km, over a total distance of 70 km, and sited at different terrain elevations provide an opportunity to characterize the winds at each site, to quantify the site-to-site wind-flow variability and to better understand physical processes related to the terrain effects, seasonality, or diurnal cycle and to capture cases of unusual wind flow such as wind speed and direction ramps, cold fronts, or marine pushes. High-resolution lidar measurements provide a comprehensive dataset to validate NWP model forecasts at each site and to obtain model-error metrics.

Acknowledgments: The WFIP-2 study was sponsored by the U.S. Department of Energy, Energy Efficiency and Renewable Energy Office and by the NOAA Boundary Layer Research Program.

Synchronized Agile Beam Scanning of Coherent Continuous-Wave Doppler Lidars for High-Resolution Wind Field Characterization

Mikael Sjöholm, Nikolas Angelou, Michael Courtney, Ebba Dellwik, Jakob Mann, Torben Mikkelsen, and Anders Tegtmeier Pedersen
Technical University of Denmark, DTU Wind Energy, Campus Risø,
Frederiksborgvej 399, DK-4000 Roskilde, Denmark
misj@dtu.dk

Abstract: Developments of coherent continuous-wave lidar technologies for detailed studies of the three components of wind fields in the three-dimensional space has been pursued at DTU Wind Energy during the last decade with a concept called WindScanner. Since a single lidar only measures the wind component along the laser beam, the concept builds on synchronized agile beam-scanning of several lidars. The technological developments including synchronized double-prism-based scanning and an all-fiber in-phase and quadrature direction detection scheme are described. Several different instrument implementations have been realized and application examples from wind energy, structural design, aviation, boundary-layer meteorology, and recently also forestry and detailed studies in wind tunnels are highlighted. The wind lidar developments has not only proven to provide useful remote-sensing tools for wind field characterization at high spatial and temporal resolution but also holds prospects as tools for calibrating traditional point-monitoring anemometric devices which currently is investigated.

Keywords: Doppler wind lidar, Continuous-wave lidar, Synchronized multiple beam scanning, Double-prism scanner, WindScanner

1. Introduction

DTU Wind Energy has been leading the development of various scanning lidars utilizing the multiple-beam WindScanner concept which is about measuring the three wind components remotely by combining the line-of-sight speeds obtained from three steerable lidar beams crossing within the same volume [1]. One of the developments has been on mirror-based scanners for long-range pulsed lidars that in collaboration with the lidar manufacturer Leosphere has become an integral part of commercially available lidars. Another development has been on prism-based agile scanners for coherent continuous-wave lidars in collaboration with the lidar manufacturer ZephIR and this is the focus of this communication, which will review the technological developments including a synchronized double-prism-based scanning technology and an all-fiber in-phase and quadrature direction detection scheme. Several different instrument implementations have been realized and application examples from wind energy, structural design, aviation, boundary-layer meteorology, and recently also forestry and detailed studies in wind tunnels are highlighted.

2. Synchronized double-prism-based scanning for coherent detection of wind

The continuous-wave short-range WindScanner is originally building on the concepts of the ZephIR profiling lidar which scans the laser beam on a conical surface by rotating a single wedge-shaped prism. For agile scanning of the lidar laser beam, a double-prism scanner as depicted in Figure 1 has been developed which is similar to the classical Risley prism configuration but with the rotation axes of the two prisms not aligned but rather at the specific angle that allows the laser beam to pass through both prisms at the angle of minimum deviation in order to maintain a tight focus which is crucial for obtaining confined sampling volumes in the atmosphere. The position of the measurement location along the laser beam is obtained by a linear translation stage moving the light emitting and receiving fiber relative to the focusing lens in the transceiver telescope.

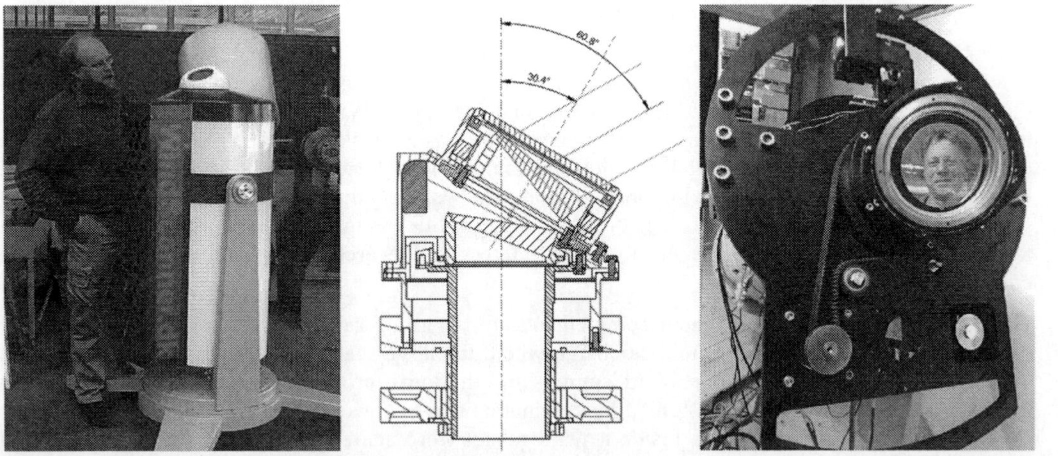

Figure 1. (Left) The exterior of the first generation short-range double-prism WindScanner. (Middle) The double-prism arrangement with individual prism rotation illustrated for the case of total maximum deviation of about 60°. (Right) The details of the scanner-head with one of the rubber belt transmissions visible.

The prisms are steered individually by one servomotor for each axis which allows the laser beam to be directed within a conical volume with a full opening angle of about 120°. All motors in all the lidars comprising the 3D WindScanner system are synchronously controlled by a Delta Tau Programmable Multi-Axis Controller system (PMAC) interconnected by optical fiber-based Motion And Control Ring Optical (MACRO) fieldbus technology. The focusing mechanism is driven by a linear motion motor and both stable ultrasonic piezo-ceramic motors and agile voice coil motors have been utilized.

A special rapid-scanning variety of the short-range WindScanner called a SpinnerLidar has also been created. The main difference is that it only has one single motor for driving both of the two prisms. The motor is operated at a constant but selectable rotation speed with a hardware-fixed gear ratio between the two prism axes such that a rosette-like scanning pattern can be completed within one second. The SpinnerLidar has not only been used from inside a wind turbine spinner but also from the top of wind turbine nacelles, either measuring the inflow or the wake [2].

Since the state-of-the-art wind turbines are becoming larger and larger, there is a drive for lidars that can measure at longer distances with small enough sampling volumes. Therefore, the WindScanner fauna recently has been populated by a new WindScanner creature using 6 inch optics instead of the 3 inch optics used in the original implementation.

During the course of the development, the coherent detection system has evolved from a heterodyne system where an acousto-optic modulator was used for shifting the zero Doppler frequency to about 27 MHz into a more robust homodyne system using an in-phase and quadrature detection scheme realized by an optical hybrid for achieving the direction sensing functionality [3].

3. Measurement application examples

Since the beginning of the WindScanner development in 2007 several measurement campaigns, as illustrated in Figure 2, have been performed. In one of the first campaigns one single WindScanner was used for studying the flow field over a steep escarpment of the westerly side of the Bolund peninsula in Roskilde Fjord in Denmark [4] and later a similar study was performed at a similar escarpment at the Prince Edward Island in Canada [5].

Two-dimensional synchronized measurements using two WindScanners with the scanning beams inside the measurement planes have subsequently been done for studying the downwash flow below rescue helicopters [6]. A similar approach has also been used for measuring inside wind tunnels [7] -

[8] as well as for studying the coherence of the inflow towards the Lysefjord bridge [9] where also the wake [10] of the bridge deck was investigated.

The main application field for the WindScanners is the field of wind energy and several campaigns for assessing both inflow to and wakes behind wind turbines have been performed using complete three-dimensional measurements from three synchronously sampling WindScanners. For instance, inflow towards a Vestas V27 horizontal axis turbine [11] as well as the wake behind a Nenuphar vertical axis turbine [12] have been measured by the short-range WindScanners and compared to models. Furthermore, the wake behind a wind break [13] and the wake behind a single Oak tree [14] have been studied by the WindScanners and the measurement capabilities in urban environments as illustrated in a recent overview paper [15] seems to be of interest to the architecture and pollution modeling communities.

The synergetic benefit of instrument complementarity regarding sampling volume and measurement distance was demonstrated in a field campaign where the large-scale flow over two mountain ridges was studied by pulsed long-range WindScanners and the local turbulent flow around a wind turbine located on one of the ridges was studied by continuous-wave short-range WindScanners [16]. Similar benefits on the very small scales inside a wind tunnel were achieved by combining a small easily handled Lidic telescope with the larger double-prism WindScanners [8].

Traditionally within the wind energy community mast-mounted in-situ point monitoring anemometers have been the standard measurement device against which wind lidars have had to be calibrated. However, currently DTU Wind Energy is attempting to reverse this situation by verifying the lidar Doppler speed estimation against a spinning wheel with a traceably calibrated speed [17] and then using the lidar for field calibration of cup and sonic anemometers [18]. This is possible by using continuous-wave lidars focused at short distances such that the sampling volumes become comparable or even smaller than the sampling volumes of the in-situ instruments.

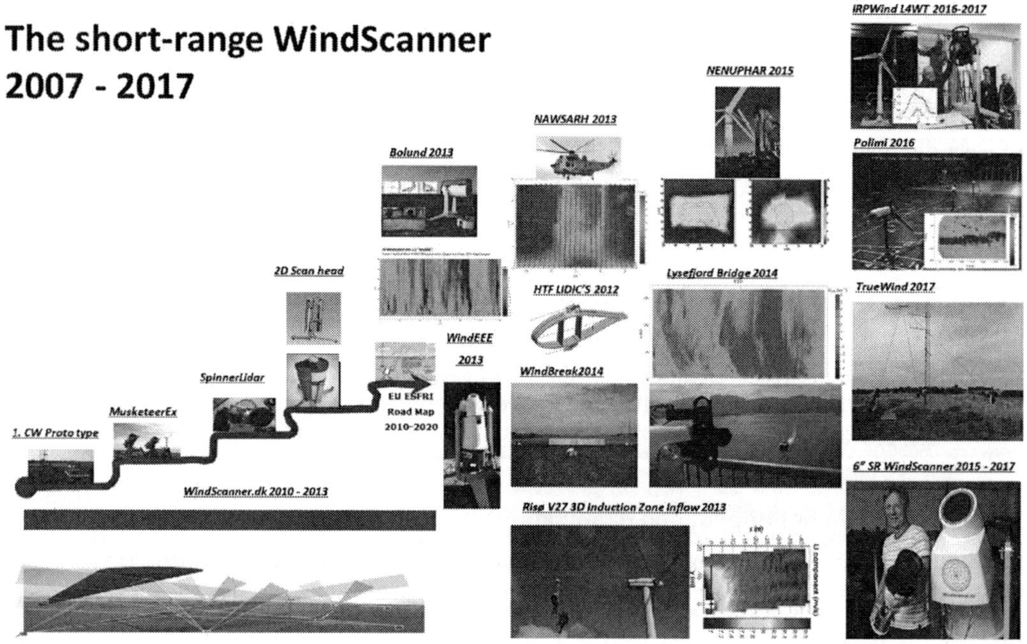

Figure 2. A flavor of activities with the DTU short-range WindScanners.

4. References

[1] T. Mikkelsen, S. Siggaard Knudsen, M. Sjöholm, N. Angelou and A. T. Pedersen, "WindScanner.eu - a new Remote Sensing Research Infrastructure for On- and Offshore Wind Energy," in *Proceedings of the International Conference on Wind Energy: Materials, Engineering, and Policies (WEMEP-2012)*, Hyderabad, India, 2012.

[2] T. G. Herges, D. C. Maniaci, B. T. Naughton, T. Mikkelsen and M. Sjöholm, "High resolution wind turbine wake measurements with a scanning lidar," *Journal of Physics: Conference Series,* vol. 854, p. 012021, 2017.

[3] C. F. Abari, A. T. Pedersen and J. Mann, "An all-fiber image-reject homodyne coherent Doppler wind lidar," *Opt. Express,* vol. 22, pp. 25880-25894, 2014.

[4] J. Lange, J. Mann, N. Angelou, J. Berg, M. Sjöholm and T. K. Mikkelsen, " Variations of the Wake Height over the Bolund Escarpment Measured by a Scanning Lidar," *Boundary-Layer Meteorology,* vol. 159, pp. 147-159, 2016.

[5] D. Parvu, J. LoTufo and H. Hangan, "Shear Layer Effects of Cliff Edge: PEIWEEE Experiment 2015," in *Book of Abstracts - WindTech 2015*, London Ontario, 2015.

[6] M. Sjöholm, N. Angelou, P. Hansen, K. H. Hansen, T. Mikkelsen, S. Haga, J. A. Silgjerd and N. Starsmore, " Two-Dimensional Rotorcraft Downwash Flow Field Measurements by Lidar-Based Wind Scanners with Agile Beam Steering," *Journal of Atmospheric and Oceanic Technology,* vol. 31, no. 4, pp. 930-937, 2014.

[7] M. F. van Dooren, F. Campagnolo, M. Sjöholm, N. Angelou, T. K. Mikkelsen and M. Kühn, "Demonstration and uncertainty analysis of synchronised scanning lidar measurements of 2-D velocity fields in a boundary-layer wind tunnel," *Wind Energy Science,* vol. 2, pp. 329-341, 2017.

[8] M. Sjöholm, A. Vignaroli, N. Angelou, M. B. Nielsen, J. Mann, T. K. Mikkelsen, H. C. Bolstad, K. O. Merz, L. R. Sætran, F. V. Mühle, M. Tiihonen and V. Lehtomäki, "Lidars for Wind Tunnels - an IRPWind Joint Experiment Project," *Energy Procedia,* vol. 137, pp. 339-345, 2017.

[9] E. Cheynet, J. B. Jakobsen, J. Snæbjörnsson, T. K. Mikkelsen, M. Sjöholm, J. Mann, P. Hansen, N. Angelou and B. Svardal, " Application of short-range dual-Doppler lidars to evaluate the coherence of turbulence," *Experiments in Fluids,* vol. 57, no. 12, pp. 184-200, 2016.

[10] E. Cheynet, J. B. Jakobsen, J. Snæbjörnsson, N. Angelou, T. K. Mikkelsen, M. Sjöholm and B. Svardal, " Full-scale observation of the flow downstream of a suspension bridge deck," *Journal of Wind Engineering and Industrial Aerodynamics,* vol. 171, pp. 261-272, 2017.

[11] E. Simley, N. Angelou, T. K. Mikkelsen, M. Sjöholm, J. Mann and L. Y. Pao, " Characterization of wind velocities in the upstream induction zone of a wind turbine using scanning continuous-wave lidars," *Journal of Renewable and Sustainable Energy,* vol. 8, no. 1, p. 013301, 2016.

[12] A. Immas, J. Kluczewska-Bordier, P. Beneditti, D. H. S. Pitance, N. Parneix, F. Silvert, P. Deglaire, C. de Jouette, M. Sjöholm, N. Angelou, K. Hansen and T. Mikkelsen, "Study of large-scale vertical axis wind turbine wake through numerical modelling and fullscale experiments," in *EWEA Annual Conference and Exhibition*, Paris, 2016.

[13] A. Pena Diaz, A. Bechmann, D. Conti and N. Angelou, "The fence experiment – full-scale lidar-based shelter observations," *Wind Energy Science,* vol. 1, no. 2, pp. 101-114, 2016.

[14] N. Angelou, E. Dellwik and J. Mann, "Characterization of a tree wake using three short-range WindScanners," in *European Lidar Conference*, Greece, 2018.

[15] T. K. Mikkelsen, M. Sjöholm, N. Angelou and J. Mann, " 3D WindScanner lidar measurements of wind and turbulence around wind turbines, buildings and bridges," *I O P Conference Series: Materials Science and Engineering,* vol. 276, no. 1, p. 012004, 2017.

[16] N. Vasiljevic, J. M. Palma, N. Angelou, J. C. Matos, R. Menke, G. Lea, J. Mann, M. Courtney, L. F. Ribeiro and V. M. Gomes, "Perdigão 2015: Methodology for atmospheric multi-Doppler lidar experiments," *Atmospheric Measurement Techniques,* vol. 10, no. 9, pp. 3463-3483, 2017.

[17] M. Courtney, N. Angelou, K. Enevoldsen, P. Hansen, T. Mikkelsen, M. B. Nielsen and R. Wagner, "Calibrating lidars using a flywheel," in *WESC2017 – Wind Energy Science Conference*, Lyngby, 2017.

[18] E. Dellwik, J. Mann, M. Sjöholm, N. Angelou and T. Mikkelsen, "Comparison of three-dimensional wind measurements by wind-lidars and a sonic anemometer," in *EMS Annual Meeting Abstracts*, Prague, 2014.

Development and testing of a triple-beam Doppler lidar system for 3D wind measurement

Matthew S. Warden (a), Simon T. Sørensen (a), John Macarthur (a), Mark Silver (b), Theodore C. Holtom (c), Craig McDonald (d), Peter J. M. Clive (d), Henry T. Bookey (a)

(a) Fraunhofer Centre for Applied Photonics,
Technology and Innovation Centre, 99 George Street, Glasgow G1 1RD, UK
(b) Thales UK, 1 Linthouse Road, Glasgow, G51 4BZ, UK
(c) Wind Farm Analytics Ltd., 2/1, 3 Queens Park Avenue, Glasgow, G42 8BX, UK
(d) Wood, Floor 2, St Vincent Plaza, 319 St Vincent Street, Glasgow, G2 5LD, UK
Lead Author e-mail address: matthew.warden@fraunhofer.co.uk

Abstract: We present results from our recent development of a triple-beam Doppler lidar system for accurate 3-dimensional wind measurements. The eye-safe all-fibre system consists of a central unit and three spatially separated remote units. The central unit contains a single seed laser that is shared between the remote unit. Each remote unit has its own optical amplifier, Risley prism beam steerer, and heterodyne detector. The photodiode signals are digitised and analysed in the central unit. The three lidar emitters were verified against each other and a calibrated meteorological mast at Myres Hill near Glasgow. To test the converging beam methodology, the three remote units were arranged in an equilateral triangle with an 8 m distance from each to the central unit and the three beams were brought to intersect at a height of 60 m. Steps towards wind turbine blade integration will be outlined.

Keywords: Doppler lidar, Wind energy

1. Introduction

Doppler wind lidars are widely used to make remote measurements of wind speed and direction for the wind energy industry. Typical measurement cases include assessing the available wind resource at sites prior to installing new turbines, and providing a measure of the wind speed at an operational turbine that allows a comparison of the actual output energy of the turbine against the specified output energy at that wind speed.

Existing commercial lidars use a single transmitter with beams pointing (either sequentially or simultaneously) in different directions to determine the wind speed and direction. This requires assumptions about the homogeneity of the wind field throughout the sensed volume. Using multiple separated transmitters with beams converging upon a single point, it is possible to make a more spatially resolved measurement of wind speed and direction, without requiring assumptions of large scale homogeneity of the wind field [1, 2, 3]. We have developed a three beam lidar system designed specifically for this purpose.

2. Lidar system design

A high level overview of the main lidar components is shown in Figure **1**. The system consists of a central unit which has connections to three spatially separated remote units. Each remote unit is connected to an adjacent lidar head. The central unit contains components that are able to share their functionality between the remote units and heads. This includes a control PC, digitiser, pulsed laser source and power supply electronics. The remote units each contain control electronics for the components in the heads, an optical amplifier and a balanced receiver. The heads each contain a dual Risley prism beam steerer, a shutter, 50 mm diameter telescope and focus adjustment stage. The shutter is used periodically to block the beam and enable background noise measurements to be made. The focus adjustment stage was used to maximise signal strength over the ranges of interest by locating the beam focus at the most favourable range (typically at the mid-point of the ranges of interest).

The laser in the central unit consists of a single continuous wave 1550.12 nm narrow linewidth seed laser (TeraXion NLL) that is modulated into 200 ns pulses at 20 kHz repetition frequency with an acousto-optic modulator, and amplified in a core pumped Er-Doped Fibre before the output is divided and transmitted to the three remote units. A 10 m fibre was inserted between the splitter and each of the remote units. The remote units each have their own power amplifier that is based on a cladding pumped Er/Yb-Doped Fibre. Their output power is limited to 120 mW due to Stimulated Brillouin Scattering in the ~1 m between the amplifier and telescope. The backscattered signal is mixed with a local oscillator in a heterodyne detection scheme [1]. The system was fitted in weatherised and thermally controlled enclosures for the outdoor testing.

All lidar functions are coordinated by the control PC housed in the central unit. This includes beam steering, shutter operation and telescope focusing. This allows for synchronisation of lidar operations such as adjusting the point in space the lidar beams are directed at. A single digitiser (with multiple channels) is housed in the central unit. This receives analog signals from the balanced detectors in each remote unit. The timing of laser pulses is determined by the pulsed laser source housed in the central unit. The combination of a centralised source of laser pulses and a centralised digitiser allows for accurately synchronised data acquisition from the three remote units.

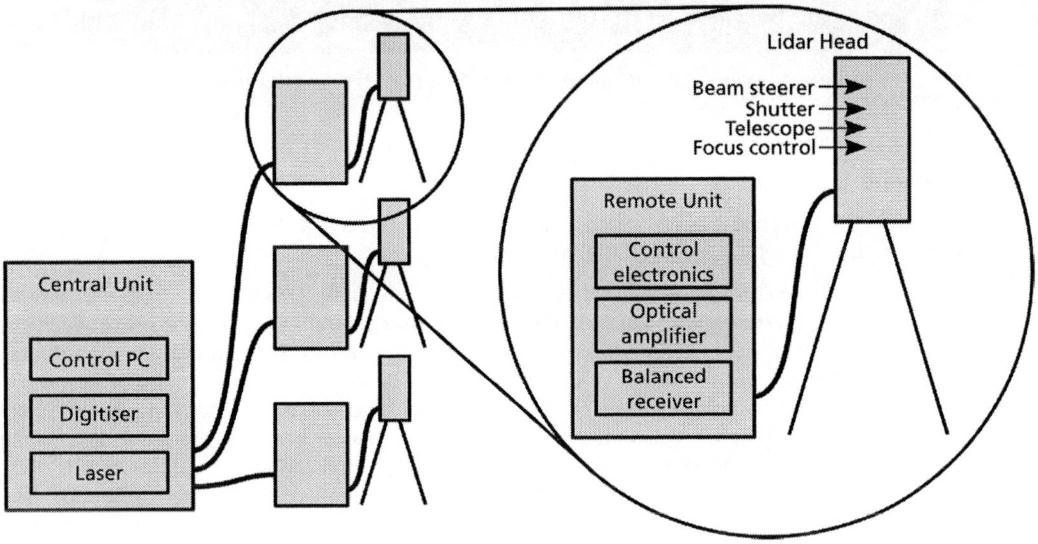

Figure 1. High level overview of main system components

Table 1. Lidar key parameters

Parameter	Value
Wavelength	1550 nm
Pulse duration	200 ns
Pulse energy	6 μJ
Pulse repetition rate	20 kHz
Telescope aperture diameter	50 mm
Maximum beam steering angle	60 deg

Figure 2. Photograph of lidar testing at Myres hill, near Glasgow, UK

3. Results

The lidar system performance was evaluated at Myres Hill near Glasgow. We report the results of tests made with two different arrangements of the lidar heads.

For the first arrangement, the three lidar heads were verified against each other and a calibrated meteorological mast. The three lidar heads were arranged with a less than 2 m spacing between the heads, with parallel beams directed so that they would pass near an anemometer on a calibrated meteorological mast. The measured LOS velocities from the lidar heads were 10-minute averaged and the meteorological mast wind speed projected onto the lidar's LOS for comparison. An excellent correlation was observed between the three lidar heads and also between the lidar heads and meteorological mast, as can be seen in Figure **3**. It should be emphasized that similar results were obtained for all three lidar heads. We ascribe the -0.54 m/s offset compared to the meteorological mast to an inaccuracy in the modulator frequency. As this is a constant offset, once the source has been confirmed a calibration procedure can be used to compensate for this.

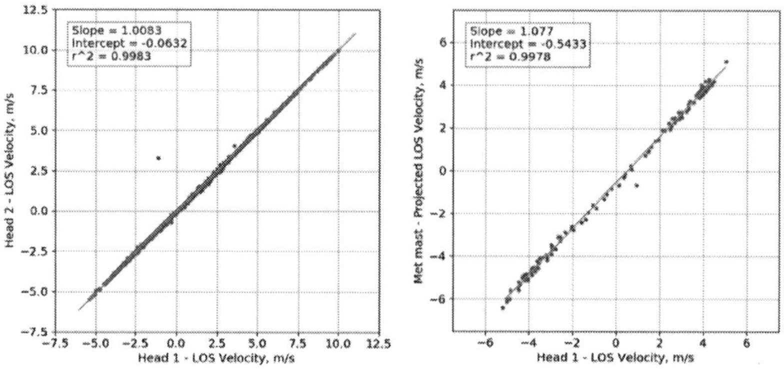

Figure 3. Correlation between lidar head 1 and lidar head 2 (left) and correlation between lidar head 1 and meteorological mast (right)

In the second arrangement of the lidar heads, to test the converging beam methodology, the three lidar heads were arranged in an equilateral triangle with an 8 m distance from each lidar head to the centre and the three beams were brought to intersect at a height of 60 m. The relative position and orientation of the lidar heads was determined from distance measurements between the heads, sighting and along reference edges on each head, and level measurements along references edges on each head. The global orientation of the system was determined from compass bearings between pairs of lidar heads.

An example of data recorded from the converging beam arrangement is shown in Figure **4**. This shows how a change in wind direction was accurately resolved with the intersecting beam lidar.

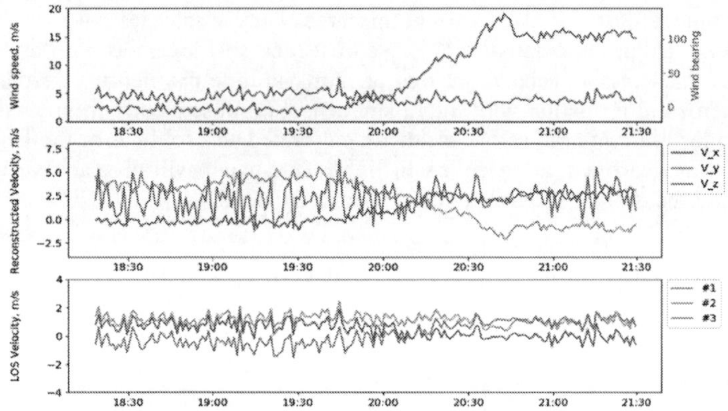

Figure 4. (bottom) Measured 1-minute averaged LOS velocities of the converging lidar beams. (middle) Reconstructed velocities in the (east, north, vertical) reference frame and (top) wind speed and direction.

4. Conclusion

We have developed a pulsed Doppler lidar system with three spatially separated emitters. The emitters share a single narrow linewidth seed laser that is amplified in multiple stages. The system shows excellent correlation with a calibrated meteorological mast and accurate 3-dimensional wind measurements were demonstrated. It is noted that use of continuous wave lidar can provide smaller probe length at short range. The compact and distributed design of the system makes it suitable for integration into large wind turbines for implementation of rotor based intersecting beam.

This work was funded by InnovateUK.

5. References

[1] Peña, A., *et al.*, "Remote Sensing for Wind Energy," DTU Wind Energy E-Report, no. 0084(EN) (2015). (1997).

[2] Mann, J., *et al.* "Comparison of 3D turbulence measurements using three staring wind lidars and a sonic anemometer." *IOP Conference Series: Earth and Environmental Science*. Vol. 1. No. 1. (2008).

[3] Vasiljević, N. and Courtney, M. "Accuracy of dual-Doppler lidar retrievals of near-shore winds" *WindEurope Resource Assessment Workshop 2017, Edinburgh, UK*

Airborne wind and backscatter measurements during the A-Life campaign in Cyprus

Stephan Rahm (a), Fernando Chouza Keil (a,b), Oliver Reitebuch (a), Benjamin Witschas (a)
(a) German Aerospace Center, DLR
(b) University of Vienna
Stephan.Rahm@dlr.de

Abstract: During the A-LIFE campaign 2017 in Cyprus the well-known DLR 2-μm Doppler Lidar has been integrated into the research aircraft Falcon F20 together with an extensive number of in situ sensors from DLR and the University of Vienna. Main objective of the campaign was the investigation of aerosol optical properties, mainly mineral dust and black carbon mixtures. The Doppler lidar was used to gather information about the extension of those dust (aerosol) layers as well as to estimate the mass transport. At each flight one or more altitude profile(s) were performed to gather aerosol information with the in situ sensors. Subsequently in order to calibrate the Doppler Lidar in term of backscatter a calibrated aerosol lidar of the University of Munich was overflown at nearly each flight. This paper will give an overview of all flights and focus on one Saharan dust outbreak for details.

Keywords: Coherent Laser Radar, A-Life, SALTRACE.

1. Introduction

A coherent Doppler lidar is usually not the instrument of choice to provide an estimate for the backscatter coefficient. Due to the heterodyning with a local oscillator it only detects the fundamental transverse mode of the backscattered light. This leads to a much higher speckle noise compared to a direct detection lidar. The comparably long pulse duration of a coherent Doppler lidar which is necessary to obtain an accurate velocity estimate also reduces the geometrical resolution of the backscatter estimate. However the accurate wind estimates, 3 dimensional vector as well as line of sight (LOS) made the airborne DLR 2 micron Doppler lidar to a well-suited instrument for meteorological campaigns for almost 20 years all over the world. Since the SALTRACE campaign in 2013 the question arose if the Doppler lidar can provide additional data like backscatter beside the velocity estimates. One advantage of coherent detection is the much lower dynamic range compared to direct detection. As a result the 8 bit digitizer used in this system is sufficient. The procedure necessary to obtain a calibrated backscatter was described by Chouza in 2015 at the SALTRACE campaign [1,7]. All of those lessons learned were implemented into the software so that at the campaign A-LIFE in CYPRUS uncalibrated but throughout the campaign consistent backscatter profiles were calculated on an operational basis. The next step will be the calibration at overflights of the POLIS Lidar which was also located in Cyprus during the A-Life campaign.

2. Two micron Doppler lidar and signal processing

The 2 μm Doppler lidar has been described in several publications [2-5]. The main parameters of the system are summarized in Table 1. During the years of operation two main operation modes have been established to cover most of the required measurements:

One mode is a fixed nadir with a control loop to compensate for platform (aircraft) movements. This mode is used to acquire vertical wind phenomena like gravity waves and also for high resolution backscatter profiles. The measurement volume is determined vertically by the pulse length and horizontally by aircraft speed (100 - 200 m/s depending on altitude) and the accumulation time (1-2 s).

The accuracy of a LOS estimate was calculated to be in the order of 0.2 m/s typically [6], depending on the SNR.

The other scan pattern is a conical 20° off nadir step and stare scan with 18 positions and one second accumulation time at each stare position. Together with the time needed to move between those positions one scan needs about 40 s. This pattern is used to estimate the 3D horizontal wind. The vertical measurement volume is 100 m as in fixed nadir mode, again determined by the pulse width. The horizontal volume varies with distance from the aircraft. It is roughly 8 km square near the ground with an aircraft altitude of 10 km.

Table 1: Parameter of the 2 micron lidar

Parameter	Value
Wavelength	2,02245 µm
Energy per pulse	1 – 1,5 mJ
Pulse duration FWHM	500 ns
Spectral width FWHM	0,75 MHz
Repetition rate	500 Hz
Average optical power	0,5 – 1 W
Offset frequency Transmitter - LO	100 MHz
Telescope diameter	11 cm
Double wedge scanner max deviation	30° off nadir
Data acquisition: single shot; resolution:	8 bit
Sampling rate	500 MS/s

At signal processing stage the digitized heterodyne signal is divided into 100 m altitude range gates. As the number of samples in the range gate is usually not a power of two zero padding is applied in order to be able to use an efficient FFT algorithm. This zero padding has to be corrected to obtain comparable power estimates for different range gates in pointing directions. The power spectrum obtained by the FFT has to be corrected also by the frequency response of the system: detector, amplifier, digitizer. For this the assumption is used that the system is dominated by shot noise from the local oscillator that is spectrally white noise. After passing detector, preamplifier, amplifier and A/D converter the resultant noise spectrum carries all of the information needed about the frequency response and local oscillator power. At an airborne lidar the part of the signal below the ground return contains definitely only noise. Thus the spectrum of this noise is used to correct for the frequency response by dividing the FFT from the signal by the noise spectrum and subtracting 1 for all frequency channels. The resultant signal power has then only to be corrected by the transmitted energy which can easily be accessed by measuring the pulse built up time (BUT) of the laser pulse. With an R^2 correction the resulting power represents the optical back scattered power multiplied by all of the instrument parameters like transmission of the optics and heterodyne efficiency which are assumed to be constant.

3. A-LIFE campaign

A-LIFE (Absorbing aerosol layers in a changing climate: aging, **life**time and dynamics) is a project that investigates the optical properties of aerosols. Details about the project can be found at https://www.a-life.at/ . Central part of the project was the field campaign contucted from Cyprus in April 2017. The DLR research aircraft Falcon F20 was equipped with the 2 micron Doppler Lidar and an extensive range of in-situ instruments collecting the aerosol data. Ground-based in-situ and remote sensing instruments were deployed at Cyprus as well. The interesting instrument for the synergy with the coherent Doppler lidar is the POLIS backscatter lidar that can be used to “calibrate” the coherent lidar in term of backscatter [1].

One idea of the combination of remote sensing and in-situ instruments is that atmospheric layers with aerosol / dust are investigated in detail with the in-situ particle counter by performing an altitude profile with the in-situ instruments collecting data on aerosols at different altitudes. With the Doppler lidar it may be possible to extend the point measurements over a wider area passing at a higher altitude over the volume investigated by in situ. As Saharan dust and other aerosols were investigated over the Mediterranean Sea at stable cloud free weather condition it can be assumed that those air masses are stable at least over the observation time which ranges from one to four hours. A nice example of this strategy can be found on the research flight to Malta on April 19. The received optical power (range corrected) can be seen in Figure 2. Figure 1 shows all of the A-Life flight tracks including the transfer flight from Oberpfaffenhofen. The main region of investigation was located close to Cyprus, but three cross sections of Saharan dust were investigated on two flights to Crete and one to Malta and back.

Figure 1: Overview of all A-Life flights.

The obvious idea to search with the lidar for an interesting dust laden air mass and to investigate this area with the in-situ sensors was not possible due to flight planning constraints. Fortunately the forecast for Saharan dust was accurate enough to identify the regions for an altitude profile and to consider this already at the flight planning stage.

Figure 2: Range corrected backscatter power at flight from Cyprus (right edge) to Malta (left) on April 19.

Figure 2 shows a rare example of nearly full coverage from aircraft level down to the ground. Two layers of dust can easily be seen. The high backscatter region close to the sea surface (below 500 m altitude) may be caused by the ship traffic to and from the Suez Canal. This volume did show a rather high black carbon concentration according to the in situ sensors. Further were some optically thin clouds present at this flight lag. Those can be identified as white areas with a virtually reduced backscatter below. At a further processing step it is planned to compensate for this loss in transmission through the clouds.

4. Conclusion and further outlook

The lidar worked well during the whole A-Life campaign. The processing software was modified in order to be able to have a quantitative estimate of the backscatter coefficient. The comparison, calibration and synergy with the in-situ instruments are pending. During the local flights close to Cyprus the same pattern was performed at different flight levels. Here the volume of Doppler lidar measurements was investigated by the in-situ sensors.

5. Acknowledgements

The project A-LIFE has received funding from the European Research Council (ERC) under the European Union ' s Horizon 2020 research and innovation program under grant agreement No. 640458 (A - LIFE). The German Aerospace Center (DLR) provided funding for a significant amount of additional flight hours and aircraft allocation days for the A-LIFE aircraft field experiment. Furthermore, two EUFAR projects were clustered with A - LIFE which provided funding for 16 flight hours.

6. References

[1] F. Chouza, O. Reitebuch, S. Groß, S. Rahm, V. Freudenthaler, C. Toledano, B. Weinzierl "Retrieval of aerosol backscatter and extinction from airborne coherent Doppler wind lidar measurements", Atmos. Meas. Tech., 8, 2909-2926, (2015).

[2] F. Köpp, S. Rahm, I. Smalikho, " Characterization of Aircraft Wake Vortices by 2-µm Pulsed Doppler Lidar" J. Atmos. Oceanic Technol., 21, 194-206 (2004)

[3] M. Weissmannn, R. Busen, A. Dörnbrack, S. Rahm, O. Reitebuch, "Targeted observations with an airborne wind lidar " J. Atmos. Oceanic Technol., 22, 1706-1719 (2004)

[4] U. Schumann et al, "Airborne observations of the Eyjafjalla volcano ash cloud over Europe during air space closure in April and May 2010" Atmos. Chem. Phys., 11, 2245-2279 (2011)

[5] B. Witschas, S. Rahm, A. Dörnbrack, J, Wagner, M. Rapp " Airborne Wind Lidar Measurements of Vertical and Horizontal Winds for the Investigation of Orographically Induced Gravity Waves" J. Atmos. Oceanic Technol., 34, 1371-1386 (2017)

[6] Chouza, F., Witschas, B., and Reitebuch, O., "Heterodyne high-spectral-resolution lidar", Appl. Opt. **56**, 8121-8134 (2017)

[7] B. Weinzierl et al. "The Saharan Aerosol Long-Range Transport and Aerosol Cloud Interaction Experiment", Bull. Am. Meteorol. Soc. **98** / 7, 1427-1451 (2017)

Overview of Demonstration Project of JAXA Onboard Turbulence Detection System

Shigeru Machida, Hamaki Inokuchi, Naoki Matayoshi
Aeronautical Technology Directorate, Japan Aerospace Exploration Agency, Tokyo, Japan

Abstract: Aircraft Accident and Incident Reports published by Japan Transport Safety Board show that turbulence is the leading cause of in-flight injuries during the past 20 years. In order to conduct R&D to reduce injuries penetrating a clear air turbulence, the JAXA R&D project team was formed in 2014. The team had successfully developed the onboard clear air turbulence detection system that weighs only 83.7 kilogram. In 2017, the system has been demonstrated using an experimental small jet aircraft that the LIDAR has a capability to observe air turbulence at a distance of 17.5 km (Ave.) which is the world's longest detection range. The advisory function has been also demonstrated, which can provide turbulence information to pilots. This innovative solution is selected as one of technologies to be tested on the ecoDemonstrator program 2018 by The Boeing Company. In this paper, project overview, flight test and next steps are described.

Keywords: Doppler LIDAR, Onboard, System, Flight Demonstration

1. Introduction

Once an aircraft accident is occurred, it would cause not only human injury or loss of lives but also adverse effect psychologically and economically. According to Aircraft Accident and Incident Reports[1] published by Japan Transport Safety Board, more than 50% of domestic airline accidents over the past 20 years were caused by turbulence. A turbulence with water drops in rain clouds in front of passenger aircraft can be detected to some degree by weather radar and/or by consulting weather forecast in advance, but it is difficult to look a turbulence without rain clouds, which is known as clear air turbulence. If it is possible to find a clear-air turbulence in front of an aircraft somehow, a pilot can be informed an early warning from turbulence detection system and turn on the seat-belt sign to reduce those accidents.

JAXA had conducted basic research on clear-air turbulence detection technology since 1998 and the JAXA R&D project, SafeAvio Project, team was formed in 2014 based on previous research results. The mission of SafeAvio Project is to develop onboard Turbulence Detection and Advisory System which can detect clear-air turbulence and conduct flight demonstrations to verify its function and performance of the system.

2. Project Overview

Prior to the project, effectiveness of an onboard Doppler LIDAR technology was evaluated using prototype devices during the experimental flight tests in 2012. In the SafeAvio Project, a compact and high-performance Doppler LIDAR was developed as a turbulence detection system based on the prototype one. An advisory system was also developed to inform pilot before entering into a turbulence based on forward airflow information from Doppler LIDAR. To demonstrate the performance of the developed onboard Turbulence Detection and Advisory System, flight demonstration was conducted to show providing information to pilot for go-around judgement during landing approach and for seat belt sign judgement during cruising and descending. A maximum observation range and measurement accuracy of the wind speed of the system were evaluated by flight tests.

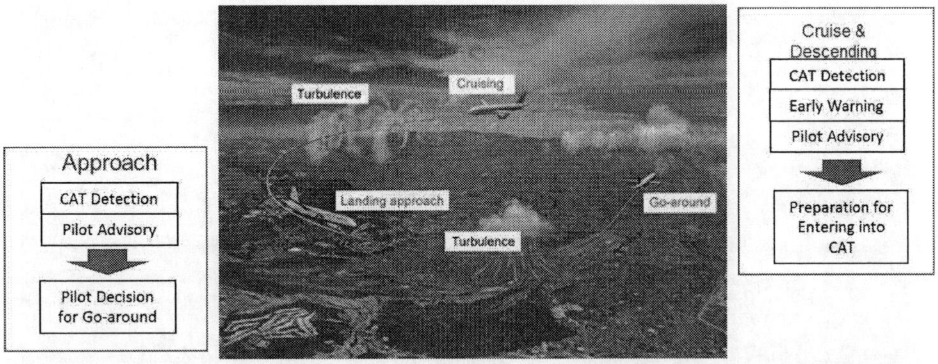

Figure 1. Operation Concept of the onboard Turbulence Detection and Advisory System

3. Developed Technologies

The developed onboard Turbulence Detection and Advisory System using Doppler LIDAR Technology [2] is shown in Figure 2. The laser pulse signal generated by the optical transmitter is amplified by optical amplifiers incorporated in the optical antenna and radiated into the atmosphere from optical telescopes. The optical antenna was equipped with a large telescope with a 150-mm aperture for observation for long distance detection and a small telescope with a 50-mm aperture for detecting short range wind velocity.

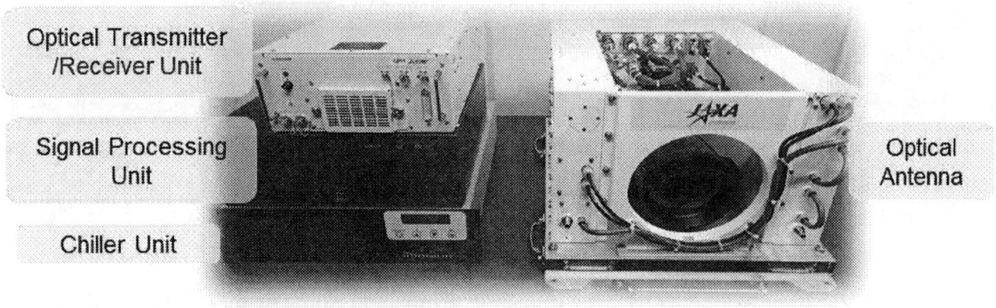

Figure 2. Developed onboard Turbulence Detection System using Doppler LIDAR Technology

The emitted laser beam from the optical antenna would be scattered by small aerosol particles present in an atmosphere (fine water drops and dust, etc.), and the scattered light is received by optical telescopes and translated by optical receiver unit and signal processer unit.

Scattered light from small aerosol particles in front of an aircraft has an information of relative moving speed and turbulence. Since the wavelength of the received light shifts due to the Doppler effect according to the moving speed of the aerosol particles, the wind speed can be calculated. And analyzing light wavelength variation caused by the Doppler effect, a Doppler LIDAR can determine transitions in airflow which is known as turbulence. This makes it possible to detect clear-air turbulence that onboard weather radar on most aircraft have can not find.

One of the functions of the developed onboard Turbulence Detection and Advisory System is to provide early warning information related to turbulence in front an aircraft and predictive airspeed information to pilots. [3] Figure 3 shows concepts of lidar-detected windshear and moderate/severe turbulence direction and range, and time to encounter on existing displays in a cockpit. A concept of multiple predicted airspeeds on speed tape at different prediction timings (in 5/10/20 seconds) which correspond to different distances ahead based on LIDAR information is shown in Figure 4.

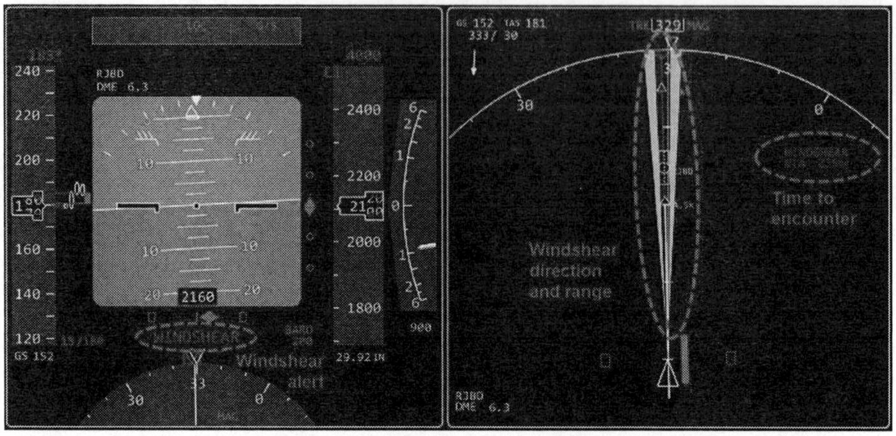

a) Lidar-detected Windshear

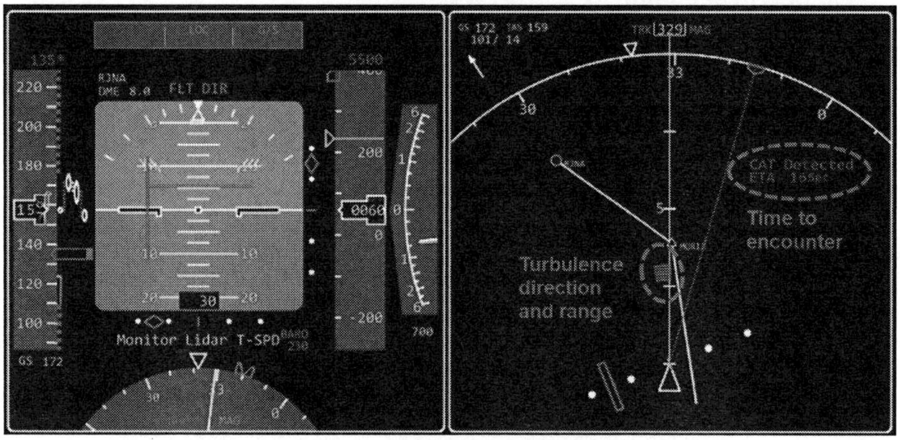

b) Lidar-detected Turbulence

Figure 3. Display Concepts of lidar-detected Windshear and Turbulence Information

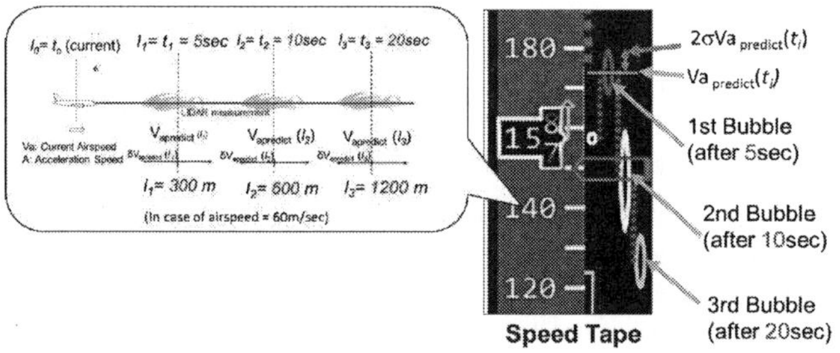

Figure 4. Display Concept of Multiple Predicted Airspeeds on Speed Tape

4. Flight Demonstration

The flight demonstration was conducted from December 2016 and January 2017 in Japan. The optical antenna device was mounted underneath front part of the fuselage of the experimental small jet aircraft to emit laser beam in front of the aircraft. A double wedge prism type scanner was installed in front of the laser radiation part of the optical antenna device for adjusting the optical axis direction of the laser beam. The beam direction can be adjusted inside the cabin and it is possible to look a wished direction. All other devices were installed in the cabin. (Figure 5)

Figure 5. Experimental Aircraft with Developed onboard Turbulence Detection and Advisory System

It was demonstrated that detection range was 17.5 kilometers (Average) in detecting clear-air turbulence ahead of the aircraft with the developed the onboard system that weighs only 83.7 kilogram, which is equivalent to a passenger with one baggage. A 17.5 kilometer, which is the world's longest range as onboard turbulence detection system, means roughly seventy seconds for cruising aircraft, during which pilots can turn on the seatbelt sign to make advanced warning for passengers and crews to prepare for hazardous shaking.

Addition to that, the developed Advisory System to provide early warning information related to turbulence in front an aircraft and predictive airspeed information was demonstrated for the first time in the world.

5. Current and Future Activities

After completion of the successful flight demonstration with the onboard Turbulence Detection and Advisory System, JAXA is to work with the Boeing Company to conduct flight tests with the same system during ecoDemonstrator program 2018 in March and April 2018.

It would be a great chance to evaluate the system from aircraft manufacturers' point of view. And from R&D point of view, JAXA would like to demonstrate how the clear-air turbulence detection system can be fitted in to large commercial aircraft and find any technical issues to be solved toward a practical application. The program will help promote the effectiveness of the system on commercial aircraft among airline companies and manufacturers. And it is also expected that it will lead to bring the importance and value of the system to the attention of standardization organizations. JAXA will continue making further R&D efforts in developing safety technology to prevent aircraft accidents.

Reference

[1] Japan Transport Safety Board, Aircraft accident reports, http://www.mlit.go.jp/jtsb/.

[2] Inokuchi H, Akiyama T, and Machida S. *Development of an Onboard Safety Avionics System using a Doppler Lidar.* Proc 18th Coherent Laser Radar Conference, Boulder, USA, 2016.

[3] Iijima T, Uemura T, Matayoshi N, Entzinger J O, Matsumoto J, Ueda S and Yoshikawa E. *Development and Evaluation of a New Airspeed Information System utilizing Airborne Doppler Lidar*. 36th Digital Avionics Systems Conference, St. Peters berg, U.S.A, 2017.

Heterodyne high-spectral-resolution lidar

Fernando Chouza (a), Benjamin Witschas (a), Oliver Reitebuch (a)
(a) DLR, Institute of Atmospheric Physics, Oberpfaffenhofen, Germany

Lead Author e-mail address: fchouza@gmail.com

Abstract: This study presents a novel lidar technique to conduct high-spectral-resolution measurements of the atmospheric backscatter. The proposed method, based on a heterodyne detection receiver, allows the separation of the molecular and the aerosol component of the atmospheric backscatter, as well as the analysis of the spectral shape of the Rayleigh–Brillouin line. As in the case of the direct-detection high-spectral-resolution lidars (HSRLs), the separation of the different scattering processes allows an independent system calibration and aerosol extinction measurements. The proposed measurement technique was successfully tested on the DLR airborne Doppler wind lidar system based on data collected during different campaigns and under different atmospheric conditions. In light of these results, further ideas for the implementation of a dedicated heterodyne HSRL (HHSRL) will be discussed.

Keywords: High-spectral-resolution, aerosols, temperature, pressure, heterodyne detection.

1. Introduction

Several types of atmospheric lidars rely on accurate measurements of Rayleigh-Brillouin (RB) scattering for the retrieval of atmospheric parameters like wind, aerosols properties or temperature. These systems, based on the direct detection technique, make use of optical filters to perform spectrally resolved measurements of the atmospheric backscatter.

While heterodyne detection of RB scattering has already been used in some laboratory measurements for the characterization of gases and liquids, e.g., [1], its application to atmospheric measurements was only theoretically discussed [2,3] and limited to the retrieval of atmospheric temperature by relating the measured RB spectrum to RB line shape models.

In this work, measurements conducted by the airborne doppler wind lidar (DWL) of the Deutsches Zentrum für Luft- und Raumfahrt (DLR) during different field campaigns (SALTRACE, DEEPWAVE and A-LIFE) are used to demonstrate the feasibility of a novel lidar measurement technique based on Rayleigh-Brillouin (RB) scattering measurement by a heterodyne detection lidar. The retrieval of high-spectral-resolution measurements of the atmospheric backscatter allows, as in the case of direct detection HSRLs, the separation of the different scattering processes, and thus, an independent system calibration and aerosol extinction measurements.

2. Method

The DLR airborne coherent DWL was built based on an instrument from CLR Photonics, now Lockheed Martin Coherent Technologies, with a scanning and acquisition system developed by DLR that provides airborne wind measurement capabilities [4]. The lidar operates at a wavelength of 2.02254 µm with a pulse full width at half-maximum (FWHM) of 400 ns, a pulse energy of 1–2 mJ, and a repetition frequency of 500 Hz.

The lidar detection is based on a double-sideband receiver with an IF shift of about 100 MHz and a standard bandwidth of 200 MHz. Although first evidence of RB scattering measurements by our system were found with the standard configuration, in order to resolve the complete RB spectrum, which has a half width at half-maximum (HWHM) of about 400 MHz at a 2-µm wavelength, the data acquisition configuration was modified during two flights to conduct high-bandwidth measurements with a sampling frequency of 2 GHz.

Heterodyne lidar signal processing is typically done in the frequency domain. Temporal signals corresponding to each shot are first divided in range gates and the power spectral density (PSD) of each range gate calculated. In order to reduce variability due to speckle and improve the signal-to-noise (SNR), a temporal average of the power spectra corresponding to several shots is calculated. It can be shown [5] that assuming shot noise limit operation, a good approximation of the average measured PSD P_s as a function of range r and frequency f can be written as

$$\langle P_s(r,f)\rangle = H(f)\left[4G^2R^2\eta_h(r)P_{LO}P_a(r,f-f_{IF}) + N_{atm}(r,f) + N_{bkg}(f)\right] \quad (1)$$

where H is the frequency response of the whole receiver chain, G is the receiver chain trans-impedance, R is the receiver photodiode sensitivity η_h is the heterodyne efficiency, P_{LO} is the local oscillator (LO) power, P_a is the average PSD of the atmospheric backscatter, N_{atm} is shot noise dominated by the Mie scattering heterodyne current and N_{bkg} is the background noise dominated by the contribution of the LO shot noise ($N_{bkg} = 2qG^2RP_{LO}$, with q the electron charge).

Dividing Eq. 1 by the PSD corresponding to range gates r_N where no atmospheric return was received (range larger than aircraft altitude), a normalized dimensionless PSD can be written as

$$R(r,f) = \frac{P_s(r,f)}{P_s(r_N,f)} - 1 = \frac{2R\eta_h(r)P_a(r,f-f_{IF})}{q} + \frac{N_{atm}(r,f-f_{IF})}{2qG^2RP_{LO}} \quad (2)$$

where the first term is proportional to the atmospheric backscatter (Mie and RB) and the second term is proportional to the noise induced by the first term. Due to the different nature of the Mie and RB scattering process, the Mie scattering corresponds to a narrow-band signal (about 1 MHz) while the RB scattering corresponds to a broad band signal (about 400 MHz HWHM). Based on this difference, the contribution of the RB scattering P_{Ray} can be obtained by integrating Eq. 2 over a broadband window (BB) centered around IF where the Mie scattering contribution is zero (lidar is vertical pointing and no deviation from IF is expected from wind contribution). In a similar way, the contribution of the Mie scattering can be obtained by integrating over a narrow-band window (NB).

$$P_{BB}(r) = \sum_{f\in BB} R(r,f) \approx k_{det}(r)k_f P_{Ray}(r) + k_n N_{atm}(r) \quad (3)$$

$$P_{NB}(r) = \sum_{f\in NB} R(r,f) \approx k_{det}(r)P_{Mie}(r) \quad (4)$$

where $k_{det}(r) = 2R\eta_h(r)/qB$ and $k_n=1/(2qG^2RP_{LO}B)$, with B the detection bandwidth. Since depending on the detection bandwidth not all the RB scattering bandwidth is covered, a correction factor k_f is included to take this into account. In the case of the Mie scattering, and since the integration bandwidth is small, the contribution of the Mie induced noise can be neglected.

Finally, based on the lidar equation, P_{ray} and P_{mie} can be written as a function of the atmospheric molecular and aerosol backscatter

$$P_{BB}(r) \approx k_{Cal}(r)k_f\beta_m(r)T^2 + k_n N_{atm}(r) \quad (5)$$

$$P_{NB}(r) \approx k_{Cal}(r)\beta_a(r)T^2 \quad (6)$$

where $k_{Cal}(r) = k_{det}(r)P_t k_i(r)A/r^2$, with P_t equal to the transmitted power and A the receiver area, β_m is the molecular backscatter coefficient, β_a the aerosol backscatter coefficient and T the atmospheric transmission.

3. Results

Figure 1 presents an example of measurements performed with the high bandwidth configuration, under low aerosol load conditions, during the A-LIFE campaign. Although the flight altitude was 9.9 km for this case study, due to limitations of the acquisition system, only measurements down to 5.9 km were recorded.

The normalized PSD R(r,f) presented in Fig. 1(a) was calculated from a FFT of 2048 samples, equivalent to range gates of about 150 m and an averaging time of 210 s. For the background noise correction process, the average of the spectra corresponding to range gates from 20 (r > 3.5 km) and onwards were used after replacing the samples between 95 and 115 MHz, containing the Mie spectrum, by a linear interpolation.

Figure 1(b) show the RB scattered power as a function of range resulting from integrating the received power for each range gate and the NB channel (indicated by green lines in Fig. 1(a)), after correction by Mie signal induced noise (k_nN_{atm}).

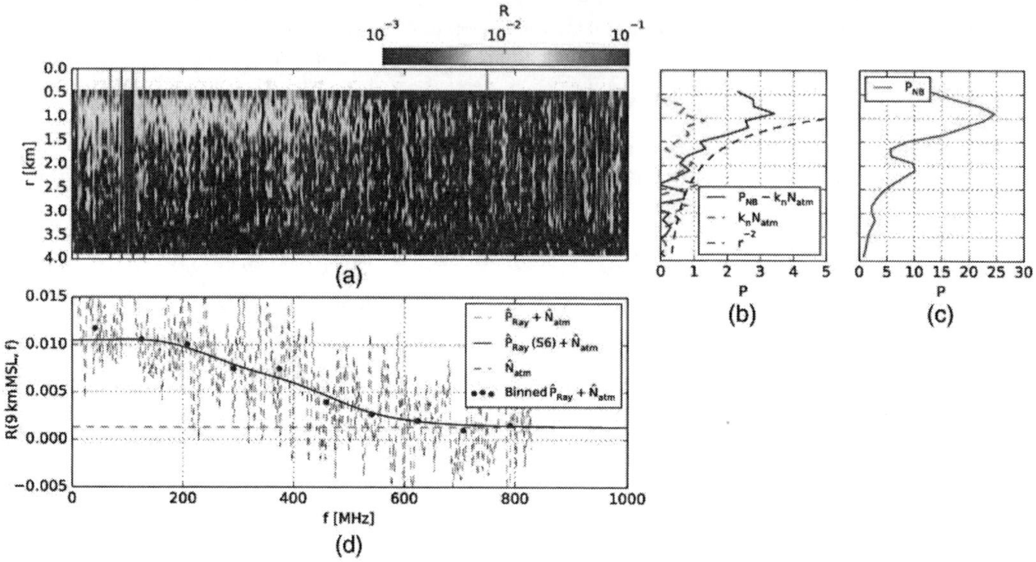

Figure 1. Analysis on data acquired during the A-LIFE campaign using the high-bandwidth receiver configuration.

The analysis presented in Fig. 1(d) show the PSD for a range gate 900 m below the aircraft together with the expected RB spectral shape derived from the Tenti S6 model and considering the spectral folding effect due to the small IF. The agreement between the measurements and the model provide further evidence about the RB origin of the broadband signal observer by the lidar.

In order to provide further evidence about the molecular origin of the observed broadband signals, an analysis of the change in the received power from a range gate at a fixed distance from the aircraft during descending or ascending flight sections was conducted. In this way, no characterization of the heterodyne efficiency is required, and the range gate with the maximum SNR can be selected. Furthermore, and since a range gate close to the aircraft can be chosen, the effects of the unknown atmospheric extinction can be minimized. In this way, the Rayleigh backscatter power is proportional to the molecular backscatter coefficient corresponding to the range gate altitude.

Figure 2 presents the results derived from Eqs. 5 and 6 based on measurements carried out during the SALTRACE campaign. By fitting the vertical profile of P_{BB} (after correction by Mie induced noise) obtained with the previously described techniques to a vertical molecular backscatter profile calculated based on in-situ temperature and pressure measurements, k_{Cal} was calculated (Fig. 2(a)). This calibration constant was then applied to retrieve a vertical aerosol backscatter profile (Fig. 2(b)). In order to perform a validation of the results, the retrieved profile was compared with an aerosol backscatter profile derived by the ground-based aerosol lidar POLIS. Since the DWL operates at 2 um and detects only one polarization component, a wavelength conversion coefficient and depolarization extinction correction was applied. The results show a good agreement between both retrievals.

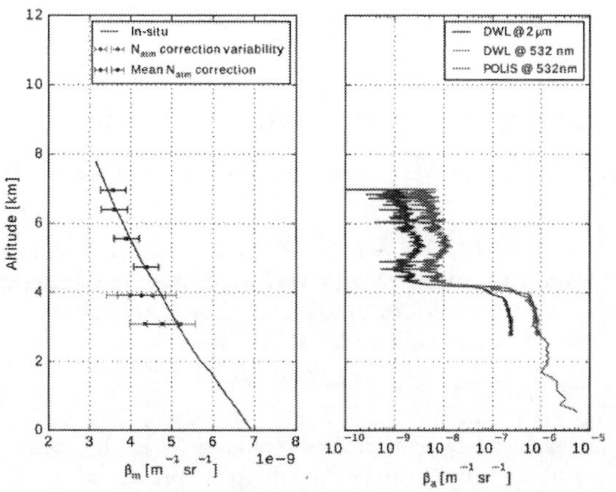

(a) (b)

Figure 2. Retrieved vertical profiles of molecular and aerosol backscatter profiles during SALTRACE.

4. Conclusions

To our knowledge, first-time measurements of RB scattering by an heterodyne detection lidar are presented in this work. Additionally, a novel calibration technique that doesn't require of additional information from aerosol ground-based lidars [6] was briefly described. Despite a series of instrumental limitations mainly related to the original purpose for which the instrument used in this study was developed (e.g. limited wavelength and power, measurement of only one polarization component), the bases for a novel detection architecture for the retrieval of temperature and aerosol optical properties were settled. This proof of concept for a HHSRL shows the potential and requirements of a system based on this technique and guides the future implementation of a system with a large range of applications, including temperature, wind, and aerosol measurements.

Acknowledgment. The authors thank Silke Groß for providing the POLIS data used in this study, Martin Wirth for the support in the modification of the DWL data acquisition system and the helpful comments on this work, Stephan Rahm for the assistance in the DWL data processing, and Joseph Gasteiger for the simulations of the Saharan dus optical properties at 2 μm.

5. References

[1] H. Tanaka and T. Sonehara, "Superheterodyne light beating spectroscopy for Rayleigh–Brillouin scattering using frequency-tunable lasers," Rev. Sci. Instrum. 73, 1998–2010 (2002).

[2] B. J. Rye, "Molecular backscatter heterodyne lidar: a computational evaluation," Appl. Opt. 37, 6321–6328 (1998).

[3] C. F. Abari, X. Chu, J. Mann, and S. Spuler, "A micropulse eye-safe all-fiber molecular backscatter coherent temperature lidar," in EPJ Web of Conferences (EDP Sciences, 2016), vol. 119, 25005.

[4] B. Witschas, S. Rahm, A. Dörnbrack, J. Wagner, and M. Rapp, "Airborne wind lidar measurements of vertical and horizontal winds for the investigation of orographically induced gravity waves," J. Atmos. Oceanic Tech. 34, 1371–1386 (2017)

[5] F. Chouza, B. Witschas, and O. Reitebuch, "Heterodyne high-spectral-resolution lidar," Appl. Opt. 56, 8121-8134 (2017)

[6] F. Chouza, O. Reitebuch, S. Groß, S. Rahm, V. Freudenthaler, C. Toledano, and B. Weinzierl, "Retrieval of aerosol backscatter and extinction from airborne coherent Doppler wind lidar measurements"Atmos. Meas. Technol. 8, 2909–2926 (2015).

Onboard wake vortex tracking with an all fibered Coherent 1.5µm Doppler LIDAR for aircrafts in formation flight

Dolfi-Bouteyre Agnès, Goular Didier, Augère Beatrice, Planchat Christophe, Fleury Didier, Lombard Laurent, Valla Matthieu, Besson Claudine.
ONERA/ DOTA Université Paris Saclay
F 91123 Palaiseau - France
Agnes.dolfi-bouteyre@onera.fr

Abstract: An onboard LIght Detection And Ranging (LIDAR) sensor designed to track wake vortex created by aircrafts in formation flight is presented. It is based on a 1.5-µm pulsed fibered laser and has a very short range detection capability (<30 m) compared to previous sensor designs. I was tested on board an aircraft and was able to detect the two-dimensional (2-D) velocity map of the wake vortex created by another aircraft flying ahead. In addition, it includes a real-time display of the 2D velocity map which enables the operator to locate the vortex center. Data postprocess allowed determining the position of the vortex cores within less than 2 m.

Keywords: Coherent Laser Radar, Wake-vortex, Formation flight.

1. Introduction

Studies of birds have shown that the V formation can greatly enhance the overall aerodynamic efficiency by reducing the drag and, therefore,increasing the flight distance that they can cover [1].Formation flight for aircraft has been extensively studied during the last 10 years as technology have provided more precise navigation and control systems to maintain formations [2]. Nowdays, close formation flight (such as the ones used by birds or military aircraft) is not possible for commercial flight due to collision hazards. However, distant formation flight with spacing between 10 to 40 wingspans would still be beneficial because of the persistence of cruise wake vortices (WV) of large commercial aircraft. The induced drag reduction was estimated to be up to 30% [2]. At these distances, aircraft tracking error is the most significant contribution to the variation in drag reduction.

For the study reported in this article, formation flight tests have been carried out, where a first leading aircraft seeded its wake vortex to enable the pilot of a second following aircraft to visually locate the vortex. We present a scanning coherent pulsed LIDAR installed in the second aircraft that have been developed to precisely locate the vortex core position relative to the aircraft's wing tip. The seeding of the vortex allowed the LIDAR to detect the WV with a high probability.

2. LIDAR design

The main LIDAR requirement is its ability to measure air flow at a distance down to 25 m from the sensor (that correspond to length of the wing) to be able to locate the wake vortex center relative to the wing tip. The technical solutions have been driven by the system performance objective and by technical requirements for airborne operations. Most coherent Doppler LIDAR for wind measurements use long laser pulses of 400 ns to 800 ns FWHM resulting in high spectral accuracy allowing the wind velocity to be precisely measured . However, for these pulse lengths, the optical reflections of the laser pulse on the output optics of the LIDAR generate a strong signal on the detector that makes a blind zone superior to 150 m. When using LIDAR with fiber laser, shorter pulses can be designed reducing the blind zone length. In addition, reducing the pulse length increases the accuracy in the measurement of the wake core position by reducing the spatial integration of the recorded signal [3]. To maintain a spectral resolution compatible with accurate wake vortex characterization, 250 ns FWHM laser pulses are typically used leading to a blind zone of 40 to 50 m. In our case, the blind zone must be shorter than 25 m, so that pulse length shorter than 160 ns with steep edge had to be used (Gaussian pulse shape could not be used as the foot could compromise the first range gates).

In Fibered Coherent Doppler LIDARs developed at ONERA, laser sources are based on "Master Oscillator Power Fiber Amplifier" (MOPFA). The master oscillator is a 1545-nm low–phase-noise narrow-linewidth Laser Module and the pulses are generated by a Fiber Coupled Acousto-Optique Modulator (AOM). The pulse length is usually limited by the rising/falling time of the AOM. In Wind LIDARs, AOM with 100-ns rising time are typically used to generate 200 ns to 800 ns pulses. In this study, we used a 10-ns rising/falling time AOM to generate the short pulses. The fiber amplifier is a custom made amplifier adapted to amplify such short pulses.

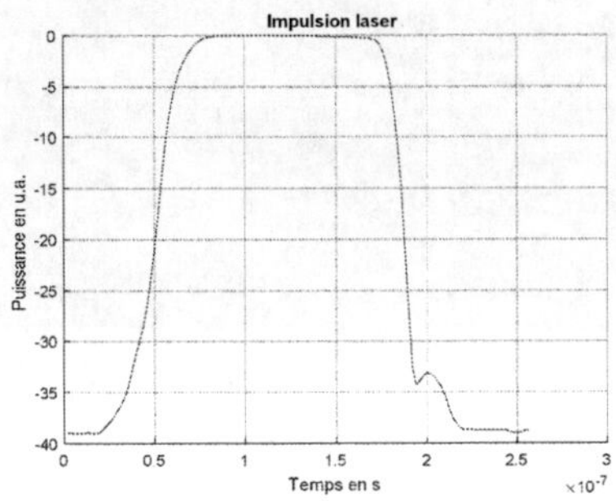

Figure 1 : Temporal profile of the laser pulse (ordinate is in dB)

The generated pulse is 150 ns long at -30 dB after amplification (Figure 1).The LIDAR blind zone is shorter than 23 m.

In order to optimize measurements from 25 m to 250 m, a LIDAR aperture diameter of 40 mm has been chosen combined with a focalization distance of about 100 m.

Another critical point of the LIDAR design is the difference in speed between the aircraft and the air that can reduce the measurement dynamics. The LIDAR has been designed to measure the wake vortex position in flight with a side slip angle up to 5 degrees (in absolute value), to determine the projection of the aircraft velocity on the LIDAR axis between about ∓20 m/s. During flying tests, the velocity range chosen for real time display was [-30 m/s +30 m/s] to account for the aircraft velocity projection.(Figure 3)

Because of the short pulse, the accuracy in velocity measurement was low (+/-1.5 m/s at CNR(Carrier to Noise Ratio) = -20 dB to +/- 0.5 m/s at CNR= -14 dB). However, this accuracy was good enough to detect and locate the wake vortex using an appropriate signal processing program to determine its core from the WV velocity signatures.

3. Fligth tests

The optical head, composed of the laser and all the optical components, were installed in a 73 cm x 44 cm x18 cm box suitable for flight certification (Figure 2). The laser electrical power and control, the scanner power and control, a rackable computer dedicated to acquire the signal, the real time signal processing and display were installed in the flight test bay.

Five flights were set with two aircrafts in formation flight configuration (left picture in figure 2) the lead aircraft seeding the vortex. Measurements were also possible without seeding the vortex when contrails where presents. Different parameters were varied: the distance between the two aircrafts varied from 0.5

NM to 2 NM, the spacing between the closest vortex and the LIDAR (horizontal axis) varied from 30 m to 200 m and the estimated vortex core height with respect to the aircraft wing (vertical axis) varied from +20 to -20 m. Real time display of the CNR and of the velocity maps were available for the flight test engineer. This real time display enabled to give directions to the pilot to precisely position the aircraft wing tip relatively to the vortex core. The LIDAR was operated during 12 tests flight hours. During the flight tests, approximately 11000 scans were recorded. More than 5000 scans contained detectable wake vortices.

Figure 2: left: Formation flight test configuration; right: LIDAR for formation flight installed in the test aircraft.

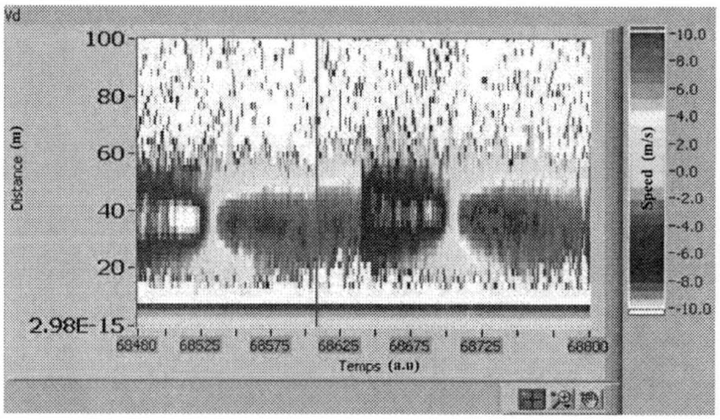

Figure 3: An example of an onboard real-time display (which corresponds to two successive scans in each window created by Labview®) is presented. The horizontal and vertical axis corresponds to the distance along the laser axis and the time respectively. The color scale represents the airspeed in m/s.

4. Vortex core tracking

A post processing algorithm has been developed to detect and locate the wake vortex center using averaged air motion velocity maps (an example is presented in figure 3). It has been optimized for artificially seeded wake vortices. This algorithm work in three steps: (i) the wake vortex detection, (ii) the core angle position and calculation, and (iii) the vortex core distance on the laser axis calculation. Figure 4 shows two examples of post processing output, including the location of the vortex core (68 m and 31 m from the LIDAR respectively).

To estimate the bias and error associated with this signal processing, 600 simulations of the lidar output signals were conducted for various vortex positions and using random LIDAR noise. Figure 5 shows the histogram of the wake vortex distance error for the closest vortex (blue) and the furthest vortex (red).

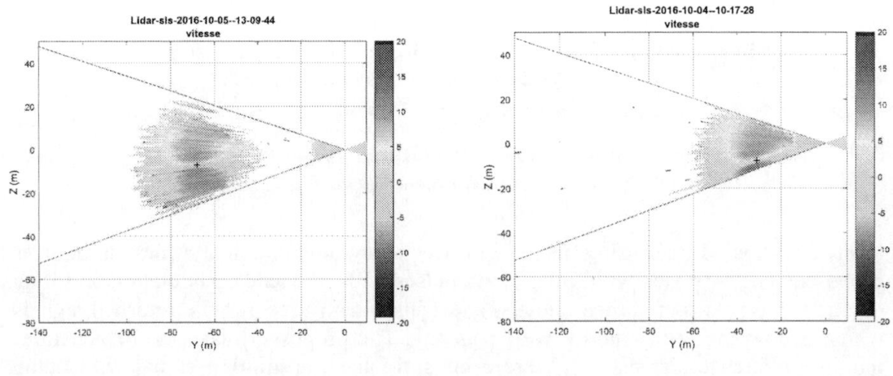

Figure 4 : Two examples of post processing outputs. The determined vortex core positions are shown (black crosses).

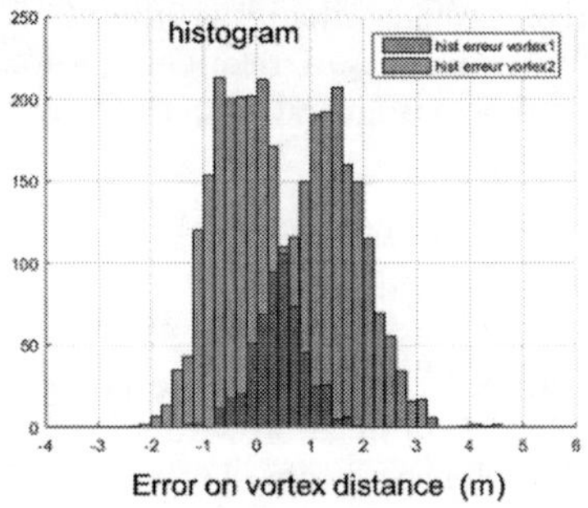

Figure 5 : Histogram of wake vortex distance error for the closest vortex (blue) and the furthest vortex (red).

The averaged bias is equal to 1.3 m for vortices closer than 100 m and equal to -0.3 m for vortices further than 100 m. The associated standard deviations are approximately 0.7 m and 0.6 m respectively. These results fulfill the requirements.

5. Conclusions

An on-board LIDAR sensor for wake vortex tracking for aircrafts flying in formation flight is presented. The sensor, based on pulsed 1.5-µm fibered laser, has a very short range detection capability (<30 m). A real time display of the velocity maps enables the flight test engineer to give directions to the pilot to precisely position the vortex core relatively to the aircraft wing tip. The LIDAR has been operated during 12 tests flight hours and has provided a huge amount of data that are still being processed.

6. References

[1]. Lissaman, P. B. S., & Shollenberger, C. A. (1970). Formation flight of birds. Science, 168(3934), 1003-1005).

[2]. Ning, S. A., Flanzer, T. C., & Kroo, I. M. (2011). Aerodynamic performance of extended formation flight. Journal of aircraft, 48(3), 855-865.

[3]. Dolfi-Bouteyre, A., Augere, B., Valla, M., Goular, D., Fleury, D., Canat, G., ... & Cariou, J. P. (2009). Aircraft wake vortex study and characterization with 1.5 µm fiber Doppler lidar,. AerospaceLab Journal, AL01-07.

Analytical Feasibility Study of Wind Lidar with Long-Duration Frequency-Modulated Pulse

Eiichi Yoshikawa(a), Tomoo Ushio(b), Hiroshi Yamasuge(b)
(a) Japan Aerospace Exploration Agency
6-13-1 Osawa, Mitaka, Tokyo, 181-0015, Japan
(b) Tokyo Metropolitan University
6-6 Asahigaoka, Hino, Tokyo, 191-0065, Japan
yoshikawa.eiichi@jaxa.jp

Abstract: A wind lidar using pulses with frequency modulation and long duration is proposed. The proposed wind lidar emits pulses with frequency modulation and long duration, receives its returned signals, and consequently records the returned signals which is converted to a microwave frequency. The proposed lidar a-priori generates multiple reference signals which are replicas of the transmitting signal with being shifted by presupposed Doppler frequencies, and parallelly implements convolutions between a returned signal and each of the reference signals. The convolution results correspond to range profiles at each of the presupposed Doppler shifts. That is, the multiple convolutions derive a range-velocity profile of received signal. While range and velocity resolution are in a trade-off relation in a traditional wind lidar, in the proposed wind lidar, they can be designated independently on each other.

Keywords: Coherent Laser Radar, Doppler, Frequency Modulation.

1. Introduction

Wind lidars have recently become utilized in various practical purposes since they have been reasonable to be purchased. The author has worked for aviation weather safety using a wind lidar, such as wake vortex detection [1][2], low-level turbulence advisory [3], and development of airborne wind lidar [4]. Through these researches, the author has heard many users making complaints against wind lidars about their slow scanning speed, short observation distance, poor accuracy of measurements, and coarse range resolution. Although traditional wind lidar needs to shorten its transmitting pulse in order to improve range resolution, shortening pulse duration enlarges Doppler velocity resolution. Since range resolution and velocity resolution are thus in a trade-off relation, lidar system design is non-flexible and it is only way for improving signal-to-noise ratio (SNR) to enhance peak power of laser pulses.

The authors proposed a wind lidar using pulses with frequency modulation and long duration, that can designate range resolution and velocity resolution independently [5]. Simply speaking, the pulse compression technique, that is frequently utilized in the radar engineering field for both point-targets' and distributed-targets' detections [6], [7], is applied with a modification as below. In principle, the radar's pulse compression technique does not work in a case with a large Doppler frequency shift, and does not work on lidar. A returned signal with a large Doppler frequency shift does not have a correlation with a reference signal that is normally a replica of a transmitting signal. The proposed lidar generates multiple reference signals which are with presupposed Doppler frequency shifts, and convolutes a returned signal with each of the reference signals. The convolution results correspond to range profiles at each of the presupposed Doppler shifts. That is, the multiple convolutions derive a range-velocity profile of received signal. This paper describes an analytical feasibility study of the proposed wind lidar.

2. Hardware

Figure 1 shows an example of the proposed wind lidar. A laser light of continuous wave is mixed with microwave pulses with frequency modulation and long duration by an acousto-optic modulator, then modulated laser pulses are emitted to the air. After a returned light is down-converted to microwave by a heterodyne detector, the microwave signal is recorded by a analog-to-digital converter.

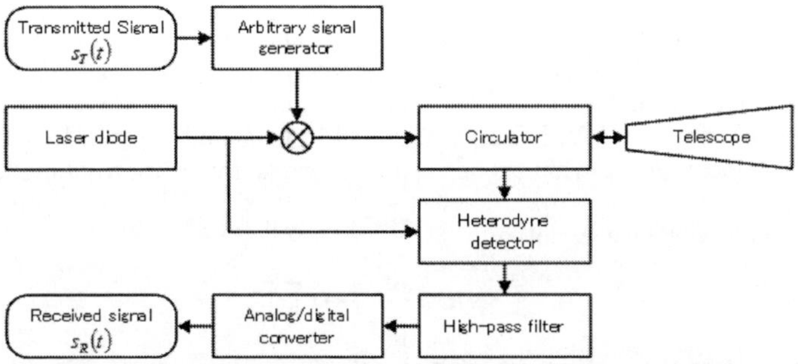

Figure 1: Hardware example of the proposed wind lidar [5]

3. Software

Figure 2 shows a calculation flow of the proposed wind lidar. Multiple reference signals are generated in advance of observation by applying presupposed Doppler frequency shifts to a digital waveform of transmitting signal. Multiple convolutions of a received signal with each of the multiple reference signals are parallelly performed, and calculation results are rearranged to form a range-velocity profile.

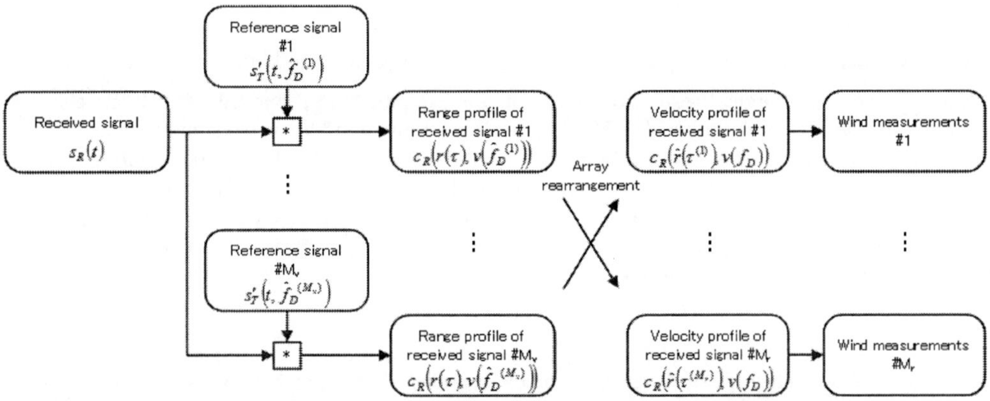

Figure 2: Software example of the proposed wind lidar [5]

4. Analytical Performance

See [5] about detailed derivation for the analytical performance. Because of the software of the proposed lidar, ambiguity function for ranging is equivalent to an autocorrelation function of the transmitting signal. For example, when the transmitting signal is with a liner frequency chirp, its autocorrelation can be approximated to a sinc function. Range resolution accomplished by a modulated signal is expressed as

$$\Delta_r \approx \frac{c}{2B}. \tag{1}$$

where Δ_r, c, and B are range resolution, speed of light, and bandwidth of modulation, respectively. Meanwhile, ambiguity function for velocimetry is represented by cross-correlations of a single frequency wave (at a Doppler frequency shift due to target) to ones at presupposed frequency shifts. As a result, it is also a sinc function whose velocity resolution is as

$$\Delta_v \approx \frac{c}{2Tf}. \quad (2)$$

where Δ_v, T, and f are velocity resolution, pulse duration, and frequency of a carrier light, respectively. As obviously shown in Eqs. (1) and (2), range resolution is independent on velocity resolution in the proposed wind lidar. Since a wind lidar receives returned signals from aerosol particles, a signal intensity at a desired range-velocity is resulted from a summation of signals returned from aerosol particles existing in a range-velocity lattice which is defined by range and velocity resolution. A signal-to-noise ratio can be estimated as a ratio of the received signal intensity and noise (a convolution result of background noise and a reference signal),

$$SNR \approx \frac{c^2}{8} \frac{a_t^2}{a_n^2 Bf} n\overline{|x|^2}. \quad (2)$$

where a_t and a_n are amplitude of transmitting signal and receiving noise, respectively. $\overline{|x|^2}$ is mean of absolute square of aerosol particles' scattering amplitude. Note that $\overline{|x|^2}$ additionally contains a gain of telescope, attenuations due to roundtrip propagations, and so on. n is the number of aerosol particles in the range-velocity lattice defined by a range-velocity resolution.

5. Conclusion

This paper proposes a new wind lidar which emits a laser pulse with frequency modulation and long duration. In the proposed wind lidar, multiple reference signals supposing possible Doppler frequency shifts is generated a-priori, and multiple convolutions between a returned signal and each of the reference signals are simultaneously implemented. Since a convolution result corresponds to a range profile at a Doppler velocity, the set of convolution results is equivalent to a range-velocity profile of aerosol particles. Two-dimensional range-velocity ambiguity function of the proposed wind lidar is approximately represented by a (two-dimensional) sinc function whose range and velocity resolution are defined independently, while they are dependent in a traditional wind lidar.

In order to accomplish range resolution of 30 m, for example, the proposed wind lidar uses a long duration pulse with a bandwidth of 5 MHz. While a traditional wind lidar needs to equip a laser source with a narrow linewidth of a few hundreds of kHz and with pulse duration of 0.2 micro seconds, the proposed wind lidar alleviate those constraints to a linewidth of several MHz and pulse duration of several tens of micro seconds. That is, the proposed wind lidar can be designed simply with a semiconductor laser, and, if a high-power semiconductor laser is developed, it will solve the traditional wind lidars' problem due to low power.

[1] E. Yoshikawa and N. Matayoshi, "Wake vortex observation campaign by ultra fast-scanning lidar in Narita airport, Japan," Proceedings of 29th Congress of the International Council of the Aeronautical Sciences (ICAS 2014), St. Petersburg, Sept., 2018.

[2] E. Yoshikawa and N. Matayoshi. "Aircraft Wake Vortex Retrieval Method on Lidar Lateral Range–Height Indicator Observation", AIAA Journal, Vol. 55, No. 7 (2017), pp. 2269-2278. https://doi.org/10.2514/1.J055224

[3] T. Iijima, N. Matayoshi, E. Yoshikawa, "Development and evaluation of low-level turbulence advisory display for aircraft operation," Proceedings of 29th Congress of the International Council of the Aeronautical Sciences (ICAS 2014), St. Petersburg, Sept., 2018.

[4] H. Inokuchi, H. Tanaka, and T. Ando, "Development of an Onboard Doppler Lidar for Flight Safety," J. Aircr. 46(4), 1411–1415 (2009).

[5] E. Yoshikawa and T. Ushio, "Wind ranging and velocimetry with low peak power and long-duration modulated laser," Optics Express, OSA, Vol. 25, Issue 8, pp. 8845-8859 (2017) https://doi.org/10.1364/OE.25.008845

[6] P. Z. Peebles, Jr., Radar Principles, (Wiley Interscience Publications, 1998).

[7] E. Yoshikawa, T. Ushio, Z. Kawasaki, T. Mega, S. Yoshida, T. Morimoto, K. Imai, S. Nagayama, "Development and initial observation of high-resolution volume-scanning radar for meteorological application," IEEE Transactions on Geoscience and Remote Sensing, Vol. 48, Issue: 8, Aug. 2010.

The Scheimpflug Lidar Technique and Its Recent Progress in Atmospheric Remote Sensing

Liang Mei(a), Zheng Kong(a) and Peng Guan(a)
(a)School of Optoelectronic Engineering and Instrumentation Science, Dalian University of Technology, Dalian 116024, China

Abstract: The Scheimpflug lidar technique has been recently developed for atmospheric remote sensing. The entire laser beam that is transmitted into atmosphere can be clearly focused over km-range on a tilted image sensor, if the optical configuration satisfies the Scheimpflug principle. The range-resolved lidar signal is in this case angular-resolved and obtained from pixel intensities of the image sensor. The special imaging principle enables the usage of low-cost, compact high-power CW laser diodes and highly integrated CMOS/CCD image sensors. This paper reports our recent progress of the Scheimpflug lidar technique in the applications of atmospheric aerosol and NO_2 monitoring.

Keywords: Scheimpflug lidar, aerosol, NO_2, extinction coefficient, depolarization.

1. Introduction

Lidar techniques have been of great interest for atmospheric aerosol [1], trace gases [2], temperature [3] and wind speed [4] monitoring for decades. Recently, the Scheimpflug lidar technique has been proposed for atmospheric remote sensing [5-7]. The laser beam that is transmitted into atmosphere can be clearly focused on a tilted image sensor, if the optical layout of the lidar system satisfies the Scheimpflug principle, as shown in Figure 1. Infinite depth-of-focus is achieved while employing large aperture optics. Thus, the range-resolved atmospheric backscattering signal can be retrieved from recorded pixel intensities. The pixel information can be transformed to measurement distance according to geometrical optics. The Scheimpflug lidar technique significantly reduces the complexity of laser sources and detection units, the requirement on system maintenance, and the overall cost of an atmospheric lidar system by utilizing high-power continuous-wave (CW) laser diodes as light sources and highly integrated CMOS/CCD sensors as detectors [8]. Moreover, the wide availability of high power laser diodes from 375 to 1550 nm provides a great possibility for developing multi-wavelength Scheimpflug lidar system, which is of great interest for precise particle sizing. This paper reports recent progress of the Scheimpflug lidar technique in the applications of atmospheric aerosol and trace gas monitoring in Dalian University of Technology (DLUT).

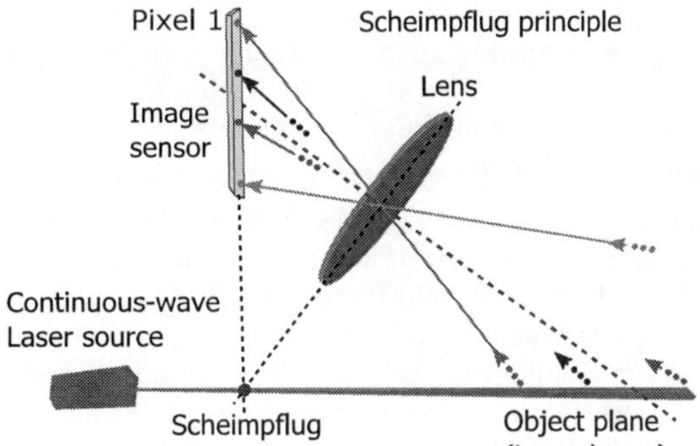

Figure 1. Principles of the Scheimpflug lidar technique. Infinite depth of focus can be achieved when the image plane, lens plane and the object plane intersect into a single line according to the Scheimpflug principle.

2. Atmospheric aerosol sensing

The schematic of the typical Scheimpflug lidar system as well as the photograph are shown in Figure **2**. High power CW laser diodes are often employed as the light sources, which are housed by homemade aluminum mounts with large-capacity thermoelectric cooler to stabilize the case temperature and thus the center emission wavelength. The laser beam is collimated by a refractor telescope (F6). Additional cylindrical lens pairs may be used to reshape the laser beam for the Scheimpflug lidar system operating at e.g., 405 nm and 520 nm [9]. The sunlight background is suppressed by interference filters with 1.7-10 nm full width at half maximum. The backscattering light of the laser beam is collected by a Newtonian telescope with approximately 806 mm separation to the refractor telescope. Both the refractor telescope and the Newtonian telescope are mounted on an aluminum bar, carried by an equatorial mount. A 2D CMOS camera is mounted with 45° titled to the optical axis of the receiving telescope. A trigger signal generated by the camera is fed to a Johnson ring counter, and the output is utilized to switch the laser diode on and off through the driving current. The CMOS camera records the "laser on" and "laser off" images alternatively. The background signal can then be subtracted from the recorded raw images. Each image pairs are vertically binned to extract the "on" and "off" curves. After background subtraction, signal averaging, and digital filtering, etc, a single lidar curve can be obtained in 45 s (1000 times signal averaging). The pixel-distance relationship is calibrated by measuring backscattering echo from a hard target with known distance.

Recently, single-band Mie scattering Scheimpflug lidar systems operating at various wavelengths, e.g., 405 nm, 450 nm, 520 nm and 808 nm, have been implemented and employed for real-time, large area atmospheric aerosol remote sensing [9,8]. Atmospheric measurements were performed in urban area to investigate the performance of the lidar systems. All-time operation is feasible for the 405-nm and 808-nm Scheimpflug lidar system due to relatively lower sunlight background and the narrow bandwidth of interference filters (less than 3 nm). However, the 450-nm and 520-nm Scheimpflug lidar systems, where 10-nm FWHM interference filters are often used, could be saturated when the lidar system is facing to the direction of the sun. The maximum measurement distance of the Scheimpflug lidar system can reach up to 7-10 km during a sunny and clean weather conditions, while the measurement distance is limited during severe haze particularly for the 405 nm lidar system due to severe atmospheric attenuation. The noise level of the raw lidar signal is dominated by the sunlight background noise and the photon-response non-uniformity (PRNU) noise of the image sensor [10]. During nighttime measurements when the sunlight background noise is negligible, the signal-to-noise ratio (SNR) of the raw lidar signal is inverse proportional to the PRNU noise ratio of the image sensor, which is about 0.5% for the CMOS camera. The maximum SNR is thus limited to about 200:1. Thus, the Savitzky–Golay filter is utilized for signal de-noising, and the maximum SNR can then reach up to 1000:1 during nighttime. The Klett-Fernald inversion algorithm was also employed for atmospheric extinction coefficient retrieval. Figure 3 shows the distribution of aerosol extinction measured by the 405 nm Scheimpflug lidar system. In several case studies, the trends of the extinction coefficients retrieved from the Scheimpflug lidar system were in good agreement with the PM10/PM2.5 concentrations measured by a conventional air pollution monitoring station, which successfully demonstrates the feasibility of the Scheimpflug lidar technique for atmospheric remote sensing. Practical applications of the Scheimpflug lidar technique in atmospheric pollution monitoring or tracking can be expected in the near future.

A polarization Scheimpflug lidar system based on the Scheimpflug principle has also been developed by employing two linearly polarized 808-nm laser diodes and a CMOS image sensor [11]. The polarization of one laser diode is rotated 90° by a half-wave plate. The two laser beams with orthogonal polarizations are combined by a polarization beam splitter, and then transmitted into atmosphere. The corresponding parallel and perpendicular polarized backscattering echoes are detected by the 45° tilted CMOS sensor using a time-division multiplexing scheme. 24-hour continuous atmospheric vertical profiling of depolarization ratio successfully demonstrated that the polarization Scheimpflug lidar system has a potential for the depolarization studies of atmospheric aerosols. However, the SNR of the perpendicular polarized backscattering signal should be improved for daytime measurements.

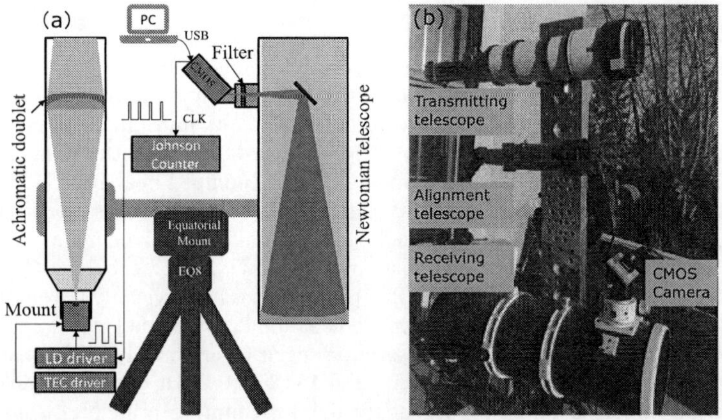

Figure 2. (a) Principle and (b) photograph of the Scheimpflug lidar system developed in DLUT [8].

Table 1. Primary specifications of the Scheimpflug lidar techniques.

	Model	**Specifications**
Laser source	Continuous-wave high power multi-mode laser diodes	Wavelength: 405 nm, 450 nm, 520 nm, 638 nm, 808 nm, 850 nm, 1064 nm; Output power: 1-5 W
Collimator	F6 refractor	Focal length: 600 mm, Diameter: 100 mm
Receiver	Newtonian telescope	Focal length: 800 mm; Diameter: 200 mm
Detector	2D CMOS Sensor, e.g., CMV2000	Tilt angle: 45°; Area: 2048×1024 Pixels Frame rate:170 fps; Bit depth: 12/8 bit Exposure time: 20-500 ms
Filters	Interference filters + color filters	1.7-10 nm FWHM

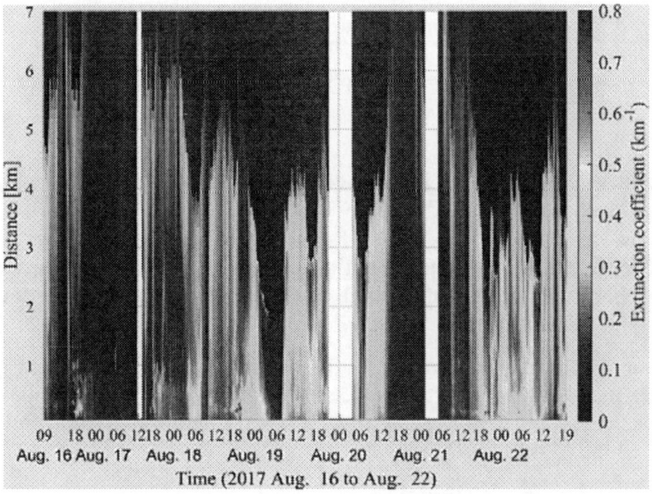

Figure 3. Distribution of extinction coefficient retrieved from all-time atmospheric measurements employing the 405-nm Scheimpflug lidar system. The PM2.5/PM10 concentrations were about 60 μg/m^3, and the relative humidity is 80% during the severe haze on August 19th, 2017.

3. Atmospheric NO_2 monitoring

Differential absorption lidar (DIAL) technique provides unique capability of remotely monitoring urban/rural area localized NO_2 emissions and profiling tropospheric vertical NO_2 distribution. By measuring and evaluating atmospheric backscattering echoes of transmitted laser pulses at on-line wavelength and off-line wavelength of NO_2, range-resolved NO_2 concentration can be retrieved. Conventional pulsed lidar techniques often employ Nd:YAG pumped dye lasers [12] or Raman lasers [13], sum frequency generation based on solid state lasers [14], etc. Nevertheless, the laser sources mentioned above are extremely complicated and require intensive maintenance during the measurements, which prevents practical applications of the NO_2 DIAL technique. A continuous-wave (CW) NO_2 DIAL system based on the Scheimpflug principle has been recently developed by employing a compact high-power multimode 450-nm laser diode as the laser source [15]. Laser emissions at the on-line and off-line wavelengths of the NO_2 absorption spectrum are readily implemented by tuning the injection current of the laser diode, i.e., 448.6 nm and 452.1 nm. Lidar signals are detected by an area CCD image sensor, which is tilted 45° to satisfy the Scheimpflug principle. Range-resolved NO_2 concentrations on a near-horizontal path are obtained by the NO_2 DIAL system in the range of 0.3-3 km during nighttime, and show good agreement with those measured by a national air pollution monitoring station. A detection sensitivity of ±0.9 ppbv in the region of 0.3-1 km is achieved with 15-minute averaging and 700 m range resolution, which allows accurate concentration measurement of ambient NO_2. The low-cost and robust DIAL system opens up many possibilities for field NO_2 remote sensing applications in the near future.

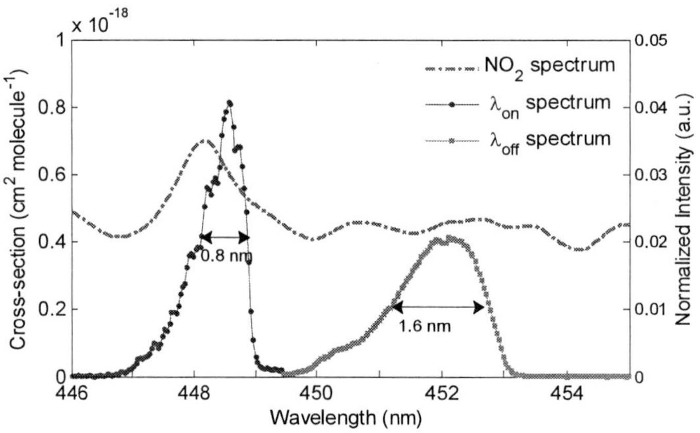

Figure 4. NO_2 absorption spectrum and the on- and off-wavelengths of the NO_2 DIAL technique.

4. Conclusions

The Scheimpflug lidar technique has been successfully demonstrated for atmospheric extinction coefficient retrieval, depolarization studies and NO_2 monitoring. It has shown a great potential for practical atmospheric applications with the advantages of being low cost, less complexity, and low maintenance. Scheimpflug lidar systems operating at various wavelengths are also developed and validated for atmospheric remote sensing, which are of great interest for the development of multi-wavelength Scheimpflug lidar system in the near future.

This work was supported by the National key research and development program of China (2016YFC0200600), the National Natural Science Foundation of China (61705030), the Fundamental Research Funds for the Central Universities (DUT18JC22), and the Natural Science Foundation of Liaoning Province, China (201602163).

5. References

[1] H. Baars, P. Seifert, R. Engelmann, and U. Wandinger, "Target categorization of aerosol and clouds by continuous multiwavelength-polarization lidar measurements," Atmos. Meas. Tech. **10**, 3175-3201 (2017).

[2] C. Weitkamp, R. Baumgart, A. Behrendt, P. Bisling, J. Glauer, U. B. Goers, S. Kohler, W. Lahmann, and J. Reichardt, "Lidar ozone measurements in the marine and terrestrial atmosphere from the ground to the tropopause," Igarss 2001: Scanning the Present and Resolving the Future, Vols 1-7, Proceedings, 1023-1025 (2001).

[3] J. Hildebrand, G. Baumgarten, J. Fiedler, and F. J. Lubken, "Winds and temperatures of the Arctic middle atmosphere during January measured by Doppler lidar," Atmos. Chem. Phys. **17**, 13345-13359 (2017).

[4] S. H. Wu, J. P. Yin, B. Y. Liu, J. T. Liu, H. W. Zhang, X. Q. Song, and K. L. Zhang, "Observations of Wind Profile of Marine Atmosphere Boundary Layer by Shipborne Coherent Doppler Lidar," Epj Web Conf **119**, (2016).

[5] L. Mei and M. Brydegaard, "Atmospheric aerosol monitoring by an elastic Scheimpflug lidar system," Opt. Exp. **23**, 247841 (2015).

[6] L. Mei and M. Brydegaard, "Continuous-wave differential absorption lidar," Laser & Photon. Rev. **9**, 629-636 (2015).

[7] M. Brydegaard, A. Gebru, and S. Svanberg, "Super resolution laser radar with blinking atmospheric particles - Application to interacting flying insects," PIER **147**, 141-151 (2014).

[8] L. Mei, P. Guan, Y. Yang, and Z. Kong, "Atmospheric extinction coefficient retrieval and validation for the single-band Mie-scattering Scheimpflug lidar technique," Opt. Express **25**, A628-A638 (2017).

[9] L. Mei, Z. Kong, and P. Guan, "Implementation of a violet Scheimpflug lidar system for atmospheric aerosol studies," Opt. Express **26**, A620-A274 (2018).

[10] L. Mei, L. Zhang, Z. Kong, and H. Li, "Noise modeling, evaluation and reduction for the atmospheric lidar technique employing an image sensor," Optics Communications (to be published), (2018).

[11] L. Mei and P. Guan, "Development of an atmospheric polarization Scheimpflug lidar system based on a time-division multiplexing scheme," Opt. Lett. **42**, 3562-3565 (2017).

[12] T. Nayuki, T. Fukuchi, N. Cao, H. Mori, T. Fujii, K. Nemoto, and N. Takeuchi, "Sum-frequency-generation system for differential absorption lidar measurement of atmospheric nitrogen dioxide," Appl. Opt. **41**, 3659-3664 (2002).

[13] S. Hu, H. Hu, Y. Zhang, J. Zhou, G. Yue, K. Tan, Y. Ji, and B. Xu, "A new differential absorption lidar for NO_2 measurements using Raman-shifted technique," Chinese Optics Letters **1**, 435 (2003).

[14] R. Toriumi, H. Tai, and N. Takeuchi, "Tunable solid-state blue laser differential absorption lidar system for NO_2 monitoring," Opt. Eng. **35**, 2371-2375 (1996).

[15] L. Mei, P. Guan, and Z. Kong, "Remote sensing of atmospheric NO_2 by employing the continuous-wave differential absorption lidar technique," Opt. Express **25**, A953-A962 (2017).

Ultrahigh accuracy distance measurements with self-correction of the air-refractive index using optical frequency combs

Kaoru Minoshima[(a,b)]
(a) The University of Electro-Communications
(b) JST, ERATO MINOSHIMA Intelligent Optical Synthesizer
k.minoshima@uec.ac.jp

Abstract: High-accuracy long-distance measurements are essential for many applications. For practical distance measurements in atmosphere, the air refractive index is one of the most important sources of uncertainties. However, correcting the air refractive index along the exact optical path under measurement is difficult. By length measurements using lasers of two different colors, dispersion relation for air refractive index can be used for the self-correction of the air refractive index, i.e., two-color method. In this talk, ultrahigh-accuracy two-color method using optical frequency comb is presented. Using the fundamental and second-harmonic wavelengths generated from a Er:fiber comb, two-color self-correction was performed with 10^{-9} uncertainty, exceeding the accuracy provided by the empirical equations.

Keywords: optical frequency comb, air refractive index, distance measurement, environmental fluctuation correction

1. Introduction

The correction of the refractive index of air is one of the most important sources of uncertainty in optical distance measurements. Conventionally, correction of air refractive index is performed by calculations based on environmental parameters using an empirical equation such as Ciddor's equation [1]. However, the correction accuracy is limited due to spatial and/or temporal refractive index variations and ultimately the accuracy of the equation itself (on the order of 10^{-8}). Therefore, a self-correction method i.e., based on direct optical distance measurements, is desired to perform highly accurate optical distance measurements. The two-color method is a self-correction technique, in which two light beams with different wavelengths measure the same target distance simultaneously to determine the geometrical distance D, which can be given by $D = D_1 - A(D_2 - D_1)$, where D_1 and D_2 are the optical distances corresponding to the two wavelengths, and A represents the dispersion relation for the refractive index of air. Since A is quite insensitive to environmental changes under normal environmental conditions, D is determined only by the measured optical distances at the two wavelengths. This is very powerful tool for practical measurements since no precise environmental parameter measurements are needed. However, the measurement uncertainty is enhanced by A, which is a large number, for example, $A = 140$ in our experimental conditions; therefore, an ultrahigh-accuracy optical measurement method is required to achieve high-accuracy two-color correction. In our previous study, by using an optical frequency comb, we demonstrated correction with an accuracy of 10^{-8} order [2,3]. However, the accuracy of the evaluation of the developed method itself was limited by the variation of D during the measurements due to its thermal expansion though the applicability of the developed method was not limited by the thermal expansion, rather it can be precisely measured. In this study, we constructed a compact and thermally stable long-path delay line for the target and demonstrated 10^{-9}-order self-correction of air refractive index, which exceeds the uncertainty of the empirical equation. With the developed method, displacement and shape could be measured with nanometer-level uncertainty over several tens of meter distance without correction of air refractive index.

2. Experimental results and discussion

Fig.1 shows the experimental setup. The light source is an Er:fiber-based, fully phase-stabilized frequency comb with repetition frequency, f_{rep} = 54 MHz. The fundamental outputs from EDFA and SHG have wavelengths of 1560 nm and 780 nm, respectively, and were used as the two colors. Each beam pass through a reference arm with acousto-optic modulators (AOMs) for heterodyne detection. In the measurement arm, a fixed delay line forms a pulse-to-pulse interferometer with an unbalanced path length of 61 m, which corresponds to the interference between two pulses with a separation equal to 11 times the pulse interval. To eliminate the thermal instability caused by the AOMs, a monitor arm with a length equal to that of the reference arm is introduced. The heterodyne interference signal at 4 MHz of each color between the reference and monitor arms was detected by PD and used as the reference signal, while those between the reference and measurement arms was used as the measurement signal of the lock-in-amplifier. Consequently, the optical path difference between the monitor and measurement arms of each color was obtained.

In this study, in order to demonstrate an ultrahigh accuracy better than that of the empirical equation, we designed a compact and stable delay line with a multiple-pass-cavity (MPC), placed it on a breadboard with a low nominal thermal expansion coefficient of α = 0.05 × 10^{-6}/K, and kept the geometrical distance constant. In this way, we could directly evaluate the uncertainty of the developed two-color method by evaluating the variation of the corrected geometrical distance D. The distance variation was obtained by using only values measured by the two-color interferometer as $\Delta D/D = \Delta D_1/D - A\Delta(D_2 - D_1)/D$, where other terms with ΔA were estimated to be more than two orders of magnitude smaller in our measurement conditions with a standard deviation of ΔA = 0.0007 for A = 141.2132.

As mentioned in the introduction, the measurement accuracy of $D_2 - D_1$ is required to be extremely high. In the developed system, the fundamental interference fringe is stabilized by changing f_{rep} of the frequency comb; therefore, the dispersion effect, i.e., the optical path difference between the fundamental and SHG wavelengths, $D_2 - D_1$, can be obtained directly and precisely by measuring the SHG interference fringe signal. Simultaneously, the variation of the fundamental optical distance D_1 can be obtained precisely by the frequency measurement of f_{rep} change as the error signal for fringe stabilization. Fig.2(a) shows the short-term stabilities of the measured interference signals during 500 s. In such a short time period, the environmental refractive index variation is negligible, so the system resolution could be evaluated by using 0.03 and 0.27 nm standard deviations for the fundamental and SHG fringe signals, respectively. Although the SHG fringe signal stability already shows that of the two-color difference, the difference between the two fringe signals was further calculated to eliminate the residual common noise. As a result, the measurement resolution of the two-color difference, i.e., $D_2 - D_1$, was found to be 0.27 nm, corresponding to 4.4 × 10^{-12} of the 61 m measured path length. This result demonstrates that our developed phase-stabilizing system using frequency comb yields extremely precise two-color difference measurements which is required for two-color method.

Next, we performed long-term measurements for 10 h. In this case, due to environmental fluctuation, air refractive index varied over time for 10^{-6} order, thus optical length of each color varied by the same amount. For the evaluation, we installed environmental sensors to measure the temperature, pressure, and humidity in the air around the MPC delay line and the air refractive index was calculated using Ciddor's equation simultaneously to the two-color interferometric measurement. As a result, the measured two-color difference $D_2 - D_1$ agreed very well with the calculated value with standard deviation of 3.8 × 10^{-11} for 10 h suggesting the long-term precision of the developed method even with large refractive index change. For the D_1 measurement, the standard deviation of the difference between the experimental and calculated values was 5.5 × 10^{-9}, suggesting ultrahigh precision. However, such high precision as 10^{-9} and 10^{-11} already exceeds uncertainty of the empirical equation, thus we cannot confirm the high accuracy measurement by comparing to the empirical equation. Therefore, finally, self-correction of the air refractive index was performed by applying the two-color method to the measured values of D_1 and $D_2 - D_1$, and the obtained geometrical distance was directly compared to the reference distance. In this experiment, since we could expect the constant distance, i.e., ΔD = 0, the obtained distance could be directly evaluated whether it is constant over long period of time. Fig.2(b) shows the obtained corrected geometrical distance D, which is constant with a standard deviation 4.1 × 10^{-9} for 10 h with changes in temperature, pressure, humidity, and air refractive index of ΔT = 0.06 K, ΔP = 2.8

hPa, $\Delta H = 0.3\%$, and $\Delta n_1 = 0.8 \times 10^{-6}$, respectively. This result demonstrates the ability of the proposed method to achieve extremely high-accuracy self-correction of the air refractive index exceeding that of the widely used empirical equation. Here we emphasize that the air refractive index correction was performed only by using optical distance measurements of the target. Moreover, the evaluation does not rely on the empirical equation, but rather on the "constant" geometrical distance D, which was made possible by introducing a stable and compact long-path MPC delay line on a breadboard made of a low-thermal-expansion material. Here the constant D is required only for the evaluation purpose but not for the real application of the developed distance measurement method with self-correction of air-refractive

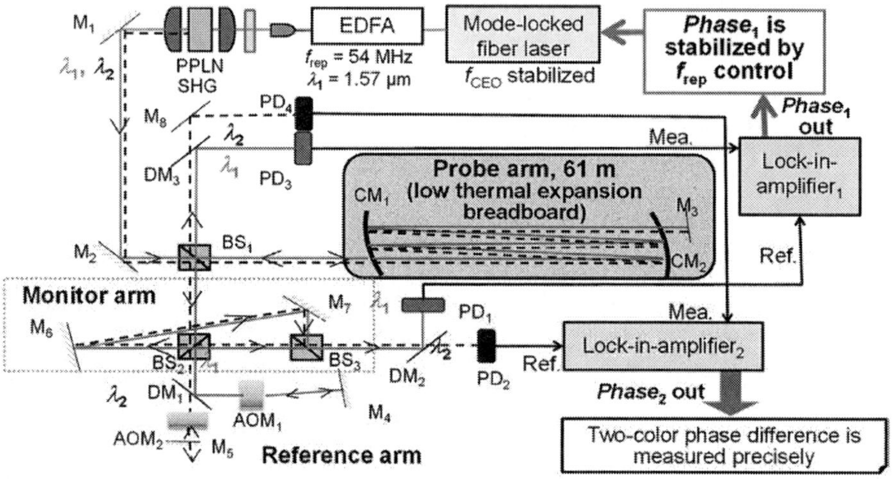

Fig. 1. Experimental set up for two-color self-correction of air refractive index with optical frequency comb. M: mirrors, BS: beam splitter, DM: dichroic mirror, CM: curve mirror, AOM: acousto-optic modulator, PD: photo detector, SHG: second harmonic generation.

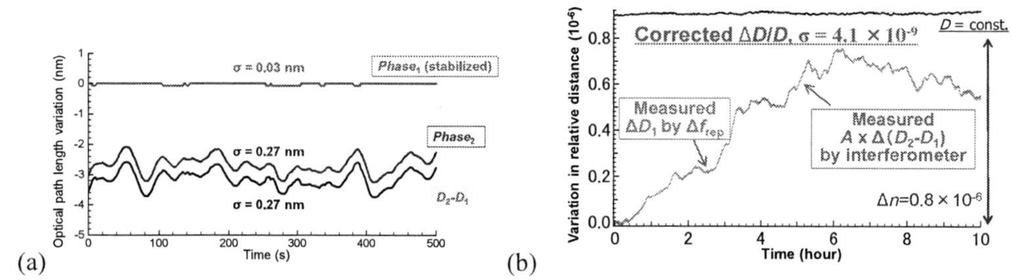

Fig. 2. (a) Phase stability of two-color interferometer. The fundamental interference signal is stabilized by feedback to the f_{rep}. (b) Long-term variation of the corrected geometrical distance based on two-color method, $\Delta D/D$, with experimentally obtained one-color distance $\Delta D_1/D$ and two-color difference $A\Delta(D_2 - D_1)/D$.

index. On the contrary, such thermal distance change can be precisely measured by the developed method since the two-color difference is much less sensitive to the real distance change due to the common-path nature.

In order to show applicability of the method in actual distance measurement, we applied small displacement to the end mirror of the MPC, M_3 in Fig. 2. PZT was attached to the mirror, and continuous sinusoidal displacement of amplitude of 230 μm at 0.1 mHz was applied for 30 hours. During the measurement, due to the environmental fluctuation, the optical distance measured with one color showed large variation of 5×10^{-6}. However, after applying the developed two-color method with optical frequency comb, air refractive index variation was automatically corrected, thus actual displacement was obtained. By comparing with displacement sensor attached to the PZT, 250 nm uncertainty (for 3 hours) was achieved in displacement measurement over 61 m optical path, which corresponds to 4.1×10^{-9}.

3. Conclusion

We demonstrated extremely high-accuracy self-correction of the refractive index of air by using a two-color interferometer of frequency comb. An ultrahigh accuracy of 10^{-11} in the two-color difference measurements, which is essential in the high-accuracy two-color method, was achieved by precisely controlling the frequency parameters of optical frequency comb. In a 10-h-long measurement, the corrected geometrical distance was constant with a standard deviation of 4.1×10^{-9}, demonstrating extremely high accuracy of the developed method which exceeds that of Ciddor's equation. Moreover, actual displacement measurement was performed and achieved nanometer-level uncertainty over 61 m optical path with self-correction of the environmental fluctuation. These results prove that the developed method will be a powerful tool in practical distance metrology and shows the potential to improve the accuracy of the air refractive index equation in the future.

This work was supported by the JST through the ERATO Intelligent Optical Synthesizer (IOS) Project (JPMJER1304) and partly JSPS and NSFC under the Japan-China Scientific Cooperation Program.

4. References

[1] P. E. Ciddor, Appl. Opt. 35, 1566 (1996).

[2] K. Minoshima, K. Arai, and H. Inaba, Opt. Express. 19, 26095–26105 (2011).

[3] G. Wu, M. Takahashi, K. Arai, H. Inaba, and K. Minoshima, Sci. Rep. 3, 1894 (2013).

Low Coherence Doppler Lidar with High-power DFB-LD

Tatsuo Shiina
Graduate School of Engineering, Chiba University
1-33 Yayoi-cho, Inage-ku, Chiba-shi,263-8522 Japan
shiina@faculty.chiba-u.jp

Abstract: A compact Doppler lidar was developed with a unique DFB-LD. This DFB-LD is a bulk device with the CW power of 2.7W[max]@3A by itself and the center wavelength of 974nm. Its spectrum width is about 10pm[max]. The coherence length and output power was optimized to 0.8m and 1.76W, respectively by adjusting the device temperature and the forward current. The low coherence Doppler lidar system was designed with this unique device. The concrete lidar setup was fabricated with an optical fiber interferometer and a differential amplifier. The Doppler echo of the wind at the measurement point, which is matched with the reference path distance, is detected and interfered with the reference light. In the experiment, a rotating reflector was utilized as wind. The rotating speed of 1 – 10m/s was measured with the accuracy of 0.03m/s. The mist flow was measured with this lidar system. the system sensitivity and the detectable maximum distance were estimated.

Keywords: DFB-LD, lidar, low coherence, Doppler, wind, sea wave

1. Introduction

The compact Doppler lidar is the successful innovation in the wind measurement market. Compared with the current wind measurement apparatuses such as anemometer, radiosonde, and Doppler soda, Doppler lidar is remote sensing, putting in a position and covering a certain direction up to 1km. The typical compact Doppler lidar has a laser light source which consists of Distributed Feedback Laser Diode (FDB-LD) and an optical fiber amplifier. It has the optical output power of a few Watts. The beam keeps its coherency within a few hundred – kilo meters.[1]-[3] There are plural commercialized products, while most of them have a blind area in near range and its range resolution is not so high (a few tens meters).

The low altitude wind has a quick activity especially just on the ground surface. Eddies of wind around structures and wind on the sea and river surface have such strong and rapid dynamics. Interaction between the wind and sea wave motion causes the complex behavior of boundary mixing. The sea wave observations were conducted by a lidar for trial.[4]-[6] It is not enough to observe only the sea wave motion; the atmospheric dynamics should be monitored. Wind measurement is adequate for that parameter.

In this study, firstly, we developed the LED mini-lidar to observe the sea wave motion with sallow angle.[7]-[9] The sea wave dynamics could be visualized with the atmospheric activity on the sea surface. The interaction between them is vivid and suggestive for the new acknowledge. The direct wind measurement with remote sensing is ideal following to the rapid wind motion with high accuracy. Here, we proposed the low coherence Doppler lidar. It has a unique DFB-LD as a beam transmitter. It has the optical output of 2.7W[max] by itself, while its spectrum width is <10pm. With this low coherence light source, we started to build up the Doppler lidar. We will report the concept of this low coherence lidar and concrete set-up.

2. Concept of low coherence Doppler lidar

Low coherence interference occurs when the measurement path matches with the reference path in an interferometer. The reference path length becomes an off-set range in this lidar set-up. At an arbitrary distance (off-set range), the wind measurement is possible by changing the reference path length. The interference width (distance) is equal to the coherence length and it depends on the spectrum width. Laser diode has the characteristics to broaden the spectrum width when the optical output increases. The bulk type DFB-LD we used has a peak in the change of the spectrum width. Its output power is 2.7W[max]. The spectrum width will be changed by the forward current and device temperature. To enlarge the coherence length on the condition of keeping a certain output power, the forward current and the temperature are adjusted.

The air cell size depends on the altitude, especially as a target of the low altitude atmosphere. Within the altitude of a few meters on the ground surface, this air cell size is of the order of 1 – a few meters. When the coherence length is of the order of 1 – a few meters, the air cell is regarded as the interference signal source. On contrary, the air cell flows with the wind, that is, its flow motion becomes rapid owing to the low altitude. Therefore, the low coherence Doppler lidar requires the necessary conditions that the coherence length of the lidar light source should match with the air cell size and that the lidar summation time to get the data should follow to the quick motion due to the low altitude atmosphere.

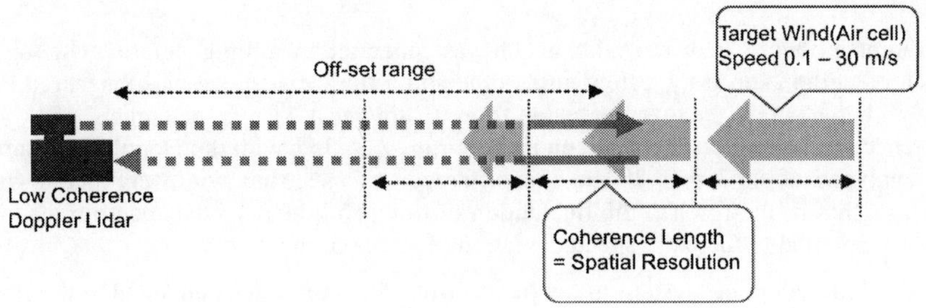

Fig. 1 Concept of DFB-LD based Doppler lidar.

3. Bulk type DFB-LD control

Bulk typed DFB-LD we used has the output of 2.7 W[max] with the wavelength of 974 nm. It is produced by Hamamatsu Photonics K.K. The nominal feature is that its spectrum width is 10 pm[max] at the device temperature of 25 ℃. We adjusted the forward current and the device temperature to enlarge the coherence length with keeping the high output power of more than 1W. The coherence length is measured by a fundamental interferometer. The result shows in Figure 2 and 3. The output optical power had a linear increase forward current from 0.4 A (Threshold) to >2A. In that range of the forward current, the spectrum width had a maximum (10 pm), that is, it is narrow cross the maximum. The coherence length depends on its feature. The spectrum width becomes narrower, the coherence length gets longer. Figure 3 shows such a behavior of the coherence length on each temperature. As a result, we decided the control condition of the bulk type DFB-LD of the forward current of 2A and the temperature of 35℃. Its output optical power was 1.76W and the coherence length was 0.8m.

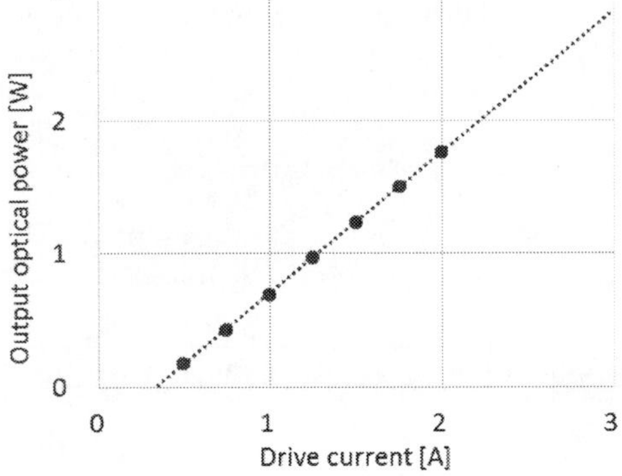

Fig. 2 Optical output of DFB-LD.

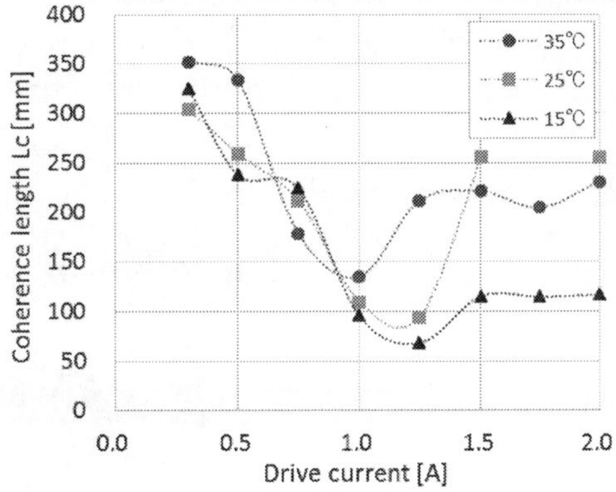

Fig. 3 Coherence of DFB-LD.

4. Fundamental measurement and lidar setup

For the Doppler measurement, optical system was assembled as shown in Figure 4. The output beam was enlarged to 30mmϕ and collimated with the beam divergence of 3 mrad. The interferometer was structured with a beam splitter and a reference path. The first target was a rotating retro-reflector. It is an imitated moving target of wind speed. The photodiode was used as a detector with an iris and a focusing lens. The high-speed spectrum analyzer was utilized to deduce a signal frequency. The rotating retro-reflector reflected the beam in its rotation partially, and the spectrum analyzer took the trigger signal to synchronize with rotating the reflector.

Figure 5 shows the measured velocity of the rotating retro-reflector analyzed by the detected Doppler frequency. Comparing with actual rotating speed, the result had high linearity with an error of +/- 0.03m/s. The target changed more diffusive ones, too. The system had the same result as shown in Figure 5.

Next, the target shifted to the water mist, which was generated by a fog generator. The fog was forced to follow with a certain speed by a flow guide tube and a fun. Mist speed was measured in 6 minutes with an anemometer as reference. It fluctuated from 0.38m/s to 1.21 m/s. Average speed was 0.72 m/s. The interference signal could be detected on its spectrum with 35 s with the Doppler measurement. The result was analyzed as 0.63 m/s. It had the difference of spectrum width compared with a retro-reflector's case, that is, the fog had the fluctuation of flow, and the detected spectrum had its width of 0.2MHz. It is converted to the wind speed deviation of +/- 0.05 m/s.

Then we replaced the optical system to the lidar set-up. The bulk optics changed to the fiber optics to reduce the background light and to improve stability and compactness. The reference path of the fiber optics was led by a glass plate in front of the transmitting beam unit. The reference path length was adjusted by the optical fiber length. It was tested from 2m – 11m. The receiver optics was utilized a Maksutov Cassegrain telescope of the aperture of 90mmϕ. A balance detector (C12668-01, Hamamatsu Photonics K. K.) was installed into the optics to increase the signal-to-noise ratio. The optical fiber was fixed to the telescope with a coupling lens. The same experiment with the fundamental one was conducted and obtained the same quality result.

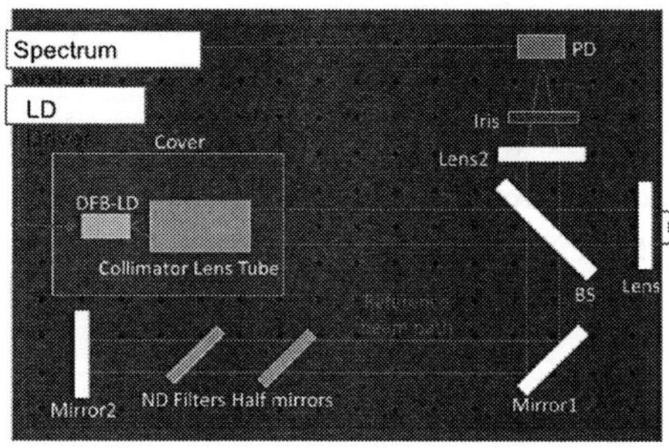

Fig. 4 Fundamental Experiment of Doppler lidar.

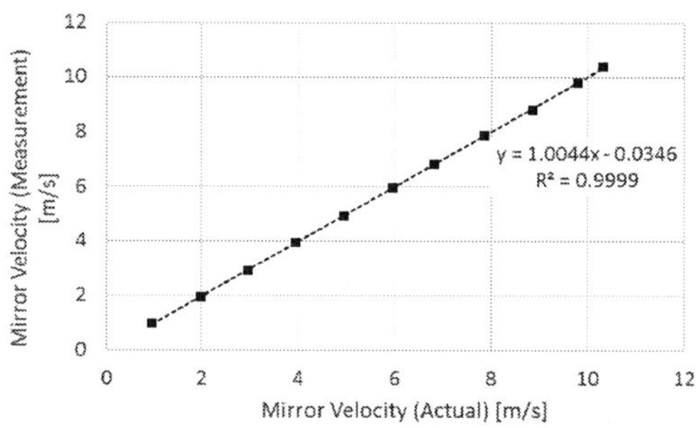

Fig. 5 Velocity measurement of rotating reflector.

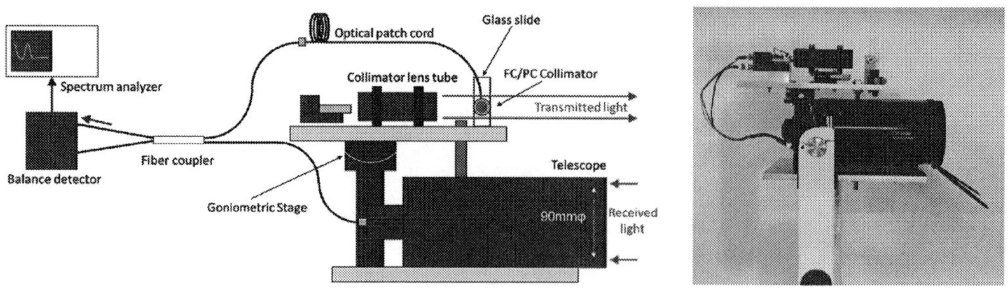

Fig. 6 Lidar setup with DFB-LD transmitter.

5. Discussion

The fundamental experiment of the low coherence Doppler lidar was successfully performed. Not only the moving hard target and soft target such as water mist flow were detected, too. The transmitting and receiving efficiency was evaluated with this lidar setup. As the receiving intensity was decreased by a ND filter, its decay trend of the interference intensity was measured. In the case of the threshold of 0 dBm, the approximate line indicated that the minimum detectable signal was 0.7 nW. The near range detection is also considered. The combination of Cassegrain telescope and coupling lens to the single mode optical fiber restricts the observation range. Especially in the lab-experiment, the observation range is so near (<10m), and its transmitting and receiving efficiency is low (2%). When the optical fiber coupling efficiency is improved to 30%, the minimum detectable signal of 0.7 nW is equivalent to the far small signal of 180m ahead. The observation range has an ability to enlarge up to 1km because of the optical output power.

Actual wind measurement is still undergoing. The current magnification of the signal is not enough. We now make proceeded the theoretical approach for electronics. Signal-to-noise ratio from the atmosphere is estimated with the balance detector and an additional amplifier. The bulk type DFB-LD is convenient for small device, easy access, and design free. On contrary, the beam intensity distribution is unevenness. Its divergence is ellipse, too. We considered that the influence of its unevenness was small for the Doppler wind measurement, while the interference intensity will be influent to its low coherency.

In this low coherence Doppler lidar, the measurement range is a point, that is, restricted in a coherence length of about 1m. To shift the measurement point or enlarge the measurement range, the reference path length of the interferometer should be changed. To make it realize, the optical fiber selector is available. Depending on the channel number, the measurement range can be increased.

The single processing and visualization of the wind data should be specialized. The commercialized spectrum analyzer is utilized for current analysis. The sweep time and repetition frequency is restricted under the wider measurement frequency range. It will also be a benefit for compactness and light weight.

We plan to apply this low coherence lidar into the surface atmosphere measurement on sea surface. Interaction between the sea wave motion and the surface atmosphere is interested. On a fisher boat and a treasure boat, the shallow angle measurement is possible. The other application is interesting, too. The eddies of wind around structures, updraft from evaporation on soil and urban car activity on load cause the small and rapid wind dynamics. They are the target of this lidar.

6. References

[1] M. L. Aitken, M. E. Rhodes, and J. K. Lundquist, "Performance of a wind-profiling lidar in the region of wind turbine rotor disks", Journal of Atmospheric and Ocean Technology, **29**, 347-355, (2012).

[2] K. Mizutani, T. Itabe, S. Ishii, M. Sasano, T. Aoki, Y. Ohno, and K. Asai, "Space-borne coherent Doppler lidar" Journal of the Communications Research Laboratory, **49**, 47-53, (2002)

[3] Liu Jiqiao, Z. Xiaopeng, Z. Jun, "Development of a coherent Doppler lidar to measure atmosphere windshear" Proceeding of IQEC/CLEO Pacific Rim 2011, 843 - 845 (2011)

[4] M. R. Belmont, J.M. K. Horwood, R. W. F. Thurley, and J. Baker, "Shallow angle wave profiling lidar", Journal of Atmospheric and Ocean Technology, **24**, 1150 – 1156, (2007).

[5] B. D. Reineman, L. Lenain, D. Castel, and W. K. Melville, "A Portable Airborne Scanning Lidar System for Ocean and Coastal Applications", Journal of Atmospheric and Ocean Technology, **26**, 2626 – 2641, (2009).

[6] M. Birkebak, F. Eren, S. Pe'eri and N. Weston, "The effect of surface waves on airborne lidar bathymetry (ALB) measurement uncertainties", remote sensing, **10**, 453, (2018).

[7] T. Fukuchi and T. Shiina Eds., *Industrial application for laser remote sensing*, (Bentham Science Publishers, 2012), Chapter 2.

[8] Y. Mori, S. Shimada, T. Shiina, H. Baji, and Sae Takemoto, "Dynamic analysis of sea wave data measured by LED lidar" SPIE proceedings of SPIE 10006, 10060G, (2016).

[9] A. Nikiforov Eds., *Photon Counting fundamentals and Applications*, (IntechOpen,2018) Chapter 11.

Micro crack analysis of optical fiber by specialized TD-OCT

Masayuki Tanaka and Tatsuo Shiina
Graduate School of Science and Engineering, Chiba University
1-33 Yayoicho Inageku Chiba 263-8522 Japan
e-mail address: tanaka.masayuki@greenkonnect.co.jp

Abstract: We have developed a method to evaluate absolute value of return loss from crack on optical fiber in a optical connector by using the specialized TD-OCT. The absolute measurement of the return loss could be accomplished with an error of ± 3 dB in the range from 10 to 100 dB. As a result, the micro crack of the optical fiber could be evaluated quantitatively. Furthermore, the state of the micro crack which could not be detected in the past could be analyzed without taking the product apart. If the value of the return loss and the elapsed time from its manufacturing are known, the angle of micro crack of the optical fiber and presence or absence of the gap between the fracture surfaces will be estimated.

Keywords: absolute measurement, reflection, OCT, micro crack, optical connector

1. Introduction

These days, the performance of the optical fibers has been improved therefore, it has been utilized in the various fields such as medical and sensing fields; however, the main materials of the optical fibers are glass, and the fear of micro crack still remains.

The product with the optical fiber often installs it into the main part of the system. Therefore, if there is a micro crack, the performance of the product would be deteriorated. Even the products need to be repaired, some products cannot be repaired by reinstalling the new fiber and it has a risk to cause serious accident such as communication disturbance. For these reasons, the manufacturers have to be careful for the quality of the optical fibers even more.

The detection of the micro crack has done mainly by measuring insertion loss and return loss. It is hard to detect the micro crack by this way because if there is a part which has the return loss away from the crack part in the measuring path, it is impossible to distinguish the crack part from the other reflections. Additionally, the position has been specified by checking a leaked red laser light with visual observation; however, it cannot be distinguished either the micro crack or a leaked light when a leaked light by bending is crossed to the crack part.

To detect the micro crack, it is necessary to measure the position and the return loss of the crack. OCT (Optical Coherence Tomography) can detect them, so it is suitable for micro crack detector.

However, in terms of detecting the micro crack, the return loss of the absolute measurement is needed for inferring the inside condition[(1)~(3)]. The absolute measurement means recognizing the same return loss as the same value even if the measuring date, place and device are different.

As for the absolute measurement, TD (Time Domain)-OCT is superior to the other OCT method. In the cases of FD (Fourier Domain)-OCT or SS (Swift Source)-OCT, they have the nonlinear signal processing in viewpoint of interference intensity because of Fourier Transformation. On the other hand, when TD-OCT is used, it becomes linear because of its simple signal amplification, therefore, the linearity of the measurement value is high. Moreover, TD-OCT can carry out long-depth measurement, so it enables to check the inside of the optical connecter, which depth is approximately 20mm.

We have developed the TD-OCT systems with the rotation mechanism for several years, and now it is possible to conduct absolute measurement within the error of ±3 dB and the return loss range of 10～100 dB. Therefore, this system is possible to detect the existence of the micro crack. The micro crack indicates the small crack which cannot be detected by the current return loss and insertion loss testing system. Optical fiber with initial stage of the micro crack can be used as usual; however, as the time elapses, the return loss and the insertion loss increase significantly, and the performance of the product declines remarkably. In this research, we utilized our developed system and analyzed the crack to predict the condition of the micro crack inside.

2. OCCR (Optical Component Coherence Reflectometer)

TD-OCT is a low coherent optical interferometer. Figure 1 depicts a principle of fundamental OCT using a Michelson interferometer. The beam from the low interference light source is divided into two paths by an optical beam splitter, and their interference occurs only when each optical path length reflected from a sample and a reference mirror matches. This is because of a low coherent light source. The optical device of this system is configured by a optical fiber assembly.

Table 1 shows specification of this device based on our developed TD-OCT. SLD (Super Luminescent Diode) light source is λ=1310 nm and its power is 10mW to improve the sensitivity. To make the system stable in the temperature range of 10～40 °C, cooling device is attached.

The semi-diameter of rotation reflector [(4) ~ (8)] of the optical path length variable mechanism is 20mm [(9)] to ensure the measurement length of 20mm. Additionally, considering the Doppler shift, Rotation speed is 1.1 sec/rotation to improve the strength of interference signal. As the only reference light gets the influence of Doppler shift, the interference intensity is weaken due to the rotation speed. The measurement light is taken inside the system though the optical fiber. Figure 2 shows the screen shot of measurement software when the system is connected to PC. The measurement result is shown in a graph, the horizontal axis is the distance from the fiber connecting point (0~20mm), and the vertical axis is the return loss (10~100dB). It is possible to set the threshold level of the return loss, and pass or fail judgement based on the threshold level is indicated. Maximum 6 points of detail information from the abnormal points are displayed. Figure 3 is the measurement result of the return loss obtained by this measurement system and an optical attenuator. It is possible to conduct absolute measurement of the return loss in the range of 10~100dB with less than ±3dB of error.

Figure 4 is the picture of the optical fiber micro crack part detected by the OCCR. The optical fiber micro crack parts seemed the mirror surface and had some degrees.

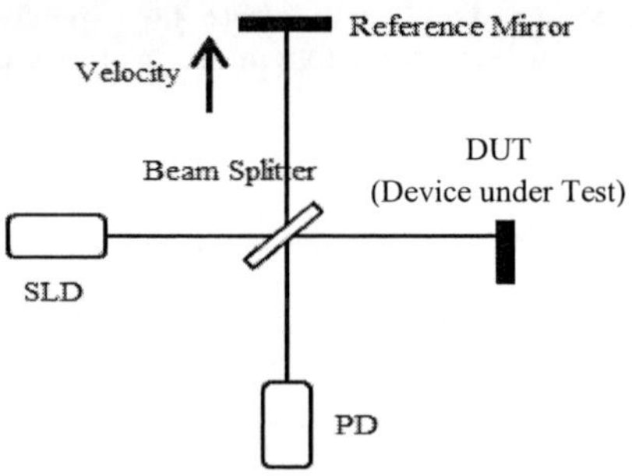

Figure 1. Principle of OCT.

Table1. Specification of OCCR

Item	Specification
Wavelength	λ=1310nm Δλ=40nm
Fiber	SM (10/125)
Scanning Range	20mm
Resolution	1.25μm
Return Loss	10～100dB
Return Loss Accuracy	+/-3dB
Measurement Time	3 seconds
Size	431×89×350mm

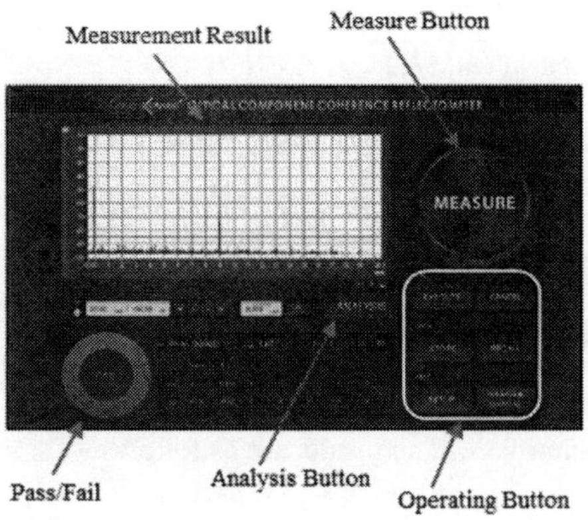

Figure 2. Measurement screen of OCCR PC software.

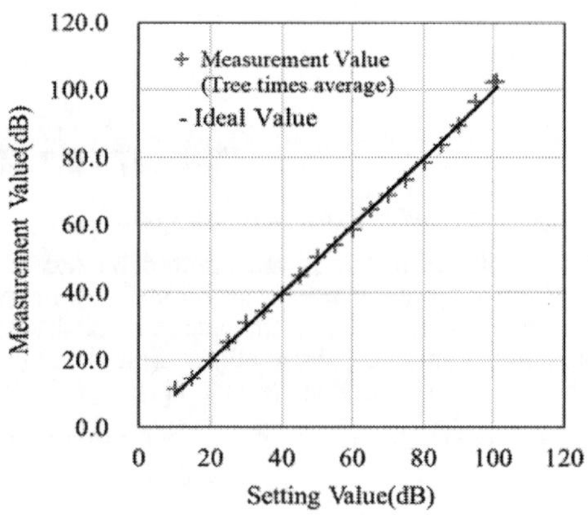

Figure 3. Linearity of rotating mechanism type TD-OCT.

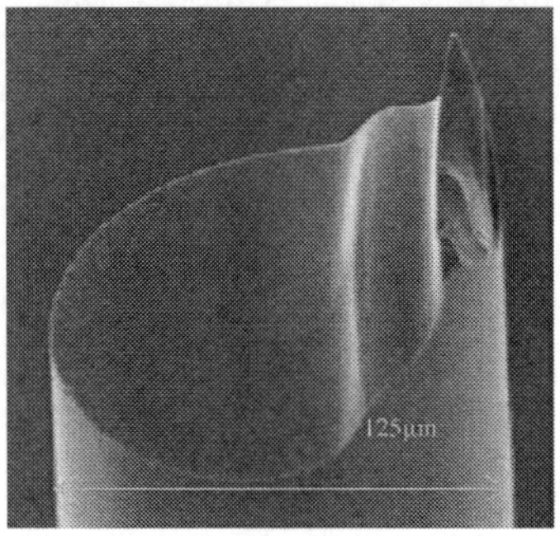

Figure 4. SEM image of fiber micro crack in optical connector.

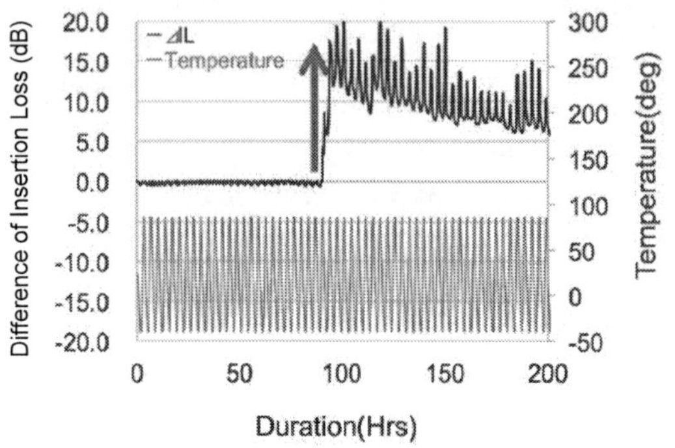

Figure 5. Temperature cycle of fiber micro crack products ΔIL@1.31μm

Additionally, we have done temperature character test with the optical connector sample which is detected the optical fiber micro crack by the OCCR measurement in the range of -40～85°C. The wavelength is λ=1.31μm. As a result, after 27 temperature cycles, Insertion Loss increased greatly, approximately 10dB. After that, insertion loss was around 10~20dB, and it did not return normal value. Figure 5 shows that state. The adhesive is used between a ferrule and a fiber; therefore, the gap between micro crack surface would become bigger as the time passes, because the temperature cycle appeared in the measurement values.

3. Micro crack of optical fiber

The formula of the return loss RL (1) [10], when depends on the degree of the surface of the fiber micro crack, is below.

$$RL(1) = RL_0\, exp\left[-\left(\frac{2\theta_b}{\theta e}\right)^2\right] \tag{1}$$

$$RL_0 = \frac{(n-1)^2}{(n+1)^2} \qquad \theta e = \frac{\lambda_0 n}{\pi \omega_0}$$

here, RL_0: return loss of vertical incidence, λ_0: wavelength in a vacuum, n: Refractive index, ω_0: semi diameter of mode field.

In addition, in the case that the surfaces of a fiber micro crack confront with a gap between them, the formula of the return loss RL (2) is represented by multiple interference as below.

$$RL(2) = \frac{4R_{gap}\, sin^2(\delta/2)}{(1-R_{gap})^2 + 4R_{gap}\, sin(\delta/2)} \tag{2}$$

$$R_{gap} = \frac{(1-n)^2}{(1+n)^2} \qquad \delta = \frac{4\pi g\, cos\, \theta_b}{\lambda\, cos\{sin^{-1}(n\, sin\, \theta_b)\}}$$

here, g: gap length between micro crack.

The actual return loss is represented by the combination of RL (1) and RL (2). We simulated the conditions of optical fiber micro crack using the FC/APC connectors with different polish degrees, (Figure 6). Furthermore, the gap caused by the time elapse was simulated that surfaces of two connecters contacted physically at first (RL (1) +RL (2)), and that the infinite gap is represented by single surface of a connector (RL (1)).

Figure 7 shows the results of the simulation and the experiment. The results are as follows.

1) The experimental results matched with the theoretical formula (1) and (2) until the degree of the surface angle of θ_b <8°.

2) When θ_b is over 8°, the measured value and the theoretical value did not match. The experiment results were stable in the range of 90～100dB.

3) The return loss difference between the physical contact results RL (1) + RL (2) and non-physical contact RL (1) was correlated with the surface angle as follows.

When the degree of the micro crack θ_b is small, the surface's gap is big because of large return loss. When θ_b is large, the gap is small. When θ_b is larger than 8°, the difference is almost none.

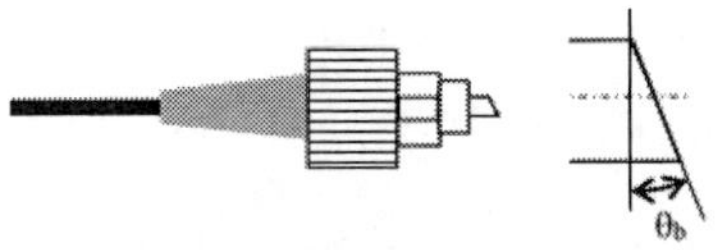

Figure 6. FC/APC Connector with different polishing angle θ_b

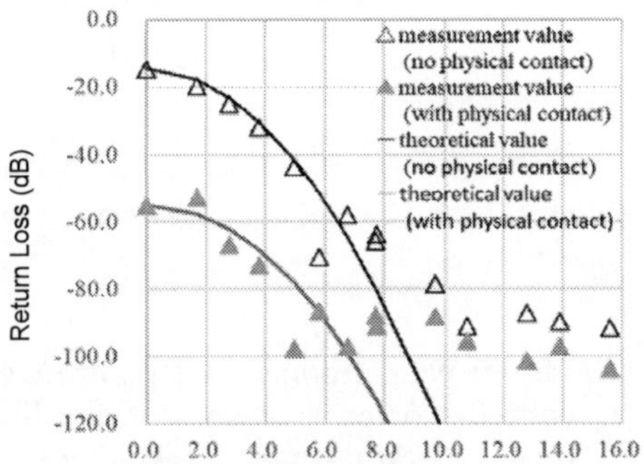

Figure 7. Return Loss by end face angle of FC/APC connector

Table 2. Specification of FC/APC sample simulating micro crack.

Item	Specification
Connector Type	FC/APC Connector
End Face Angle: θ_b	0～15°
Fiber	SM (10/125)

4. Consideration

As the result of analyzing the optical fiber micro crack using the developed TD-OCT system which can conduct absolute measurement of its return loss, it is found that it is possible to predict the condition of the optical fiber micro crack without dismantling. If the micro crack angle is within 8 degrees, the micro crack angle can be predicted.

The return loss follows the theoretical formula RL (1) +RL (2) when the time elapse is short after the fiber micro crack occurs. It also follows theoretical formula RL (1) when the time elapse is long. In addition, when it is hard to predict the length of time elapse, by remeasuring it after running in temperature cycle -40～85°C, it would be more similar to the theoretical formula RL (1). In brief, the return loss of the fiber micro crack is different depending on the time elapse.

This phenomenon can be explained as follows. The epoxy-based adhesive is used for fixing of optical connector; therefore, the adhesive repeats expansion and contraction as the time passes, and the gap would become bigger. The biggest return loss in the condition of physical contact and the surface angle of micro crack θ_b=0 was -60dB. It is lower than the return loss of the usual fiber connection point. It is exceedingly hard to detect in the general RL testing. By getting the position information of the micro crack, this system can detect the fiber micro crack accurately.

5. References

[1] A.D.Kersey, M.J.Marrone, M.A.Davis, Electronics Letters 14th March 1991 Vol.27 No.6, pp518-520(1991)

[2] K. Takada and M. Horiguchi, Member, IEEE, J Lightwave Technol, Vol.12, No.4, April 1994, pp658-663(1994)

[3] K. Takada and M. Horiguchi, Electronics Letter 8th July 1993 Vol.29 No.14, pp1273-1275(1993)

[4] T. Shiina, Y. Oritani, M. Ito, and Y. Okamura, Applied Optics Vol.42, Issue19, pp.2795-3799, 2003

[5] T. Shiina, S. Miyazaki, T. Honda The transactions of the Institute of Electrical Engineers of Japan. C, A publication of Electronics, Information and System Society 129(7), 1276-1281, 2009-07-01 (written in Japanese)

[6] K. Ojiri, T. Shiina, IEE Japan IM-15-017, pp25～30(2015) (written in Japanese)

[7] K. Akimoto, T. Shiina, IEE Japan IM-13-5, pp21~24(2013) (written in Japanese)

[8] S. Tabuchi, T. Shiina, IEE Japan IM-14-060, pp15~20(2014) (written in Japanese)

[9] Wolfgang Drexler, James G. Fujimoto, Springer-Verlag Berlin Heidelberg 2008

[10] M. Takahashi Doctoral thesis of Shizuoka University (1997) (written in Japanese)

A comb heterodyne receiver on a geosynchronous satellite? for long-path absorption measurements of atmospheric trace gases: a feasibility study

Nobuo Sugimoto
National Institute for Environmental Studies
16-2 Onogawa, Tsukuba 305-8506 Japan

Abstract: We performed an experiment on laser long-path absorption measurements using a retroreflector on a polar-orbit satellite (the Retroreflector in Space Experiment with ADEOS in 1996). It was an exciting technical demonstration, but it was not an efficient method for atmospheric observations. Our concept of earth-to-satellite laser long-path absorption measurements however was originally the method using a detection system on a satellite. It is much more efficient compared with the method using a retroreflector, and a large-diameter satellite-tracking telescope is not required in the ground station. Various measurement methods can be applied with a detection system on the satellite. One of the possibilities is applying dual-comb spectroscopy (DCS). This paper discusses the feasibility of applying symmetric or asymmetric DCS to the earth-to-satellite long-path absorption measurements. In the asymmetric earth-to-satellite DCS, a comb heterodyne receiver is used on the satellite.

Keywords: Satellite laser long-path absorption spectroscopy, Trace gas, Dual-comb spectroscopy

1. Introduction

The concept of the earth-to-satellite laser long-path absorption method is already described in the book edited by Hinkley in 1976 [1]. The method using lasers has potential advantages compared with the passive methods using solar radiation, such as ground-based Fourier Transform Spectrometers [2, 3], in accuracy, temporal resolution, measurement frequency, and obviously in the nighttime measurement.

We performed an experiment using the Retroreflector in Space (RIS) on the polar-orbit satellite ADEOS in 1996 [4]. In the experiment using a retroreflector in a low altitude orbit, we need to consider large velocity aberration in the direction of the reflected beam and large Doppler shift in the reflected laser wavelength. We also need a large-diameter accurate satellite-tracking telescope. In the RIS experiment, the orbital altitude of ADEOS was 800 km, and the ground speed was 7 km/s. We designed a special 0.5-m diameter single-element hollow cube-corner retroreflector having a slightly curved mirror surface [5] to compensate the velocity aberration (~60 μrad at maximum). After studying various spectroscopic methods (including a heterodyne method [6]), we adopted the method for measuring the spectrum of the target gasses utilizing the Doppler shift of the reflected beam which is dependent on the angle to the satellite (0-1.3 GHz at 10 μm). We developed single-longitudinal mode frequency agile TEA CO_2 lasers for the experiment [7]. We used two TEA CO_2 lasers for the signal and the reference, and the pulses from the two lasers are transmitted with an interval of 200 μs. The pulse repetition was 50 Hz, and the pulse energy was 100 mJ. We also used a second-harmonic Nd:YAG laser for tracking the satellite. We performed the experiment using a 1.5-m Contraves satellite-tracking telescope at the Communications Research Laboratory (current National Institute of Information and Communications Technology) in Koganei, Tokyo. We successfully obtained return signals from the satellite and performed spectroscopic measurements of ozone in the 9.6-μm region [4]. However, the experiment was discontinued in 1997 because the operation of ADEOS was stopped due to the trouble of the solar paddle.

Though the RIS experiment was an exciting technical demonstration, it was not an efficient method for measuring atmospheric trace gasses. The measurement was possible in both day and night, but the frequency of the measurement was limited from the satellite orbit and thc geometrical condition. Also, the efficiency of the round-trip optical transmission was low. It was 10^{-11} to 10^{-8} in the measurement depending on the range and angle to the satellite, using the 1.5-m transmitter/receiver telescope with a transmitting beam divergence of 100 μrad.

Our concept of earth-to-satellite laser long-path absorption measurements however was originally the method using a receiver on a satellite [8, 9]. Receiving the beam from the ground station on the satellite is much efficient than the system using a retroreflector. Also, we do not need to consider the Doppler shift even with a low altitude satellite, and we do not need a large satellite-tracking telescope. The tracking telescope with a diameter less than 10 cm can be used to transmit the laser beam to the satellite. In this paper, we discuss the feasibility of the earth-satellite long-path absorption system using a detection system on a satellite, especially the system based on Dual-Comb Spectroscopy (DCS) [10-12].

2. Earth-Satellite Long-Path Absorption Method Using a Receiver on a Satellite

We studied feasibilities using a direct detection receiver on a geosynchronous satellite in our previous study and obtained positive results [9]. The method using double pulses like in the RIS experiment is feasible. In that case the signal laser may be scanned over an absorption line of the target molecule.

The efficiency of the optical transmission to the satellite is the key parameter that determines the sensitivity of the measurement. Table 1 lists rough estimates of the efficiency for a geosynchronous satellite at a distance of 36,000 km and a low altitude satellite at a distance of 1000 km.

Table 1. Efficiency of Optical Transmission to the Satellite

Distance to Satellite	Transmitted Beam Divergence (full angle), and Diameter of Satellite Receiver	Optical Efficiency
36,000 km	100 μrad, 20 cm	$\sim 3 \times 10^{-9}$
1,000 km	100 μrad, 10 cm	$\sim 1 \times 10^{-6}$

The efficiency listed in Table 1 seems low, but it is not low when compared with lidars using a hard target. The efficiency of $\sim 3 \times 10^{-9}$ for a geosynchronous satellite is comparable to the reflection of a Lambertian hard target with 100% reflectivity at a distance of 1 km received with a 10-cm diameter telescope. The efficiency of $\sim 1 \times 10^{-6}$ for a low-orbit satellite is comparable to the reflection of the same target at 50 m received with the same receiver. Consequently, high sensitive measurements are expected with the methods using pulsed lasers.

A pointing system to direct the receiver telescope to the ground station is required on the satellite. The field-of-view (FOV) of the receiver can be rather wide in the direct detection system. In that case, background radiation from the earth surface and reflected solar radiation will be a major noise source. A satellite tracking system having a sufficient accuracy and sufficient speed (for a low-orbit satellite) is required in the ground stations.

3. Possibility of Applying Dual-Comb Spectroscopy

Figure 1 illustrate the concept of applying Dual-Comb Spectroscopy (DCS) to the earth-to-satellite long-path absorption measurements. There are two types, symmetric DCS and asymmetric DCS [10]. The symmetric DCS is equivalent to Fourier Transform Spectroscopy. The asymmetric DCS is a multi-heterodyne detection system using optical combs. In the asymmetric DCS-based system, Comb 2 is on the satellite.

A higher sensitivity is expected for the asymmetric DCS-based system when the signal intensity is low because it is a multi-heterodyne detection system. However, very accurate pointing to the ground station is required (a few microradian with 20 cm telescope). Also, Comb 2 must be controlled and monitored with sufficient accuracy on the satellite. In the symmetric DCS, the satellite receiver is simply a time-resolved direct detection system, and the requirement for the pointing accuracy is much relieved because a wider FOV can be used.

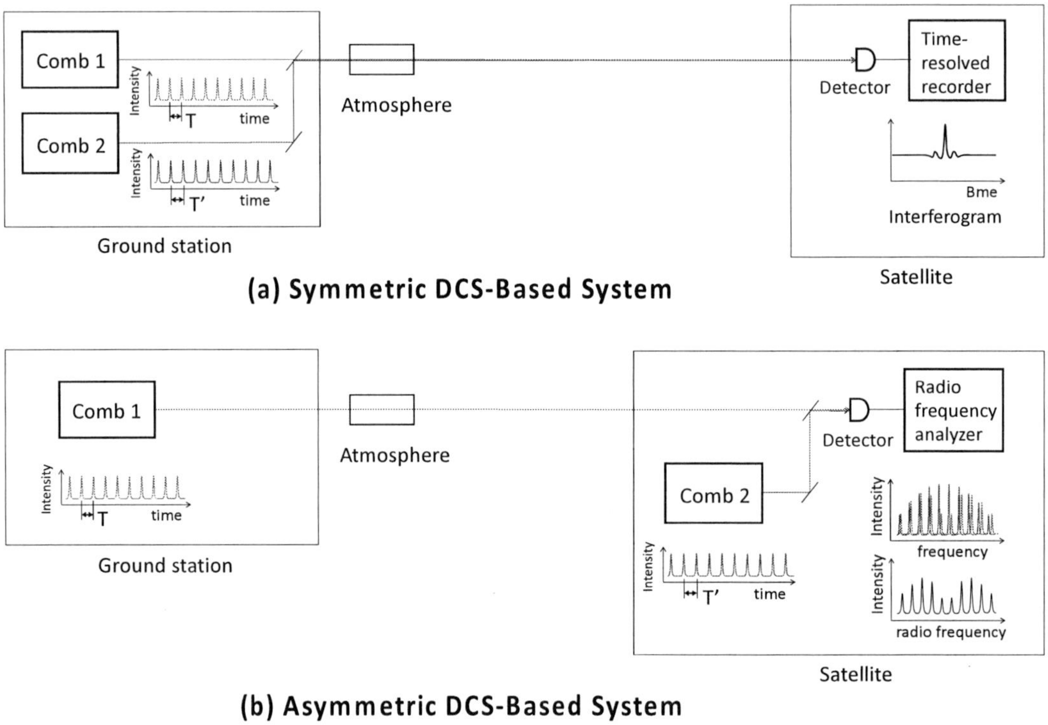

Figure 1. Concept of Earth-to-Satellite Long-Path Absorption System Based on DCS.

Open-path symmetric DCS measurements of CO_2, CH_4, and H_2O in 1.57-1.66 μm using a retroreflector on a flying multicopter is reported by Cossel et al. [12]. In their measurement, the efficiency of round-trip transmission to the retroreflecter at a distance about 1 km was about 1 x 10^{-2}. The output power of the transmitted comb beam was about 5 mW. If we try to perform the measurement using a receiver on a low altitude satellite with the same sensitivity, we require 10^4 times larger transmitter power considering the transmission efficiency in Table 1. We need more power with a geosynchronous satellite. However, we may also consider the fact that most of the teeth of the comb is actually not used to determine the concentration of the target gasses. High-power comb having an optimum envelope spectrum is ideal.

One of the advantages of using DCS is that the measurement protocol can be standardized more simply. The symmetric DCS-based system is realistic if high-power combs are available. The asymmetric DCS-based system is more sensitive, but a high-power comb is still required for the ground station because the transmission efficiency is low. A rough estimation shows sufficient photon number is received if transmitted power of ~mW is available per tooth for the system with a geosynchronous satellite. The pointing accuracy of the satellite receiver is the problem with the asymmetric DCS-based system. It would be not realistic to have an asymmetric system with a low-orbit satellite where fast tracking to the ground station is required for the satellite receiver. An asymmetric system with a geosynchronous satellite would be technically possible.

4. Conclusions

The optical transmission efficiency from a ground to a satellite is not low. It is comparable to that for hard target reflection. Consequently, the measurements using pulsed lasers like hard target lidars are feasible with both low-orbit and geosynchronous satellites. The systems based on DCS are attractive. However, high-power optical combs (~mW/tooth) are required. Experiments on symmetric DCS using hard target reflection would be a good feasibility demonstration for the earth-to-satellite measurements. For further technical demonstration, it would be useful to have a time-resolved direct-detection receiver system on a small satellite in a low altitude orbit, which can be used for testing various methods including the double-pulse method and the symmetric DCS-based method.

5. References

[1] E. D. Hinkley ed., *Laser Monitoring of the Atmosphere* (Springer-Verlag, 1976), p. 288.

[2] D. Wunch, et al., "The total carbon column observing network," Phil. Trans. R. Soc. A **369**, 2087–2112 (2011).

[3] M. De Mazière, et al., "The Network for the Detection of Atmospheric Composition Change (NDACC): history, status and perspectives," Atmos. Chem. Phys. **18**, 4935–4964 (2018).

[4] N. Sugimoto, et al., "Earth-satellite-earth laser long-path absorption experiment using the Retroreflector in Space (RIS) on the Advanced Earth Observing Satellite (ADEOS)," J. Opt. A: Pure Appl. Opt. **1**, 201-209 (1999).

[5] A. Minato, N. Sugimoto and Y. Sasano, "Optical design of cube-corner retroreflectors having curved mirror surfaces," Appl. Opt. **31**, 6015-6020 (1992).

[6] N. Sugimoto, A. Minato, Y. Saito and A. Nomura, "Heterodyne spectroscopy using spectral spread of short laser pulse," Jpn. J. Appl. Phys. **33**, 11B, L1602-L1603, (1994).

[7] R. J. Nordstrom, L. J. Berg, A. F. Desimone, and N. Sugimoto, "Time-gated gain cell for frequency-stable, single-longitudinal-mode operation of a transverse, electric, atmospheric CO_2 laser," Rev. Sci. Instrum. **64**, 6, 1663-1664 (1993).

[8] N. Sugimoto, "Atmospheric environment monitoring system based on an earth-to-satellite Hadamard transform laser long-path absorption spectrometer: a proposal," Appl. Opt. **26**, 763-764 (1987).

[9] N. Sugimoto, A. Minato, K. Ozawa, Y. Saito, and A. Nomura, "Theoretical evaluation of earth-to-satellite laser long-path absorption measurement of atmospheric trace species in the infrared region," Jpn. J. Appl. Phys. **34**, 5A, 2329-2334 (1995).

[10] I. Coddington, N. Newbury, and W. Swann, "Dual-comb spectroscopy," Optica **3**, 4, 414-426 (2016).

[11] G. B. Rieker, et al., "Frequency-comb-based remote sensing of greenhouse gases over kilometer air paths" Optica **1**, 5, 290-298 (2014).

[12] K. C. Cossel, et al., "Open-path dual-comb spectroscopy to an airborne retroreflector," Optica **4**, 7, 724-728 (2017).

Coherent Laser Radar in Japan, Past, Present and Future

Kazuhiro Asai
Tohoku Institute of Technology

Abstract: A research activity on coherent laser radar in Japan has begun in 1974. Kenya Goto, Tokyo Shibaura Electric Co.Ltd., had applied for a patent on "Doppler radar using laser" to United States Patent and Trademark Office(USPTO) and successfully acquired it. Since then, more than 200 papers written not only in Japanese and but also in English have been published by Japanese researchers until now. This paper is to describe on coherent laser radar activities in Japan, past, present and future.

Active Optical Remote Sensing Sensors and Instrumentations for NASA's Future Earth Science, Planetary Science and Mars Explorations Measurements/Missions

Upendra N. Singh (a)
(a) NASA Langley Research Center

Abstract: Active optical (Laser/Lidar) is critical for future NASA Earth, Planetary Science, and Mars Explorations. Eight out of fourteen missions recommended by 2017 NRC Earth Science Decadal Survey require active optical remote sensing for successful mission implementation. Given the cross-cutting synergies in critical Active Optical measurements for NASA Science and Exploration, a strategic approach in needed to identify areas where NASA can lead, leverage and partner with existing national and international capabilities to meet its future needs. This presentation will focus on efforts being planned to assess the current capabilities of NASA for the Active Optical and identify opportunities and synergies with industry, academia, government agencies and international partners for NASA to address its critical gap areas and work with external capabilities to meet NASA Science and Exploration future needs for space-based active optical remote sensing applications.

Investigating Space-based Lidar Performance for Decadal Survey Objectives

R. Michael Hardesty (a), Sara C. Tucker (b), Sunil Baidar (a), Benjamin Witschas (c), Oliver Reitebuch (c), and Stephan Rahm (c)

(a) Cooperative Institute for Research in Environmental Sciences
University of Colorado Boulder, 216 UCB, Boulder, Colorado 80309 USA
(b) Ball Aerospace
1600 Commerce Street, Boulder, CO, 80301 USA
(c) German Aerospace Center (DLR)
Institute of Atmospheric Physics, Münchner Strasse 20, 82234 Oberpfaffenhofen-Wessling, Germany

r.michael.hardesty@colorado.edu

Abstract: The recent NRC Earth Sciences and Applications from Space Decadal Survey ranked 3-dimensional atmospheric winds as one of a number of "most important" observables. The report notes that winds would be transformative for weather and air quality forecasting in the boundary layer and free troposphere. Several space-based instrument performance simulations of notional instruments have been performed; however, most used theoretical models of atmospheric properties and system performance. Here, we reexamine performance models for direct-detection and coherent Doppler lidar systems, but also analyze observations from the Ball airborne OAWL and the DLR airborne coherent lidar to predict space-based system performance for Decadal Survey wind measurement scenarios. Data from flight campaigns are analyzed to investigate measurement capability, then scaled to predict performance of a space-based instrument seeing the same atmosphere. Based on these results we evaluate the suitability of similar instruments to address Decadal Survey needs.

Keywords: Coherent Laser Radar, Wind Observations. Space Measurements, Aircraft Measurements

1. Introduction

Measurement of wind profiles from space has remained an elusive goal for more than 3 decades. Observing system simulation experiments indicate that wind observations from space-based platforms assimilated into numerical prediction models would likely produce significant improvements in short to medium range weather forecasts. Recognizing the impact of global wind observations, the World Meteorological Organization stated in 2012 that "Development of satellite-based wind profiling systems remains a priority for the future global observing system".

Responding to this need, the European Space Agency is planning to launch Aeolus during the late summer of this year. Employing direct detection in the ultraviolet, Aeolus will provide full profiles of the line of sight wind at a single fixed pointing angle orthogonal to the spacecraft velocity vector in regions not obscured by clouds. Because Aeolus is a three year mission, several international efforts are underway to define and plan the next generation of space-based wind missions.

In the US, the National Research Council recently completed its Earth Sciences and Applications from Space (ESAS) Decadal Survey [1], designed to shape science priorities and guide agency investments into the next decade. The ESAS report focuses on major science questions and potential space-based observations to address these questions. In particular, the report specifies linking boundary layer processes to weather, hydrological, and air quality forecasts and extending environmental predictions of weather and air quality by improving observations of "natural, low-frequency modes of weather/climate variability" as two major questions for which better wind observations are required.

Although Aeolus will provide winds throughout the troposphere and lower stratosphere, the instrument was not designed to address a focus on boundary layer processes, which will likely require higher spatial resolution than that available from the Aeolus data set. As such, when developing a strategy for deploying next-generation space instruments, two questions arise: 1) What boundary layer information can a direct detection lidar instrument designed to provide full tropospheric winds provide, and 2) would a coherent lidar instrument developed to specifically measure in the boundary layer where aerosol scattering is high provide sufficient additional scientific return to justify the mission expense?

Numerous modeling studies have been performed to predict performance of different space-based lidar designs. However, because modeling studies typically incorporate parameterized estimates of backscatter, they do not generally reflect capability of instruments under specific measurement scenarios. In recent years, several airborne Doppler lidar instruments employing both direct and coherent detection have been developed and flown during field campaigns at various locations around the world. By analyzing the results of such field campaigns and scaling the results using validated radiometric models to predict performance of similar designs in space, we can develop insight into the performance of space-based instruments to investigate various phenomena or to provide measurements at certain resolution, precision and accuracy.

Here we illustrate this approach by applying it to investigate space-based performance based on airborne measurements from two instruments: the Ball Aerospace 532 nm Optical Autocovariance Wind Lidar (OAWL) and the German Aerospace Center (DLR) 2μm heterodyne coherent detection system that has been deployed on the DLR Falcon aircraft during a number of field campaigns. In this paper the problem is simplified to answer the question "How would one design a space-based instrument to provide similar measurement capability as the airborne system (i.e., collect equivalent photons per measurement interval at the same pulse rate), assuming equivalent instrument performance?" In the future our intention is to expand the analysis to investigate impacts on system design of changing measurement requirements such as resolution or precision; this more intensive investigation requires specification of the relationship between measurement precision and system carrier-to-noise ratio.

2. General approach

Our goal here is to select actual measurement data from field campaigns during which the airborne lidars provided scientifically useful measurements and scale the measurements to specify space-based designs that would provide the same results when observing the backscatter environment probed by the airborne lidars. We assume that the relevant design parameter is received photons per second, and scale the space-based specification of transmit power, receiver diameter, range resolution and horizontal resolution to achieve similar results. Implicit in this approach is the assumption that a space-based instrument has identical sensitivity in terms of velocity error versus received photons as the aircraft instrument. Assuming that the system pulse repetition frequency (prf) and the number of pulses averaged remains the same (note that this reduces the horizontal resolution of the space-based measurement due to the high orbit velocities), and for the same range gate accumulation (vertical resolution) similar performance between airborne and space-based systems is achieved when

$$\frac{D_s^2 J_s}{R_s^2} = \frac{D_a^2 J_a}{R_a^2}$$

where D is telescope diameter, J is energy per pulse, R is slant range to the measurement volume, and the subscripts *s* and *a* refer to the space-based and aircraft-based measurements, respectively. Then, by choosing a measurement range to the scattering volume of interest for the airborne system, we can compute the equivalent space-based system needed to give similar precision for the same range and accumulation time. Additional along-range accumulation (e.g. larger range gates) that are typically required for space-based operation would decrease the space-based system requirements.

3. Scaling OAWL measurements

The OAWL design [2], based on a Cat-eye-based Quadrature Mach-Zehnder interferometer, offers a different approach to direct detection wind measurements from that employed in Aeolus. While OAWL uses some of the same lidar technologies (Cassegrain-type telescope, Nd:YAG lasers, etc.) being implemented on Aeolus, there are some additional advantages of the OAWL approach. One such advantage is the elimination pf laser pulse-to-pulse frequency stability requirements and platform pointing requirements, since the OAWL interferometer accounts for the outgoing laser frequency and pointing on a per-pulse basis; each pulse is measured separately with no on-detector accumulation and returns are phase-adjusted in real-time FPGA processing prior to pulse accumulation. Another OAWL advantage stems from the all-reflective, field-widened, Cat-eye interferometer that provides high throughput efficiency and enables use of large telescopes with relaxed FOV and co-boresighting requirements.

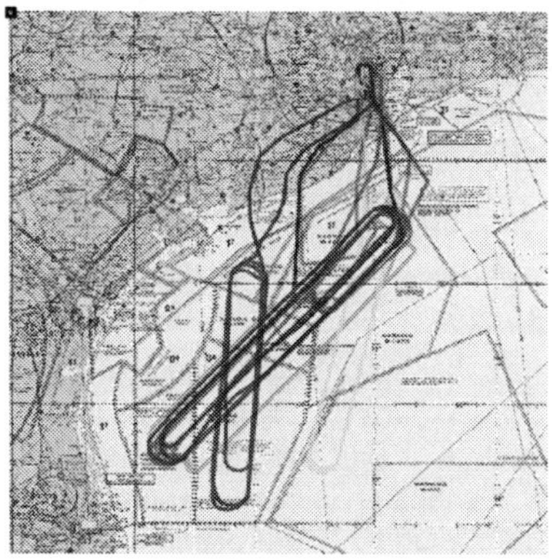

Figure 1. GrOAWL test flight tracks over the Gulf of Mexico.

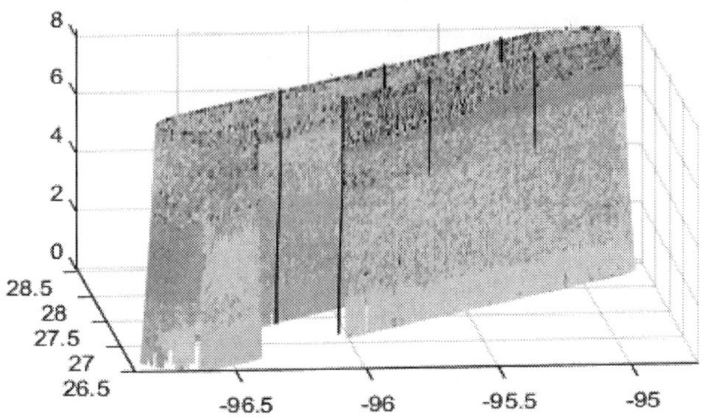

Figure 2: Line of sight wind speeds vs. longitude (y-axis), latitude (x) and altitude (z) measured by GrOAWL with dropsonde deployments (black lines).

A "Nested OAWL" system is currently under development at Ball to add a UV molecular channel to the OAWL system design by nesting a UV molecular bandwidth resolving interferometer inside a 532 nm aerosol bandwidth resolving interferometer. The advantage of using the all-reflective QMZI approach for the molecular channel is that, unlike for the Double Edge Fabry Perot approach to molecular return wind measurements, there is no aerosol-presence-induced bias or contamination in the OAWL winds retrieval. The same performance models validated under the OAWL and GrOAWL development efforts are being used to predict space-based performance of such a system operating at both the 532 nm (for aerosol returns) and 355 nm (for molecular returns) wavelengths. Initial model runs indicate that a system using a laser like that being developed under ESTO funding at Fibertek would provide returns from the full atmosphere.

The Green-OAWL (GrOAWL) airborne system, built under NASA Venture Technology funding and managed by NASA ESTO, provided a demonstration of OAWL 532nm aerosol performance which was used to evaluate potential measurement capability of a space-based version [2,3]. The effort also demonstrated that returns gathered from two azimuthally-orthogonal pointed beams may be combined to produce representative vector wind measurements. The system was deployed on a NASA WB-57 aircraft for a series of test flights over the Gulf of Mexico during the spring and summer of 2016. Figure 1 shows the flight pattern for the series of OAWL flights, while Figure 2 shows an example of line-of-sight measurements from the two telescopes.

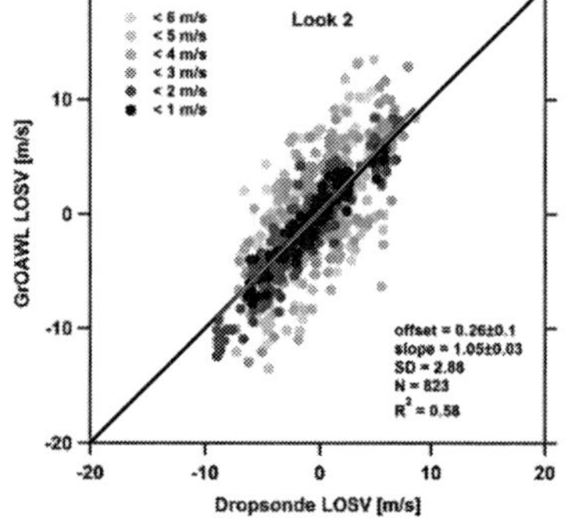

Figure 3: GROAWL and dropsonde comparison (10 s average)

A dropsonde system deployed during the flight provided comparison information for evaluating the precision and accuracy of the OAWL measurements [3]. Figure 3 shows a scatter diagram comparing GrOAWL line-of-sight wind measurements with the nearest dropsonde observations as projected on to the GrOAWL line-of sight direction. Although the dropsonde measurements did not always overlap in space and/or time with the GrOAWL measurements and, though they could be separated by as much as 10 km, the measurements agree quite well. In addition, the GrOAWL data, meet the WMO specification stating that wind measurements should have uncertainty of less than 3 m s^{-1} to be useful for assimilation into numerical forecast models.

Applying the basic scaling from Equation 1 and accounting for different range gate sizes, we find that a GrOAWL design including a ~1-m telescope, ~50 mJ of pulse energy at a prf of 200 Hz (10w), and a vertical resolution of 150 m could provide equivalent measurements to those obtained during the Houston campaign. As noted previously, scaling in this manner implies that the measurement time remains the same; thus the 10 second accumulation time applied for the GrOAWL measurements would correspond to about 70 km horizontal resolution of the scaled space-based instrument. This is within the spatial resolution needed for useful assimilation (Aeolus will have a spatial resolution of 100 km) but not sufficient to resolve boundary layer processes.

4. Scaling the DLR Coherent Lidar Measurements

The DLR airborne coherent Doppler lidar [4] has been deployed for a number of field campaigns at various locations around the globe, including the North Atlantic near Iceland, Southern Ocean off New Zealand, Africa, and Europe. This extensive range of locales and aerosol environments offers the opportunity to assess potential coverage of a space-based coherent lidar. The instrument provides excellent sensitivity and measurement precision in regions where aerosol loading is sufficient. In responding to the Decadal Survey identification of boundary layer phenomena as a key focus, observations from the DLR instrument provide a useful example of how well these phenomena can be observed by coherent lidar and how to design a space-based instrument to provide these observations.

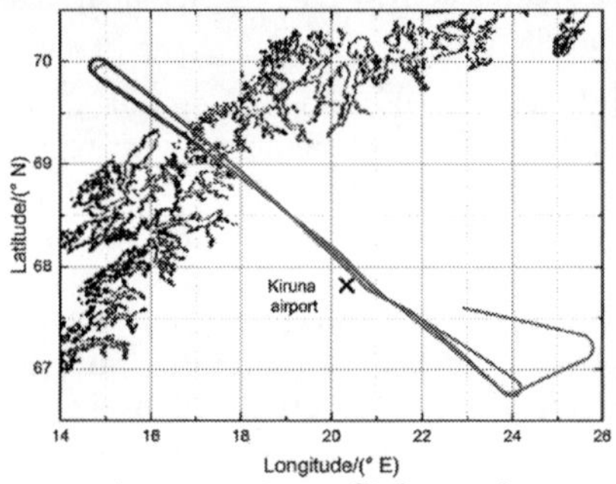

Figure 5: Kiruna flight track

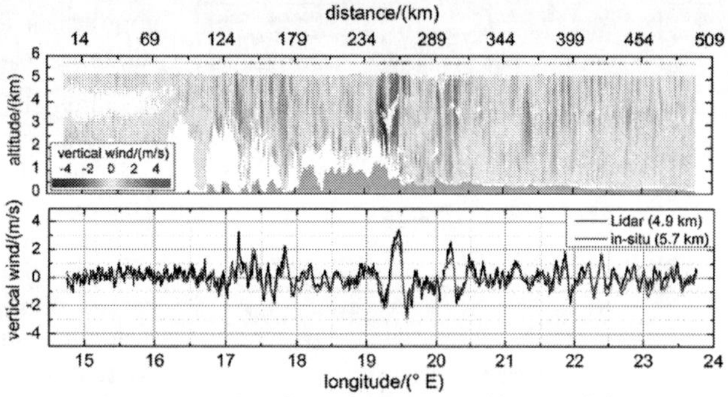

Figure 6: Vertical winds measured by DLR lidar

As an example of observations from the DLR Falcon, we show data taken during an investigation of internal gravity waves near Kiruna, Sweden. Figure 4 shows a flight track from the Falcon, while Figure 5 shows measurements of the vertical wind measured by the lidar. Also shown in Figure 5 is a comparison between an in situ sensor and a near-range lidar gate. One can see that the coherent lidar provides excellent coverage of winds in the aerosol layer near the surface. As above, we pose the problem as computing the power-aperture product necessary to obtain similar measurements from a coherent lidar on an orbiting platform at 400 km with the same temporal resolution (1 s) and vertical resolution (100 m) as the airborne lidar. Assuming the space-based system looks off-nadir at 35° (identical to Aeolus), and using 10 km as the nominal height for many Falcon measurements, we estimate that we would need about 16.5 W of transmitter power using a 1 m telescope to provide measurements equivalent to those obtained by airborne system at 10 km altitude, except with 7 km (1 s) resolution. Note this would provide very high measurement precision, on the order of a few tens of cm s^{-1} in many cases, which would probably be more than is needed for measurements averaged over 7 km, where variability will dominate the measurement error.

This scaling exercise provides a notional design for a space-based coherent lidar that would observe phenomena similar to the cases studied during DLR field deployment of the Falcon-based system. Because many of the boundary layer phenomena identified in the decadal survey are associated with smaller scales than the 7 km horizontal resolution for this notional design, in developing a design for a space-based instrument it may be desirable to improve resolution by relaxing the precision specification for the lidar, i.e., accumulating fewer pulses before providing an estimate. Several trades among the key variables of aperture size, transmit energy, pulse repetition frequency, averaging time, and vertical resolution are possible to optimize the design to meet specific observational objectives. These trades will be discussed during the presentation

5. Discussion

The extremely simplified scaling discussed here is useful for extending results from real instruments observing the real atmosphere to develop notional space-based designs. This exercise provides an indication of the specifications, and potential tradeoffs, required to provide useful measurements as demonstrated by both the GrOAWL and the DLR coherent lidar. It does indicate the potential difficulty of obtaining measurements over length scales of a few hundred meters to a few km, because of the speed of the spacecraft and the need to accumulate pulses to improve signal to noise ratio, which may be needed to address many boundary layer issues noted in the Decadal Survey.

Data from other airborne Doppler lidars could also be examined to broaden the range of applications and to investigate the extent to which the sensitivity of the two systems used is representative of typical performance. The DLR instrument has been flown together with the Aeolus Airborne Demonstrator, providing very valuable information from two instruments observing the same phenomena. NASA has deployed their DAWN coherent instrument [5] to study phenomena in polar regions and the tropics. Investigation and comparison of results from the different instruments will be important as designs and approaches for Aeolus follow-on Doppler lidar wind missions are considered.

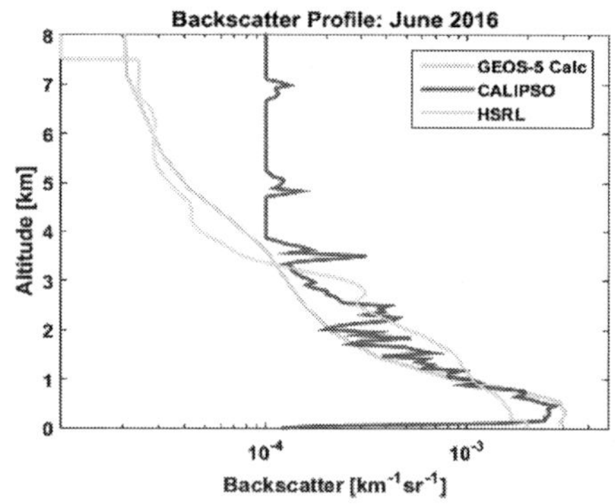

Figure 6: Backscatter profile estimates

A more detailed study would also consider the aerosol environment during the measurement campaigns in order to extend the results to other areas of the globe. This can done by applying analysis/forecast models that include aerosol information, such as the NASA GEOS-5. Figure 6 shows GEOS-5 estimates of the backscatter profile during the GrOAWL measurement campaign near Houston, compared with estimated backscatter from CALIPSO and from lidar measurements taken during a campaign at the same time of year but a year earlier. The agreement provides confidence that model results, if cautiously applied, can be used to extend results from field campaigns to other regions of the globe. Additionally, results from airborne could be used to verify earlier simulation studies.

6. Acknowledgements

We thank our colleagues at Ball and DLR for their contributions to the flight campaigns and data analysis, and Ball Aerospace, NASA Earth System Technology Office, DLR, and the German Ministry of Education Research for their support of the research activities described here.

7. References

[1] National Academies of Sciences, Engineering, and Medicine, *Thriving on Our Changing Planet: A Decadal Strategy for Earth Observation from Space.* Washington, DC: The National Academies Press. https://doi.org/10.17226/24938, 2018).

[2] S. C. Tucker, R. M. Hardesty, S. Baidar, and C. Weimer, "The ATHENA-OAWL Venture Tech instrument". *18th Coherent Laser Radar Conference and the Lidar Working Group on Space Based Winds, CLRC 2016,(2016).*

[3] S. Baidar, S. Tucker, and R. M. Hardesty, "Preliminary results from the ATHENA-OAWL Venture Tech airborne mission", *18th Coherent Laser Radar Conference, CLRC 2016 (2016).*

[4] B. Witschas, S. Rahm, A. Dörnbrack, J. Wagner, and M. Rapp, "Airborne Wind Lidar Measurements of Vertical and Horizontal Winds for the Investigation of Orographically Induced Gravity Waves", *J. Atmos. Oceanic Technol.,* **34**, 1371–1386, https://doi.org/10.1175/JTECH-D-17-0021.1 (2017).

[5] M. J. Kavaya, J.Y. Beyon, G.J. Koch, M. Petros, P.J. Petzar, U.N. Singh, B.C. Trieu, and J. Yu, "The Doppler Aerosol Wind (DAWN) Airborne, Wind-Profiling Coherent-Detection Lidar System: Overview and Preliminary Flight Results", *J. Atmos. Oceanic Technol.,* **31**, 826–842 (2014).

Space-Based Coherent Lidar for Wind Measurements with High-Percentage Tropospheric Coverage

Sammy Henderson and Don Jacob
Beyond Photonics, Boulder, CO, 80301, sammy@beyondphotonics.com

Abstract: We describe the potential for a 2-micron wavelength space-based coherent lidar system with power-aperture product and receiver efficiency sufficient to provide high percentage coverage tropospheric wind vector measurements along two continuous stripes in the atmosphere below the satellite track.

Keywords: Coherent Lidar, Wind Measurement

1. Introduction

Global measurements of altitude-resolved vector winds from an Earth-orbiting platform are of interest for both Earth science and weather forecasting applications. The potential for global wind measurements using lidar systems has been recognized for many years [1-3], and the technology has now advanced such that the first global wind measurement mission using direct detection Doppler lidar technology is scheduled to launch in late 2018 [4]. In recent years coherent lidar systems have primarily been considered for space-based wind measurements only in regions of the atmosphere where the aerosol backscatter is relatively high, e.g., the boundary layer and elevated aerosol layers higher in the troposphere. We describe the potential for a coherent detection wind lidar that has sufficient backscatter sensitivity to allow for high percentage coverage throughout the troposphere and lower stratosphere.

2. Coherently Detected Signal Photons and Diversity

As described in Frehlich's paper and elsewhere [2,5], the total number of backscattered photons that are detected by the coherent receiver over M_p pulses from a range gate of duration T_g is

$$\phi_T = \frac{\eta_{eo}\eta_a E T^2 \lambda \beta A_r M_p T_g}{2hR^2}. \qquad (1)$$

In this equation, ϕ_T is the total number of coherently detected photons in the measurement, E is the pulse energy from the transmitter, T^2 is the two-way transmission to and from the target range, λ is the lidar wavelength, β is the aerosol backscatter coefficient at the lidar wavelength, A_r is the receiver aperture area, h is Planck's constant, R is the range to the center of the range gate, and $M_p = f_p T_m$ is the number of laser pulses used in the measurement, with f_p being the pulse repetition rate and T_m the total measurement time. The total efficiency of the coherent receiver is given by $\eta_{eo}\eta_a$, where η_{eo} is the electrooptic efficiency and η_a is the antenna efficiency. The electrooptic efficiency is $\eta_{eo} = \eta_{to}\eta_{ro}\eta_q\eta_{sn}$, and contains all the transmit and receive optical path losses other than the antenna efficiency loss, where the terms on the right-hand side of the equation are the efficiencies of the transmit and receive optics, detector quantum efficiency, and coherent detection shot noise efficiency respectively. The antenna efficiency contains losses due to transmit beam truncation, and mismatch between the wavefront of the received signal field and that of the local oscillator. Reduction of antenna efficiency occurs due to misalignments, wavefront aberrations in the lasers or optics, and atmospheric refractive turbulence.

Equation 1 can be inverted to solve for the backscatter coefficient. To determine the minimum detectable backscatter, we must determine the minimum number of coherently detected photons required to detect the signal in the presence of noise. Given that the effective noise of a coherent receiver is one coherently detected photon per diversity mode (aka, independent sample) [2], it is important to determine the total diversity in the measurement.

As the pulse travels through the atmosphere the coherent signal return exhibits well known temporal speckle fluctuations and the diversity, or number of statistically independent speckles, of the return in the range gate duration T_g is given by [6,7]

$$M_g = \left\{ \frac{\tau_{cg}}{T_g} \text{erf}\left(\sqrt{\pi}\frac{T_g}{\tau_{cg}}\right) - \frac{1}{\pi}\left(\frac{\tau_{cg}}{T_g}\right)^2 \left[1 - exp\left(-\pi\frac{T_g^2}{\tau_{cg}^2}\right)\right]\right\}^{-1}, \tag{2}$$

where $\tau_{cg} = \int_{-\infty}^{\infty} |\mu_A(\tau)|^2 \, d\tau$, with τ_{cg} being the signal coherence time defined by Goodman [7] and $\mu_A(\tau)$ being the normalized signal autocorrelation function. For a Gaussian signal spectrum, which we assume for this analysis, the signal coherence time is given by $\tau_{cg} = 1/2\sqrt{\pi}\sigma_f$, where the total rms spectral width is $\sigma_f = \left(\sigma_{fT}^2 + \sigma_{fv}^2\right)^{1/2}$, with $\sigma_{fT} = \sqrt{\ln 2}/(\sqrt{2}\,\pi\Delta T)$ being the rms spectral width of the assumed Gaussian temporal laser pulse with FWHM duration of ΔT, and σ_{fv} being the rms spectral width due to wind fluctuations (shear and turbulence) in the measurement volume.

In the limit that $T_g \gg \tau_{cg}$ the range gate diversity becomes $M_g \approx T_g/\tau_{cg}$, and if $T_g \ll \tau_{cg}$ it becomes $M_g \approx 1$. In many measurement scenarios the approximation $M_g \approx T_g/\tau_{cg}$ is sufficiently accurate and the total diversity of the measurement is $M_T = M_p M_g \approx f_p T_m T_g/\tau_{cg} = T_T/\tau_{cg}$, where T_T is the total signal collection time for the measurement.

The detectability of the coherent lidar signal depends on the detectivity signal to noise ratio

$$SNR_d = \frac{\phi_T}{\sqrt{M_T}} = \sqrt{M_T} \cdot \frac{\phi_T}{M_T} = \sqrt{M_T} CNR_{MF}, \tag{3}$$

which is the mean height of the accumulated signal divided by the rms fluctuations of the accumulated noise floor. The matched filter carrier to noise ratio, CNR_{MF}, is the number of coherently detected photons per diversity mode. Using these relations, Equation 1 can be rewritten as

$$\beta = \frac{2hR^2 SNR_d}{\eta_{eo}\eta_a T^2 \lambda A_r \sqrt{T_m \cdot T_g \cdot \tau_{cg}} \cdot E\sqrt{f_p}}, \tag{4}$$

As will be shown below, the significance of SNR_d is that for moderate and high diversity levels the SNR_d required to detect the signal successfully is nearly constant.

3. Minimum Detectable Backscatter

The expression for backscatter illustrates that backscatter sensitivity is minimized by increasing the pulse energy, aperture area, signal coherence time, range gate time, and measurement time while decreasing the range. Higher pulse energy is more effective than higher pulse repetition rate. In order to determine the minimum detectable backscatter level that the lidar can sense, we must determine the minimum SNR_d at which the detection meets a given reliability, or probability of good estimates.

As described in the literature, detection of coherent lidar signals results in good detections at the actual signal location, and also detections that are randomly (uniformly) distributed across the search bandwidth (bad estimates) [5]. The minimum detectable backscatter depends on the fraction of uniformly spread bad estimates that is acceptable for the measurement. An integral equation providing an estimate of the fraction of bad estimates, P_b, is

$$P_b = \left(\frac{N_S}{N_S-1}\right)(1 - P_{NA}) = \left(\frac{N_S}{N_S-1}\right)\left\{1 - \int_{-\infty}^{\infty} pdf_{s+n}(y)\left[\int_{-\infty}^{y} pdf_n(x)\, dx\right]^{N_s-1} dy\right\}. \tag{5}$$

We developed this and associated equations independently, but similar equations have been developed by others [8]. In this equation pdf_n and pdf_{s+n} are the probability density functions of the noise only and the signal plus noise respectively. At low to moderate total diversity the pdfs are both Gamma functions appropriate for accumulated coherently detected signals. At higher total diversity the signals are well approximated by Gaussian signals. The mean of the noise only pdf is unity and that of the signal plus noise is $1 + CNR_{MF}$ with the variance for both being $1/M_T$. When Gaussian pdfs are appropriate (i.e., at large total diversity) the integral equation is greatly simplified. Other terms in the integral equation are the number of search bins, N_S, which we define as the search bandwidth divided by the signal FWHM bandwidth, and P_{NA} which is the probability that the detection is in the signal bin, i.e., it is not an anomaly, which are any detections outside the signal bin. The $N_s/(N_s - 1)$ factor accounts for the fact that not all detections in the correct signal bin are actually good estimates.

An example of requirements that can be derived using Equation 5 are shown in Figures 1 and 2.

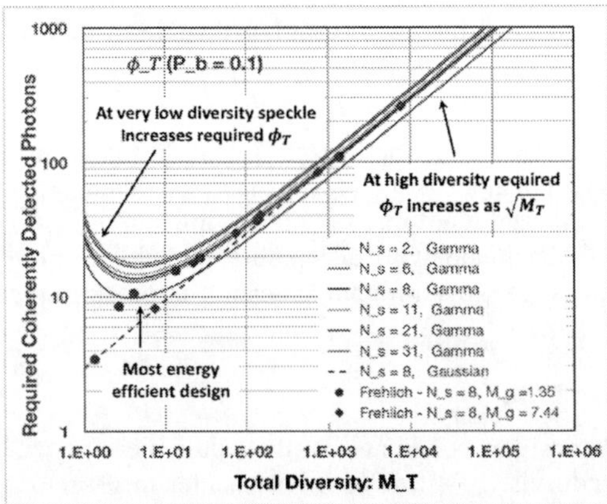

Figure 1. Coherently detected photons required to achieve 90% good estimates vs total diversity.

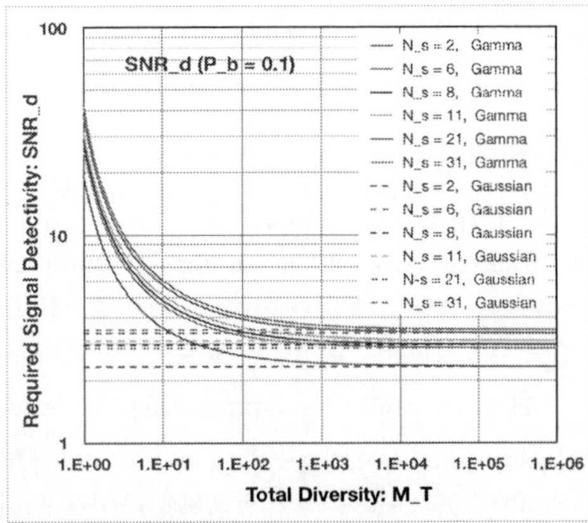

Figure 2. Signal detectivity required to achieve 90% good estimates vs. total diversity.

The lowest number of signal photons required to achieve $P_b = 0.1$ occurs for total diversity of 3-6, with the exact number depending on the number of search bins. Therefore, the total integrated transmitter energy required to make the measurement occurs at low diversity, with diversity sufficient to reduce speckle fades but not so high that more noise is introduced into the measurement. The $\sqrt{M_T}$ increase at higher diversity shown in the figure is a consequence of the effective coherent receiver noise of one detected photon per diversity mode. In general, coherent lidar systems are most energy efficient when a few coherently detected signal photons are detected from each signal coherence time, τ_{cg}, or diversity mode. But in many measurement scenarios this drives the required transmitter pulse energies to very high levels that are not easily obtained, so lower transmitter energy at higher pulse repetition rate is used with the penalty that the overall transmitted energy expended for the measurement increases.

Also shown in Figure 1 are the number of coherently detected photons required as calculated using the complicated analytic fit equations that Frehlich developed [5]. At higher total diversity (> 10) these calculated points using the Frehlich equations match very well the values calculated from the simpler Equation 5. At lower total diversity there is significant disagreement which indicates that the Frehlich analytic fit equation are not valid at very low diversity, which is reasonable as he indicates in the paper that simulations with 10 pulse averaging were the lowest considered in his analysis.

The SNR_d required to achieve $P_b = 0.1$ parametric in the number of search bins is shown in Figure 2, using both the exact Gamma statistics and the Gaussian approximation. This again illustrates that the Gaussian approximation is accurate for high diversity but cannot be used at lower diversity. When the total diversity is above about 500 the required SNR_d flattens to the fixed values predicted by the Gaussian pdfs. The invariance of the required SNR_d level at higher diversities illustrates that it is a valuable fundamental parameter for determining the performance of coherent lidar systems.

Similar requirements like those shown in Figures 1 and 2 and others can be generated at any bad estimate probability value using Equation 5. Once the required SNR_d to achieve a given probability of bad estimates is determined, and the other lidar parameters are defined, Equation 4 can be used to determine the minimum backscatter level consistent with this level of bad estimates.

4. Velocity Precision

A lower bound of velocity measurement error for the good estimates (not including the uniformly distributed bad estimates) is given by the Cramer Rao Lower Bound (CRLB) [2]. Written in terms of the total number of coherently detected photons and total diversity the CRLB is given by

$$var[V_r] \geq \left(\frac{\lambda}{2}\right)^2 \sigma_f^2 \left[\frac{2}{\phi_T} + \frac{2M_T}{\phi_T^2} + \frac{1}{2M_T}\right] . \quad (6)$$

When the coherently detected photons per diversity mode is small (i.e., $\phi_T \ll M_T$), then the middle term dominates, and the CRLB is well approximated by $var[V_r] \geq \left(\frac{\lambda}{2}\right)^2 \sigma_f^2 \frac{2M_T}{\phi_T^2} = \frac{\lambda^2 \sigma_f^2}{2SNR_d^2}$.

With well-designed frequency (velocity) estimation algorithms optimized for low signal levels achieving performance close to that predicted by Equation 6 at low signal levels has been demonstrated. For typical coherent lidar designs considered for space-based wind measurement applications, as long as the backscatter is sufficient for measurements to be made with acceptable probability of bad estimates, the velocity precision will be 1-2 m/s. Of course, even higher precision is possible if the backscatter is well above the minimum detectable levels.

5. High-Sensitivity Space-Based Coherent Lidar

Our objective in this investigation was to conceptually design a coherent lidar that could measure winds from a space-based platform at very low backscatter levels. Past aerosol backscatter measurements imply that a 2 μm wavelength coherent lidar with backscatter sensitivity of $\sim 10^{-10} (m \cdot sr)^{-1}$ would be able to make useful measurements in the upper troposphere and lower stratosphere a large fraction of the time (possibly as high as 80-90% of the time).

In order to measure horizontal wind vectors multiple look directions are required. For space-based wind measurements, overlaid fore and aft tracks produce robust radial wind measurements along each line of sight (LOS) direction and therefore can be combined to reliably produce horizontal vector measurements along one track in the atmosphere. To allow for measurements along two tracks below the flight path, four look directions are required. Many system concepts assume multiple telescopes looking in each LOS direction, which is reasonable for small optics size but is not desirable for meter class telescopes. A feasible strawman design for a system that can achieve high backscatter sensitivity over four look directions is summarized below and will be described in more detail in the presentation.

The transmitter is an in-band-pumped Ho:LuLF laser operating at a high atmospheric transmission wavelength near 2 μm. The laser front end produces 80 mJ, 180 ns FWHM pulses at 100 Hz PRF (similar laser is currently under development under NASA's Wind-SP program). Alternate pulses from the front-end laser are optically switched into two paths with each path having a high-efficiency in-band-pumped amplifier operating at 50 Hz PRF. The high energy amplifiers require development to optimize efficiency and beam quality but similar performance has previously been demonstrated in Ho:YLF [9]. Following alternate pulse recombination, the final transmitter output is 240 mJ at 100 Hz. With optimization, the overall high-power transmitter wall plug efficiency is expected to be ~3%, so the overall spacecraft prime power required by the transmitter is <1 kW.

Lidar sensitivity is significantly enhanced by using a 1.5 m aperture diameter and 320 km orbit altitude (*same as ESA Aeolus*). An EO-tunable aperture at the telescope output based on pulse-to-pulse polarization control and liquid crystal polarization gratings, toggles the lidar line of sight at the laser PRF sequentially across four lines of sight (fore and aft views for each of the two stripes) with each line of sight operating at 25 Hz PRF. In this concept the polarization gratings are in large beam at the output of the telescope. The polarization gratings are very efficient and have been demonstrated in coherent lidar systems [10], but development is required to scale the technology to large apertures.

To predict the minimum backscatter sensitivity, we also assume a high-efficiency lidar design with $\eta_{eo}\eta_a = 16\%$, rms wind turbulence and shear in the measurement volume of 2 m/s, LOS angles that are ~ 28.4° from nadir (results in 60° elevation wrt to Earth surface), and a LOS velocity search space of ~ 60 m/s ($N_s = 11$). Achieving 16% total lidar efficiency requires that optical transmission and detector quantum efficiency be maximized, wavefront aberrations be minimized, and lag angle due to platform rotations be removed or greatly minimized. The approach planned to maximize the lidar efficiency is described in a separate paper [11]. The lidar model we use is very detailed, including the atmospheric attenuation, refractive turbulence, and antenna efficiency variation with range.

Example model results are shown in Figures 3. The minimum detectable backscatter is shown for probability of bad estimates of 0.5, 0.1, and 0.02 and for both high and low measurement resolutions. For the high-resolution measurement (red solid squares under lines) the height resolution is 250 m and the horizontal resolution is 11 km (1.5 s accumulation). For the low-resolution measurement (black squares under lines) the height resolution is 2 km and the horizontal resolution is 88 km (12 s accumulation). Also shown for comparison are various measurement-based backscatter coefficient models we use in our lidar modeling. The clean air model, solid blue line with open blue circles, represents very clean air with past measurements indicating that aerosol backscatter will be above this level 80-90% of the time. The lidar described above allows for accurate wind measurements to minimum backscatter coefficient of ~ 10^{-10}/m/sr for 1 km vertical x 88 km horizontal resolution, providing high-percentage coverage at acceptable resolution for the mid- and upper-troposphere; and to minimum of ~10^{-9}/m/sr for 250 m vertical x 11 km horizontal resolution, allowing for high resolution measurements in the boundary layer and in enhanced aerosol layers in mid- and upper-troposphere. Of course, at high backscatter levels, even higher resolution measurements are possible with the coherent lidar. For example, with only 3 shot averaging, measurements can be made to 50 m vertical x 0.88 km horizontal resolution as long as the aerosol backscatter is above ~10^{-8}/m/sr. So very high-resolution boundary layer wind measurements are possible even under very clear air conditions. The lidar architecture and additional modeling will be described in more detail in the presentation.

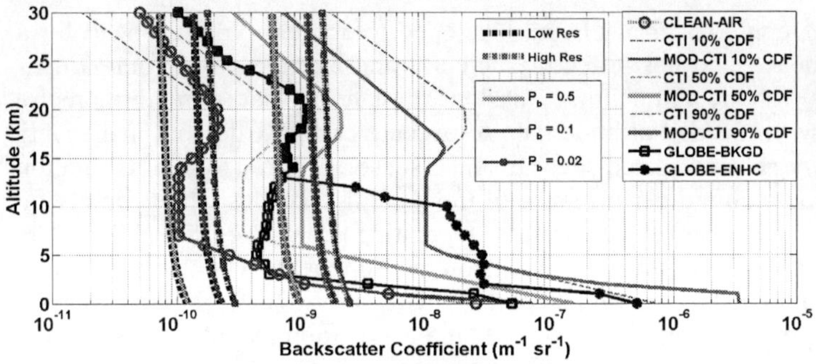

Figure 3. Predicted minimum detectable backscatter levels and measurement-based model backscatter profiles.

Figure 4. Nonmechanical beam scanning concept.

6. Acknowledgments

Michael Kavaya provide the predictions from the Frehlich paper that were included in Figure 1.

7. References

[1] R.M. Huffaker, et.al., "Feasibility studies for a global wind measuring satellite system (Windsat): Analysis of simulated performance," Appl. Opt., 23, 2523-2536, 1984

[2] S.W. Henderson, et.al, "Wind Lidar" in *Laser Remote Sensing,* T. Fujii and T. Fukuchi, Eds., CRC Taylor and Francis, 469-722 (2005)

[3] W. E. Baker, eta.al., "Lidar-Measured Wind Profiles: The Missing Link in the Global Observing System," Bulletin of the American Meteorological Society, Vol 95, No 4, (2014)

[4] A.M Dabas, et,al., "ESA wind lidar mission Aeolus", 19th Coherent Laser Radar Conference, paper Mo2, Okinawa, Japan (2018)

[5] R. Frehlich, "Velocity Error for Coherent Doppler Lidar with Pulse Accumulation," Journal American Meteorological Society, Vol 21, 905-920, (2004)

[6] D. Jacob and P. Gatt, "Defining CNR, Diversity, Window Efficiency, and Resolution of Coherent Lidar for Optimizing System Performance", in proceedings of 16th Coherent Laser Radar Conference

[7] Goodman, J.W., Statistical Optics: An Introduction, New York: Wiley, 1984.

[8] Shapiro, J.W. et.al., "Performance analysis of peak-detecting laser radars," SPIE Vol. 663, pp 38-56, (1986)

[9] H.J Strauss, et.al., "330 mJ single-frequency Ho:YLF slab amplifier," Opt. Lett., Vol 38, 1022 (2013)

[10] S. Serati, et.al., "Large-aperture, wide-angle nonmechanical beam steering using polarization gratings", *Opt. Eng*. 56(3), 031211 (Oct 27, 2016). Also see paper We7 in this conference.

[11] S.W. Henderson, "Antenna Efficiency Optimization in Coherent Lidar Systems" 19th Coherent Laser Radar Conference, paper Tu2, Okinawa, Japan (2018)

Development of a coherent Doppler lidar for precision maneuvering and landing of space vehicles

Farzin Amzajerdian(a), Glenn D. Hines(a), Diego F. Pierrottet(b), Bruce W. Barnes(a), Aram Gragossian(b), Mitchell J. Davis(a), Tak-kwong Ng(a), Alexander D. Scammell(a), Adam Ben Shabat(a), Larry B. Petway(a), and John M. Carson III(c)

(a) NASA Langley Research Center
Hampton, Virginia 23693, USA
(b) Coherent Applications, Inc.
Hampton, Virginia 23693, USA
(c) NASA Johnson Space Center
Houston, Texas 77058, U.S.

Abstract: A coherent Doppler lidar has been developed to address NASA's need for a high-performance, compact, and cost-effective velocity and altitude sensor onboard its landing vehicles. Future robotic and manned missions to planetary bodies require precise ground-relative velocity vector and altitude data to execute complex descent maneuvers and safe, soft landing at a pre-designated site. This lidar sensor, referred to as a Navigation Doppler Lidar, meets the required performance of the landing missions while complying with vehicle size, mass, and power constraints. Operating from over five kilometers altitude, the lidar obtains velocity and range precision measurements with 2 cm/sec and 2 meters, respectively, dominated by the vehicle motion. After a series of flight tests onboard helicopters and rocket-powered free-flyer vehicles, the Navigation Doppler Lidar is now being ruggedized for future missions to various destination in the solar system.

Keywords: Coherent Lidar, Doppler Lidar, Precision Navigation, FMCW Lidar

1. Introduction

Global Positioning System (GPS) is commonly used in terrestrial navigation for vehicle position and velocity knowledge. In the absence of a GPS signal, past landing missions to planetary bodies relied on radar to provide the necessary data to execute descent and landing maneuvers.[1,2] The Navigation Doppler Lidar (NDL) offers several critical advantages compared to radar, including significantly higher precision with reduced size, weight, and power. In addition, the laser-based NDL sensor does not suffer from measurement perturbation from terrain features or signal ambiguity from transmitted side lobes, and is far less susceptible to signal clutter, such as returns from the lander structure or jettisoned vehicle components such as heatshields. The higher quality data provided by the NDL will enable both a more precise navigation towards the designated landing site and a well-controlled touchdown (with greater stability and lower impact loads). This translates into lower fuel reserve and smaller leg span for a landing vehicle and thus further reduces vehicle mass, therefore the NDL can potentially reduce the overall cost and risk of landing missions and enable new capabilities for planetary exploration missions including missions to the Moon, Mars, asteroids, and planetary moons.

2. NDL System Description

The principle of the lidar operation is illustrated in Fig. 1 showing the modulation waveform consisting of three segments: up-ramp chirp, constant frequency, and down-ramp chirp. The resultant returned waveform from the target is delayed by t_d, the light round trip time. When the target or the lidar platform is not stationary during the beam round trip time, the signal frequency will be also shifted up or down, depending on the velocity direction, due to the Doppler effect. When mixing the two waveforms at the detector, an interference signal is generated whose frequency is equal to the difference between the transmitted and received frequencies. In absence of velocity along the laser beam, the signal frequency during the "up-ramp" and "down-ramp" periods are equal and their magnitude is directly proportional to the distance to the target. When the vehicle is moving, the up-ramp and down-ramp frequencies will not be equal and their difference is related to the Doppler velocity. The target range and magnitude of the velocity component along the laser beam are determined through the following simple equations:

$$R = \left(\frac{TC}{2B}\right)\left(\frac{f_{IF}^{+} - f_{IF}^{-}}{2}\right) \qquad V = \left(\frac{\lambda}{2}\right)\left(\frac{f_{IF}^{+} + f_{IF}^{-}}{2}\right)$$

where f_{IF}^{+} and f_{IF}^{-} are the intermediate up-ramp and down-ramp frequencies, B is the modulation bandwidth, T is the waveform period, C is the speed of light, and λ is the laser wavelength. The constant frequency segment also produces a Doppler velocity that can be used for eliminating data dropouts when either up-ramp or down-ramp frequency is very close to zero, and allows for eliminating the measurement ambiguities that may arise in certain scenarios.

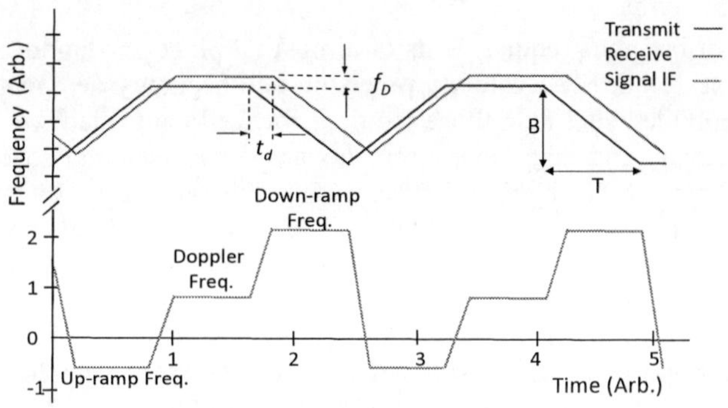

Figure 1. Linearly frequency modulated transmitted beam and returned signal, and the resulting intermediate frequency (IF) of the homodyne signal.

The NDL transmits three laser beams at different pointing angles toward the ground to measure range and velocity along each beam using a frequency modulated continuous wave (FMCW) technique.[1] The three line-of-sight measurements are then combined in order to determine the three components of the vehicle velocity vector and its altitude relative to the ground. Figure 2 shows the NDL prototype system developed for a series of flight tests to demonstrate its capabilities and viability for future landing missions to planetary bodies. The prototype consists of an electronic chassis and an optical head. All the lidar components including the transmitter laser, receivers, and signal processor, are housed in the electronic chassis. The optical head consists of three transmit/receive lenses connected to the chassis via a long fiber optic cable.

Figure 2. Prototype NDL.

3. Projected Performance

The NDL performance were characterized over different phases of its development through ground tests, helicopter flight tests, and onboard autonomous rocket-powered test vehicles (Fig. 3) while operating in open and closed-loop with a guidance, navigation, and control (GN&C) system.[2,4] The results of these tests helped optimize the NDL design leading to the development of a compact and robust prototype unit shown in Fig. 2.

Figure 3. Morpheus vehicle (left), built by NASA-JSC, and Xodiac vehicle (right), built by Masten Space Systems, used for demonstration of NDL capabilities.

The NDL meets the performance requirements of almost all planetary landing missions being considered for lunch over the next decade. The velocity precision and accuracy demonstrated in the flight tests are about an order of magnitude better than those required for well-controlled soft landing. The resolutions of line-of-sight (LOS) velocity and range measurements are 0.2 cm/sec and 25 cm, respectively, limited by the FFT processing algorithm. The NDL measurements precision, defined as random noise about its mean, was determined experimentally using the flight tests data. The LOS velocity and range precisions are about 1.7 cm/sec and 2.2 m, which were predominantly due to vehicle vibration and rapid angular motions. The NDL can measure velocities up to 200 m/sec along the LOS of its beams. The actual vehicle velocity can be significantly higher than 200 m/sec since there will always be an angular separation between the laser beams and the vehicle flight path. As a result, the NDL can measure vehicle velocities of well over 1000 km/hr.

The maximum operational range of the NDL is harder to define since it is highly dependent on the atmospheric environment and the surface albedo. To better assess the NDL performance in different terrestrial and planetary environments, a set of operational range measurements were performed on the Langley Air Force Base runway which provides a relatively long path length. For these tests, the NDL was placed at one end of the runway pointing one of its beams toward a calibrated target on the back of a moving box-car truck. The targets used for these measurements have a diffuse Lambertian surface and placed nearly orthogonal to the NDL beam's optical axis. Range, velocity, and return signal magnitude data were collected continuously as the truck travelled from 250 meters range to 3.8km.

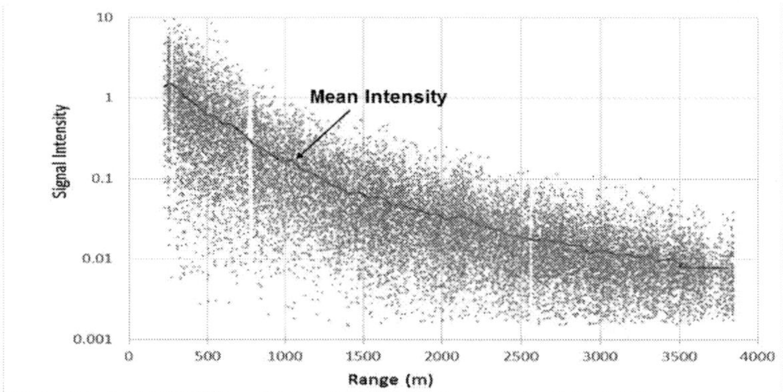

Figure 4. Signal intensity versus range.

Fig. 4 provides an example of signal intensity data as a function of LOS range using a 50% reflectance target. The blue dots are the individual measurements, showing fluctuations due to speckle noise, and the solid line shows the mean of signal intensity data. Fig. 5 shows the lidar equation fit to measured data indicating a terrestrial maximum operational range of over 4 km in most atmospheric conditions and terrains. This is a conservative estimate considering that the measurement were performed over a horizontal path of approximately 1.5m above the runway's concrete surface and in 70% humidity. Using these results, the projected maximum LOS operational range in the Moon and Mars, under nominal atmospheric and surface albedo, were estimated to be 7.5 km and 5.7 km, respectively.

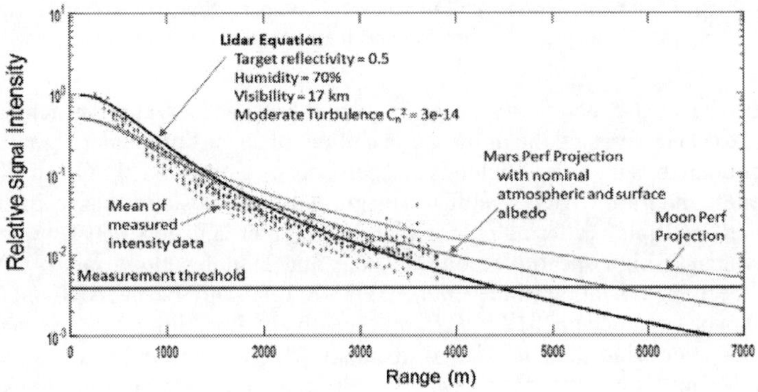

Figure 5. Estimated maximum operational range for operation on Earth, Moon, and Mars.

4. Technology Advancement towards Spaceflights

The prototype lidar described above was designed with the vision of its eventual integration and operation from a planetary landing vehicle. The design of the prototype unit considered the size and mass constraints of landing vehicles and the harsh thermal and vacuum environments of space while meeting the performance requirements for a wide range of landing scenarios. This prototype unit is now serving as a testbed for the development of a spaceflight Engineering Model (EM) of the NDL intended to reduce the risk and cost of building spaceflight units. The EM is being designed to meet the thermal, vacuum, and radiation environments of space and be able to tolerate the launch vibration loads. Efforts are underway to build the components of the EM using space-grade or spaceflight-qualifiable parts and validating their functionality, robustness, and reliability for operation in space. Once the EM is fully assembled, it will be subjected to extensive spaceflight-qualification tests. The EM will be then integrated with other landing vehicle avionics for a series of tests that will probably include aircraft and/or a rocket-powered free-flyer test vehicle prior to building the spaceflight units.

5. References

[1] F. Amzajerdian, D. F. Pierrottet, G. D. Hines, L. B. Petway, B. W. Barnes, J. M. Carson III, "Development and Demonstration of Navigation Doppler Lidar for Future Landing Mission," AIAA Space Forum, 10.2514/6.2016-5590 (2016)

[2] V. E. Roback , D. F. Pierrottet, F. Amzajerdian, B. W. Barnes, G. D. Hines, L. B. Petway, P. F Brewster, K. S. Kempton, and A. E. Bulyshev, "Lidar sensor performance in closed-loop flight testing of the Morpheus rocket-propelled lander to a lunar-like hazard field," Proc. of AIAA Science and Technology Forum and Exposition (2015)

[3] Pollard, B. D., and Chen, C. W., "A Radar Terminal Descent Sensor for the Mars Science Laboratory Mission," IEEE Aerospace Conference, paper no.1597 (2008).

[4] Pierrottet, D. F., Amzajerdian, F., Petway, L. B., Hines, G. D., and Barnes, B. W., "Field Demonstration of a Precision Navigation Lidar System for Space Vehicles," AIAA GN&C Conference, 10.2514/6.2013-4717 (2013).

Next-generation-synergetic-observation-system for the unified analysis of CALIPSO, ADM-Aeolus and EarthCARE-ATLID

Hajime Okamoto(a), Kaori Sato(a), Shoken Ishii(b), Makoto Aoki(b), Tomoaki Nishizawa(c) Nobuo Sugimoto(c), Yoshitaka Jin(c), Yuichi Ohno(b), Hiroaki Horie(b), Masahiro Fujikawa (a)

(a) Research Institute for Applied Mechanics, Kyushu University
Kasuga Fukuoka 816-8580 Japan
(b)National Institute of Information and Communications Technology, Tokyo 184-8795 Japan.
(c) National Institute for Environmental Studies, Ibaraki 305-8506 Japan
okamoto@riam.kyushu-u.ac.jp

Abstract: The next-generation-synergetic-ground-based-observation-system is being developed to (1) understand the information content of space borne lidar signals and (2) construct continuous data sets of clouds and aerosols by using CALIPSO and CloudSat, ADM-Aeolus and EarthCARE satellites. The ground-based system consists of multiple-field-of-view multiple-scattering polarization lidar, direct- and coherent-Doppler lidars, multi-wavelength high spectral resolution lidar and Doppler cloud radar. CALIPSO lidar has been in operation since 2006. ADM-Aeolus and EarthCARE-ATLID are scheduled to be launched in 2018 and 2020, respectively. In addition, future space-borne coherent Doppler wind lidar is studied in Japan. The new ground-based system will offer unique opportunity to fill the gaps in different observation conditions among space-borne lidars. We also discuss the retrieval algorithms of clouds and air motion for the ground-based and space borne systems.

Keywords: Doppler lidar, multiple scattering lidar, high spectral resolution lidar, radar, EarthCARE

1. Introduction

Cloud-Aerosol Lidar and Infrared Pathfinder Satellite Observations (CALIPSO) has started its observations in June 2006 with other satellites such as 94GHz cloud profiling radar on CloudSat in A-Train orbit. These active sensors have provided the three-dimensional picture of clouds, aerosols and radiative properties, which are the large source of uncertainties in climate sensitivity projections. ESA's ADM-Aeolus will be launched in August in 2018. It will carry the Atmospheric Laser Doppler Instrument (ALADIN) that is a direct detection Doppler wind lidar operated at 355nm. It will provide the first Doppler wind information at angle of 35 degrees off-nadir and also provide extinction and backscattering coefficient. JAXA-ESA joint satellite mission EarthCARE is expected to be launched in 2020. It carries the Atmospheric Lidar (ATLID), the Cloud profiling radar (CPR) with Doppler function, the Multi-Spectral Imager (MSI), and the Broad Band Radiometer (BBR) [1]. ATLID will be the HSRL with depolarization ratio at 355nm and CPR will be the first 94GHz Doppler cloud radar in space. These active and active sensors are expected to provide much more reliable signatures of clouds and aerosols with dynamical features of atmosphere.

Characteristics of cloud signals appears to be different for different ground-based lidars and space-borne lidar observations. Conventional ground-based lidar cannot penetrate optically thick clouds. CALIPSO has a much larger footprint size (~90m) compared to that of conventional ground-based lidar (smaller than a few meters at 1km height). Large foot print size of space-borne lidar enables to collect more multiple scattered signals and can penetrate optically thicker clouds. There are also differences in foot print sizes among CALIPSO lidar, ADM-Aeolus ALADIN and EarthCARE ATLID. These differences should produce different lidar signals for the clouds with same macroscale and microphysical properties. It is therefore not sufficient to reproduce spaceborne lidar signals by conventional ground-based lidars. In order to overcome the situation, we recently have introduced a lidar system that can detect backscattering coefficients and depolarization ratios similar to those observed by space-borne lidar. It has been extended to be 10 channel-system.

The next-generation-synergetic-ground-based-observation-system is being developed. The ground-based system consists of multiple-field-of-view multiple-scattering polarization lidar, direct- and coherent-Doppler lidars, multi-wavelength high spectral resolution lidar and Doppler cloud radar. We will conduct synergetic observations in Koganei, Tokyo. These data sets will provide unified view for analyzing these three different satellites and will be used to develop and evaluate algorithms for CALIPSO and Cloudsat, Aeolus ALADIN and EarthCARE ALTID-CPR.

2. Results from the next generation observation system

We first describe the multiple-field-of-view multiple-scattering polarization lidar (MFMSPL) system. It has originally eight channels (four parallel, four perpendicular) with detectors. Each receiver channel has a 10 mrad FOV and a total FOV is 70mrad with polarization function [2]. Odd-numbered channels detected parallel signals. The tilt angles of Ch.1, 3, 5 and 7 from the vertical direction were 0, 10, 20 and 30 mrad, respectively. Even-numbered channels were used to detect perpendicular signals. The angles of Ch.2, 4, 6, and 8 were 0, 10, 20 and 30 mrad, respectively. Observations of attenuated backscattering coefficient β_{att} of low-level water clouds by channels 1 and 3 are shown in Fig. 1. Cloud top height detected by off-beam channel (Ch.3) is larger than that by on-beam channel because the off-beam could collect more multiple scattered photons. Synergetic observations with cloud radar revealed that the agreement in cloud top height with cloud radar can be achieved with off-beam channel than with on-beam channel. It is also noted that the cloud top height by the conventional lidar is lower than that by the MFMSPL.

Depolarization ratio for the four tilt angles was estimated. Only the results of depolarization ratio for 0 and 10 mrad are shown in Fig. 1. Depolarization ratio from off-beam channel for 10 mrad is generally larger than that for 0 mrad. The value tends to be smaller near cloud bottom and gradually increased as height increased. The system will be extended for 355nm and 1064nm. Thus the extended version of the MFMSPL would be used to develop the algorithms for ADM-Aeolus, ATLID and CALIPSO.

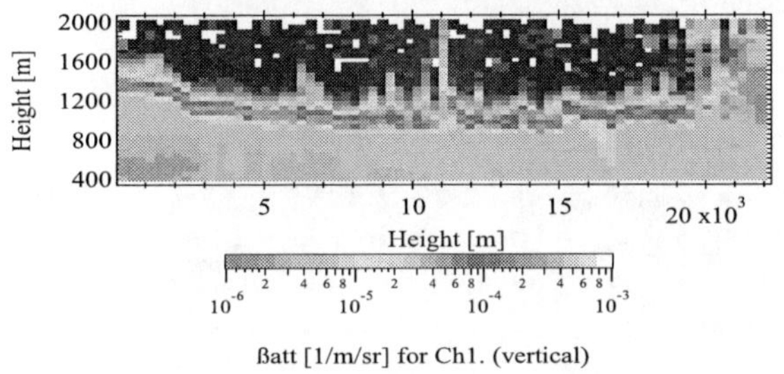

(a)

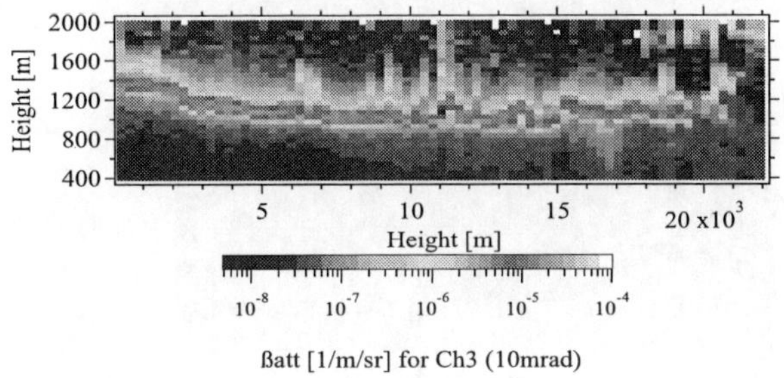

(b)

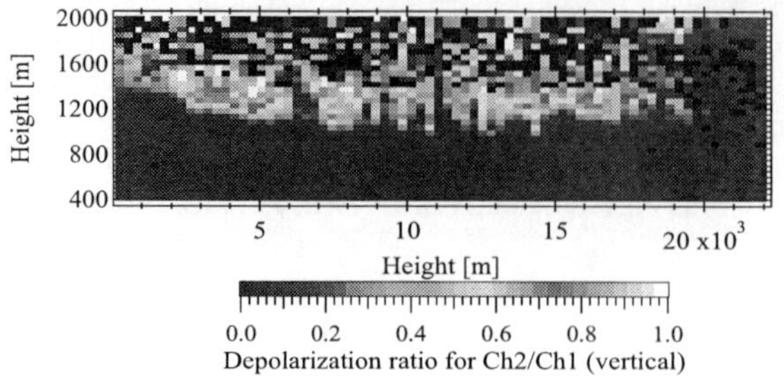

(c)

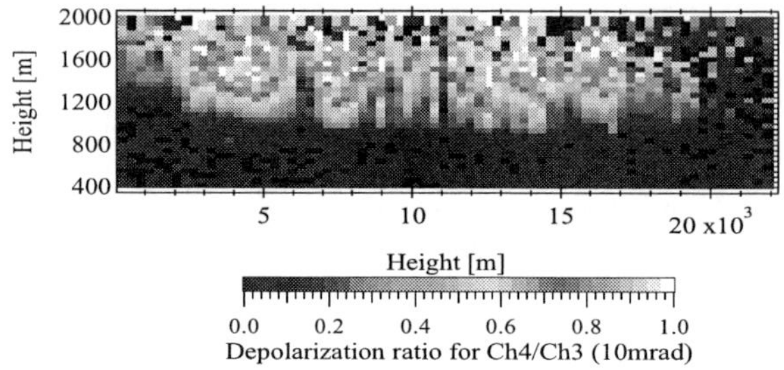

(d)

Figure 1. Attenuated backscattering coefficient (β_{att}) and depolarization ratio (δ) at 532nm detected by the Multiple-field-of-view Multiple-scattering Polarization Lidar (MFMSPL). (a) β_{att} for Channel 1, (b) β_{att} for Channel 2, (c) δ estimated by Channels 1 and 2, and (d) δ estimated by Channels 3 and 4.

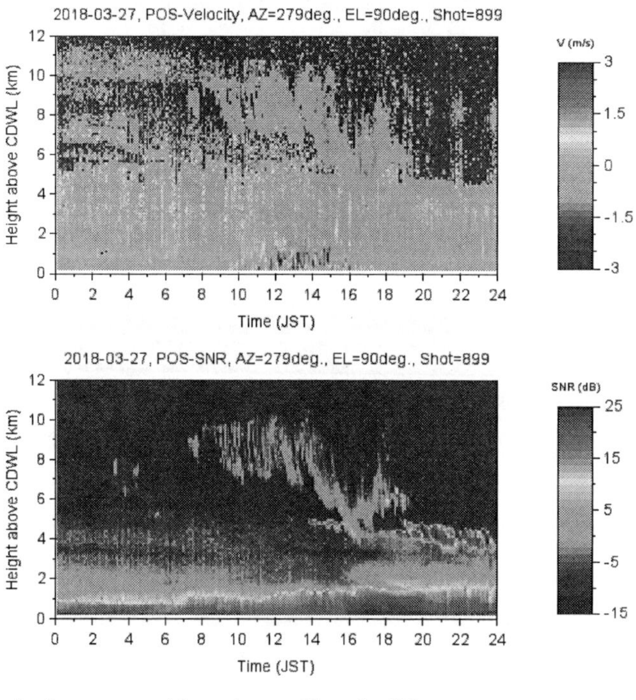

Figure 2 Doppler velocity measured by coherent Doppler lidar.

ALADIN on ADM-Aeolus and CPR on EarthCARE will provide Doppler information at 355nm and at 94GHz, respectively. The vertical air motion and terminal velocity of cloud and precipitation particles are expected as JAXA L2 products for EarthCARE. Retrieval of the two values may be achieved by simultaneous retrieval of cloud microphysics and vertical air motion [3]. Fig. 2 shows vertical velocity of atmosphere, aerosols and clouds by coherent Doppler lidar developed by NICT. The dual Doppler measurements by Doppler lidar and cloud radar would be suitable to evaluate and characterize errors in the space-borne algorithm errors. We plan to conduct Doppler direct and coherent lidar observations with Doppler cloud radar HG-SPIDER for the purpose.

Treatment of the multiple scattering in the lidar and radar signals can be performed by Physical Model (PM) [4]. Applicability of the PM was examined by the backward Monte Carlo method [5]. The PM is implemented in the satellite algorithms for the retrieval of cloud microphysics.

3. Summary

Satellite algorithms consist of (1) Cloud and aerosol mask [6], (2) Cloud and aerosol particle type[7,8], (3) Cloud and aerosol microphysics [9], and (4) vertical air motion for EarthCARE. We test (1) by the extended version of the MFMSPL. Evaluation of the (2) can be tested by using multi-wavelength HSRL. Evaluation of (3) can be achieved by the synergy use of MFMSPL, multi-wavelength HSRL and HG-SPIDER. We plan to evaluate (4) by direct detection Doppler lidar, coherent lidar and HG-SPIDER.

4. Acknowledgements

This work is supported by JSPS kakenhi JP17H06139 and JP18K03745; the Japan Aerospace Exploration Agency for EarthCARE Research Announcement; The Arctic Challenge for Sustainability (ArCS); Collaborated Research Program of Research Institute for Applied Mechanics, Kyushu Univeristy.

5. References

[1] A. Illingworth et al., "The EarthCARE Satellite: The Next Step Forward in Global Measurements of Clouds, Aerosols, Precipitation, and Radiation" BAMS **96**(8), 1311-1332 (2015).

[2] H. Okamoto et al., "Development of a multiple-field-of-view multiple-scattering polarization lidar: comparison with cloud radar" Opt. Express **24**(26), 30053-30067 (2016).

[3] K. Sato et al., "95-GHz Doppler radar and lidar synergy for simultaneous ice microphysics and in-cloud vertical air motion retrieval" J. Geophys. Res. **114**, D03203, 1-17 (2009).

[4] K. Sato, H. Okamoto and H. Ishimoto, "Physical model for multiple scattered space-borne lidar returns from clouds" Opt. Express **26**(6), A301-A319 (2018).

[5] H. Ishimoto and K. Masuda, "A Monte Carlo approach for the calculation of polarized light: application to an incident narrow beam", J. Quant. Spectrosc. Radiat. Transf. **72**(4), 467-483 (2002).

[6] Y. Hagihara, H. Okamoto and R. Yoshida, "Development of a combined CloudSat-CALIPSO mask to show global cloud distribution" J. Geophys. Res., **115**, D00H33, 1-17 (2010).

[7] R. Yoshida, H. Okamoto, Y. Hagihara and H. Ishimoto, "Global analysis of cloud phase and ice crystal orientation from Cloud-Aerosol Lidar and Infrared Pathfinder Satellite Observation (CALIPSO) data using attenuated backscattering and depolarization ratio" J. Geophys. Res., **115**, D00H32, 1-12 (2010).

[8] M. Kikuchi et al., "Development of algorithms for discriminating hydrometeor particle types with a synergestic use of CloudSat and CALIPSO" J. Geophys. Res., **122**, 10.1002/2017JD027113, 1-23 (2017).

[9] H. Okamoto, K. Sato and Y. Hagihara, "Global analysis of ice microphysics from CloudSat and CALIPSO: incorporation of specular reflection in lidar signals", J. Geophys. Res., **115**, D22209, 1-20 (2010).

Nonmechanical Step and Stare Telescope for Coherent Laser Radar Applications

S. Serati[1], C. Hoy[1], D. McKnight[1], L. Hosting[1], K. Kluttz[1], J. Stockley[1] and C. Hale[2]

[1]Boulder Nonlinear Systems, 450 Courtney Way, Lafayette, CO, USA 80026
[2]Beyond Photonics, 1650 Coal Creek Dr., Suite B, Lafayette, CO, USA 80026

Abstract: Step and stare operation is desirable for several laser radar applications, since it allows the sensor system to dwell and collect returns from multiple laser shots, reducing the pulse energy requirements on the laser transmitter. Also, a static look angle through transmit and receive eliminates the need for lag-angle compensation. For most applications, it is also desirable to maximize dwell time and minimize transition time between look angles. For any application, the scanning telescope needs to provide sufficient aperture and scan over the required field of regard. This paper discusses the development of low-mass, low-power, step-and-stare telescopes that use low-voltage, low-frequency electro-optic switching (i.e. no moving parts) to steer a large effective aperture (potentially greater than 15 cm) over wide angles (± 45 degrees or larger). Currently, the technology is being integrated into different coherent laser radar systems. Results from these efforts are presented.

Keywords: Nonmechanical beam steering, switchable diffractive optical element, random access steering, step and stare scanning, coherent lidar

1. Introduction

Step and stare operation is reported to be the best method for measuring line-of-sight wind speed from multiple look angles to determine the wind speed and direction in a target volume from a space-borne platform [1-3]. This is based on requirements analysis and performance trade studies conducted over the years. One obvious advantage to this type of operation is the ability to dwell and collect multiple shots from one look angle, which through averaging the multiple returns improves the signal-to-noise ratio (SNR). For example, a system that collects and averages the returns from 60 shots will improve SNR by approximately eight times if the returns come basically from the same volume sample. This SNR improvement significantly reduces the pulse energy requirement on the laser transmitter and thus reduces the size, weight and power (SWaP) and cost of the laser transmitter. The drawback is that step and stare scanning is a SWaP and cost issue itself, since quickly accelerating and deaccelerating large, massive optics takes considerable power and these scanning motions require mechanisms on the platform to counterbalance the forces generated. To mitigate some of these scanning problems, techniques such as using multiple separate telescopes [2] have been considered. Another approach is to combine multiple holographic or diffractive optical elements (HOEs or DOEs) into a single large aperture optic to discretely steer a beam [3,4]. The Shared Aperture Diffractive Optical Element (ShADOE) Telescope developed by Goddard Space Flight Center is one such device.

The technique discussed in this paper also uses DOEs, but it uses a different addressing method than the one used in the ShADOE Telescope. In the ShADOE approach, multiple DOEs are patterned into a single optical element, which are then angle addressed using a relatively small rotating mirror to direct an expanding beam into the DOEs from different directions [3]. The DOE's then collimate the beam and redirect it into a 40? cone angle. In our approach, we use DOEs produced using UV holography to create a polarization grating (PG) film. These solid liquid crystal (LC) films are a few microns thick and produce nearly continuous phase shifts that emulate glass wedges or lenses, but the PG film produces a negative phase shift for left circularly polarized light and a positive phase shift for right circularly polarized light. These PG properties allow us to combine film layers to make a nonmechanical, low-SWaP, step and stare telescope. In such a telescope, the expanding beam is brought in on-axis where a large-aperture PG lens collimates the beam and a series of low-power LC switches and uniaxial PGs deflect the beam into different steer angles (see Fig. 1). This technique eliminates all moving parts and simplifies the beam-delivery optical train. Also, liquid crystal polarization grating (LCPG) scanners produce beams without introducing much wavefront distortion, which allows them to be used in coherent laser radar systems (see Section 3). However, fabrication of large-aperture LCPG components is still in its infancy, and nonmechanical beam scanners (NMBS) with clear apertures of 10 - 15 cm (Fig. 2) are the current state of the art [4].

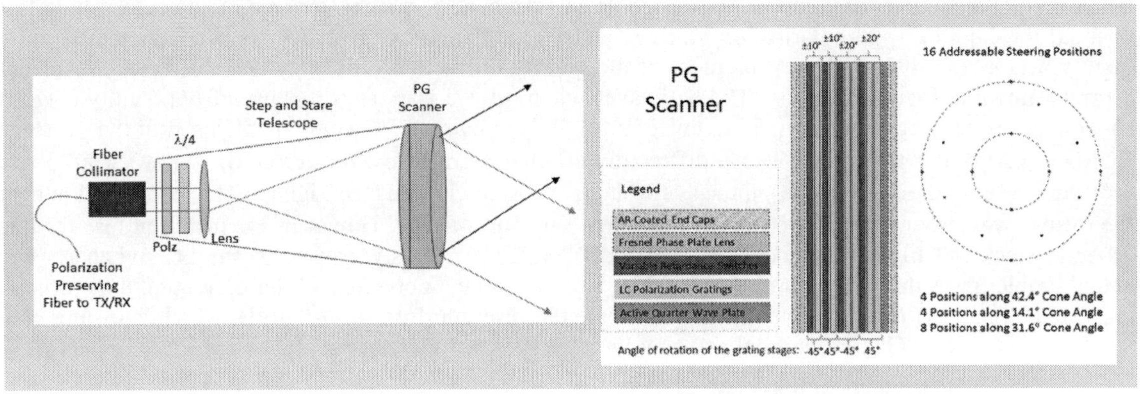

Figure 1. Nonmechanical step and stare telescope using a PG lens to further reduce SWaP.

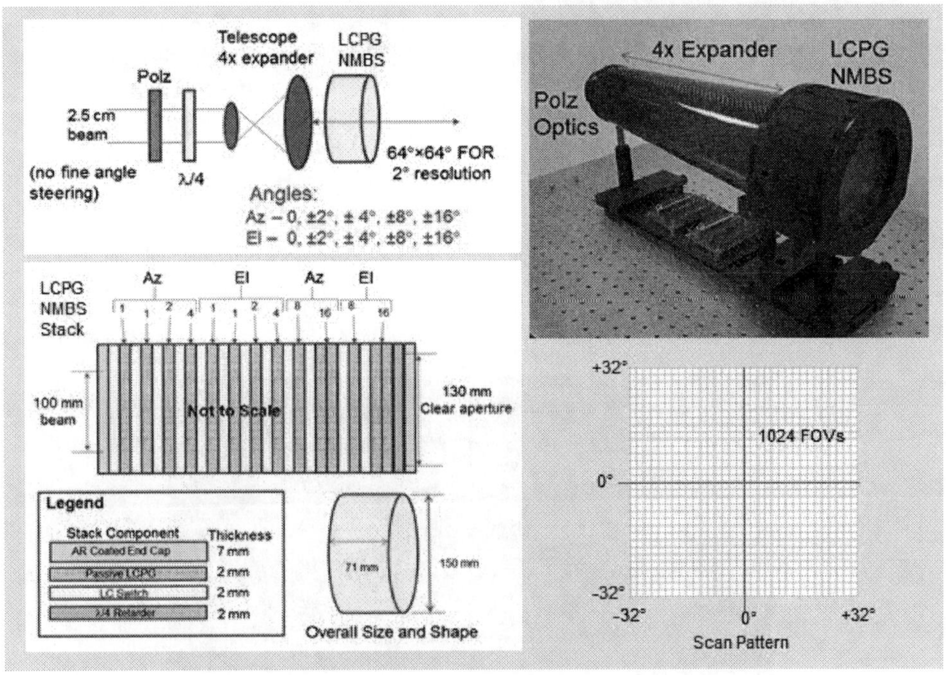

Figure 2. Large aperture (13 cm), wide angle (64°×64°) LCPG NMBS prototype.

2. Large-aperture, Wide-angle LCPG NMBS Development

The LCPGs have successfully been implemented in several steering systems providing precision wide-angle beam control for active and passive systems. Because of the significant improvement in SWaP, this technology is strongly being considered for a variety of applications where platform constraints limit the overall size of the optical system, but large-aperture, wide-angle capability is still essential. To demonstrate this type of capability, we developed a 1550 nm demonstration beam deflector system, which has a13-cm clear aperture and steers over a 64° x 64° field of regard with 2° resolution (see Fig. 2).

The goal was to provide large-aperture performance without significantly sacrificing optical quality. To better predict the impact of the LCPG components on system performance, a range of characterization test on both the liquid crystal polarization gratings and the halfwave LC switches were conducted. Grating quality was assessed through measurement of the grating diffraction efficiency and diffracted wavefront error as shown in Figures 3 and 4. The halfwave switches were also tested for wavefront quality (Fig. 4) as well as response time and switching uniformity. The individual PG components used in the 12-stage LCPG stack all had better than 99.5% diffraction efficiency and wavefront error was between λ/25 and λ/50 rms. For all the LC switches, the wavefront error was negligible (see Fig. 4). This combination kept the overall wavefront error of the steered beam to ~λ/10 rms or less. However, the insertion loss for this 12-stage stack was higher than expected (a few dB), which was mostly caused by the LC switches. The actual problem was due to scatter in the switches, which was later corrected. After correcting the problem, a 3-stage large-aperture stack was assembled where the insertion loss was ~0.4 dB, which is in-line with smaller aperture LCPG stack assemblies being built today.

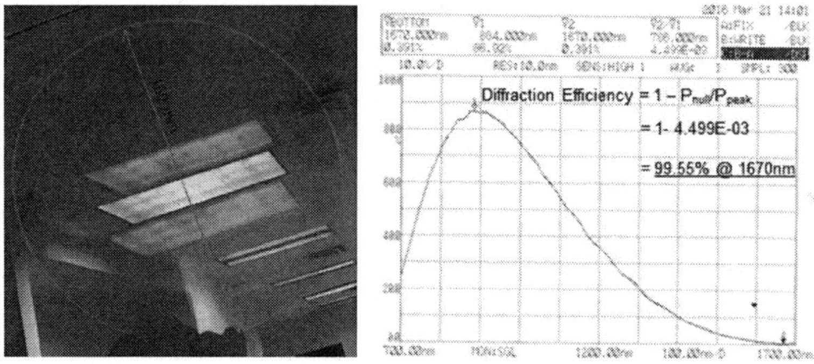

Figure 3. Large aperture LCPG on 15 cm substrate.

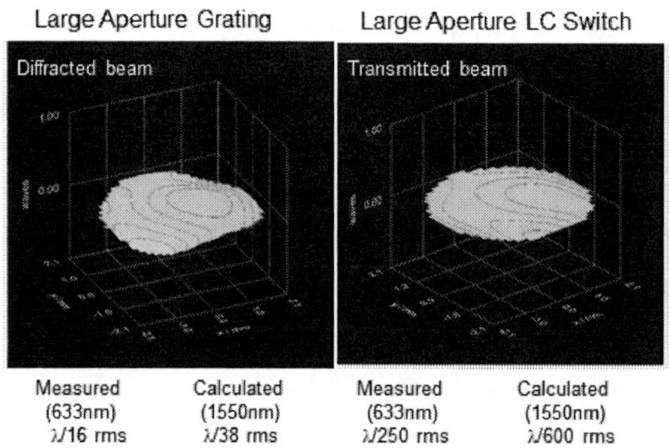

Figure 4. Wavefront measurements of 15 cm LCPG and LC switch.

3. Coherent Laser Radar Compatibility

A smaller LCPG beam steering stack, as shown in Fig. 5, was used to test scanner performance as part of a monostatic coherent laser radar system. This stack consisted of two ±1° LCPGs oriented orthogonally to achieve four accessible steering angles in two axes. Each grating is paired with a LC switch, with a third LC switch added to the exit aperture and tuned to behave as a quarter wave plate. Anti-reflection coated end caps were added to each end face to reduce Fresnel losses. All components were aligned, epoxied at points along the circumference to mechanically mate the components and then back-filled with index-matching optical adhesive and cured.

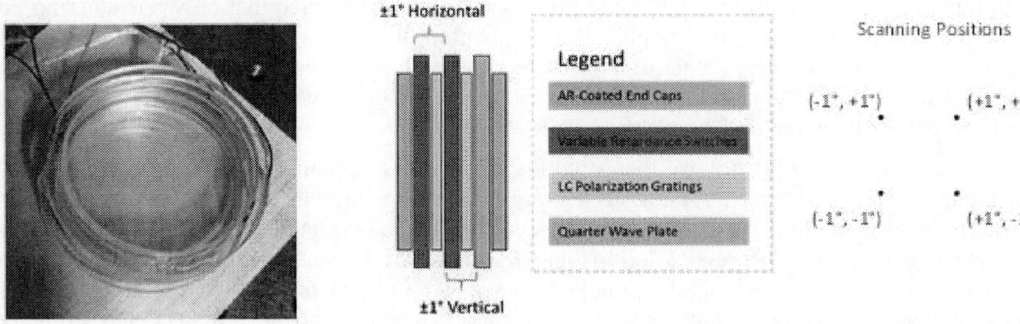

Figure 5. LCPG stack used to test beam steering performance in a monostatic coherent laser radar.

First, the wavefront of the steered beams were analyzed with the Shack-Hartmann wavefront sensor, and the measured RMS wavefront error at 1550 nm was found to vary between λ/61.2 to λ/84.5 depending on the steering angle. This variation in wavefront error arises because the wavefront errors of each individual grating are partially correlated. This correlation occurs because the wavefront error is written into the grating during the fabrication process and the gratings were fabricated with the same patterning beam configuration. As a result, when each grating is steering to its +1 order, they introduce very similar wavefront distortions and the overall wavefront error is more than the single-grating error. Conversely, when one grating steers to its +1 order and the other to its -1 order, the wavefront errors partially cancel and the overall wavefront error is less than the single-grating error. Because the measured errors are centered about the measured single-grating error of λ/76.5, we conclude that the index-matching and end caps do not introduce any significant wavefront error of their own. Importantly, the RMS wavefront errors of λ/61.2 to λ/84.5 indicate that the complete beam steering stack have negligible impact on the heterodyne detection efficiency of coherent laser radar systems.

In addition to the wavefront error, the stack was tested for switching speed and insertion loss. The 10%-90% rise and 90%-10% fall switching speeds were measured to be 5 ms and 1 ms respectively, which is sufficiently short compared to the ~1 s averaging times usually used in 3D wind sensing applications. The insertion loss, given by the amount of light steered into a desired order divided by the amount of incident light, was measured to be 0.85 dB for this stack. This insertion loss is likely due primarily to absorption in the ITO transparent conductor and residual Fresnel reflections due to the difference in refractive index between the ITO and the glass substrates. Since the fabrication of this device, we have improved our fabrication procedures and are now building stacks with four stages (i.e. 16 steering positions) that have ~0.6 dB insertion loss. We believe there is still some improvement to be had by reducing ITO absorption and Fresnel reflections at the ITO layer.

Hard target Doppler testing of the beam steering stack was conducted by focusing the laser beam coming out of LCPG stack onto a spinning disc target at a distance of 1.27 m. Since the lidar was configured for monostatic operation, the return signal from the target was steered back on axis by the LCPG stack and into a fiber-coupled heterodyne receiver. Individual measurements of 4000-sweep averages were taken with the beam steered into each of the four available steering angles and were compared with a 4000-sweep measurement of the target without the beam steering stack in the system (baseline control measurement).

The CNR measurements at the four available steering angles were 28.9 dB, 29.1 dB, 29.8 dB, and 29.4 dB, which are very close to the 29.7 dB CNR measured without the LCPG stack. The CNR measurements through the beam steering stack are somewhat higher than expected given the insertion loss measured in the laboratory, which we attribute to either an overly conservative loss measurement in the laboratory (possibly due to poor degree of polarization) and/or scatter in the measured CNR values due to measurement precision and repeatability.

Poorly polarized light can result in "off-angle light" leaving the stack at one of the undesired steering angles. This corresponds to the undesired polarization component in the light following an orthogonal steering path. In this experiment, the steering angles of the LCPG stack were such that only one steering position could be incident on the target for any given target position. In other words, if the target was positioned such that it was hit by any given steering angle, steering the beam to any of the other 3 available angles would steer the beam off the target. This allowed us to rule out the possibility of low-level light in any of the "off" angles contributing to the measured CNR.

To investigate the concept further, we measured the CNR with the beam steered onto the target and then measured the CNR with the beam steered off the target. In this manner, the contribution of any "off-angle" light that is incident on the target should be apparent. In this experiment, we could identify no Doppler signal associated with any "off-angle" light (Fig. 6). This result demonstrates an important feature of the LCPG technology, in that the highly polarization-sensitive nature appears to provide discrimination eliminating contributions from any spurious low-energy beams. This is not necessarily the case in conventional wedge scanners, where low-energy ghost reflections can contaminate the signal.

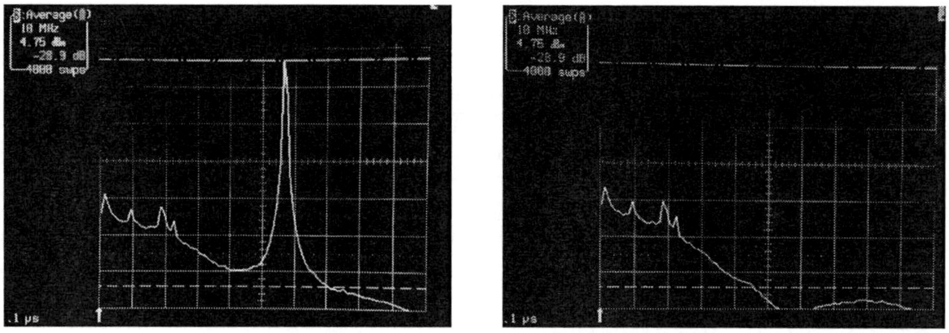

Figure 6. CNR measurements through the two-axis LCPG stack with beam steered on and off target.

4. Conclusion

Large-aperture, wide-angle LCPG step and stare scanners with good optical quality can be built and this beam steering technology shows no measurable degradation of the Doppler lidar CNR beyond what may be expected from the minimal insertion loss of these high-efficiency components.

5. References

[1] P. Hays, M. Dehring, L. Fisk, P. Tcoryk, I. Dors, J. Ryan, J. Wang, M. Hardesty, B. Gentry and F. Hovis, "Space-based Doppler Winds LIDAR: A Vital National Need," RFI response to National Research Council (May 2005).

[2] U.N. Singh, "Two Micron Laser Technology Advancements at NASA Langley Research Center," http://geodesy.unr.edu/hanspeterplag/library/IGARSS2010/pdfs/4229.pdf

[3] G. Schwemmer, B. Gentry, C. Trout-Marx, P. Dogoda, S. Lee and R. Rallison, "Shared Aperture Diffractive Optical Element (ShADOE) Telescope for LIDAR," https://www.researchgate.net/publication/228584697 uploaded May 2014.

[4] S. Serati, C. Hoy, L. Hosting, K. Kluttz, J. Stockley and C. Hale, "Large-aperture, wide-angle nonmechanical beam steering using polarization gratings," *Opt. Eng.* **56**(3), 031211 (March 2017).

High Power, Uncooled InGaAs Photodiodes with High Quantum Efficiency for 1.2 to 2.2 Micron Wavelength Coherent Lidars

Shubhashish Datta and Abhay M Joshi
Discovery Semiconductors, Inc.
119 Silvia Street, Ewing, New Jersey 08628, USA
sdatta@discoverysemi.com

Abstract: Coherent lidars require photodiodes having high quantum efficiency (QE), which operate with a high local oscillator power for shot noise limited detection. We report an uncooled, lattice-mismatched InGaAs photodiode module for lidars operating at 1.2 μm to 2.2 μm wavelength. The photodiode is coupled to a 105 μm diameter core fiber to maximize its free-space coupling efficiency. The fiber-coupled module exhibits a responsivity of 1.0 A/W (80% QE) at 1550nm wavelength and 1.2 A/W (73% QE) at 2050 nm wavelength. This device demonstrates a linear performance up to a photocurrent of 65 mA, and can deliver up to +14 dBm of continuous-wave RF output power, i.e. 3.1 V peak-to-peak amplitude. Thus, no further RF amplification is needed to maximize the effective number of bits for the backend digitizers. These photodiodes have a -3 dB bandwidth of 1.7 GHz and are adequate for frequency modulated continuous wave and nanosecond-scale pulsed lidars.

Keywords: extended InGaAs, high power, multimode

1. Introduction

Coherent lidars enabled by high power lasers promise better range and spatio-temporal resolution than their direct detection counterparts. These systems have leveraged Erbium doped fiber lasers and amplifiers, which were developed for 1550 nm wavelength telecom industry. Innovations in Thulium and Holmium doped high power lasers and optical amplifiers have opened up the 2 micron wavelength applications [1]. These lasers are available in both continuous wave (CW) and pulsed formats, and support a diversity of lidar designs [2 - 5]. Fully exploiting the coherent gain and speed afforded by these laser technologies requires a fast, high-power photodetector that demonstrates high quantum efficiency (QE) at these wavelengths.

An additional requirement for photodetectors used in lidar system is maximizing the optical collection area to boost the free-space coupling efficiency. Large-area photodiodes packaged with a free space optical interface have been historically used in such systems. These lidars require mechanically stabilized optical elements, including lenses and mirrors, to ensure a consistent optical coupling efficiency into the photodetector. The stability requirements are especially stringent when a symmetrical optical illumination profile is needed to boost the power handling and linearity of photodiodes [6, 7]. Photodiodes packaged with a large-core multimode fiber interface offer an attractive alternative by providing a consistent optical coupling efficiency and illumination profile without requiring stabilized free-space optical elements. The consequent reduction in the size and weight of the lidar is especially crucial for aerospace applications, which are not only sensitive to SWaP considerations, but also require sub-systems to withstand elevated levels of mechanical shock and vibration.

We report an uncooled, lattice-mismatched InGaAs photodiode module for lidars operating at 1.2 μm to 2.2 μm wavelength. The photodiode is coupled to a 105 μm diameter core multimode fiber to maximize its free-space coupling efficiency. The fiber-coupled module exhibits a responsivity of 1.0 A/W (80% QE) at 1550nm wavelength and 1.2 A/W (73% QE) at 2050 nm wavelength, with a -3 dB bandwidth of 1.7 GHz. This device demonstrates a linear performance up to a photocurrent of 65 mA, and can deliver up to +14 dBm of continuous-wave RF output power. This device can also produce

200 ps wide impulses that are linear up to 3.1 Vpp RF output, and thus enable both continuous wave and nanosecond-scale pulsed coherent lidars.

2. Device Description

Conventional InP / InGaAs photodiodes, commonly used in 1550 nm wavelength systems, utilize $In_{0.53}Ga_{0.47}As$ absorber layer that is lattice-matched to the InP substrate (lattice constant = 5.87 Å). The cut-off wavelength of these devices is limited to 1650 nm by the bandgap of $In_{0.53}Ga_{0.47}As$ material (Eg = 0.74 eV). In this work, we utilized a lower bandgap material, namely Indium-rich $In_{0.72}Ga_{0.28}As$ (Eg = 0.57 eV) as the absorber layer, to extend the photodiode's cut-off wavelength to 2200 nm at ambient room temperature. The epitaxial structure of the lattice-mismatched $In_{0.72}Ga_{0.28}As$ / InP photodiode is compared with the conventional lattice-matched $In_{0.53}Ga_{0.47}As$ / InP photodiode in Fig. 1.

The primary challenge in incorporating Indium-rich absorption layer is to minimize the dislocation defects that may arise from the lattice mismatch between $In_{0.72}Ga_{0.28}As$ (lattice constant = 5.94 Å) and the InP substrate. These defects act as generation-recombination centers and increase the photodiode's dark current. Also, these defects may provide interstitial sites for material impurities and elevate unintentional background doping level in intrinsic $In_{0.72}Ga_{0.28}As$ absorption layer [8]. As a result, larger reverse bias is needed in a material with higher defect density to achieve a targeted depletion width and device capacitance for ultra-fast operation. The photodiode structure shown in Fig. 1a utilizes n-doped graded $InAs_yP_{1-y}$ buffer layer that is lattice-matched to the (100) oriented, n^+-doped InP substrate in the bottom and the $In_{0.72}Ga_{0.28}As$ absorber on the top. This buffer layer contains compositionally abrupt interfaces that minimize the propagation of lattice defects into the subsequently grown intrinsic $In_{0.72}Ga_{0.28}As$ absorption layer. We have previously reported uncooled photodiodes and photoreceivers having high QE in 1.2 μm to 2.2 μm wavelength range for ultra-fast applications [9].

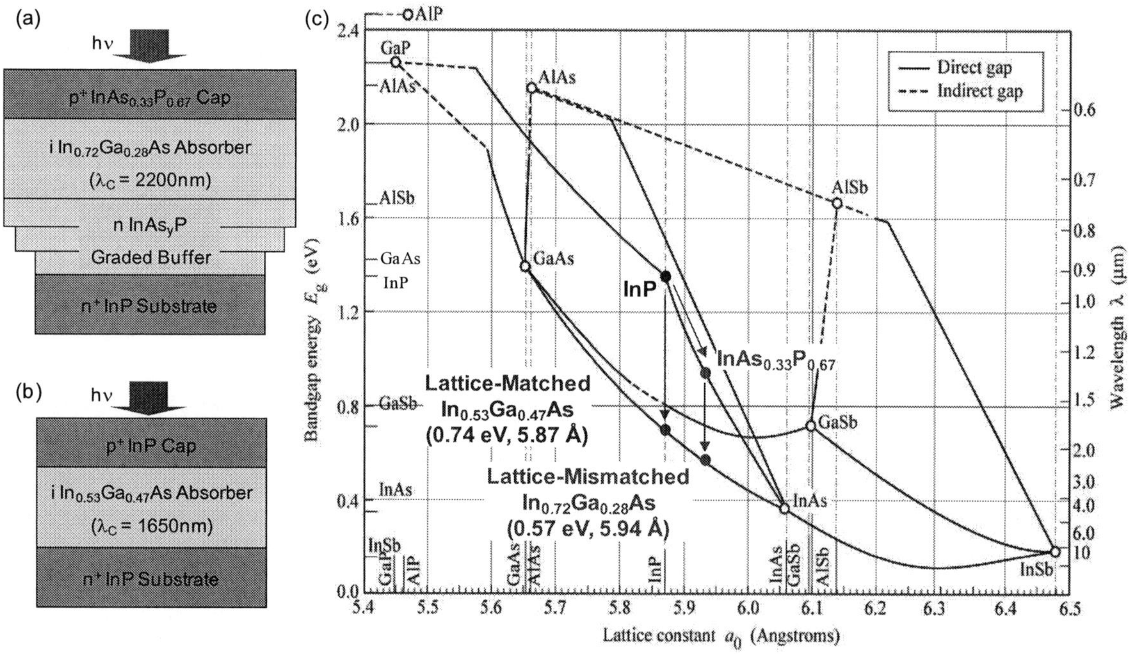

Figure 1. Epitaxial structure of (a) lattice-mismatched $In_{0.72}Ga_{0.28}As$ / InP photodiode used in this work is compared with (b) conventional lattice-matched $In_{0.53}Ga_{0.47}As$ / InP photodiode. (c) Bandgap and lattice constant of III-V semiconductors with materials of the two photodiode structures is annotated.

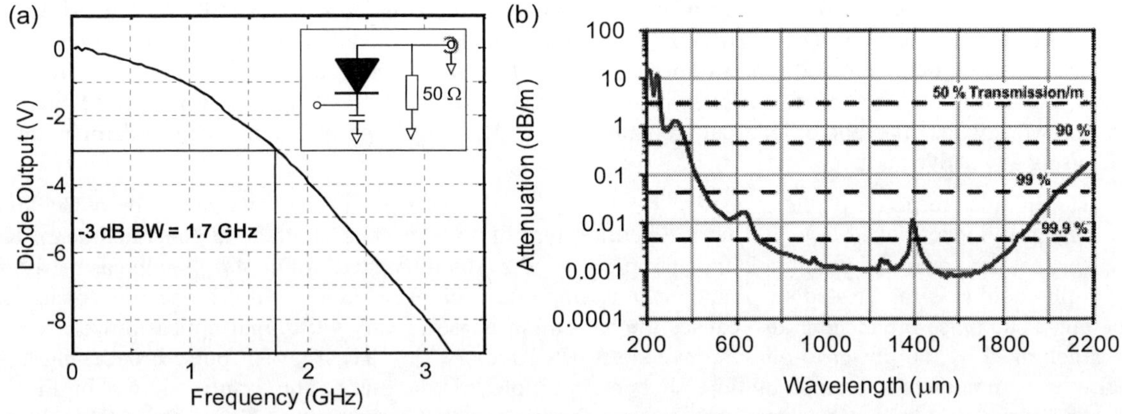

Figure 2. (a) Normalized RF response of fiber-pigtailed 150 μm diameter $In_{0.72}Ga_{0.28}As$ photodiode module. (Inset) The photodiode module includes an internal 50 Ω resistive termination. (b) Optical attenuation of the 105 μm core diameter multimode fiber used for the photodiode module (Source: Om-6 Technologies, Quebec, Canada).

For this work, a 150 μm diameter $In_{0.72}Ga_{0.28}As$ photodiode was terminated with an internal resistive 50 Ω load and assembled in a microwave package. The photodiode's active area was optically coupled to a 1 m long multimode fiber pigtail having a 105 μm diameter core and a numerical aperture of 0.22. The packaged photodiode demonstrated a DC responsivity of 1.0 A/W at 1550 nm wavelength and 1.2 A/W at 2050 nm wavelength with a -3 dB bandwidth of 1.7 GHz, as shown in Fig. 2a. The optical propagation loss for the fiber pigtail was less than <0.2 dB over the photodiode's spectral range, as shown in Fig. 2b.

3. Device Results

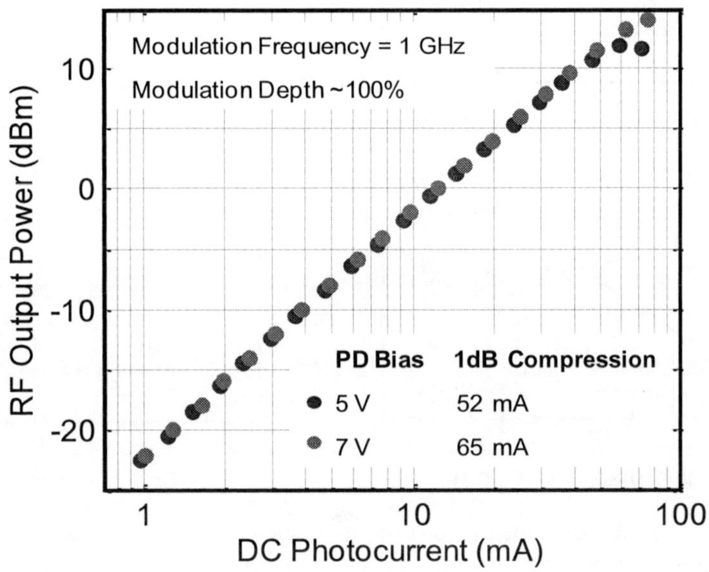

Figure 3. RF output power of 100 μm diameter $In_{0.72}Ga_{0.28}As$ photodiode for continuous-wave (CW) stimulus with 1 GHz sinusoidal modulation at 5 V and 7 V reverse bias. These results were obtained at an ambient room temperature of 20 °C without any active cooling.

The RF output power of the packaged 150 μm diameter $In_{0.72}Ga_{0.28}As$ photodiode was measured at different optical power levels for a 1 GHz sinusoidally modulated, continuous wave (CW) signal. As shown in Fig. 3, the photodiode shows linear behavior up to a DC photocurrent of 52 mA at 5 V reverse bias. Increasing the photodiode's reverse bias to 7 V enhances the 1 dB compression point to 65 mA. It is noteworthy that the photodiode can reliably deliver RF output power of up to +14 dBm without requiring any active cooling.

The impulse response of the uncooled $In_{0.72}Ga_{0.28}As$ photodiode was characterized using a 1550 nm wavelength, mode locked laser having a repetition rate of 1 GHz. The photodiode generated sharp RF pulses that are contained within a 200 ps temporal window, as illustrated in Fig. 4a. Nonlinear behavior of a photodiode is manifested simultaneously as amplitude compression and pulse broadening of its RF output. The pulse broadening also causes the RF output phase to vary with input optical power, and is characterized by the power-to-phase conversion (PPC) factor [7]. Excessive RF pulse broadening also introduces nonlinear crosstalk in time division multiplexed systems. The results shown in Fig. 4 demonstrate that this device can generate linear RF output up to 3.1 Vpp with a PPC <10 rad/W for input optical pulse energy up to 18 pJ.

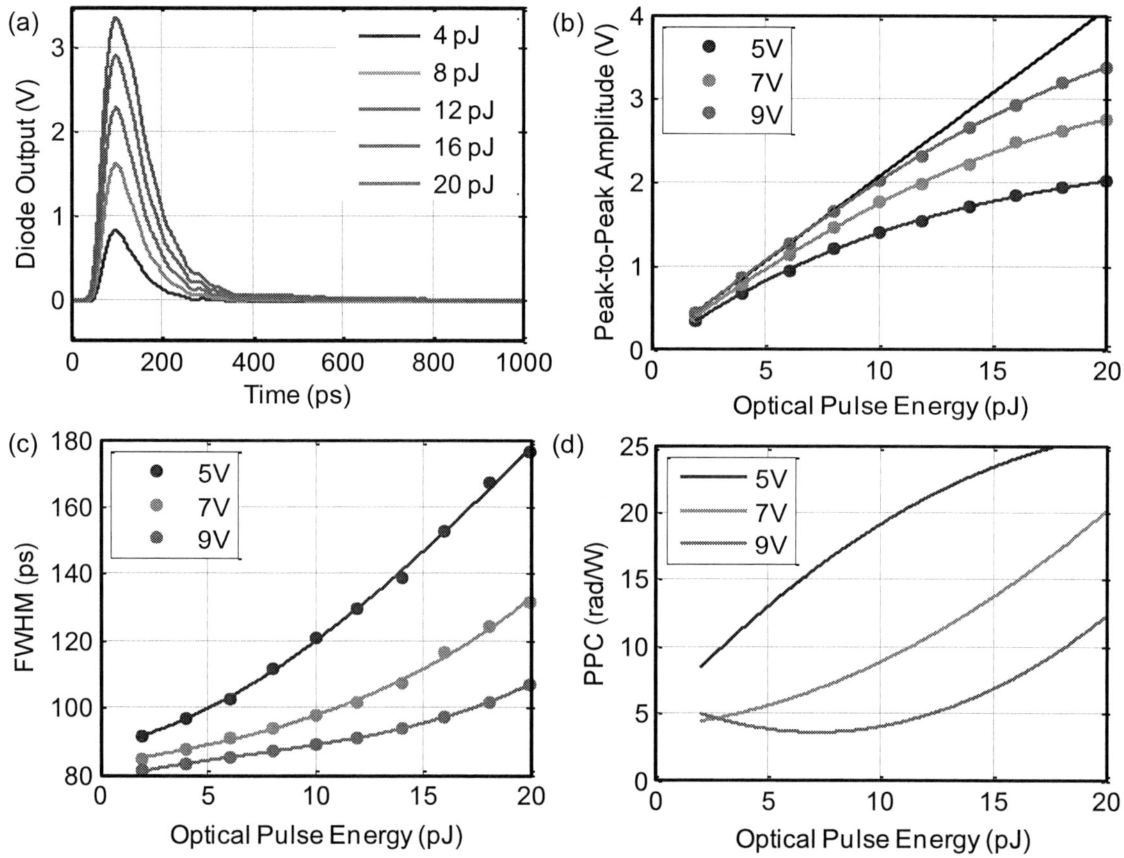

Figure 4. (a) Impulse response of uncooled, 150 μm diameter $In_{0.72}Ga_{0.28}As$ photodiode at 9 V reverse bias for different optical pulse energy levels. (b) Peak-to-peak amplitude, (c) Full-Width Half Maximum (FWHM), and (d) Power-to-Phase Conversion (PPC) calculated from the impulse response.

4. Conclusion

In summary, we have demonstrated an uncooled, lattice-mismatched InGaAs photodiode module having high quantum efficiency at 1.2 µm to 2.2 µm wavelength range and a -3 dB bandwidth of 1.7 GHz. This device generates linear continuous-wave RF output up to +14 dBm, and RF pulses up to 3.1 Vpp. Such power handling not only ensures shot noise limited operation in coherent lidar, but also eliminates the need for any further RF amplification in back-end electronics. The utility of this photodiode in both CW and pulsed coherent lidars is further bolstered by its 105 µm diameter core, multimode fiber interface, which ensures consistent optical coupling without requiring mechanically stabilized free-space optical elements.

5. References

[1] U. Singh, B. Walsh, J. Yu, M. Petros, M. Kavaya, T. Refaat, and N. Barnes, "Twenty years of Tm:Ho:YLF and LuLiF laser development for global wind and carbon dioxide active remote sensing," Optical Materials Express, vol. 5, pp. 828-837, 2015.

[2] F. Amzajerdian and D. Pierrottet, "Fiber-based coherent LIDAR for target ranging, velocimetry, and atmospheric wind sensing," in Conference on Lasers and Electro-Optics/Quantum Electronics and Laser Science Conference and Photonic Applications Systems Technologies, Technical Digest Optical Society of America, paper CPDB9, 2000.

[3] S. Kameyama, T. Ando, K. Asaka, Y. Hirano, and S. Wadaka, "Compact all-fiber pulsed coherent Doppler lidar system for wind sensing," Applied Optics, vol. 46, pp. 1953 – 1962, 2007.

[4] C. Wang, H. Ya, M. Shangguan, Y. Wu, L. Wang, L. Zhao, J. Qiu, and R. Zhang, "1.5 µm polarization coherent lidar incorporating time-division multiplexing," Optics Express, vol. 25, pp. 20663 – 20674, 2017.

[5] L. Lombard, G. Canat, A. Durecu, A. Dolfi-Bouteyre, M. Valla, B. Augere, D. Goular, and C. Besson, "High peak power single-frequency MOPFA for lidar applications," Proc. 18th Coherent Laser Radar Conference, Boulder, CO, June 26-July 1, 2016, pp. 220 - 225.

[6] S. Datta and A. Joshi, "High-Power Highly-Linear Photodiodes for High Dynamic Range LADARs," Proc. 18th Coherent Laser Radar Conference, Boulder, CO, June 26-July 1, 2016, pp. 252 - 256.

[7] A. Joshi and S. Datta, "Dual InGaAs Photodiodes having High Phase Linearity for Precise Timing Applications," IEEE Photonics Technology Letters, vol. 21, pp. 1360-1362, 2009.

[8] A. Joshi, G. Olsen, S. Mason, M. Lange, and V. Ban, "Near-Infrared (1-3 µm) InGaAs Detectors and Arrays: Crystal Growth, Leakage Current, and Reliability," Proc. SPIE, vol. 1715, pp. 585-593, 1992.

[9] A. Joshi, S. Datta, and M. Lange, "2.2 micron, uncooled, InGaAs photodiodes, and balanced photoreceivers up to 25 GHz bandwidth," Proc. SPIE, vol. 8704, pp. 87042G1-8, 2013.

History of 1.5µm laser source for the coherent Doppler LIDAR in Mitsubishi Electric.

Takayuki Yanagisawa, Takeshi Sakimura, Kenichi Hirosawa, Takuya Takasaki, Narito Samejima, Masashi Furuta, Hisamichi Tanaka, Shumpei Kameyama, Toshiyuki Ando, Kimio Asaka, and Yoshihito Hirano
Mitsubishi Electric Corporation, Information Technology R&D Center,
5-1-1 Ofuna, Kamakura, Kanagawa, 247-8501
Yanagisawa.Takayuki@cw.MitsubishiElectric.co.jp

Abstract: We have been developing 1.5-µm coherent Doppler LIDARs and its laser sources for 20 years. For pulsed lasers used in CDL, stable single-frequency oscillation, a relatively long pulse width, linearly polarized output, and a high output energy are required. A single-frequency diode-pumped Q-switched Er,Yb:phosphate glass laser with injection-seeding was used as a laser source in 1st stage of our CDL. To improve stability of output energy, we applied OPA system using Yb:YAG disk laser and a periodic-poled $LiNbO_3$ in 2nd stage. However it was difficult to realize high stability in the field using a laser oscillator because of its high sensitivity to the temperature and vibration. Therefore finally we applied a waveguide amplifier using Er,Yb:phosphate glass for CDL. In this paper, we will introduce the history of laser sources and the CDLs in Mitsubishi Electric.

Keywords: Coherent Laser Radar, etc.

1. Introduction

A coherent Doppler lidar (CDL) is now recognized as a promising sensor for wind velocity detection in clear air condition, and is applied in the fields of air-safety precautions for air traffic and wind power. For pulsed lasers used in long range CDL, stable single-frequency oscillation, a relatively long pulse width, linearly polarized output, high energy, and a high repetition rate are required. In early works, the possibilities of the CDLs were demonstrated by use of pulsed CO_2 laser [1]. Diode-pumped 1-µm and 2-µm pulsed lasers such as Nd:YAG, Ho,Tm:YAG and Tm:YAG lasers were used to improve the efficiency and compactness of CDL systems[2,3]. The wavelength range of 1.5 µm has the following attractive characteristics; the single-pulse maximum permissible exposure to the 1.5-µm range for human eyes is 30 times higher than that for the common 1-µm lasers; there are few absorption lines for molecules in the atmosphere in the 1.5-µm range; optical fiber components, laser source, and detectors for the optical fiber communications are easily available.

To realize mJ class high energy 1.5-µm laser sources, we have developed three types of laser generation methods. In the developing stage of CDLs in our lab in late 1990's as 1st generation, we chose a diode-pumped Q-switched Er,Yb:phosphate glass (Er,Yb:glass) laser with injection-seeding [4,5]. Unfortunately, a thermal loading in the glass host of Er,Yb:glass glass prevents the laser from working at the repetition rate higher than 15 Hz. This low repetition rate limited the measurement speed of the CDL, and consequently, limited the system performance. Therefore in order to improve the performance of 1.5-µm CDL, we replaced pulsed Er,Yb:glass laser with PPLN parametric amplifier in combination with all-fiber type transmitter as 2nd generation [6,7]. However it was difficult to realize high stability in the field using a fundamental laser oscillator because of its high sensitivity to the temperature and vibration. Therefore finally we developed a planar waveguide laser amplifier as 3rd generation using Er,Yb:glass for direct amplification of 1.5-µm laser [8,9,10]. Table 1 shows characteristics of 3 types of the laser sources.

Table 1 Laser characteristics of Mitsubishi CDLs [4,6,8,9]

Parameter	1st Gen.	2nd Gen.	3rd Gen.
Configuration	Er,Yb:glass Q-SW laser oscillator	Yb:YAG + PPLN Parametric amplifier	Er,Yb:glass Waveguide amplifier
Energy	11 mJ	0.8 mJ	3 mJ (Max. 9 mJ)
Repetition rate	15 Hz	4,000 Hz	500 - 4,000 Hz
Pulse width	228 ns	195 ns	450 ns
Optical gain	-	26 dB	26 dB
Max. range	> 6 km	> 8 km	34 km

2. Diode-pumped Q-switched Er,Yb:glass laser [4]

Figure 1 shows a schematic diagram and an internal photo of the Q-switched Er,Yb:glass laser. The optical path of the 2-m-long cavity is folded eight times to realize a compact assembly. A β-barium borate Pockels cell is used for the electro-optic (EO) Q-Switch. The polarizer-output-coupler composed of a cube-type polarizer (P1) and a quarter-waveplate (QWP1) is used to obtain linearly polarized laser output. The cavity-length tuner is composed of two roof prisms and a piezoelectric transducer that can vary the spacing between two prisms. A commercial external-cavity single-frequency laser diode is used as an injection-seeding light. The maximum output energy of the laser was 10.9 mJ with Q-switched operation. A stability of the beat frequency was ±1.9 MHz standard deviation.

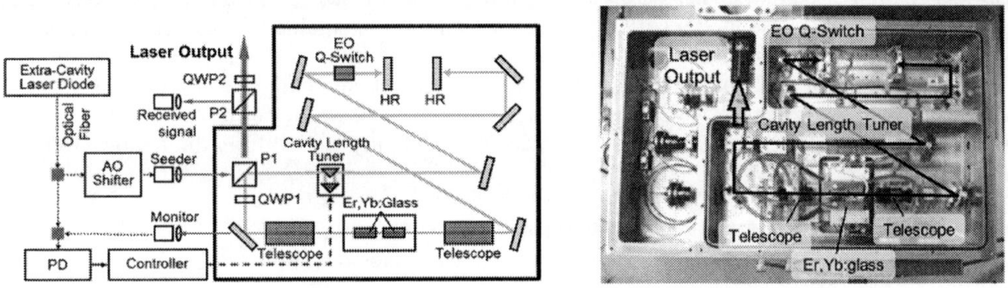

(a) Schematic diagram (b) Internal photo

Fig. 1 Diode-pumped Q-switched Er,Yb:phosphate glass laser [4]

3. Optical parametric amplifier (OPA) system using Yb:YAG disk laser and a periodic-poled $LiNbO_3$ [6]

Figure 2 shows a schematic diagram and an internal photo of the OPA system using Yb:YAG disk laser and a periodic-poled $LiNbO_3$. A diode-pumped Yb:YAG laser, which is easy to operate in long Q-switched pulse mode because of relatively low gain, is utilized as a pump laser source of parametric

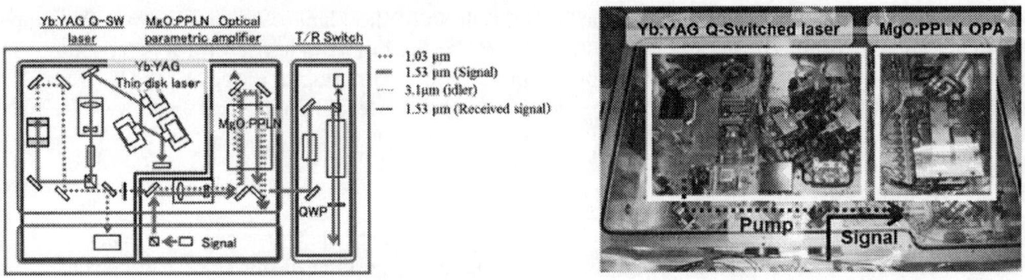

(a) Schematic diagram (b) Internal photo

Fig. 2 OPA system using Yb:YAG disk laser and a periodic-poled $LiNbO_3$ [6]

amplifier. For this pump source, a telescopic resonator with etalon filter and BBO EO Q-switch is used to obtain Q-switched single longitudinal and transverse mode operation. A MgO doped periodic-poled $LiNbO_3$ (MgO:PPLN) is selected as a nonlinear material for the parametric amplification. This material provides high photorefractive damage threshold at room temperature keeping high effective nonlinear coefficient *ref*. A seed pulse at wavelength of 1.5 μm from all-fiber type T/R optical circuit is launched from PM fiber, combined with pump pulse at wavelength of 1.03 μm, and consequently, incidented into the MgO:PPLN. Three 1-mm-thick MgO:PPLN crystals are used for amplification. The total length of amplifier is 130 mm. The 1.5-μm maximum output energy of the laser was 0.8 mJ at the repetition rate of 4 kHz.

4. Laser power amplifier using an Er,Yb:glass planar waveguide [8,10]

The laser oscillator used in our conventional CDL caused several problems. The injection-seeding method for stabilizing the laser frequency is susceptible to vibration because it is necessary to control the length of the cavity with an accuracy of less than 1 μm. Low gain and long cavity length to obtain long pulse width make the laser operation unstable. Furthermore, in spite that pulse timing management of the laser output is particularly important for TOF LIDAR to measure accurate range, laser oscillator generates tens ns timing jitter in the laser output because of build-up time. We have also applied a large mode area (LMA) fiber laser [11] to amplify the pulsed local source directly. However the output energy was limited due to nonlinear effects such as Brillouin scattering.

To solve these problems, we have developed a planar waveguide amplifier using Er,Yb:glass. Figure 3 shows a configuration of a planar waveguide amplifier. The transverse mode in the thickness direction is confined by waveguide mode, so that the beam quality is maintained. The transverse mode in the planar direction is bounced between side-surface and propagates in free space in the waveguide. The effective gain length is about 0.8 m, which is the same as active medium length of a fiber laser. The nonlinear effects is prevented by designing the beam size to decrease the laser intensity. The heat caused by pumping is generated in the whole waveguide and efficiently cooled from the bottom wide surface of the waveguide.

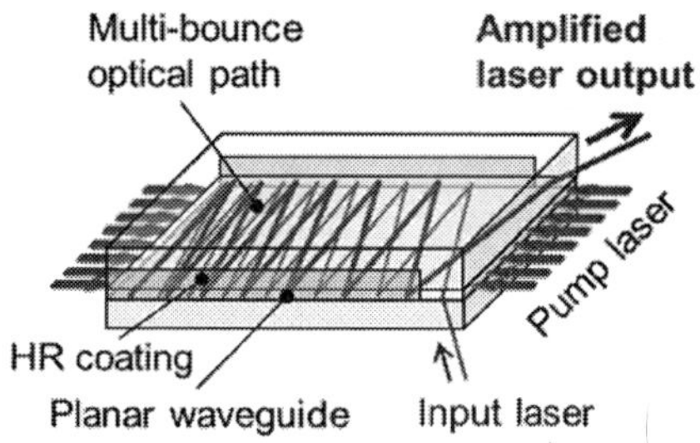

Fig. 3 Er,Yb:glass planar waveguide [*]

Figure 4 shows a schematic diagram and an internal photo of the planar waveguide amplifier system. There are no laser oscillator in the laser system except for the laser diode used for a local source. This laser is applied to our CDL installed at various airports. The 1.5-μm output energy of the laser was 3 mJ at the repetition rate of 1 kHz [8]. From recent development, we obtained 9 mJ from a single planar waveguide [9].

5. Conclusion

The history of 1.5-μm laser sources for the CDLs in Mitsubishi Electric Corporation was described. The output energy of the developed Er,Yb:glass planar waveguide applied to our CDL can be scaled from 1mJ to 9mJ and freely adjust the characteristics according to user's request. We believe that this property makes it possible to apply CDLs to various applications.

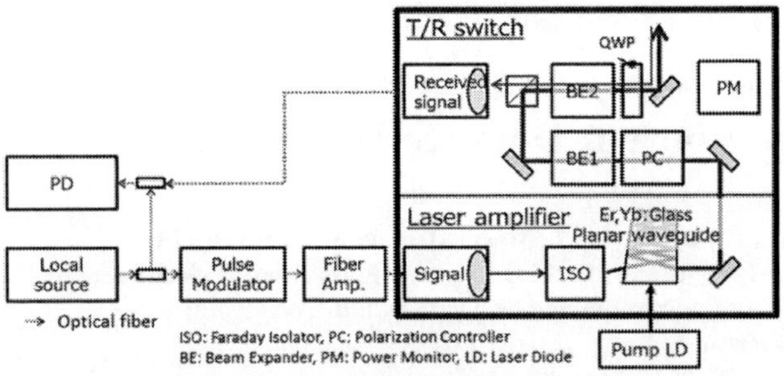

(a) Schematic diagram

(b) Internal photo

Fig. 4 Laser power amplifier using an Er,Yb:glass planar waveguide.

[1] R. M. Huffaker and A. V. Jelalian, "Laser-Doppler System for Detection of Aircraft Trailing Vortices", Proc. of IEEE, **58,** pp.322-326 (1970).

[2] T. J. Kane, W. J. Kozlovsky, and R. L. Byer, "Coherent laser radar at 1.06 μm sing Nd:YAG lasers", Optics Letters, **12,** pp.239-241 (1987).

[3] S. W. Henderson, C. P. Hale, J. R. Magee, M. J. Kavaya, and A. V. Huffaker, "Eye-safe coherent laser radar system at 2.1 μm using Tm,Ho:YAG lasers", Opt. Lett., **16**, pp.773-775 (1991).

[4] T. Yanagisawa, K. Asaka, K. Hamazu and Y. Hirano, "11-mJ, 15-Hz single-frequency diode-pumped Q-switched Er,Yb:phosphate glass laser", Optics Letters, **26,** pp. 1262-1264 (2001).

[5] K Asaka, T. Yanagisawa, Y. Hirano, "1.5-μm eye-safe coherent lidar system for wind velocity measurement", Proc. of SPIE, **4153,** pp. 321-328 (2001).

[6] T. Yanagisawa, S. Yamamoto, and Y. Hirano," Compact single-bar side-pumped Yb:YAG thin-disk laser" , in Technical Digest of CLEO2004, paper CW02 (2004).

[7] Y. Hirano, T. Yanagisawa, K. Asaka, M. Furuta, I. Kakimoto, Y. Ooga, and Y.Fujii, "Development of high-performance 1.5μm Doppler LIDAR", in 12th Coherent Laser Radar Conference (CLRC2003).

[8] T. Sakimura, Y. Watanabe, T. Ando, S. Kameyama, K. Asaka, H. Tanaka, T. Yanagisawa, Y. Hirano and H. Inokuchi, "1.5-μm high average power laser amplifier using an Er,Yb:glass planar waveguide for coherent Doppler LIDAR" , Proc. of SPIE, **8526**, 852604-1 (2012).

[9] S. Kameyama, T. Sakimura, Y. Watanabe, T. Ando, K. Asaka, H. Tanaka, T. Yanagisawa, Y. Hirano and H. Inokuchi, "Wind sensing demonstration of more than 30 km measurable range with a 1.5 μm coherent Doppler LIDAR which has the laser amplifier using Er,Yb:glass planar waveguide" , Proc. of SPIE **8526**, 85260E-1 (2012).

[10] K. Hirosawa, T. Sakimura, N. Samejima, T. Yanagisawa, and Shumpei Kameyama, "9-mJ laser amplifier at 1540 nm for a coherent lidar system with 250 Hz repetition", in 19th Coherent Laser Radar Conference (CLRC 2018), paper Mo8 (2018).

[11] S. Kameyama, T. Yanagisawa, T. Ando, T. Sakimura, H. Tanaka, M. Furuta, and Y. Hirano, "Development of wind sensing coherent Doppler LIDAR at Mitsubishi Electric Corporation ~from late 1990s to 2013," in 17th Coherent Laser Radar Conference (CLRC2013).

Development of single-frequency high-energy 2-µm laser for future space-based coherent Doppler Wind Lidar

Shoken Ishii (a), Atsushi Sato (b, a), Makoto Aoki (a), Koichi Akahane (a), Shigeo Nagano (a), Kohei Mizutani (a), Satoshi Ochiai (a), and Katsuhiro Nakagawa (a)
(a) National Institute of Information and Communications Technology
4-2-1 Nukui-kitamachi Koganei, Tokyo, 184-8795, Japan
(b) Tohoku Institute of Technology, Sendai, Miyagi, Japan
E-mail address: sishii@nict.go.jp

Abstract: The National Institute of Information and Communications Technology (NICT) is developing a single-frequency high-energy 2-µm pulse laser, 2-µm key technology and instrument. In the paper, we describe a new quantum dot laser, a Tm-doped fiber amplifier (TDFA) with an amplification of about 20, and a high-pulse energy laser emitting 125 mJ operating at 30 Hz meeting requirements for the future space-based coherent Doppler wind lidar.

1. Introduction

Global wind profiling is important to improve the initial conditions for numerical weather prediction (NWP) and meteorological studies. Non-uniform errors in NWP and its subsequent are caused by large blank regions of wind observation over oceans, the southern hemisphere, and other sparse areas. Space-based passive or micro sensors achieve a large coverage area and high temporal and horizontal resolutions (2.5 and 10 min, 0.5° × 0.5°). They can measure single-layer or a few vector winds with a low vertical resolution (1-2 km). They have the bias and root-mean-square vector difference against radiosonde wind observations [1,2]. A Doppler Wind Lidar (DWL) would be a promising approach for global wind profiling. A Doppler wind lidar (DWL) provides us with a wind profile having high vertical resolution, low bias, and good precision. A space-based DWL will, and it is necessary to fill the gap of current observations. The European Space Agency (ESA) is planning to launch the first space-borne DWL called Aeolus for global wind profiling [3,4] (ESA 1999; Stoffelen et al. 2005) in the summer of 2018. ADM-Aeolus uses a single-frequency UV laser and a direct-detection system, and it will provide us profiles of a single line-of-sight (LOS) wind speed. In Japan, we are conducting feasibility studies for a future space-based coherent DWL (CDWL) [5-7]. The impacts of the space-basede DWL wind observations on NWP have been assessed by Sensitivity Observing System Experiment (SOSE) [8-10]. The concept of the space-based CDWL is summarized in Table 1. The technical requirements of the future space-based CDWL are a single-frequency 2-µm laser producing 125 mJ at pulse repetition frequency of 30 Hz, two effective 40-cm-diameter afocal telescopes, a wide-band optical detector, a high-speed AD converter, and a systematic efficiency of 0.07. The National Institute of Information and Communications Technology (NICT) is developing a single-frequency high-energy 2-µm pulse laser [11], 2-µm key technology, and instrument meeting requirements for the future space-based coherent DWL. In the paper, we describe recent progress and results of a single-frequency 2-µm pulse laser developing at NICT.

2. 2-µm laser system

A single-frequency 2-µm Q-switched laser being developed at NICT composes of a 2-µm narrow line width single-frequency continuous-wave (CW) semiconductor laser, an optical fiber amplifier and a high-energy pulse laser. The 2-µm single-frequency CW semiconductor laser output is amplified by the optical fiber amplifier. The optical fiber amplifier output will be used as a single-frequency master laser. The output of the single-frequency master laser is divided into two portions: one portion is used for injection seeding, while the other is used as a local oscillator for heterodyne detection. Target requirements for the MO are TEM_{00} mode, wavelength tuning, optical power of >20 mW, and linewidth

Table 1. Summary of 2-μm laser and detector for future space-based CDWL.

Wavelength (μm)	2.05
Pulse energy (mJ)	125
Pulse duration (ns_{FWHM})	200
Pulse repetition frequency (Hz)	30
Telescope diameter (m)	0.4
Number of laser directions	2
Detector quantum efficiency	0.8
Detector bandwidth (GHz)	>3.8
Heterodyne efficiency	0.4
Optical efficiency	0.44
Unknown system efficiency	0.5
AD Sampling frequency (MHz)	400
Sampling points for 1 range-gated bin	256

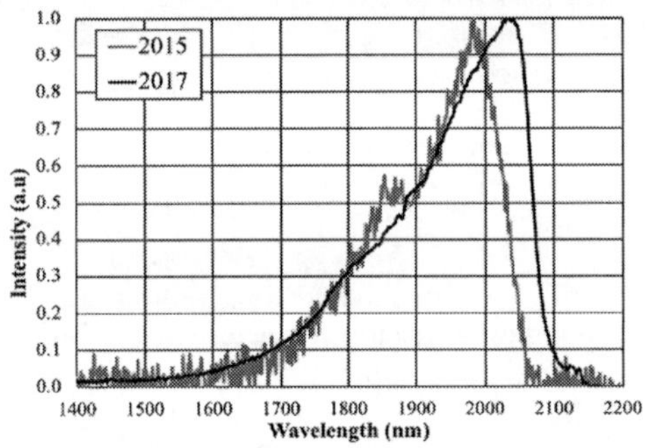

Figure 1. Photoluminescence from quantum dot laser.

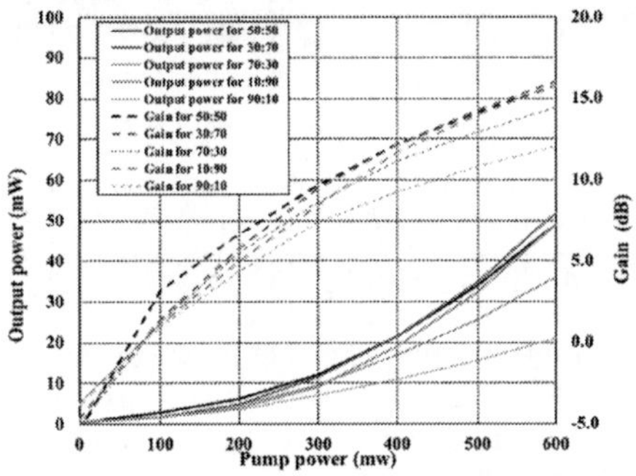

Figure 2. Output power and gain as a function of pump ratio.

of <100 KHz. The required optical power and linewidth of the 2-μm single-frequency CW semiconductor laser are 1-2 mW and <100 KHz. Figure 1 shows spectra of photoluminescence of quantum dot lasers fabricated in 2015 and 2017. The peak wavelength of the quantum dot laser fabricated in 2017 shifted toward the wavelength of longer than 2-μm and are around at wavelength of 2.03 μm. We found that the intensity of the quantum dot laser in 2017 are stronger than the one in in 2015.

To realize the MO with output power of >20 mW, the TDFA was optimized by the TDF length and the ratio between the forward and backward pump power to realize efficient amplification. The amplification efficiency depended strongly on the TDF length, and our results showed that it was highest for a 10-m TDF. Figure 2 shows output power and gain as a function of pump ratio. We define a pump ratio between forward and backward pump powers and express it as A:B in the Fig.2. The TDFA achieved a gain of over 13 dB at a pump power of 400 mW and it demonstrated an output power of >20 mW for a 2.05-μm signal input at am output power of 1.3 mW.

Target pulse energy of the single-frequency Q-switched laser is 125 mJ operating at a pulse repetition frequency of 30 Hz. The design is a master oscillator and power amplifier (MOPA) architecture based on Tm,Ho:YLF gain media operating at 2051.250 nm. Figure 3 shows top down view of a Tm,Ho:YLF laser optical layout with a master oscillator (MO) and power amplifier (AMP) configuration (left figure) and view of the Tm,Ho:YLF laser mechanical structure (right figure). The master oscillator consists of a

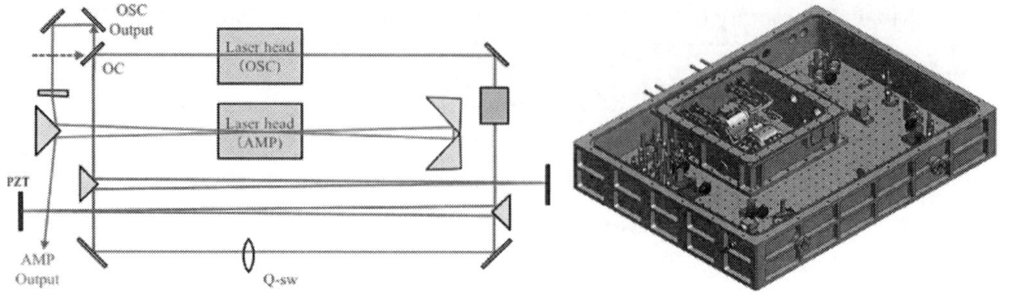

Figure 3. (left) Top down view of Tm,Ho:YLF laser optical layout with a master oscillator and power amplifier (MOPA) configuration, and (right) view of the Tm,Ho:YLF laser mechanical structure.

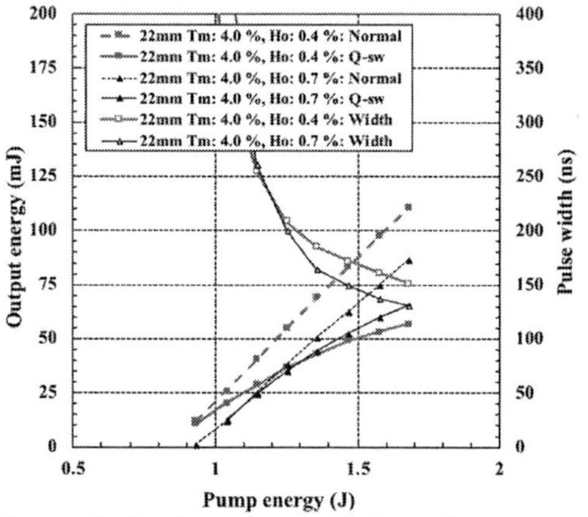

Figure 4. Lasing characteristics of a master oscillator (MO) with Tm,Ho:YLF laser rod co-doped with 22 mm in length, 4% thulium, and 0.4% or 0.7% holmium.

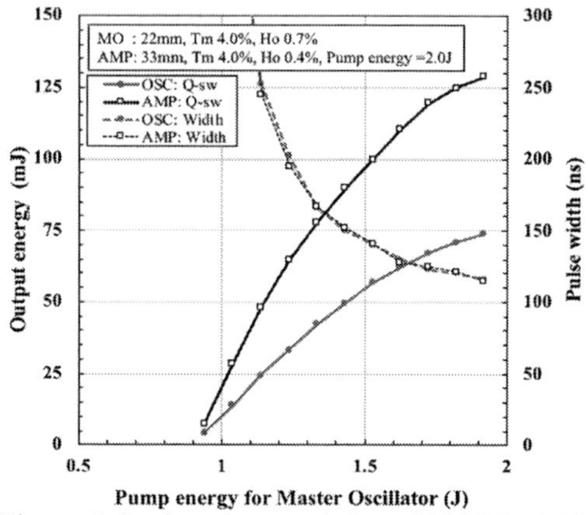

Figure 5. Lasing characteristics of MOPA. MO: Tm,Ho:YLF laser rod with 4% thulium, 0.4% holmium, and 22 mm in length. AMP: Tm,Ho:YLF laser rod with 4% thulium, 0.7% holmium, and 33 mm in length.

Q-switched, four 45-degree mirrors, two 0-dgree mirrors, two prism-type mirrors, and a lens used for compensation for thermal lens of a Tm,Ho:YLF laser rod. The power amplifier is double pass configuration with a prism-type mirrors and a wedge-type mirror.

Figure 4 shows the laser performance of the master oscillator with Tm,Ho:YLF laser rod co-doped with thulium dopant concentrations of 4%, holmium dopant concentrations of 0..4 and 0.7%, and 22 mm in length. The laser rod is pumped by InGaAs/GaAs diode arrays, each of which is composed of 3 bars that emit at wavelength of 792 nm. The laser rod and the diode arrays are conductively-cooled to -40C and approximately 25 C, respectively. The output energy in the normal-mode operation for both holmium dopant concentrations increases linearly with pump energy, while, in the Q-switched operation, the roll-off of the output energy can be seen in Fig. 4. The slope efficiency in Q-switched mode operation decreased as the pump power increased under intense pumping. The output energy in Q-switched operation obtained 50 mJ at the PRF of 30 Hz, and its pulse width were 180 and 150 ns for holmium dopant concentrations of 0.4 and 0.7%.

Figure 5 shows the laser performance of MOPA laser with an single-pass AMP. The length and holmium concentration of the laser rod used for the MO and AMP are 22 mm and 0.7 %, and 33 mm and 0.4 %, respectively. The thulium concentration of 4 % is for both the MO and AMP. The slope efficiency for the MOPA also decreases slowly due to the degradation of the slope efficiency of the MO. The output energy of 125 mJ operating at the PRF of 30 Hz in Q-switched operation is achieved obtained 125 mJ when the pump energy for MO and AMP are 1.8 and 2 J, respectively. The results indicate that the development of the laser capable of meeting the requirements for the space-based DWL is technically feasible.

3. Summary

A conductively-cooled Q-switched Tm,Ho:YLF laser is been deveopling at NICT. We achieved a output energy of 125 mJ operating at the PRF of 30 Hz in Q-switched operation and firstly demonstrated the feasibility of the development of the laser meeting the requirements for the space-based DWL. The demonstration provides us useful information to optimize lasing characteristics and to design the future space-based DWL.

4. Acknowledgements

The authors wish to thank the working group for a future Japanese space-based DWL for supporting the research activity of the working group.

5. References

[1] Hayashi M., and K. Shimoji, "Atmospheric motion vectors derivation algorithm. Meteorological satellite center technical note." **58**, (2013). (in Japanese).

[2] M. Otsuka, M. Kunii, H. Seko, K. Shimoji, M. Hayashi, and K. Yamashita, "Assimilation experiments of MTSAT rapid scan atmospheric motion vectors on a heavy rainfall event." J. Meteorol. Soc. Jpn. **93**, 459-475 (2015).

[3] ESA, "Atmospheric dynamics mission." Mission Selection Rep. ESA SP-1233(4) (1999).

[4] A. Stoffelen, J. Pailleux, E. Källén, M. J. Vaughan, L. Isaksen, P. Flamant, W. Wergen, E. Andersson, H. Schyberg, A. Culoma, R. Meynart, M. Endemann, and P. Ingmann, "The Atmospheric Dynamics Mission for global wind field measurement." Bull. Amer. Meteor. Soc. **86**, 73-87 (2005).

[5] A. Sato, M. Aoki, S. Ishii, R. Otsuka, K. Mizutani, and S. Ochiai, "7.28-W, High-Energy, Conductively Cooled, Q-switched Tm,Ho:YLF Laser," IEEE Photo. Techno. Lett. **29**, 134-137 (2017).

[6] S. Ishii, P. Baron, M. Aoki, K. Mizutani, M. Yasui, S. Ochiai, A. Sato, Y. Satoh, T. Kubota, Daisuke Sakaizawa, Riko Oki, K. Okamoto, Toshiyuki Ishibashi, T. Y. Tanaka, T. T. Sekiyama, T. Maki, K. Yamashita, T. Nishizawa, M. Satoh, and T. Iwasaki, " Feasibility Study for Future Space-Borne Coherent Doppler Wind Lidar, Part 1: Instrumental Overview for Global Wind Profile Observation," J. Meteorol. Soc. Jpn. **95**, 301-317 (2017).

[7] P. Baron, S. Ishii, K. Okamoto, K. Gamo, K. Mizutani, C. Takahashi, T. Itabe, T. Iwasaki, T. Maki, R. Oki, S. Ochiai, D. Sakaizawa, M. Satoh, Y. Satoh, T. Tanaka, and M. Yasui, " Feasibility study for future spaceborne coherent Doppler Wind Lidar, Part 2: Measurement simulation algorithms and retrieval error characterization," J. Meteorol. Soc. Jpn. **95**, 319-342 (2017).

[8] Marseille G.-J., A. Stoffelen, and J. Barkmeitjer, "Sensitivity observing system experiment (SOSE)-a new effective NWP-based tool in designing the global observing system." Tellus **60A**, 216-233 (2008).

[9] T. Ishibashi, "Observing system simulation experiments with multiple methods." Proceedings of SPIE **9265**, 926508; doi: 10.1117/12.2069087 (2014).

[10] K. Okamoto, T. Ishibashi, S. Ishii, P. Baron, K. Gamo, T. Y. Tanaka, K. Yamashita, T. Kubota, "Feasibility Study for Future Space-Borne Coherent Doppler Wind Lidar, Part 3: Impact Assessment Using Sensitivity Observing System Simulation Experiments," J. Meteorol. Soc. Jpn. **96**, 179-199 (2018).

[11] K. Mizutani, T. Itabe, S. Ishii, M. Aoki, K. Asai, A. Sato, H. Fukuoka, T. Ishikawa, and K. Noda, "Diode-pumped 2-mu m pulse laser with noncomposite Tm,Ho:YLF rod conduction-cooled down to-80 degrees C," Applied Optics **54**, 7865-7869 (2015).

Challenges past, present and future for lidar in wind energy

Michael Harris
ZephIR Lidar
Fairoaks Farm, Hollybush, Ledbury, HR8 1EU, U.K.
Michael.harris@zephirlidar.com

Abstract: Lidar made its first tentative steps in 2003 towards becoming an accepted tool in the wind industry; significant efforts were needed initially to understand the requirements and establish lidar methods. Important questions were raised, including: is lidar reliable and does it work in all conditions, particularly low backscatter? Is the accuracy sufficient? Can costs be reduced? Today, lidar is used onshore, offshore and on the turbine itself for a range of important activities including resource assessment, power performance evaluation, turbine optimisation and active control. The level of acceptance has steadily increased to the point where lidar is now considered mainstream in the growing wind industry, with the development of international standards and roadmaps to bankability. Remaining challenges include better understanding of complex terrain, wakes and turbulence, and greater acceptance by way of certification and standards. In future new applications may emerge as the technology improves with further cost reductions.

Keywords: Coherent Laser Radar, wind energy

1. Introduction

The potential for coherent wind lidar to make a contribution in the wind industry was first considered as long ago as the early 1980s [1]. The systems at that time were research tools and not suitable for commercialization; nevertheless, efforts were made to investigate lidar wind measurements for assessing the impact of turbulence on wind turbines [2], and resource assessment, wake studies and turbine control [3].

In the late 1990s and early 2000s, several groups including FOI, DERA, ONERA and Mitsubishi started to examine the potential to build a wind lidar at telecommunications wavelengths around 1.55 μm [4-7]. A major attraction of this approach was the increasing availability of cheap, reliable components developed for telecoms, with interconnection by optical fibre. The use of fibre architectures revolutionized the design of coherent lidars, allowing ease of manufacture through flexible modular construction, and paving the way for modern commercial wind lidar systems.

Wind lidars of different types are now playing a significant and increasing role in the wind industry [8] by performing the same role as conventional tall meteorological masts at greater convenience and lower cost, and also by creating a number of new applications. The essential requirement for robust, autonomous and cheap lidar systems is increasingly fulfilled, and several lidar manufacturers around the world have become established suppliers to the wind industry.

2. Initial challenges

In March 2003, a prototype ZephIR lidar from QinetiQ was mounted on the nacelle of a 2.3MW wind turbine to demonstrate the measurement of wind speed in advance of the blades (figure 1) [9]. The system was a staring continuous-wave (CW) unit configured to measure the wind from 10m to 200m in front of the turbine. While technically successful, the measurement was ahead of its time in that lidar was very much a novelty to wind energy, and the industry was far too sceptical at that point to seriously consider lidar for applications such as turbine control. A more accessible commercial application in the short term was as a mast replacement for resource assessment purposes; an important step towards lidar acceptance was the involvement of wind industry experts from Risø in Denmark. Their comparisons of lidar data against tall masts at the Høvsøre test site [10] provided a demonstration of lidar accuracy against the traditional tools for wind energy measurements, giving much-valued credibility.

Figure 1: prototype ZephIR lidar on nacelle of Nordex N90 turbine, March 2003

The next phase of development (2004-2005) required a small number of visionary "early adopters" such as Risø, who were prepared to invest in the first batch of lidar units to investigate the potential of this exciting, potentially revolutionary, new technology. The pace of development increased as other lidar manufacturers introduced their own products for the increasing wind market. At that time the most common questions from customers related to concerns about signal/backscatter levels in clear atmospheres, and reliability of the technology. These concerns, while not disappeared completely, are much diminished with the advances made in lidar design, manufacture and sensitivity over the following decade.

3. Current applications

Today, there are many hundreds, even thousands, of coherent wind lidars in use in the wind industry. This section will summarise their main uses, which contribute during all phases of a wind farm's life cycle.

Continuous-wave (CW) and pulsed coherent systems have a role to play, with both methods proven in the main applications listed below. Each of the methods has strong advantages for particular applications; e.g. the higher sensitivity of CW permits a shorter integration time, and pulsed systems are capable of operating at much longer (multi-km) ranges.

Resource assessment. The process of wind prospecting has traditionally used tall masts to assess winds around hub height, and hence to estimate the annual energy production (AEP) of a proposed wind farm. Lidar is increasingly cost-competitive in this application with the introduction of ever-taller turbines. Reduction of uncertainty in AEP is a critical part of this process, and lidar has advantages compared to a mast of measurement across the whole rotor, as well as the ability to move around the site. Traceable validation of the lidar against conventional methods currently forms an important factor for many customers. Figure 2 shows a selection of lidars undergoing performance validation at the UK remote sensing test site near Pershore in Worcestershire. The plot on the right is a correlation between the lidar and the mast: 10-minute average values of wind speed are compared at various heights above ground level.

Complex flows in hilly terrain represent a challenge for resource assessment by any method, lidar or mast; methods have been developed that model the flow over a specific site using computational fluid dynamics (CFD) in order to make appropriate adjustment to the lidar data and hence reduce the uncertainty in AEP.

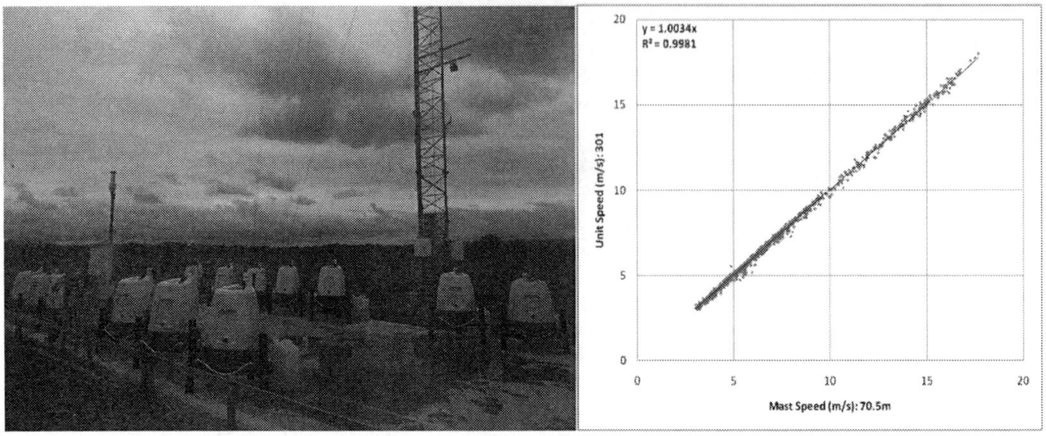

Figure 2: ZephIR 300 units at Pershore test site; typical resulting lidar/mast correlation

Floating lidar. The cost of a hub-height mast for offshore wind resource assessment is very high; lidar mounted on a floating buoy was quickly identified as a potential alternative method [11]. In recent years, the commercial case for floating lidar has become compelling, and its use has been adopted by many of the largest offshore developers. A variety of buoy solutions is in use, ranging from small waveriders to very stable spar buoys. Compensation for platform motion is sometimes required, depending on the buoy and lidar type.

Power performance assessment. Validation of turbine performance (via its "power curve") is done by measuring the dependence of generated power on input freestream wind speed. This is expensive with conventional met masts and often impractical in complex or offshore sites. Ground-based lidar can be used as an alternative to a met mast, and the unit is easily moved to other tasks after the completion of the campaign (typically several weeks).

Nacelle-mounted lidar represents an even better solution for measurement of power curves: the lidar yaws automatically with the turbine in response to changes in wind direction so that it always samples the flow upstream. Turbine-mounted lidar also can be used to measure wind yaw misalignment, which can result in inefficient operation of the turbine, with increased loads.

Turbine control

The potential for lidar to provide feed-forward input to the turbine control has been recognized since the 1980s [1,3]. Further development followed the first demonstration of turbine-mounted lidar in 2003 [9,12], and subsequent studies showed the potential for improvement in energy yield and (more significantly) load reduction. Developments in this area are now advancing rapidly, and it has the potential to generate significant growth of the lidar market.

4. Future opportunities

Development and improvement of lidar systems continues among both commercial and academic groups. The current state of the art is epitomized by DTU's Windscanner project (http://www.windscanner.dk/), which uses rapidly-scanned multiple lidars, both pulsed and CW, to characterize complex 3D flows in some detail. The project has successfully demonstrated the potential for turbine-mounted lidar mounted in the spinner or even on the blades themselves [13]. By pushing the limits of what can be achieved with current technology, the project has pointed the way towards future commercial applications. This section lists some areas of current development that are starting to become commercialized within the wind power sector.

Long-range dual-Doppler measurement has potential for wide area assessment, and is a particularly powerful method for offshore sites.

Lidar is starting to be used as a met mast replacement on operational wind farms. In addition, lidar has potential for short-range forecasting of wind conditions, which can have commercial value in the operation of a wind farm.

Turbulence measurement with lidar remains to be fully exploited, with most efforts up to now concentrating on replicating values of turbulence intensity (TI) from a mast-mounted cup. However, it is likely additional useful turbulence data can be extracted from lidar signals that can be better related than the usual TI parameter to the actual loads experienced by large wind turbines.

Wakes are recognized as a significant source of loss for downstream turbines in a large array. Lidar is well suited to the monitoring and tracking of wakes, and for assessment of their energy deficit. The first wake measurements were performed with a modified ZephIR lidar in 2004 [14], and since then a range of more sophisticated systems have been used in numerous studies.

Lidar also has a role to play in the real-time monitoring of wind conditions during wind farm construction; accurate measurement can lead to an increase in the operating limits of cranes and other machinery, with financial benefits

In summary: The wind energy sector represents a significant market for coherent wind lidar. Further increases in the market are expected as wind energy assumes greater global importance, international standards catch up with the technology, and new applications develop.

5. References

[1] J M Vaughan, Internal RSRE memo (1981)

[2] R M Hardesty & B F Weber, "Lidar measurement of turbulence encountered by horizontal-axis wind turbines", J of Atmospheric and Oceanic Technology, **4** 191 (1987)

[3] J M Vaughan & P A Forrester, "Laser Doppler velocimetry applied to the measurement of local and global wind", Wind Engineering **13** 1 (1989)

[4] M Harris, G N Pearson, J M Vaughan, D Letalick & C Karlsson, "The role of laser coherence length in CW coherent laser radar", J. Mod. Opt. **45** 1567-1581 (1998)

[5] C Karlsson, F Olsson, D Letalick & M Harris, "All-fibre multifunction continuous-wave 1.55 micron coherent laser radar for range, speed, vibration and wind measurements", Applied Optics **39** 3716-3726 (2000)

[6] B Augere & J-P Cariou, "All-fiber 1.5-um CW coherent laser anemometer for in-flight measurements", Proc SPIE. 10.1117/12.503135 (2003)

[7] T Ando, S Kameyama & Y Hirano, "All-fiber coherent Doppler lidar technologies at Mitsubishi Electric Corporation", Earth and Environmental Science **1** 012011 (2008)

[8] S Emeis, M Harris & R M Banta, "Boundary-layer anemometry by optical remote sensing for wind energy applications", Meteorologische Zeitschrift **16** 337-347 (2007)

[9] M Harris, D J Bryce, A S Coffey, D A Smith & J Birkemeyer, U Knopf, "Advance measurement of gusts by laser anemometry", J. Wind Engineering and Industrial Aerodynamics **95** 1637-1647 (2007)

[10] D A Smith, M Harris, A S Coffey, T Mikkelsen, H E Jørgensen, J Mann & R Danielian, "Wind lidar evaluation at the Danish wind test site in Høvsøre", Wind Energy **9** 87-93 (2006)

[11] D A Smith & M Harris, "Wind speed measurement apparatus and method", International patent application WO 2005/008284 (2005)

[12] M Harris, M Hand & A Wright, "Lidar for turbine control", Technical Report NREL/TP-500-39154 (2006)

[13] A T Pedersen et al, "Full-Scale Field Test of a Blade-Integrated Dual-Telescope Wind Lidar", EWEA poster ID 166 (2013)

[14] F Bingöl et al, "Fast Wake Measurements with LiDAR at Risø Test Field", Earth and Environmental Science **1** 012022 (2008)

Status of research and development for lidar-assisted wind turbine control in Japan

Yasunari Kamada(a), Hirokazu Kawabata(b), Yutaka Kajiyama(c), Nobuki Kotake(c), Tetsuya Kogaki(b) and Takao Maeda(a)
(a) Mie University
1577 Kurimamachiya, Tsu, Mie, Japan, 514-8507
(b) National Institute of Advanced Industrial Science and Technology
2-2-9 Machiikedai, Kooriyama, Fukushima, Japan, 963-0298
(c) Mitsubishi Electric Corporation
5-1-1 Ofuna, Kamakura, Kanagawa, Japan, 247-8501
Lead Author e-mail address: Kajiyama.Yutaka@dw.MitsubishiElectric.co.jp

Abstract: The status of research and development for the lidar-assisted wind turbine control in Japan is shown. Two projects have been performed using Mitsubishi Electric's 9-beam wind lidars. The overview of the projects, in addition to the configuration of the lidar, is introduced.

Keywords: Coherent laser radar, Lidar-assisted turbine control, Wind, Doppler

1. Introduction

In recent years, the application of wind lidar has made remarkable progress on the feed-forward wind turbine control and there are some demonstrations on this application to reduce both the turbine loads and yaw misalignment [1, 2]. The researches and developments for this application are now going also in Japan and are introduced here. Mie University and Mitsubishi Electric have demonstrated the power performance stabilization with blade pitch control of Mie University's test turbine using Mitsubishi Electric's 9-beam lidar data in their joint research. AIST (National Institute of Advanced Industrial Science and Technology) have tested the same lidar which was installed on the nacelle of their test turbine in their in-house project.

2. Configuration of the wind lidar and overview of the demonstration campaign

Figure 1 shows the system configuration of Mitsubishi Electric's 9-beam wind lidar. The system sequence for measuring wind data is described such as follows: The laser light source emits the light of continuous wave at wavelength 1550-nm as seed light. For the purpose of heterodyne detection, this seed light is splitted to 2 optical lines by the coupler, one optical pass is connected to (1) the acousto-optic modulator (AOM), and (2) another pass is to the 4-port coupler. At (1) optical line, the light is modulated to pulse with the AOM, amplified with the erbium-doped fiber amplifier (EDFA), passed through the circulator and optical-switch, and emitted to the atmosphere with the telescope. Then, the backscattered light from the aerosols is received by the telescope and this received light is passed through the optical-switch and circulator to the 4-port coupler. Therefore, (1) received light and (2) seed light are interfered at 4-port coupler and this coherent light is converted to an electrical signal at the balanced-receiver (BR). After this electrical signal amplified, the signal processor recognize this signal and output wind data by analyzing the frequency of coherent electrical signal as Doppler shift. As a supplement, there are 9 output ports on the optical switch and 9 telescopes, therefore the beam output direction can be changed for 9 patterns arbitrarily.

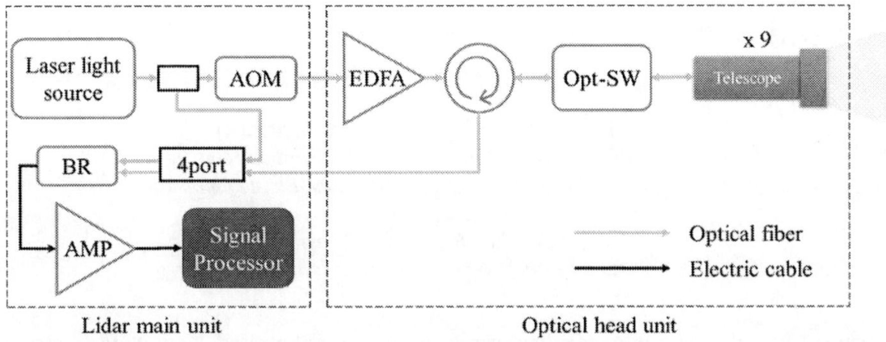

Fig. 1. System configuration.

The overview of the demonstration campaigns is the following. At Mie University's test site, this 9-beam lidar system is installed on the site located 205m behind the turbine for the purpose of blade pitch control, referred to Fig. 2. At AIST's test site, same lidar is installed on the nacelle of AIST's test turbine for the purpose of detecting yaw misalignments, referred to Fig. 3.

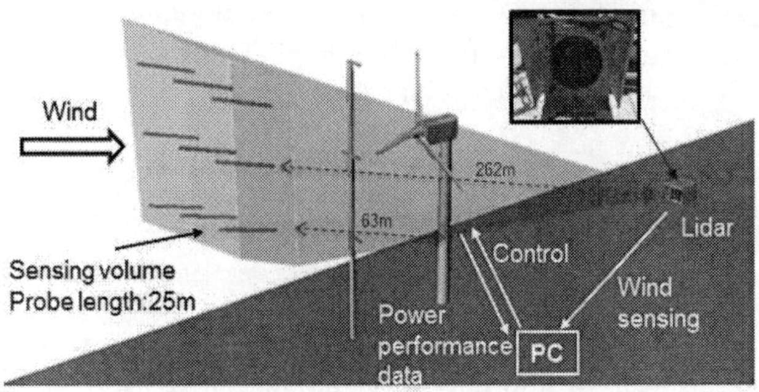

Fig. 2. Schematic of Mie University's campaign

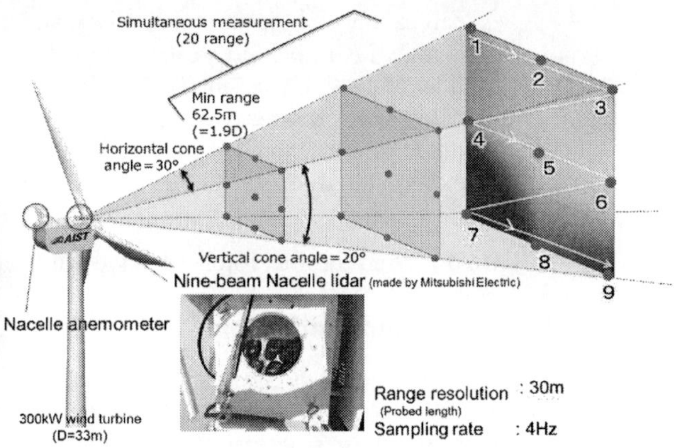

Fig. 3. Schematic of AIST's campaign

3. Results on the campaigns

In the demonstration at Mie University's test turbine, 9-beam-lidar-assisted turbine control with database produced by Mie University shows it is much effective to suppress the overshoot of turbine output power when it exceeds the target 10-kW in the range of high wind speed [3]. In the measurement campaign at AIST test turbine, with same lidar, they found out the feasibility of increasing energy capture of up to about 6 % by reducing the appearance frequency of yaw misalignment larger than $\pm$10-deg. and also of prediction of incoming extreme gust which contributes to reducing the turbine loads [4].

4. References

[1] D. Schlipf et al., Journal of Physics:Conference Series 555 (2014) 012090.

[2] A. Scholbrock et al., Journal of Physics: Conference Series 524 (2014) 012002.

[3] Y. Kamada et al., "Demonstration of lidar-assisted turbine control for stabilizing the power output", Wind Europe (2017) PO.291.

[4] http://www.aist.go.jp/fukushima/en/unit/WPT_e.html

Wind turbine rotor-effective wind speed estimated by nacelle-mounted Doppler wind lidars

Dominique P Held(a,b), Jakob Mann(a)
(a) Department of Wind Energy, Danish Technical University (DTU)
Frederiksborgvej 399, 4000 Roskilde, Denmark
(b) Windar Photonics A/S
Helgeshøj Alle 16-18, 2630 Taastrup, Denmark
Lead Author e-mail address: domhel@dtu.dk

Abstract: Over the past two decades, research has demonstrated a substantial structural load reduction potential of lidar-assisted wind turbine controllers; especially for lidar-assisted pitch control, opening the possibility of lighter rotors and cheaper wind turbines. Knowing how well a lidar can estimate the rotor-effective wind speed (REWS) is an important aspect of designing an optimal pitch controller. This contribution investigates the inter-comparison between REWS estimated by a forward-looking nacelle-mounted lidar and the wind turbine itself. We present measurements from two experimental campaigns, where an economical and robust cw Doppler wind lidar is mounted on a heavily instrumented Vestas V52 turbine. The results indicate that the coherence between lidar and turbine estimated REWS is in agreement with analytical models. Also, the time delay between the two signals conforms to the expected results. This implies that the models used to design and optimize lidar-assisted pitch controllers match with experimental results.

Keywords: Wind turbine, rotor-effective wind speed, pitch control, nacelle lidar, wind energy

1. Introduction

Lidar systems mounted on the nacelle or inside the spinner of wind turbines are able to provide in-advance information about the incoming wind field in front of a turbine. Several controllers have been proposed to improve power production and/or mitigate structural loading; for an overview see [1]. Promising simulation results were demonstrated with a relatively simple feed-forward addition to the traditional feed-back speed controller in wind turbines [2]. An illustration is shown in figure 1. Disturbances in the incoming wind field v can be estimated based on lidar measurements and a feed-forward controller $\sum_{FF}$ can derive an additional pitch demand θ_{FF} that can be added to the feed-back demand before pitch actuation of the wind turbine WT. Such a controller could ideally completely compensate any disturbances in v that are affecting WT, however due to turbine modelling and lidar measurement limitation the performance is always non-ideal. In this study we will focus on the drawbacks

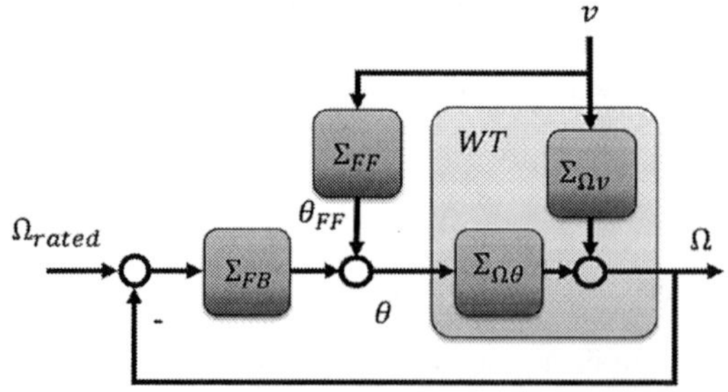

Figure 1. Example of a collective pitch feed-forward controller augmenting a traditional feed-back controller, from [2].

of estimating the rotor-effective wind speed (REWS) from lidar measurements. The REWS $v_{eff,R}$ is defined as the average of the longitudinal wind velocity u_x over the entire rotor plane

$$v_{eff,R} = \frac{1}{\pi R^2} \int_0^{2\pi} \int_0^R u_x(r,\theta) r dr d\theta \,,$$

where R is the rotor radius.

Coherent wind lidars can measure the frequency shift according to the optical Doppler effect of light backscattered on moving aerosols levitated in air. For a cw lidar the line-of-sight (LOS) measurement $v_r(\boldsymbol{r})$ at focus $\boldsymbol{r}$ is a convolution integral between the weighting function $L(s) = \frac{1}{\pi} \frac{z_R}{z_R^2 + s^2}$ and the LOS component of the wind velocity $\boldsymbol{u}$

$$v_r(\boldsymbol{r}) = \int_{-\infty}^{+\infty} L(s) \boldsymbol{n} \cdot \boldsymbol{u}((s\boldsymbol{n} + \boldsymbol{r})) \, ds \,,$$

where z_R is the Rayleigh length and $\boldsymbol{n}$ is the beam unit vector. This implies that only one component of $\boldsymbol{u}$ can be measured and a low-pass filtered signal compared to a point measurement. Together with the discreet number of measurement points and the evolution of turbulence between measurement and rotor plane, these are the imperfections of lidar measurements when estimating $v_{eff,R}$.

To evaluate the correlation between REWS estimated from the turbine and the lidar it is important to know the coherence between the two signals, which is defined as

$$\gamma_{RL}(k) = \frac{|S_{RL}(k)|^2}{S_{RR}(k) S_{LL}(k)} \tag{1}$$

with $S_{ij}(k)$ is the cross-spectrum between signal i and j as a function of the wave number $k = \frac{2\pi}{f}$; R refers to the turbine and L to the lidar. The spectra are calculated in blocks of 600 s. Further the coherence determines the minimum achievable measurement error [3]. Here we present measurements of $\gamma_{RL}(k)$ for two nacelle-mounted lidars manufactured by Windar Photonics A/S[1].

2. Theoretical Model

The coherence in eq. (1) can also be calculated analytically by using the Kaimal turbulence spectra and the exponential spatial coherence model defined in the IEC 61400-1 Ed.3 standard. Details can be found in [3].

3. Experimental Setup

An experiment using two lidars by Windar Photonics A/S was conducted at the Risø test site. The lidars were mounted on the nacelle of a Vestas V52 with 850 kW of rated power. The systems scan at discrete focus points on a cone with 2 or 4 beams, see figure 2. More information can be found in table 1.

To estimate REWS from lidar measurements no vertical components and a perfectly aligned turbine were assumed and the average of the n LOS components ($n \in \{2,4\}$) was calculated

$$\hat{v}_{eff,L} = \frac{1}{n \cos(\alpha)} \sum_{i=1}^{n} v_{r,i} \,,$$

where α is the half-cone opening angle. The sampling rate was 1Hz. The lidar measurements have also been corrected for turbine induction according to the theory of [4].

---

[1] Company webpage: http://www.windarphotonics.com/

Figure 2. Photo and illustration of the installation of the 4-beam lidar at the Risø test site.

To measure $v_{eff,R}$ from turbine data the method in [5] was followed. Measurements of pitch angle, rotor speed, low-speed shaft torque and air density were used in combination with the pre-calculated power coefficient surface to derive $\hat{v}_{eff,R}$ by applying conservation of angular moments for the drive-train.

Table 1. Overview of the measurement campaign

Quantity	2-beam lidar	4-beam lidar
Focus Distance (along LOS) L_f	37 m	62 m
Half-cone opening angle α	30°	18°
Measurement period	Mar-May 2016	Oct – Dec 2016
Nr. of useable 10Min averages	2043	1152

4. Results

The result for the measurement coherence is presented in fig. 3. It can be seen that for small wave numbers coherences for both lidars approach unity. As expected, the coherence for the 2-beam lidar drops at lower wave numbers compared to the 4-beam lidar since only measurements at two points over rotor plane are measured. It can also be seen that the measurement agree well with the theoretical calculations. The 4-beam lidar results show larger deviation from the theory, which could be attributed to the larger

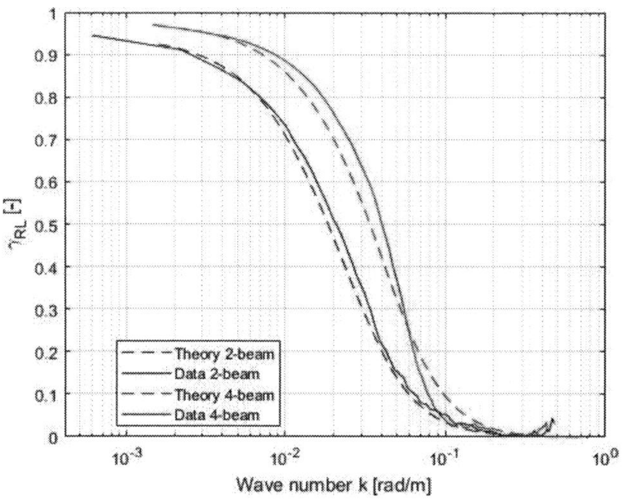

Figure 3. Coherence between lidar and turbine estimated REWS as function of wave number.

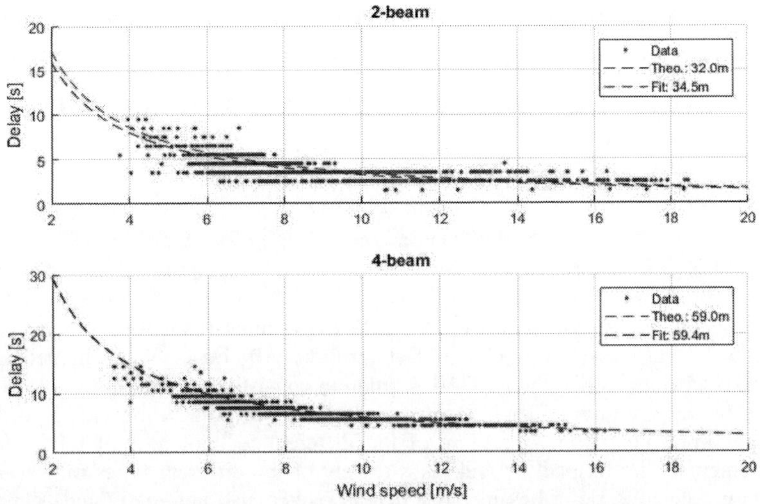

Figure 4. Measured time delay between turbine and lidar estimated REWS as function of wind speed.

distance between measurement and rotor plane (see table 1). The effect of turbulence evolution, which is not considered in the theoretical calculations, is bigger of larger distances.

Another important aspect for lidar-assisted pitch control is to predict when a disturbance that has been measured at the measurement plane will affect the turbine in the rotor plane. In order to do so, we assume that all fluctuations are advected with the mean wind speed (Taylor's frozen turbulence hypothesis). This mean wind speed can be estimated by taking a reasonable long average of the transversal wind speed component u_x. Here we used a 600 s average. The delay between the two signal Δt can then be calculated as

$$\Delta t(\overline{u_x}) = \frac{L_f \cos(\alpha)}{\overline{u_x}} \quad (2)$$

and it can also be estimated using the peak index of the cross-correlation between the two signals. The results are shown in fig. 4, where it can be seen that delays fit well to expected values. Likewise, a fit of L_f to eq. (2) shows that the delay speed can be predicted accurately by taking a 600s wind speed average.

In general, this study shows statistically how well lidar systems can estimate the REWS that affect wind turbines. The measurement results are in agreement with analytical models. This indicates that the theory models the underlying physics well and can be used to find optimal lidar configuration for lidar- assisted wind turbine pitch control.

5. References

[1] D. Schlipf, "Lidar-Assisted Control Concepts for Wind Turbines," PhD-thesis, Faculty of Aerospace Engineering and Geodesy of the University of Stuttgart (2015).

[2] D. Schlipf et. al., "Load Analysis of Look-Ahead Collective Pitch Control Using LIDAR," in Proceedings of the 10th German Wind Energy Conference, (Bremen, 2010).

[3] E. J. Simley, "Wind Speed Preview Measurement and Estimation for Feedforward Control of Wind Turbines," PhD-thesis, Faculty of the Graduate School of the University of Colorado (2015).

[4] J. Conway, "Analytical solutions for the actuator disc with variable radial distribution of load," J Fluid Mech **365**, 235-67 (1995).

[5] K. Z. Østergaard et. al., "Estimation of effective wind speed," J. Phys. Conf. Ser. **75**, 012082 (2007).

Collecting large wake vortex data with operational Doppler LIDARs for operational implementing of new Wake Turbulence regulations at major airports

Cariou Jean-Pierre, Thobois Ludovic
LEOSPHERE
14-16 rue Jean Rostand, 91400 Orsay, France

Abstract: For twenty years, coherent Doppler LIDAR systems (CDL) have been used for better understanding wake vortices (WV) behaviors at airports. All these studies helped to design new concepts of wake separations like the RECAT regulation consisting in redefining new WT categories. These new regulations are now under investigation by many airports worldwide in order to validate their operational implementation. A review of the different technologies of CDLs used for wake vortex measurements is performed, as well as a review of the different types of wake vortex (WV) algorithm. The presentation describes in more details one robust, automatic and accurate algorithm based on radial wind speeds designed for collecting large WV data for operational projects. It shows several examples of operational projects at major airports in the world like Paris-Charles de Gaulle airport or Hong Kong airport where wake vortex data collection has been started with CDLs.

Keywords: Coherent Laser Radar, Wake turbulence

1. Introduction

Major international airports are reaching more and more saturation due to the continuous and strong air traffic growth. One emerging solution is to improve the runway throughput mostly limited by the Wake Turbulence (WT) distance separation between two aircrafts, established sixty years ago by ICAO standards. WT behind aircrafts consists in Wake vortices (WV) which are a consequence of the pressure differential between the lower and upper surface of the wing moving through the air. The trailing vortices roll up into a pair of coherent, counter-rotating vortices. They can persist for more than 3 minutes after the aircraft has flown by. They can pose a potential hazard due to the rolling moment induced by the strength or the circulation of the WV to the following aircraft.
From 2000's, research LIDAR systems have been used for better understanding wake vortices behaviours [1] [2]. These studies allowed designing optimized concepts of WT distance separations like the RECAT-EU regulation developed by EUROCONTROL [7] which allows to gain 2% to 5% of airport capacity. This new WT regulation is currently planned to be implemented operationally at many congested airports worldwide and has been already successfully implemented at Paris-Charles de Gaulle airport.
This paper describes the way CDLs must be used for supporting the implementation of such optimized WT regulation at major airports. The methodology to be used for collecting huge amount of wake vortex data by scanning WINDCUBE CDLs is described as well as the associated wake vortex characterization algorithm. Finally, their operational use for the projects at Paris-Charles de Gaulle airport will be presented.

2. Instrumental set up and measurement campaigns

Before implementing new WT regulations at an airport like RECAT-EU, a safety assessment must be performed in order to quantify the risk. Its severity metric is based on the rolling moment in reasonable worst case conditions for any following aircraft encountering the WV of any leading aircraft. To do so, the local collection of WV data by scanning CDLs is mandatory to validate the statistical dimensionless decay models of the wake vortex circulation per aircraft. This decay can indeed be greatly impacted by the local air traffic management procedures, aircraft fleet, airport environment (mountains, sea, buildings) and above all local weather conditions. Such data collection should be performed for departures and arrivals and for all runways, and for the different seasons occurring at the airport. If the decay behaviors for each aircraft type are the same than the one as established in the RECAT-EU regulation the optimized WT regulation can be implemented. After this, this is still recommended to perform a risk monitoring allowing the analysis of wake vortex encounter (WVE) events reported by pilots by the ground truth database provided by the scanning CDL.

For both safety assessment and risk monitoring, CDL must provide the WV characteristics: detection, localization of their cores and the computation of their circulations. The fiber based CDL technology allows to adapt easily the parameters of the configuration like the spatial and temporal resolutions and the scanning patterns in order to monitor the wake vortices in an optimal way. Besides, the compactness of such CDL like a WINDCUBE allows to install them quickly at an airport on multiple runways, at different altitudes and during takeoff or landing. In addition, the blind zone is much lower. Figure 1 shows the typical setup used for WINDCUBE CDLs for measuring wake vortices to a runway. The CDL scans vertically (Range Height Indicator-RHI scans) and perpendicularly to the runway in order to well measure the transverse velocities induced by WV.

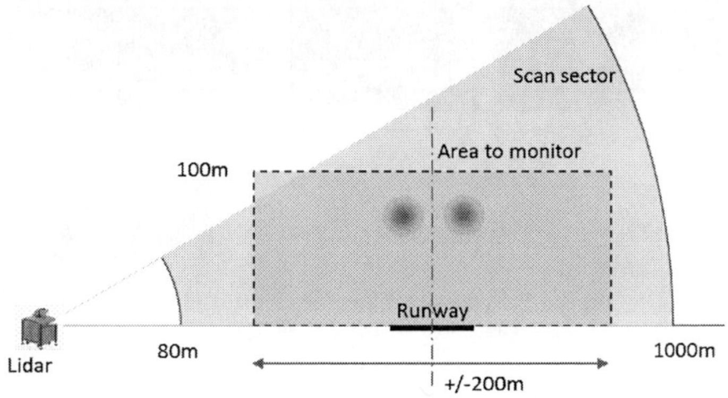

Figure 1. Typical scanning geometry for measuring wake vortices

Such a geometry is used for a WINDCUBE 200S CDL using 200ns pulses, very high resolution both along the beams (5m with overlap) and in across the beams (2m) and 8s of scan period. The scanning geometry of the WINDCUBE 200S is indeed optimized to ensure the accurate tracking and characterization of WV.

3. Signal processing

Many studies show the capabilities to track WV within the Doppler spectra [5] [6] but more recently within the mean radial wind data [7] as illustrated by Figure 2. The WV algorithm developed by LEOSPHERE for the WINDCUBE 200S is based on fitting a WV theoretical model, filtered by the lidar instrumental function, on the experimental radial wind speed map. This method is used to detect, localize and compute the circulation of WV. This last step is performed by a least mean square optimization of a Hallock-Burnham WV theoretical model on the radial wind speeds, as proposed by Smalikho [7]. This method ensures the algorithm to be adapted to short pulse CDL and to be robust and efficient when processing huge amount of data. A sketch of the WV algorithm is given on Figure 3 describing its major steps. This algorithm has been validated on real wake vortex data collected at Paris-Charles-De-Gaulle airport in 2011 and at Toulouse-Blagnac airport in 2014 [9].

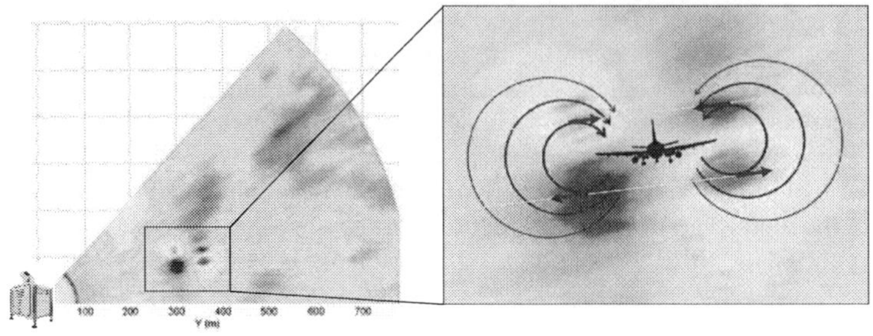

Figure 2. Example of map of radial wind speeds of wake vortices

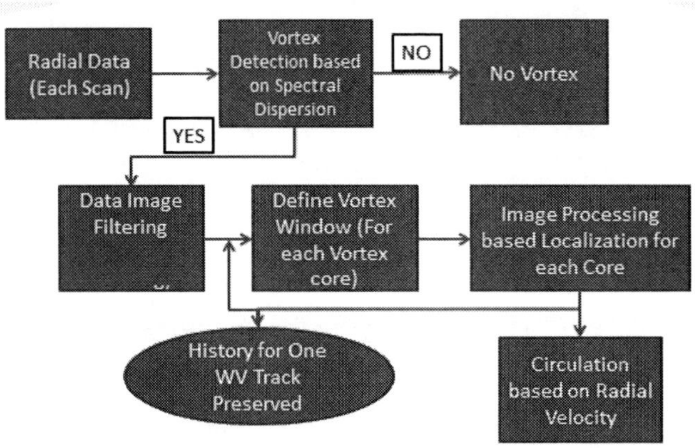

Figure 3. Synoptic of the wake vortex algorithm

4. Results and analysis

A WINDCUBE 200S CDL using the optimized scanning geometry and data processing for WV measurements are continuously collecting WV data since October 2015 at Paris Charles de Gaulle (CDG) airport. An example of WV track of a heavy aircraft is presented on the Figure 4 showing the decay in terms of circulation.

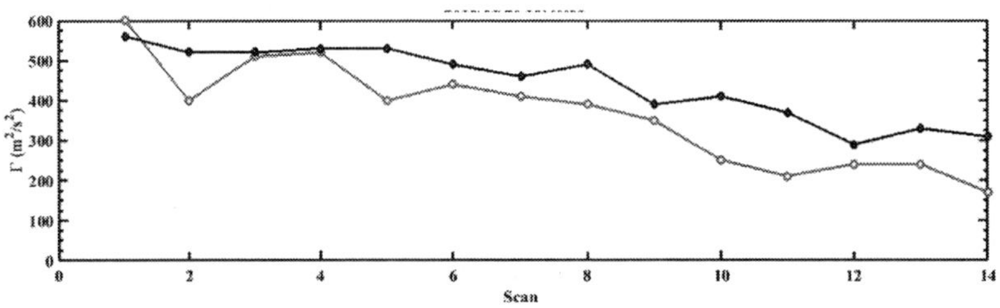

Figure 4. Time evolutions of circulations of WV of a Heavy aircraft on 20th November 2015 (red : left vortex ; black : right vortex)

In total, more than 100000 tracks have been already collected with the WINDCUBE 200S in vortex mode. Before the implementation of RECAT-EU regulation, the WV decay for each aircraft type measure at CDG has been compared to the reference database of RECAT-EU regulation. These analyses have been conducted for the CDG Database containing about 26000 tracks from 19th November 2015 to 22nd September 2016 and with the London Heathrow (EGLL) database containing about 55000 tracks collected from October 2008 until December 2010 with a WINDTRACER

LIDAR. In spite of the differences of technologies, pulse lengths and algorithms, the decay behaviors for each aircraft type between CDG and EGLL are very close as shown by Figure 5. The same satisfying results have been obtained when comparing the vortex spans. These studies show firstly the consistency between the old data collections performed with the research CDL and the WINDCUBE CDL based on optical fiber technology. Secondly, it demonstrates that the WV decay behaviors at CDG and EGLL are very close. This study, combined with others, has supported the implementation of the optimized RECAT-EU regulation at CDG in March 2016.

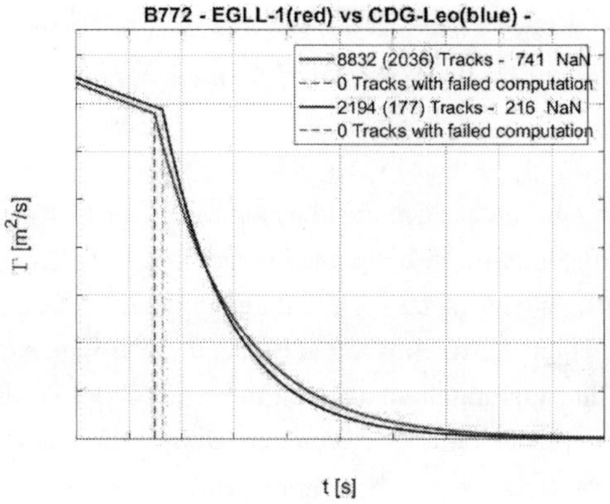

Figure 5. Wake decay comparison - B772

5. Conclusion and perspectives

The proposed methodology based on the WIDCUBE 200S optical fiber CDL in mode vortex has been developed and validated during several campaigns at airports and finally operationally used for the project of the implementation of RECAT-EU at CDG airport. The same methodology has been recently adopted by Hong Kong International airport where four WINDCUBE 200S currently under deployment will allow to collect WV during takeoff and landings at the four runway directions.

6. References

[1] "CREDOS: Crosswind-Reduced Separations for Departure Operations, EU Project Program", EU project Funded under FP6-AEROSPACE, Ref.: 30837, Record Number: 81468.

[2] "FAR WAKE Fundamental Research on Aircraft Wake Phenomena", EU project Funded under: FP6-AEROSPACE Ref.: 12238, Start date: 2005-02-01, End date: 2008-05-31.

[3] G. C. Greene, "An Approximate model of vortex decay in atmosphere", J. Aircraft, Vol. 23, July 1986, pp. 566–573.

[4] I. De Visscher et al., "Aircraft vortices in stably stratified and weakly turbulent atmo-spheres: simulation and modeling", AIAA J., Vol. 51, No. 3, March 2013, pp. 551–566.

[5] F. Köpp,et al., "Characterization of aircraft wake vortices by 2-micro pulsed doppler lidar", Journal of Atmospheric and Oceanic Technology, vol. 21, no. 2, pp. 194–206, 2004.

[6] A. Dolfi-Bouteyre et al., "1.5 m all fiber pulsed lidar for wake vortex monitoring, Conference on Lasers and Electro-Optics", Optical Society of America, 2008.

[7] I. N. Smalikho, V. A. Banakh, F. Holzäpfel, S. Rahm, "Method of radial velocities for the estimation of aircraft wake vortex parameters from data measured by coherent Doppler lidar", Optics Express 23(19) A1194-A1207 – 2015.

[8] EUROCONTROL, "European Proposal for revised Wake Turbulence Categorisation and Separation Minima on Approach and Departure", "RECAT - EU", Safety Case Report, 2013.

[9] L. Thobois et al., "Deployment of a next generation and operational LIDAR solution for monitoring wake vortices for supporting new wake turbulence regulations (Invited)", 8th AIAA Atmospheric and Space Environments Conference, AIAA Aviation, (AIAA 2016-3273)

Characterization of Aircraft Dynamic Wake Vortices and Atmospheric Turbulence by Coherent Doppler Lidar

Songhua Wu (a, b)*, Xiaochun Zhai (a), Bingyi Liu (a, b)
(a) Ocean Remote Sensing Institute, College of Information Science and Engineering, Ocean University of China, Qingdao 266100, China.
(b) Laboratory for Regional Oceanography and Numerical Modeling, Qingdao National Laboratory for Marine Science and Technology, Qingdao 266100, China.
**Email: wush@ouc.edu.cn*

1. Introduction

Wake vortex are large rolling air masses generated by aircraft as a consequence of lift. The behavior of that wake vortex is a major issue in aeronautical research [1,2]. They are strongest and most hazardous on congested airfields during take-off and landing, i.e. in close proximity to the ground. It is held that serious interactions between wake vortex and the ground can significantly affect the evolution of wake vortex, thus making them difficult to predict [3-4]. In order to investigate the real approach to mitigate wake vortex using 1.55μm coherent Doppler lidar and to study the wake vortex near ground effect (NGE), the wake vortex observation project was carried out during Jan-Mar 2017 at Beijing Capital International Airport (BCIA) [5]. This paper focuses on the NGE analyze based on 2017 BCIA experiment (2017BCIAE) observation.

2. Setup and Method

A scanning Doppler lidar and a wind profiler lidar are used during 2017BCIAE. The main specifications of these two setups are listed in Table 1. Figure 1 (a) shows the sketch map of lidar location at BCIA from 20 Jan 2017 to 20 Mar 2017. Figure 1 (b) shows the field experiment at BCIA. Since the direction of landing depends on the wind direction and the prevailing wind direction at BCIA in winter is northerly wind, the meteorological station in the south of 01L/36R runway was selected to deploy the CDL for wake vortex measurement.

Table 1. The specifications of the 3D scanning lidar and wind profile lidar

Qualification	3D scanning lidar	wind profile lidar
Wavelength	1.55 μm	1.55 μm
Data update rate	4 Hz (fastest)	1 Hz
Measurement range	40 m - 4000 m	40 ~ 240 m
Radial velocity measurement range	±37.5 ms^{-1}	±37.5 ms^{-1}
Radial range resolution	15 m - 60 m	configurable
Scanning features	VAD	5- DBS
Weight	~75 kg	45 kg
Power consumption	<300 W	90 W

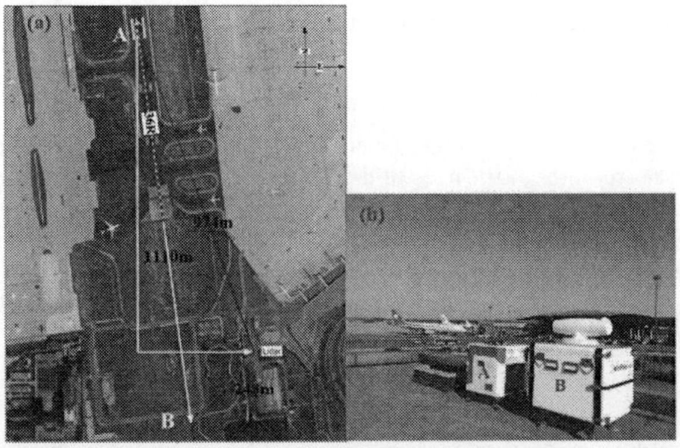

Figure 1. (a) The second outfield experiment of wake vortex NGE observation at BCIA during 20 Jan to 23 March 2017 (b) A: wind profiler lidar, B: Scanning PCDL.

Generally, the spectral width is firstly used to judge whether the wake vortex exists or not in one scanning measurement. If the maximal spectral width is less than the pre-set threshold (~10.5), it is regarded as the background wind field which will be used in the later wake vortex analysis. On the contrary, if the maximal spectral width is larger than the threshold, the wake vortex feature extraction is then processed based on specificity of wake vortex ground effect.

Let the array of Lidar estimates of the radial velocity $V(R_k,\varphi_m;n')$ contain the information about aircraft wake vortex starting from the scan number $n' = n_0 + 1$ and up to $n' = n_0 + N$. To avoid the influence of the background wind, we obtain the following data array:

$$V(R_k,\varphi_m;n) = V(R_k,\varphi_m;n_0+n) - V(R_k,\varphi_m;n_0)$$

Figure 2 (a) exemplifies the distribution of $V(R_k,\varphi_m;2)$ obtained from the data measured by CDL at the BCIA, as an A333 aircraft is scanned by CDL PPI mode at a height of 30 m. We can see that at distance from 250 m to 380 m from the CDL, the pair of wake vortex affects the radial velocity. Figure 2 (b) shows the distribution of spectral width $S(R_k,\varphi_m;2)$, and the spectral broadening due to wake vortex is obvious and thus can be an available feature to determine the existence of wake. To estimate coordinates of axes of the left and right wake vortex from the data analogous to those shown in Figure 2 (a) and (b), we introduce the function $S(R_k,\varphi_m;2)$, $D_s(R_k;n)$, $\varphi_{\max}(R_k;n)$ and $\varphi_{\min}(R_k;n)$ determined by the following algorithm:

$$\max_{\varphi}\{V(R_k,\varphi_m;n)\} = V(R_k,\varphi_{\max};n) \quad (1)$$

$$\min_{\varphi}\{V(R_k,\varphi_m;n)\} = V(R_k,\varphi_{\min};n) \quad (2)$$

$$D_v(R_k;n) = |V(R_k,\varphi_{\max};n)| + |V(R_k,\varphi_{\min};n)| \quad (3)$$

$$D_s(R_k;n) = \max\{S(R_k,\varphi_m;n)\} \quad (4)$$

where the maxima and minimums fall within the angle range φ_m for every distance R_k. Figure 2 (c) show $D_v(R_k;n)$ and $D_s(R_k;n)$ in green and blue lines, respectively. It can be seen that both $D_v(R_k;n)$ and $D_s(R_k;n)$ have two pronounced peaks. To better estimate the wake vortex coordinates, we take $a = D_s(R_k;n)/10$ as a weight factor, and introduce the function $D(R_k;n)$ determined by the following equation:

$$D(R_k;n) = a \cdot D_v(R_k;n) \quad (5)$$

which can be seen in Figure 2 (d)(e) described using black lines. So far the range bin of wake vortex can be determined exactly. Figure 2 (e) show the distribution of $\varphi_{\max}(R_{c_1};n)$ and $\varphi_{\min}(R_{c_2};n)$ in red and blue lines, respectively. Then the average of $\varphi_{\max}(R_{c_2};n)$ and $\varphi_{\min}(R_{c_2};n)$ can be regarded as the elevation of the vortex in the polar coordinate system. In Figure 2 (f), the red and black squares show the coordinates of the axes of the left and right aircraft vortex as obtained using above method.

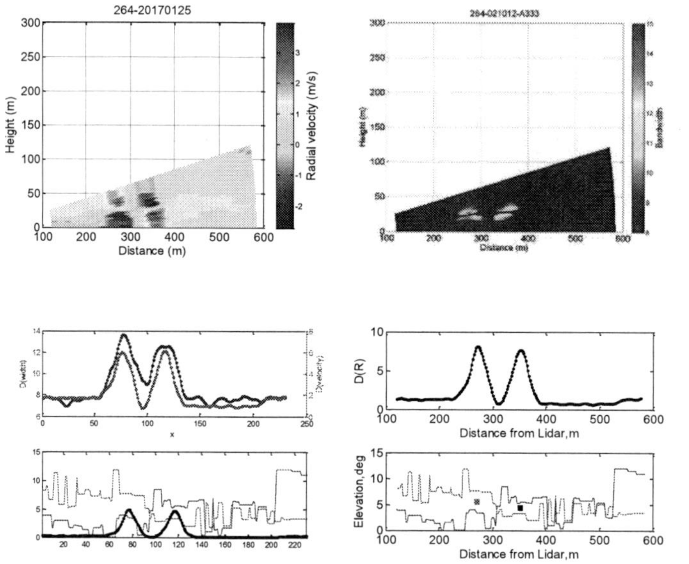

Figure 2. The CDL RHI scanning measurement (a) radial velocity (b) spectral width when an A333 crossed the scanning plane during 02:10:12 Jan 23 2017 at BCIA. (c) Corresponding function $D_v(R_k;n)$ (green curve), $D_s(R_k;n)$ (blue line) and $D(R_k;n)$ (black line) and (d) determined wake vortex position: red and black square for left and right vortex, respectively.

The circulation is defined as the mean of radius 5~15 m values. When the core position height is less than 15 m, that means the circulation in larger radius is unavailable. In this case, the continuation property is used based on Burnham-Hallock model. Figure 3 shows the distribution of wake vortex circulation with radius and corresponding B-H fitting. The fitting results match the measurement well and reasonable for circulation correction.

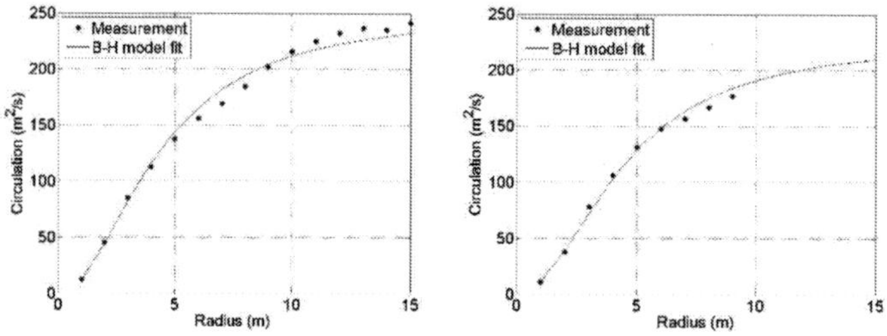

Figure 3. BH model fit using CDL measured data (a) without NGE, (b) with NGE.

3. Results

Figure. 4 shows the example of wake vortex NGE observation. It is typical for aircraft wake vortex evolution in ground proximity where the asymmetrical rebound is driven by crosswind. The results of CDL measurements of the normalized distance between the cores of the aircraft vortex, nadir angle, normalized vortex core altitude and horizontal distance, normalized vertical and horizontal component of the vortex core movement speed when A332 overflight during 17:46:17–17:49:52, 25 Jan 2017 at BCIA are shown in Figure 5 as a function of normalized time, respectively.

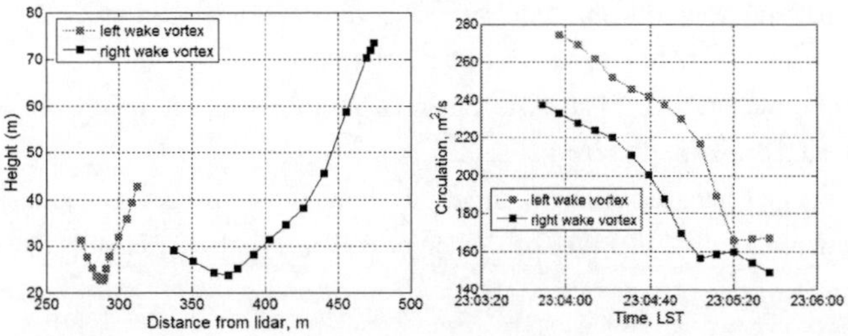

Figure 4. (a) Trajectories of left (red squares) and right (black squares) wake vortex axes and (b) evolution of wake vortex circulation generated by A333 over the BCIA at 17:46:17–17:49:52, 25 Jan 2017.

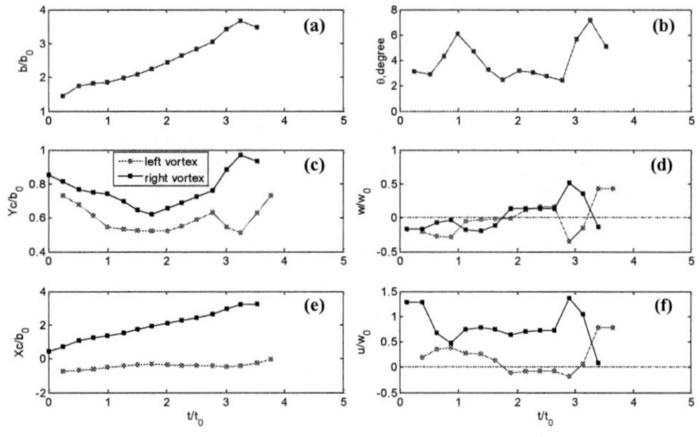

Figure 5. Measurement results of (a) the normalized distance between the cores of the wake vortex (b) the nadir angles of a pair of vortex (c) the normalized core altitude (d) normalized vertical component of the vortex core movement speed (e) the normalized transverse distance (f) normalized transverse component of the vortex core movement speed generated by A332 aircraft over the BCIA on 17:46:17 – 17:49:52 25 Jan 2017.

4. Discussion

The field experiment of wake vortex observation based on CDL quick-scanning mode in 2017 at BCIA is summarized. Different from the previous algorithm for the velocity envelope processing, a new algorithm based on the radial velocity and spectral width distribution has been developed to capture wake vortex evolution process under NEG. The circulation of wake vortex is also corrected and calculated using Burnham-Hallock model. Case studies shows the accuracy and reasonability of BH correction. A lot of representative cases under NGE have been retrieved using the batching processes based on this long-time period experimental observation. In the further study, the turbulence parameters from RHI structure function and the background wind profile from V300 wind profile measurement will be used to analyze the atmospheric condition effect on wake vortex evolution under NGE, which is fewer studied now, providing further knowledge and observational validation from simulation models.

References:

1. Köpp, Friedrich, Stephan Rahm, and Igor Smalikho. "Characterization of Aircraft Wake Vortices by 2-μ m Pulsed Doppler Lidar." Journal of Atmospheric and Oceanic Technology21.2 (2004): 194-206.

2. Smalikho, I. N., and Sh Rahm. "Lidar investigations of the effects of wind and atmospheric turbulence on an aircraft wake vortex." *Atmospheric and Oceanic Optics* 23.2 (2010): 137-146.

3. Holzapfel, Frank, and Meiko Steen. "Aircraft wake-vortex evolution in ground proximity: analysis and parameterization." *AIAA journal* 45.1 (2007): 218.

4. Robins, Robert E., and DONALD P. Delisi. "Potential hazard of aircraft wake vortices in ground effect with crosswind." *Journal of Aircraft* 30.2 (1993).

5. Wu, Songhua, et al. "Characterization of aircraft dynamic wake vortices and atmospheric turbulence by coherent doppler lidar." EPJ Web of Conferences. Vol. 176. EDP Sciences, 2018.

Wake Vortex Measurement Using Pulsed Doppler Lidar

Naoki Matayoshi(a), Eiichi Yoshikawa(a) and Masayuki Yamamoto(b)
(a) Japan Aerospace Exploration Agency
6-13-1 Osawa, Mitaka, Tokyo, Japan
(b) Mitsubishi Space Software Co., LTD.
1-6-1 Takezono, Tsukuba, Ibaraki, Japan
Lead Author e-mail address: matayoshi.naoki@jaxa.jp

Abstract: The Japan Aerospace Exploration Agency (JAXA) developed the original algorithm to detect wake vortex shed from aircraft using spectral data of pulsed Doppler lidar. The JAXA has been conducting the wake vortex measurement at two major airports in Japan using a pulsed Doppler lidar and demonstrated the capability of extracting wake characteristics including wake position and circulation strength from lidar spectral data. The acquired wake characteristics will be used to support safe and efficient aircraft operations.

Keywords: Pulsed Doppler Lidar, Wake Vortex, Aircraft Separation

1. Introduction

There is a strong demand to increase airport capacity by reducing aircraft separations to cope with increasing air traffic demand, which has been steadily expanding two-fold every 15 years [1]. The wake vortex separation minima are a major impediment to this since they require 4–8 NM separations in trail, which are greater than the radar separation minima of 2.5–3 NM that apply under radar surveillance (table 1). To reduce wake vortex separations, the International Civil Aviation Organization (ICAO) is discussing the re-categorization of wake vortex separation, a process named 'RECAT' [2]. The final goal of the RECAT is to introduce dynamic wake vortex separation according to pairs of aircraft types and weather conditions. The dynamic separation allows to reduce separations in favorable weather conditions when wake durations on flight paths become shorter. To realize this dynamic separation, the relationship between wake vortex behavior and surrounding weather condition must be clarified by developing wake vortex database measured in various weather conditions. There are several wake measurement activities using pulsed Doppler lidars to support the RECAT standardization [3][4].

In Japan, air traffic demand is also expected to grow and this demand might exceed airport capacities particularly around the Tokyo metropolitan area [5]. To increase airport capacity, the Japan Civil Aviation Bureau has been discussing the introduction of RECAT including dynamic wake vortex separation. The Japan Aerospace Exploration Agency (JAXA) has been conducting wake vortex measurements using a pulsed Doppler lidar at two major airports in Tokyo metropolitan area to support the RECAT introduction in Japan [6] [7]. The Mitsubishi Space Software supports the JAXA to analyze measured lidar data. This paper describes the JAXA's wake measurement activities. First, the JAXA's wake vortex detection algorithm using lidar spectral data is described. The lidar setup for wake measurement and example of measured wake vortex data are then presented.

Table 1. ICAO wake vortex separation minima.

Leading aircraft	Following aircraft			
	A380	Heavy	Medium	Light
A380	–	6 NM	7 NM	8 NM
Heavy (≥ 136ton)	–	4 NM	5 NM	6 NM
Medium (≥ 7ton)	–	–	–	5 NM
Light	–	–	–	–

2. Wake Vortex Detection Algorithm

The JAXA's wake vortex detection algorithm estimates wake vortex parameters including vortex core positions and circulation strength using pulsed Doppler lidar spectral data [7]. Although the pulsed Doppler lidar has an advantage that can scan wide area in short time, its spatial resolution, especially range resolution of several tens of meters, is too coarse to measure wake vortex whose core size is several meters. The JAXA's algorithm therefore uses lidar spectral data, which express probabilistic density distribution of radial winds in each lidar range bin, to estimate wake vortex parameter accurately.

The JAXA's algorithm is an iterative procedure to optimally fit wake vortex parameters to measured Doppler spectra acquired in RHI (range-height-indication) scan mode whose observation plane is perpendicular to aircraft flight path (fig. 1). We use a 2-D vortex model (eqn. 1) to model wake induced wind velocity field and assumes constant background wind.

$$v_\theta = \frac{\Gamma}{2\pi r}\frac{r^2}{r^2+r_c^2} \tag{1}$$

where v_θ: tangential wind speed, Γ: circulation strength, r_c: vortex core size, r: distance from core.

This assumption leads to a vector of sate variables that expresses wind velocity field in lidar observation plane (eqn. 2).

$$\mathbf{x} = [y \ z \ \Gamma \ r_c \ \bar{v} \ \bar{w}]^T \tag{2}$$

where y: lateral position of vortex core, z: vertical position of vortex core, $\bar{v}$: averaged lateral background wind, $\bar{w}$: averaged vertical background wind.

An optimum solution of the state vector $\mathbf{x}$ is calculated in a Bayesian scheme by minimizing the following cost function (eqn. 3).

$$J(\mathbf{x}) = \sum \frac{(y(x)-\hat{y})^2}{\sigma} \tag{3}$$

where $\mathbf{y}(\mathbf{x})$: Doppler spectral component estimated from the state vector $\mathbf{x}$, $\mathbf{y}$: measured Doppler spectral component, σ: the standard deviation of the known probabilistic property of Doppler spectral component.

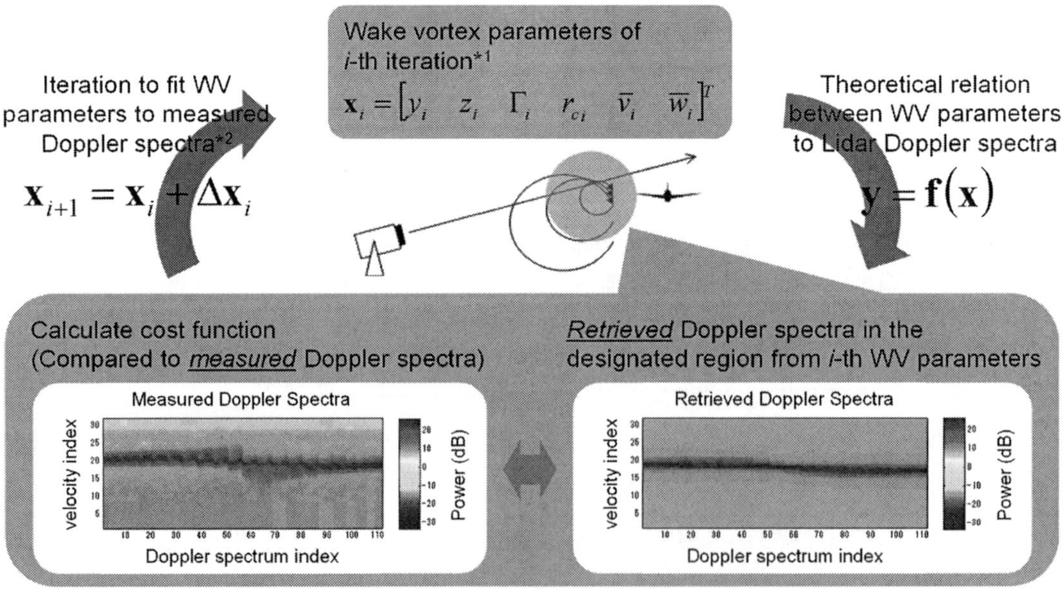

Figure 1. Flow of the wake vortex detection algorithm

3. Wake Vortex Measurement

The JAXA conducted wake vortex measurement at Narita international airport in 2013-2014 [6] and is conducting the measurement at Tokyo international airport from 2017. The JAXA uses the Leosphere WINDCUBE200S lidar for wake measurement (table 2). Figure 2 and table 3 show the lidar installation at Narita international airport. The high data output rate of 30 Hz enables rapid RHI scan of less than 10 sec while keeping enough elevation angle sampling of 0.2 deg for wake vortex detection. The lidar measured radial wind distribution in the vertical plane perpendicular to the flight path and wake vortex parameters were estimated for over 3000 landing aircraft. The collected data were mainly 'out of ground effect' wake vortex data since typical aircraft altitudes in the lidar measurement plane were about 400 ft. Whereas, the on-going measurement campaign at Tokyo international airport mainly collects 'in-ground effect' wake vortex data by locating lidar more closely to runway threshold. Typical aircraft altitudes in the lidar measurement plane are about 200 ft. As of March 2018, wake vortex data for nearly 4000 landing aircraft are collected. Figure 3 shows the example of measured 'in-ground effect' wake vortex data in calm wind condition. The wakes rebound near ground and linger long around landing path. The JAXA will develop the wake vortex database covering both 'out of ground effect' and 'in-ground effect' regions.

Table 2. Major specifications of Leosphere WINDCUBE200S

Wavelength	1543 nm
Range gate interval	1 m (minimum) with range overlapping
Number of range gates	240 (max)
Scan	0 – 360 deg (Azimuth), -10 – 190 deg (Elevation)
Pointing accuracy	0.1 deg
Radial wind speed	-30 – 30 m/sec
Measurements	CNR, radial velocity, velocity dispersion, Doppler spectrum, range, gazing angle, and time.

Table 3. Rapid RHI scan for wake vortex measurement

Range sampling	Every 5 m from 100 to 885 m
Physical range resolution	About 50 m
Elevation angle sampling	Every 0.2 deg from 0 to 40 deg
Velocity sampling	Every 3 m/sec from -30 to 30 m/s
Pulse repetition	18,000 Hz
LOS data output rate	30 Hz
Scan duration	6.7 sec (+2 sec to reset scanner head)

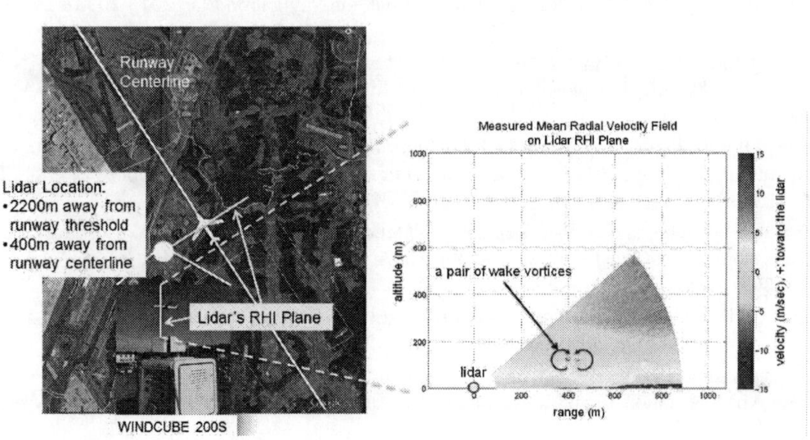

Figure 2. Lidar installation and example of measured data at Narita international airport

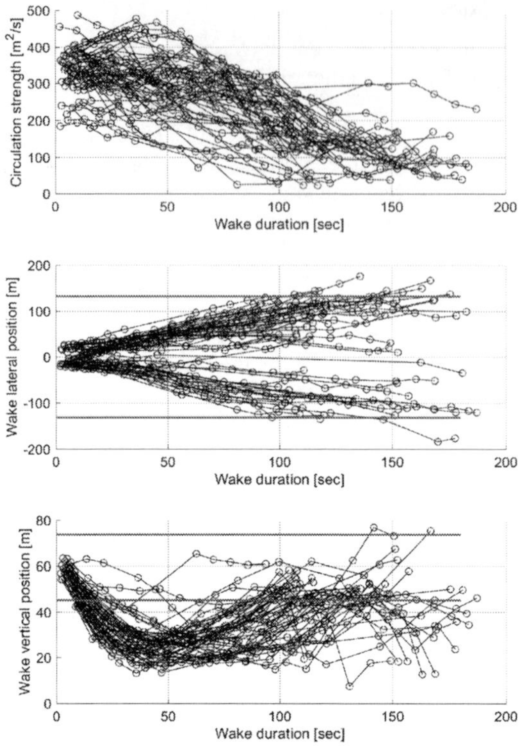

Figure 3. Example of measured wake vortex at Tokyo international airport

(lingered around landing path (indicated by red lines) for over 90 seconds)

4. Conclusions

This paper described the JAXA's wake measurement activities using pulsed Doppler lidar. The collected wake vortex data will be used to support the RECAT introduction in Japan and the standardization of dynamic wake vortex separation in ICAO.

5. References

[1] The International Civil Aviation Organization, "Global Air Navigation Plan 2016-2030", Doc 9750-AN/963 5th edition, (2016).

[2] J. Cheng, J. Tittsworth, W. Gallo and A. Awwad, "The Development of Wake Turbulence Recategorization in the United States", 8th AIAA Atmospheric and Space Environments Conference, AIAA AVIATION Forum, (AIAA2016-3434).

[3] M. J. Pruis, D. P. Delisi, D. Jacob, and D. Lai. "Summary of NASA Wake and Weather Data Collection at Memphis International Airport: 2013-2015 (Invited)", 8th AIAA Atmospheric and Space Environments Conference, AIAA AVIATION Forum, (AIAA 2016-3274).

[4] I. D. Visscher, V. Treve, and G. Winckelmans. "Characterization of Aircraft Wake Vortex Circulation Decay in Reasonable Worst Case Conditions", 54th AIAA Aerospace Sciences Meeting, AIAA SciTech Forum, (AIAA 2016-1603)

[5] The Japan Civil Aviation Bureau, "Collaborative Actions for Renovation of Air Traffic Systems (CARATS)", http://www.mlit.go.jp/en/koku/koku_fr13_000000.html (as of May 2018).

[6] N. Matayoshi and E. Yoshikawa. "Dynamic Wake Vortex Separation Combining with AMAN/DMAN Concept", 15th AIAA Aviation Technology, Integration, and Operations Conference, AIAA AVIATION Forum, (AIAA 2015-3397).

[7] E. Yoshikawa and N. Matayoshi, "Aircraft Wake Vortex Retrieval Method on Lidar Lateral Range–Height Indicator Observation", (AIAA Journal, Vol. 55, No. 7, 2017), pp. 2269-2278.

200-GHz 8-μs LFM Optical Waveform Generation for High-Resolution Coherent Imaging

Kevin W. Holman
MIT Lincoln Laboratory
244 Wood Street, Lexington, MA 02420 USA
kholman@ll.mit.edu

Abstract: Linear frequency modulated (LFM) waveforms are commonly used to provide range resolution in coherent imaging applications. Large bandwidth waveforms are needed for high range resolution, while short modulation times are needed to accommodate scene movement. In previous work on wideband swept waveform generation, the fastest reported modulation rates were on the order of 1-10 THz/s. Here we report the generation of a 200 GHz waveform with an unprecedented modulation rate of > 24,000 THz/s. We describe our unique scalable approach of modularly generating smaller-bandwidth waveforms across the optical spectrum, and then phase-stitching these sub-components together to form the full-bandwidth 200 GHz waveform in 8 μs. We also describe the use of this waveform for sub-diffraction-limited synthetic aperture imaging.

Keywords: Coherent Laser Radar, Coherent Lidar; Laser Radar; Ladar; LFM; Linear Frequency Modulation; Synthetic Aperture Ladar; SAL; Optical Waveform Generation

1. Introduction

Wideband optical waveforms are used to provide high range resolution in coherent imaging applications. When a linear frequency modulated (LFM) waveform is used, stretch processing techniques can be employed that provide range resolution commensurate with the full optical bandwidth of the waveform, with modest receiver electronic bandwidth. However, the waveform must be generated sufficiently rapidly to minimize blurring caused by movement of the scene.

In previous work on wideband swept waveform generation for coherent imaging, the fastest reported modulation rates were on the order of 1-10 THz/s [1, 2]. The waveform generation technique described here provides a 200 GHz waveform with an unprecedented modulation rate of > 24,000 THz/s. The generation technique involves modularly generating smaller-bandwidth waveforms across the optical spectrum, and then phase-stitching these sub-components together to form the full-bandwidth 200 GHz waveform in 8 μs. This waveform has applications to diffraction-limited 3D imaging with 1-mm range resolution, and sub-diffraction-limited synthetic aperture imaging, as described here.

2. Linear Frequency Modulated Waveform

The waveform commonly used for coherent imaging applications is the linear frequency modulated (LFM) waveform. This waveform enables simultaneous range and Doppler measurements, and is illustrated in Fig. 1. It consists of a linear frequency ramp over a bandwidth B, in duration T_{sweep}. It is repeated over a duration T_{CPI}. The time-delayed returned waveform (dashed blue line in Fig. 1) is combined onto a detector with a copy of the transmitted signal (red solid line) to produce the heterodyne beat frequency between these two optical signals.

This material is based upon work supported by the Assistant Secretary of Defense for Research and Engineering under Air Force Contract No. FA8702-15-D-0001. Any opinions, findings, conclusions or recommendations expressed in this material are those of the author(s) and do not necessarily reflect the views of the Assistant Secretary of Defense for Research and Engineering.

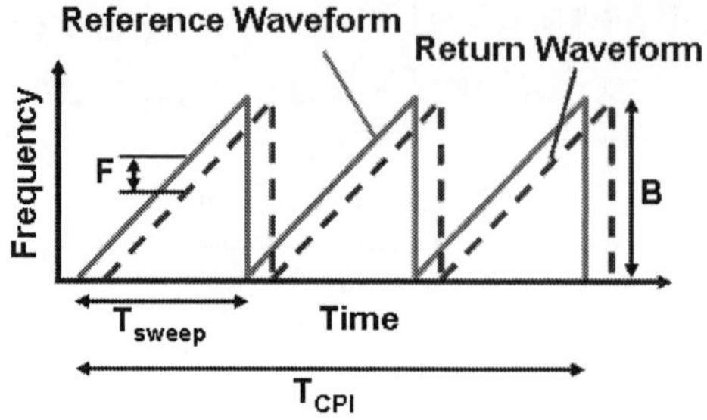

Figure 1. Linear Frequency Modulated Waveform.

The heterodyne beat frequency, F, is directly proportional to the target range, R, according to:

$$R = c\frac{FT_{sweep}}{2B} \tag{1}$$

The phase evolution of this beat signal during time T_{CPI} provides the Doppler measurement for each range resolution cell. The bandwidth of the sweep provides the range resolution, δR, and the total duration, T_{CPI}, provides the Doppler resolution, δf, according to the following equations:

$$\delta R = 1.33\frac{c}{2B} \tag{2}$$

$$\delta f = (T_{CPI})^{-1} \tag{3}$$

The factor of 1.33 in Eqn. 2 corresponds to a particular choice of windowing function used in computing the Fourier transform during range processing. Note that although the range resolution is determined by the optical bandwidth, the required detector bandwidth is determined by the much lower beat frequency F, which is related to the range to the target.

One final consideration for the waveform is that the differential optical phase change of the return light due to differential line-of-sight scene-platform motion across the scene should not exceed 2π from one sweep to the next. This sets a maximum velocity spread across a scene, v_{max}, of:

$$v_{max} = \frac{\lambda}{2T_{sweep}} \tag{5}$$

and thus a maximum rotation rate, ω_{max}, for a scene of diameter D :

$$\omega_{max} = \frac{v_{max}}{D} = \frac{\lambda}{2T_{sweep}D} \tag{6}$$

3. Generation of Broadband Optical Waveform

In generating an LFM waveform, a large bandwidth is desirable to provide high range resolution, and a short sweep time is desirable to support large velocity spreads across a scene. The waveform generation technique described here provides an LFM waveform with 200 GHz bandwidth, swept in 8 μs. This corresponds to a modulation slope of 25 THz / ms, as compared to modulation slopes of 1.25 GHz / ms and 10 GHz / ms reported in prior work [1, 2]. In addition to bandwidth and sweep time, another important consideration in waveform generation is linearity and phase distortion of the waveform, which would result in range and Doppler sidelobes for coherent imaging applications. Previous approaches to generating optical LFM waveforms for coherent imaging utilized a matched reference path to compensate nonlinearities in the sweep [1] or digital processing techniques to remove the effect of these nonlinearities [2]. The technique described here relies on an intrinsically low-noise approach to produce broad bandwidth sweeps over a sufficiently short duration, eliminating the need for a matched reference arm or digital post-processing.

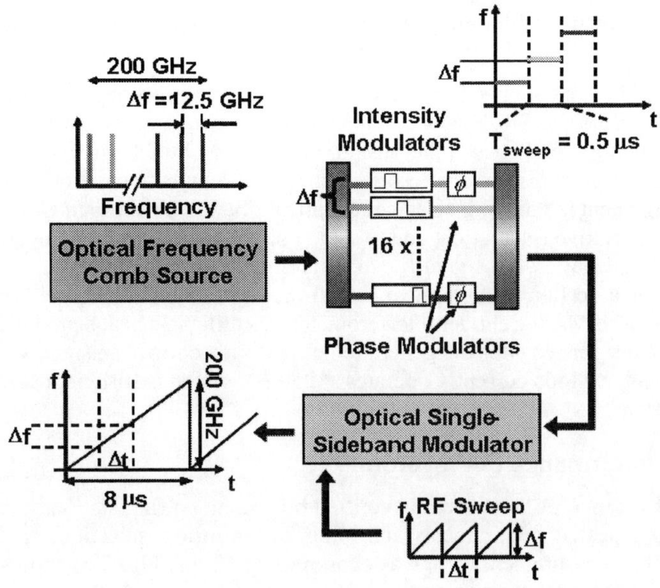

Figure 2. High-Bandwidth LFM Waveform Generation.

Our waveform generation technique relies on producing relatively small bandwidth waveforms across the optical spectrum, and then phase-stitching these sub-components together to form the full-bandwidth 200 GHz waveform. The general approach, summarized in Fig. 2, is scalable to even larger bandwidth waveforms. The building blocks of the waveform are the components of an optical frequency comb. This comb contains 16 frequency components, spanning 200 GHz and spaced by 12.5 GHz. It is generated by strong sinusoidal phase modulation of a single-frequency 1.55-μm fiber laser, followed by amplitude modulation to balance the amplitude across the comb [3]. These modulations are applied with LiNbO3 modulators external to the laser cavity. The frequency components of the comb are separated with an arrayed waveguide grating (AWG) demultiplexer, sending each frequency along a separate optical fiber. Each frequency then passes through an electro-optic intensity modulator, which passes each frequency for 0.5 μs, and then extinguishes that frequency for the remainder of the waveform generation time. The timing of the modulators is adjusted so that each of the frequency components is sequentially transmitted through the array of modulators. Upon recombination of the frequencies by an AWG multiplexer, a stair-step waveform of frequency vs. time is produced. An electro-optic phase modulator immediately after each intensity modulator provides phase control of the frequencies for compensating any relative drifts of the path lengths, as described later. The stair-step waveform is then transmitted through an optical single sideband modulator that is driven with a repetitive RF frequency-swept waveform. The RF waveform produces a 12.5-GHz sweep, centered at 10 GHz, in 0.5 μs. This introduces a sideband on each optical frequency (the transmitted carrier is nulled) that sweeps 12.5 GHz, and matches up with the beginning of the sweep of the next optical frequency's sideband. These 12.5-GHz frequency-swept waveforms are phase-coherently stitched together using the phase modulators following the intensity modulators to produce a total chirp of 16 × 12.5 GHz = 200 GHz in 16 × 0.5μs = 8 μs.

The two significant challenges in implementing this scheme are the generation of the 12.5 GHz RF LFM waveform in 0.5 μs, and the phase coherent stitching of the individual sub-chirps. The 12.5-GHz RF sweep is produced by first generating a 390.6-MHz broad sweep at baseband in 0.5 μs using an arbitrary waveform generator. Two such sweeps are produced on independent channels of the arbitrary waveform generator with the appropriate relative phase to drive an I/Q modulator, resulting in a 390.6MHz x 2 = 781.2-MHz sweep centered at 4.2 GHz. This waveform is then passed through 4 frequency doublers, resulting in a 12.5-GHz sweep bandwidth. Finally, a mixer is used to bring the center frequency of the sweep down to 10 GHz. A calibration procedure measures the amplitude and phase aberrations introduced by the 4 frequency-doubling stages, which are then pre-compensated by the arbitrary waveform generator.

The optical phase modulators are needed to control the relative optical phases of the 16 partial sweeps, such that the phase across the full bandwidth is continuous. Since the optical comb source produces components with a well-defined phase relationship, there is very little high-frequency phase noise. The dominant source of relative noise is due to differential path length fluctuations during the time when each frequency component is traveling on separate optical fibers. The phase modulators in each path are used to eliminate this relative fluctuation, and to set the phase offset of each component to be consistent with a continuous frequency-swept signal (i.e. a quadratic phase evolution). The phase deviations are measured by coupling a portion of the optical waveform into a monitor channel. This monitor channel consists of an unbalanced Mach-Zehnder interferometer, resulting in a beat signal. Digitizing and phase-demodulating the beat signal enables the generation of real-time correction signals to apply to the phase modulators. This feedback loop currently operates at 250 Hz, which is sufficient to suppress most of the path length fluctuations.

4. Imaging Performance of Waveform

The resolution performance of the system is verified by imaging a flat, stationary surface. The optical LFM waveform is repeated continuously, enabling the arbitrary selection of T_{CPI} for any given measurement. For this measurement, T_{CPI} was chosen to be 32 ms. Fig. 3(a) shows the range-Doppler image resulting from this measurement. Also shown are cuts through the Doppler and range peak locations, yielding the (b) range and (c) Doppler profiles, respectively. The range resolution is 1-mm, as expected for the 200-GHz waveform according to Eqn. 2. The sidelobes are nearly 20 dB down, and are understood to be caused by a finite reset time of the RF sweep generator of approximately 20 ns. The Doppler resolution is $(T_{CPI})^{-1}$ = 31 Hz, also as expected.

One coherent imaging application we have been exploring is inverse synthetic aperture ladar (ISAL) imaging of rotating objects. Fig. 4 illustrates the geometry for obtaining an ISAL image. In such an image, one dimension of resolution is provided by the range resolution of the waveform. The other dimension of image resolution is provided by resolving the Doppler frequency of the return light. According to Eqn. 5, a sweep time of 8 μs enables a maximum velocity spread across the rotating object of 10 cm / s for operation at λ = 1.55 μm. For a 20-cm diameter object, this corresponds to a maximum rotation rate of 500 mrad / s. (30 deg. / s).

Inverse synthetic aperture imaging of a face mask is demonstrated by rotating the mask while illuminating it with the wideband waveform. The mask is tilted backward at approximately 45 deg. enabling the range resolution of the waveform to provide the vertical resolution of the image. Rotation of the mask at 11 deg. / s provides horizontal resolution of the mask of 1 mm by resolving the different Doppler frequencies of the light scattered from the different horizontal regions of the mask. A T_{CPI} of 4 ms provides Doppler resolution of 250 Hz, according to Eqn. 3. For a rotation of 11 deg. / s, this corresponds to a horizontal displacement of 1 mm. Fig. 5(a) shows the mask being imaged, along with the reference spot indicating the size of the illumination beam. The 4-cm illumination beam is scanned across the mask, and the resulting series of range-Doppler images are summed to produce a single inverse synthetic aperture image, as shown in Fig. 5(b). The image in Fig. 5(b) is the result of 32 full-frame averages, thus reducing the speckle noise that would be evident in a single synthetic aperture image.

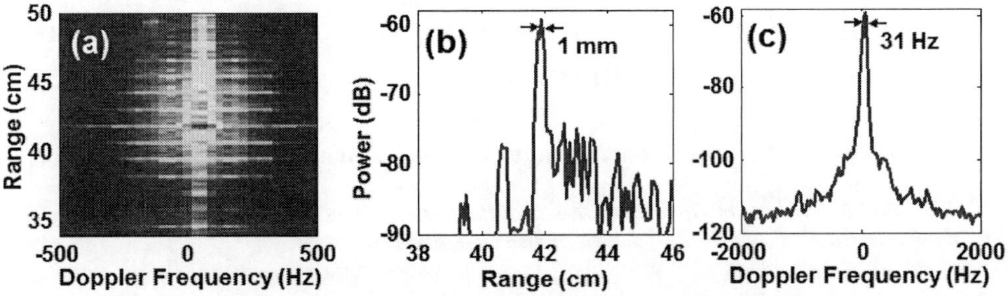

Figure 3. Range-Doppler Image of Stationary Surface.

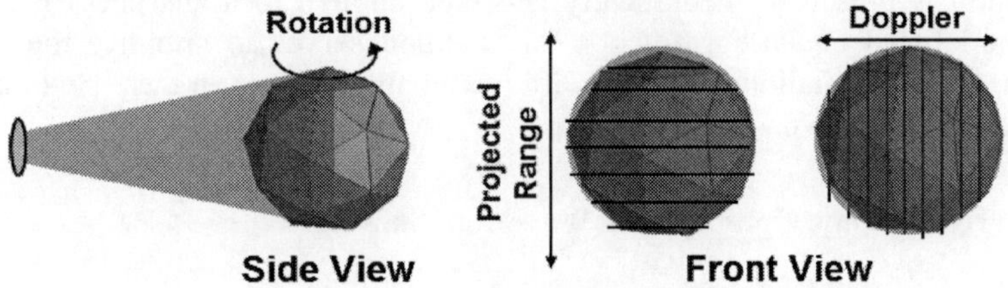

Figure 4. Inverse Synthetic Aperture Imaging Geometry

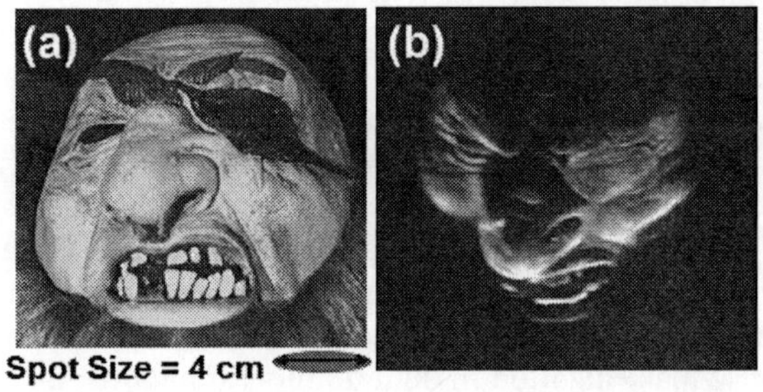

Figure 5. Inverse Synthetic Aperture Ladar Image

5. Summary

We have demonstrated an optical waveform generation technique that provides LFM waveforms suitable for making coherent images of moving objects. The 200-GHz waveform generated with a sweep rate of 25 THz / ms has sufficient bandwidth and speed to produce synthetic aperture images of rotating objects with line-of-sight velocity spreads of up to 10 cm / s, with a range resolution of 1 mm. The scalable waveform generation approach can be extended to larger waveform bandwidths while maintaining the sweep rate with the use of additional optical channels.

6. References

[1] M. Bashkansky, R.L. Lucke, E. Funk, L.J. Rickard, and J. Reintjes, *Opt. Lett.* **27** 1983 (2002).

[2] S.M. Beck, J.R. Buck, W.F. Buell, and R.P. Dickinson, *Applied Optics.* **44,** 7621 (2005).

[3] K. Mandai, D. Miyamoto, T. Suzuki, H. Tsuda, K. Aizawa, and T. Kurokawa, *IEEE Photon. Tech. Lett.* **18,** 679 (2006).

Application of Algorithmic Differentiation to Turbulence Mitigation in Digital Holography

Wesley E. Farriss (a, b), James R. Fienup (a, b)
(a) Leidos Inc.
(b) The Institute of Optics, University of Rochester, Rochester, NY, 14627, USA
wfarriss@gmail.com

Abstract: Image correction in digital holography via sharpness maximization has been an active research field for over a decade. In this work, the method of algorithmic differentiation is presented theoretically and then applied to a sharpness maximization forward model and analytic gradient. This method serves to simplify the derivation, expression, and computation of the forward model and gradient in these optimizations. A proof of concept is shown.

Keywords: Digital Holography, Optimization, Wavefront Sensing.

1. Introduction

Gradient-based optimization techniques have been applied to solve various optical engineering problems, such as phase retrieval [1], image sharpening [2], thin film design [3], and optical system aberration retrieval [4] to name a few. Similarly, work has been conducted on aberration correction in digital holography by optimizing over a small number (1-3) of phase planes with respect to maximized image sharpness metrics [5-7]. In order to ensure precision and expedite computation, explicit gradients are favored over methods such as finite differences in calculating the gradient of a sharpness metric with respect to a number of phase planes. However, gradients quickly become analytically cumbersome for more complicated forward models requiring many applications of the chain rule. This is indeed the case with turbulence mitigation/image enhancement of digital holograms using image sharpness metric maximization [6]. In this work, we present a brief discussion of the rules and advantages of algorithmic differentiation and show its implementation with respect to sharpness metric maximization in digital holographic images through turbulence.

2. Theory

Gradient-based optimizers use gradient, and possible Hessian, information as part of a process which locates a local maximum or minimum of an objective function with respect to the parameters we wish to solve for [8]. For a function $M(\vec{x})$, if any of the elements of $\vec{x}$ is sufficiently buried in mathematical operations inside of M, such as multiplication, exponentiation, linear transformations, etc., the traditional derivation process quickly becomes burdensome to track and difficult to interpret when reading. This is because the gradient requires multiple applications of the chain rule, product rule, quotient rule, inverse operator application, etc. In addition, the full analytical expression for the gradient is often difficult to isolate into components that allow for modularity in programming and often creates the requirement that certain values be calculated more than once during a single iteration of the optimizer. This motivates the implementation of a technique that addresses these issues. Reverse-mode algorithmic differentiation is a technique that has been used in the computer science field for some time, but has only recently been adopted by the optics community [9]. It is implemented by breaking the forward model and gradient down into a chain of concise operations, each of which feeds the next link in the chain. Additionally, it lends itself to a program architecture that allows values that are needed more than once to be passed and reused as a subroutine instead of re-computing them each time they are needed.

Sharpness-based image correction in digital holography is conducted under the assumption that volumetric turbulence can be well approximated as one or more 2-D transverse phase planes [10], placed along the axis of optical propagation between a target object and a detector. The sharpness of the image of the aberrated target image is assessed and the gradient (possibly also the Hessian) with respect to each the phase plane is computed [6]. This derivative information orients the direction of the line search, increasing the sharpness value of the target image during each iteration by making favorable updates to each phase plane. The power-law sharpness metric we use here is quantifies sharpness as $s = \sum_{x,y} I^{\alpha}$, where α is a real-valued number that makes dark areas in an image darker and bright areas brighter [7]. Here $I = (1/K)\sum_k^K I_k$, the incoherent average of speckle image intensities of the target. The traditional forward model for a power-law sharpness metric is [6]

$$s = \frac{1}{K} \sum_{x_{tgt},y_{tgt}} \left(\sum_k^K \left| \mathcal{P}_{Q\to tgt}\{\Psi_p \circ \mathcal{P}_{Q-1\to Q}\{\ldots \mathcal{P}_{1\to 2}\{\Psi_1 \circ \mathcal{P}_{0\to 1}\{F_{k,det}\}\}\ldots\}\} \right|^2 \right)^{\alpha},$$

$$\Psi_q(x_q, y_q) = \exp\left[i \underbrace{\left(\frac{2\pi}{\lambda} \sum_m c_{q,m} Z_m + \frac{2\pi}{\lambda} \Theta_{q,pbp} \right)}_{\Phi_q} \right], \quad (1)$$

where α is the power, K is the number of speckle realizations in an incoherent sum, $\mathcal{P}_q\{\cdot\}$ is the field propagator from the qth to the $q+1^{th}$ plane, $F_{k,det}$ is the field in the plane of the detector that has been propagated through the turbulent atmosphere, Z_m is the member of a chosen transverse basis set used to represent the turbulence, $c_{q,m}$ is the weighting coefficient associated with each Z_m at a plane q, and $\Theta_{q,pbp}$ is a point-by-point phase map at plane q, which also represents contribution of the turbulence. The symbol $\circ$ is used throughout this paper to represent element-wise products between arrays. Fig. 1 shows a graphical representation of the forward model.

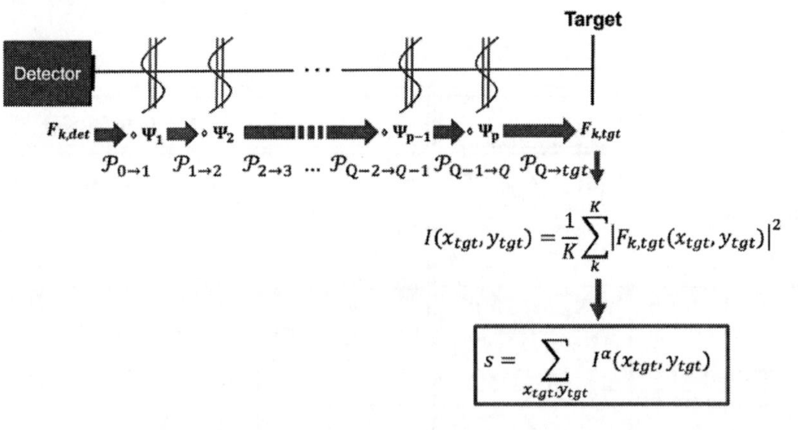

Figure 1: Sharpness maximization forward model.

The traditional gradient model is written explicitly as,

$$\frac{\partial s}{\partial c_{m,q}} = \sum_{x_q,y_q} \frac{\partial s}{\partial \Phi_q}\frac{\partial \Phi_q}{\partial c_{m,q}} = \frac{2\pi}{\lambda}\sum_{x_q,y_q}\frac{\partial s}{\partial \Phi_q} \circ Z_m \ ; \frac{\partial s}{\partial \Theta_q} = \frac{\partial s}{\partial \Phi_q}\frac{\partial \Phi_q}{\partial \Theta_q} = \frac{2\pi}{\lambda}\frac{\partial s}{\partial \Phi_q}$$

$$\frac{\partial s}{\partial \Phi_q} = \frac{2\alpha}{K}\mathrm{Im}\left\{\left[\sum_k \mathcal{P}^{-1}_{q+1\to q}\left\{\mathcal{P}^{-1}_{q\to q-1}\left\{\ldots \mathcal{P}^{-1}_{Q\to Q-1}\left\{\mathcal{P}^{-1}_{tgt\to Q}\left\{I^{\alpha-1} \circ F_{k,tgt}\right\}\circ \Psi^*_Q\right\}\circ \Psi^*_{Q-1}\right\}\ldots \circ \Psi^*_{q-1}\right\}\circ F^*_{k,q}\right]\circ \Psi^*_q\right\} \quad (2)$$

where $I = (1/K)\sum_k^K I_k$, the incoherent average of speckle image intensities of the target, $F^*_{k,q}$ is the complex-conjugate of the field value in the plane q resulting from the propagation of speckle field k during forward propagation including the effect of estimated phase screens. The reverse accumulation gradient corollary to Eq. (2) & Eq. (3) is shown in Table 1. We use overbar notation to represent the differentiation of a metric value with respect to a final dependent variable, $\bar{x}_n = \partial s/\partial x_n$, where s is the final metric and x_n is one of a number of dependent parameters that determine s. The chain rule necessitates that we find $\bar{x}_n$with respect to the sub-functions that make up s. If we consider $\vec{y} = f(\vec{x})$ to be one of these sub-functions, it's derivative is $f'_{mn}(\vec{x}) = \partial y_m/\partial x_n$. We know $\bar{x}_n = \sum_m (\partial s/\partial y_m)(\partial y_m/\partial x_n) = \sum_m \bar{y}_m f'_{mn}(\vec{x})$, where $\bar{y}_m = \partial s/\partial y_m$. The partial derivative of the metric with respect to each sub-function receives the overbar notation. Employing this convention to our problem means that $\bar{c}_{m,q} = \sum_{x_q,y_q}(\partial \Phi_{q,Z}/\partial c_{m,q})\,(\partial s/\partial \Phi_{q,Z}) = (2\pi/\lambda)\sum_{x_q,y_q}\bar{\Phi}_q \circ Z_m$. All upper-case quantities, except for K, in Table 1 are assumed to have spatial coordinates while all lower-case quantities are scalar. We see that this representation is much more concise and can be broken down into five convenient blocks for the purposes of subroutine implementation. These blocks represent one or more objects whose subroutines can be swapped out at will, with minimal effort in order to alter the functionality of the optimizer. Using the model in Table 1 also allows for easy passing of values such as I and $F_{k,tgt}$, which are computed during the computation of the objective function but used again during the gradient calculation, helping to minimize redundant computation. These values are simply stored in the modular instances of the subroutine until needed.

Table 1: Algorithmic Differentiation Model

Forward Model			Reverse Model		
Metric Block	**Propagation Block**	**Phase Component Block**	**Metric Block**	**Propagation Block**	**Phase Object Block**
$s = \sum_{x,y} I^\alpha$	$F_{k,q} = \mathcal{P}\{F'_{k,q-1}\}$	$\Phi_q = \Phi_{Z,q} + \Phi_{pbp,q}$	$\bar{I} = \alpha I^{\alpha-1}$	$\bar{F}'_{k,q-1} = \mathcal{P}^{-1}\{\bar{F}_{k,q}\}$	$\bar{\Phi}_q = \mathrm{Im}\{\bar{\Psi}_q \circ \Psi^*_q\}$
Intensity Block	$F'_{k,q-1} = F_{k,q-1} \circ \Psi_{q-1}$	$\Phi_{pbp,q} = \frac{2\pi}{\lambda}\Theta_{pbp,q}$	**Intensity Block**	$\bar{\Psi}_{q-1} = \sum_k^K \bar{F}'_{k,q-1} \circ F^*_{k,q-1}$	**Phase Component Block**
$I = \frac{1}{K}\sum_k^K I_k$	**Phase Object Block**	$\Phi_{Z,q} = \frac{2\pi}{\lambda}\sum_m c_{q,m} Z_m$	$\bar{I}_k = \frac{1}{K}\bar{I}$	$\bar{F}_{k,q-1} = \bar{F}'_{k,q-1} \circ \Psi^*_{q-1}$	$\bar{\Theta}_{q,pbp} = \frac{2\pi}{\lambda}\bar{\Phi}_q$
$I_k = \lvert F_{k,tgt}\rvert^2$	$\Psi_q = \exp[i\Phi_q]$		$\bar{F}_{k,tgt} = 2F_{k,tgt} \circ \bar{I}_k$		$\bar{c}_{q,m} = \frac{2\pi}{\lambda}\sum_{x_q,y_q}\bar{\Phi}_q \circ Z_m$

3. Results

In a test problem using algorithmic differentiation, a bar target of approximately 0.2 m in width was placed 150 m from a square detector array with a width of 0.1 m. We first simulated the 12 speckle realizations of the bar target, summing them incoherently without turbulent effects. The ideal image is shown in Fig 2 (c).

The fields producing Fig. 2 have been spatially filtered by the detector area. We choose an α-value of 1.1 for the sharpness metric. We generated the aberrated image in Fig 2. (a) by placing a $D/r_0 = 5$ simulated Kolmogorov turbulence screen in the plane of the detector and placing a different $D/r_0 = 5$ simulated Kolmogorov turbulence screen in the 75 m transverse plane and propagating the 12 speckle fields through the turbulence. Fig 2 (b) is the result of 3 rounds of optimization. The first round of optimization consisted of Chebyshev polynomials defined over the detector area and Chebyshev polynomials defined over the relevant 150 pixels of the plane halfway between the target and detector. During the first round, the middle phase plane did not have access to quadratic phase terms in order to avoid afocal telescoping/general non-physical over-sharpening of the image [5]. Polynomials were fit 25 at a time, holding all previously optimized polynomials static and then allowing all active polynomials to vary for another 8 iteration of the optimizer, a total of 160 total iterations. This process continued through ten levels of polynomial optimization with a total of 250 polynomials being defined over each phase plane (247 in the central plane when avoiding over-sharpening). After polynomials were fit, a point-by-point phase map was fit for a total of 80 iterations. Finally, polynomials were again fit, but this time including quadratic phase terms. Successful image correction is shown in the Fig 2 (b).

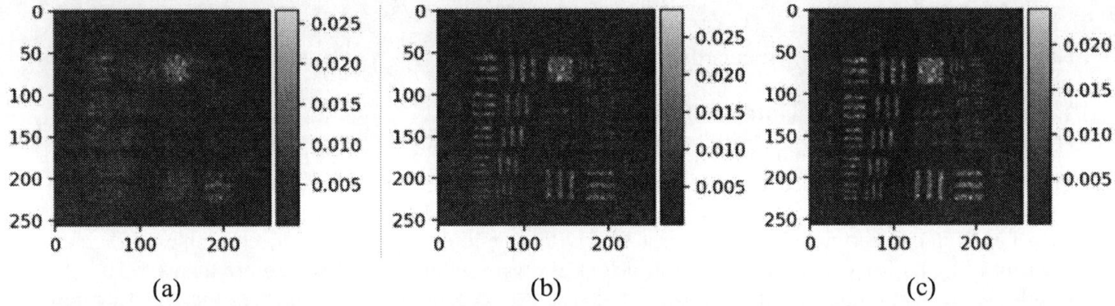

Figure 2: (a) Image aberrated with two x D/r_0=5 phase screens, (b) Sharpened Image, (c) Ideal image.

4. References:

[1] J. R. Fienup, "Phase-retrieval algorithms for a complicated optical system," Appl. Opt. 32, 1737-1746 (1993).

[2] R. G. Paxman, J. C Marron, "Aberration correction of speckled imagery with an image-sharpness criterion," Proc. SPIE 0976, Statistical Optics, (2 December 1988).

[3] A. L. Bloom, "Refining and optimization in multilayers," Appl. Opt. 20, 66-73 (1981).

[4] D. B. Moore and J.R. Fienup, "Efficient prescription retrieval from PSF data," in Frontiers in Optics 2015, OSA Technical Digest (online) (Optical Society of America, 2015), paper FTu5D.2.

[5] A. E. Tippie and James R. Fienup, "Multiple-plane anisoplanatic phase correction in a laboratory digital holography experiment," Opt. Lett. 35, 3291-3293 (2010).

[6] Abbie E. Tippie and James R. Fienup, "Phase-error correction for multiple planes using a sharpness metric," Opt. Lett.34, 701-703 (2009)

[7] S. T. Thurman and James R. Fienup, "Phase-error correction in digital holography," J. Opt. Soc. Am. A25, 983-994 (2008).

[8] J. Nocedal & S. Wright, *Numerical optimization, series in operations research and financial engineering.* (Springer, 2006).

[9] A. S. Jurling and James R. Fienup, "Applications of algorithmic differentiation to phase retrieval algorithms," J. Opt. Soc. Am. A 31, 1348-1359 (2014).

[10] R. G. Lane, A. Glindemann, and J. C. Dainty. "Simulation of a Kolmogorov phase screen." Waves in random media 2, no. 3, 209-224 (1992).

Various activities in NICT Space-Time Standards Laboratory including optical clocks and applications

Tetsuya Ido
National Institute of Information and Communications Technology
4-2-1 Nukui-kitmachi, Koganei, Tokyo 184-8795, Japan
e-mail address: ido@nict.go.jp

Abstract: Activities related to time and frequency standards in NICT are briefly presented. We not only generate and disseminate Japan Standard Time, but also develop highly-precise optical frequency standards to make an accurate frequency available in 24h/7d. A time scale generated by combining an optical lattice clock with a hydrogen maser always provides frequencies with the precision of low 10^{-16} level. We also demonstrated a fiber transfer of a standard signal to a Q-switch pulse laser system at 2μm, which is used for CO_2 LIDAR.

Keywords: optical frequency standards, time scale, LIDAR.

1. Japan Standard Time (JST)

National Institute of Information and Communications Technology (NICT) generates Japan Standard Time (JST), which is based on an atomic timescale comprising of an ensemble of eighteen Cs commercial atomic clocks (Microsemi Corporation "5071A") [1]. In this ensemble timescale, ticking rate of each clock is estimated from the last 30-day-trend and the weights for the ensemble average are determined by its frequency stability. If any clock shows a sudden rate change more than 1×10^{-14}, its weight is reduced to zero. To generate real signal synchronized to this ensemble timescale, a phase micro stepper (PMS) with reference to a hydrogen maser is employed. The PMS is automatically steered every 8 hours to trace the Cs ensemble timescale, and is manually steered to trace Coordinated Universal Time (UTC) if necessary. Atomic clocks and other critical devices are supplied with a large UPS, and a generator is equipped which has sufficient fuel to maintain power for three days. The building itself sits on quake-absorbing rubber. JST was not affected by the large earthquake in 2011. Figure 1 shows the frequency instability of JST calculated from the data during 2011 – 2015. Here, the hydrogen maser (HM#04) has a bottom of instability at low 10^{-16} level. But hydrogen masers have frequency drifts, which requires the signal to trace the Cs ensemble in longer time scale to utilize it as a timescale. Thus, the long term stability is determined by Cs ensemble to be at the 10^{-15} level.

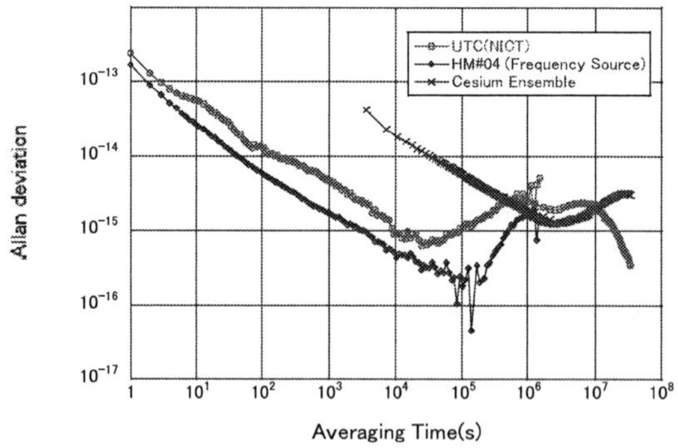

Figure 1. Allan deviation of the hydrogen maser, Cs ensemble time, and UTC(NICT). NICT generates a real-time signal equivalent to UTC, which is called UTC(NICT), of which Japan Standard Time is nine hours ahead.

NICT provides dissemination services of JST using LF radio-wave, internet, and the telephone lines. Signals from the two LF stations, Ohtakadoya-yama and Hagane-yama, mostly cover Japan including Okinawa. Both stations operate 24 hours a day. NICT also provides the public Network Time Protocol (NTP) service using a Field Programmable Gate Array (FPGA)-based NTP server which can accept up to one million NTP requests per second. NICT responds to more than 1.2 billion requests per day these days. The provision using telephone lines adopts two-way protocols that compensates the delay due to the signal transmission, realizing the accuracy of 1 ms in the user side. Companies such as broadcasting companies or rail-way companies use this service to synchronize their clocks.

2. Sr lattice clock

All time scales have so far been generated by microwave clocks. Optical clocks, on the other hand, have made rapid progress in the last twenty years, and clearly have surpassed the microwave clocks in terms of accuracy and stability.

In NICT, an optical lattice clock based on the ^{87}Sr $^1S_0$-$^3P_0$ transition has been in operation since 2011. The absolute frequency of the clock transition was evaluated in our system with uncertainty of 4×10^{-16}, which agreed with other measurements as shown in Fig. 2 [2]. The systematic uncertainty of our Sr system is 5×10^{-17}, where predominant sources of uncertainty are blackbody radiation shift, lattice Stark shift, and density shift. The absolute frequency measurements in Fig. 2 are all limited by Cs clocks as the frequency reference.

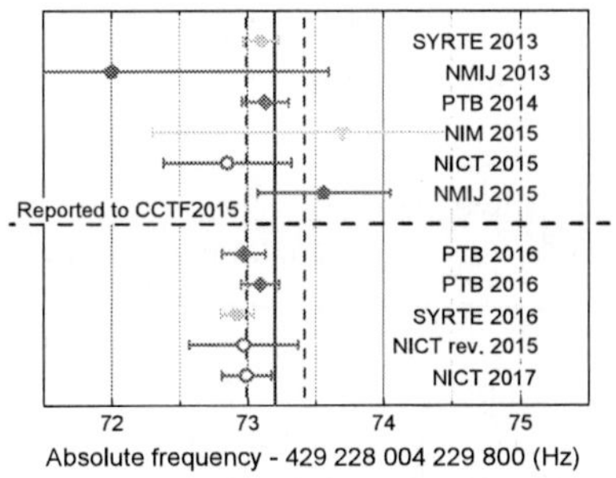

Figure 2. Absolute frequencies of the ^{87}Sr clock transition

While the small systematic uncertainties of optical clocks certainly allow more stable time scales, continuous operation of optical clocks is a difficult task. The complexity of the optical clock system still requires operators to monitor various parameters in course of operation. Thus, overnight operation is not yet realistic. On the other hand, high stability of optical clocks does not require long averaging time to reach low statistical uncertainty. Thus, short-time operations, such as 10^4 s, is sufficient to calibrate the ticking rate of a hydrogen maser, and then we can predict and compensate the frequency drift of the hydrogen maser with referring to the past intermittent calibrations.

3. Optical-microwave hybrid time scale

We operated the lattice clock steadily for half a year in 2016 [3] with the operation rate of once in a week or more, by which the steering parameter was set into the PMS. The resultant timescale TA(Sr) was as stable as UTC. As shown in Fig.3 as empty points, The TA(sr)-UTC monotonically increased to 8 ns over five months, indicating that 1 s in the TA(Sr) is longer than UTC for 0.62 fs. The accuracy of the one second in UTC is monitored by Cs fountain standards operated in metrological institutes worldwide. In every January, BIPM reviews the past UTC with respect to those result of monitoring, and publishes the further calibrated timescale, which is called TT(BIPMXY), where XY denotes the final year in the evaluation interval. In January 2017, TT(BIPM16) was issued, which is drawn in Fig. 3 as the dashed line (TT(BIPM16)-UTC). These two data indicate that the difference of 8 ns is not due to TA(Sr) but UTC. TA(sr)-TT(BIPM16) stayed in sub-ns, demonstrating a capability to generate a time scale combining an optical clock and hydrogen maser. Using the optically steered signal, we were able to evaluate the scale interval of UTC continuously for half a year.

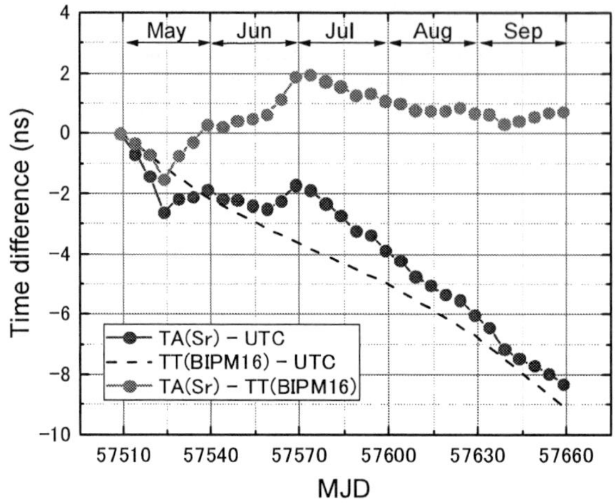

Figure 3. Time difference of the hybrid time scale with respect to UTC and TT(BIPM16). TT(BIPM16) is the most accurate timescale post-processed by BIPM. The hybrid time scale TA(Sr) keeps the difference against TT(BIPM16) within 2 ns during half a year.

State-of-the-arts Cs fountains microwave standards have accuracy below 10^{-15} these days. While some fountains including the Rb fountains are operated almost continuously, fewer new fountains have recently begun their operations. The community of time and frequency standards is instead inclined to develop optical standards. Ideally, the greatly improved accuracy and stability of optical frequency standards would be reflected to the timescale. Thus, it is productive to generate timescales using the intermittent operation of optical clocks as demonstrated here. The combination of fiber link networks and optical frequency standards will provide benefit of time scales to remote users, where we can obtain frequency standards in 24h/7d.

4. Providing a standard signal to LIDAR experiment

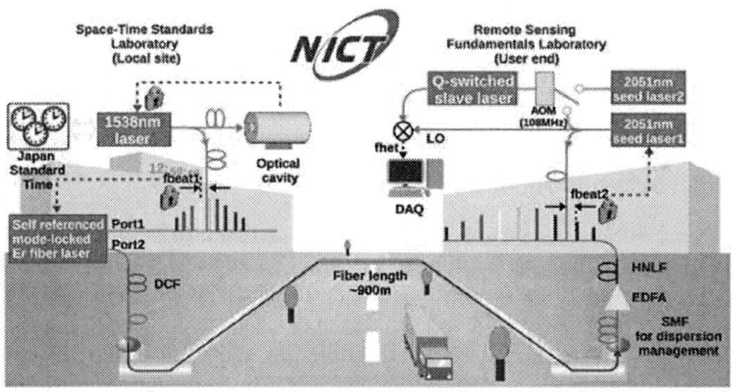

Fig. 4 Dissemination of standard signal to LIDAR system [4]

Remote sensing laboratory and space-time standards laboratory in NICT consider that standard optical signal transmitted over optical fiber may provide benefit to coherent LIDAR system. We established a fiber link between two laboratories and demonstrated the provision of standard signal as shown in Fig. 4. The optical standard signal at 1.5 μm is transferred over fiber, to which a seed laser at λ=2051 nm was phase stabilized via a frequency comb in the remote sensing laboratory. This seed laser was injected to Q-switched 2 μm pulse laser, and worked as a LO laser for the coherent light detection [4]. Here, the optical frequency of the Q-switched laser is well stabilized and is traceable to JST within an accuracy of 50 kHz [4].

[1] Y. Hanado, K. Imamura, N. Kotake, F. Nakagawa, Y. Shimizu, R. Tabuchi, Y. Takahashi, M. Hosokawa and T. Morikawa, International Journal of Navigation and Observation, vol. **2008**, p.841672, 2008.

[2] H. Hachisu, G. Petit, F. Nakagawa, Y. Hanado and T. Ido, "SI-traceable measurement of an optical frequency at the low 10^{-16} level without a local primary standard," Opt. Express **25**, 8511 (2017)

[3] H. Hachisu, F. Nakagawa, Y. Hanado and T. Ido, "Months-long real-time generation of a time scale based on an optical clock", Sci. Rep. **8**, 4243 (2018).

[4] S. Nagano, M. Kumagai, Y. Li, T. Ido, S. Ishii, K. Mizutani, M. Aoki, R. Otsuka, Y. Hanado, "Dissemination of optical-comb-based ultra-broadband frequency reference through a fiber network," Opt. Express **24**, 019167 (2016).

Fiber-based coherent DIAL/Doppler Lidar at 1645 nm for remote sensing of methane plumes

Nicolas Cézard(a), Simon Le Méhauté(a), Philippe Benoit(a), Didier Goular(a), Didier Fleury(a), Julien Le Gouët(a), Christophe Planchat(a), Matthieu Valla(a), Agnès Dolfi-Bouteyre(a), Xavier Watremez(b)

(a) Onera, the French Aerospace Lab
Chemin de la Hunière, BP80100, FR-91123, Palaiseau Cedex, France
(b) Total E&P
CST-JF, Avenue Larribau, FR-64018 Pau Cedex, France
nicolas.cezard@onera.fr

Abstract:A new coherent, fiber-based, and dual-function DIAL/Doppler Lidar is currently being developed at Onera, for remote sensing of CH_4 plume in case of industrial onshore or offshore gas leaks. The system is designed for operation at 1645 nm in pulsed mode to allow for range-resolved measurements. It relies upon a new fiber laser source developed in the lab. The laser is a semiconductor laser diode amplified and pulsed through a MOFPA architecture using two Raman amplification stages. The lidar emits 10 µJ, 100 ns, linearly polarized pulses at 20 kHz repetition rate from a monomode fiber. The pulse spectral width is 10 MHz, close from the Fourier limit, and is suitable for coherent detection. The frequency stability of emitted pulses is below 100 kHz (<0.1 m/s) for time accumulation over 20 ms. The lidar has been set in monostatic configuration for wind and CH_4 measurements tests.

Keywords: lidar, methane, CH_4, wind, fiber laser, Raman amplifier

1. Introduction

Oil and gas industry has known in the past serious methane leak crisis events, both in on-shore installations and off-shore platforms. Such accidents induce major risks for the working staff in the area and sometimes for inhabitants beyond in the case of on-shore facilities. In some cases, complete evacuation can be ordered, leaving the site unattended by human being. In such situation, remote sensing instruments are necessary to investigate the crisis scene at safety distance (typically 0.5-1 km). The critical need is to monitor the gas concentration and dispersion into the scene, in order to delineate danger zones and to characterize the leak flux rate. Of course, methane leaks also raise important environmental issues, since CH_4 is a major greenhouse gas.

Employed with favorable conditions, thermal infrared cameras can capture quickly the geometry and temporal evolution of a gas plume. However, this technique requires the existence of a thermal contrast between the gas plume and the background; otherwise the plume is invisible. Active Lidar systems have the potential to overcome this issue. A scanning and range-resolved DIAL lidar (Differential Absorption Lidar) can measure the 3D gas plume concentration directly in ppm. Hence it can reveal the gas plume geometry, quantify the leak as well, and the measurement is independent from the plume thermal contrast. The measurement time for a large scene could be quite long however. For faster measurements at local scale, Lidar systems offer an original alternative, with the possibility to measure locally the gas concentration simultaneously with the wind speed. Indeed, integrating the product of gas concentration C (in ppm or g/m^3) and wind speed V (in m/s) over a small scanning surface S, leads to the local flux rate Φ (in g/s) following the formula:

$$\Phi(\vec{r}) = \iint_S C(\vec{r})\left[\vec{V}(\vec{r}).\vec{n}\right]dS \tag{1}$$

Where r is the sensing point coordinate and n is the unitary vector normal to the local surface.

Simultaneous measurement of gas concentration and wind speed can be done by a single lidar instrument, designed for both Doppler and DIAL functions. This idea was formerly proposed by Gibert et al. for CO_2 measurements with a solid-state 2 µm laser [1] In our case, the developed instrument is based on fiber technology, which is well suited for coherent wind measurements, and hasinherent advantages such as alignment easiness, robustness to vibrations, and high versatility for temporal and spectral laser pulse shaping. However, in order to perform range-resolved CH_4 measurements, the fiber laser must deliver high peak power pulses around 1.65 µm, the main CH_4 absorption band in the silica transparency range. And unfortunately, at 1.65 µm, Erbium-doped fiber amplifiers (EDFA) – commonly used in 1.5 µm fiber Doppler lidars – are inefficient, for Erbium gain band stops around 1.62 µm. Hence it is necessary to consider another amplification process, and we have chosen non-linear Raman amplification.

We report in this paper i) the current development status of a high-power, Raman-amplification-based, fiber lidar at 1645 nm, designed for both DIAL and Doppler functions and ii) the first lidar measurements conducted with this new laser source.

2. High-power pulsed fiber Lidar at 1645 nm

The design of our combined CH_4/Doppler coherent lidar system is shown in fig.1. It employs two Distributed Feedback (DFB)laser diodes as master oscillators, running in cw mode, and set respectively at an On-line wavelength (absorbed by CH_4) and an Off-line wavelength (unabsorbed). The chosen CH_4absorption transition is located at 1645.54 nm (at peak), chosen for its high intensity, low temperature sensitivity, and low water-vapor interference [2] The On-line laser can be set at 1645.54 nm for classical two-wavelength DIAL operation, or step-scan modulated for multi-wavelength DIAL operation as shown in Figure 1. The Off-line laser remainsat a fixed frequency. Both lasers are fiber-coupled to a 2x1 optical switch, in order to alternate On and Off line sequences. While the output laser pulse repetition frequency (PRF) is 20 kHz, we only use a typical optical-switch rate of 1-10 Hz. Then the cw signal is separated into a signal channel and a Local Oscillator (LO) channel. In the signal channel, the beam first experiences an80 MHz frequency-shift using an Acousto-Optic Modulator (AOM), in order to set the detection Intermediate Frequency (IF). Then it is amplified and temporally-shaped by a pulsed Raman amplifier, itself being pumped in pulsed mode. More details on this key laser component are given below. The amplifier output pulses are then launched into the atmosphere through a free-space circulator (lenses L1, L2, a Polarizing Beam-Splitter, and a Quarter-Wave Plate). A lens telescope (L3, L4)is used to widen the beam emitted in the atmosphere up to 4 cm diameter.The emitted beam has an M^2=1 since the exit fiber is monomode. Coherent detection is performed by mixing the backscattered beam with the LO beam on a high-speed avalanche InGaAs photodetector with built-in amplifier. After data digitization and storage, signal processing methods are employed to derive the lidar raw products such as accumulated Carrier-to-Noise Ratio (CNR) and frequency barycentre, from where the target products (methane concentration and wind speed) can be derived.

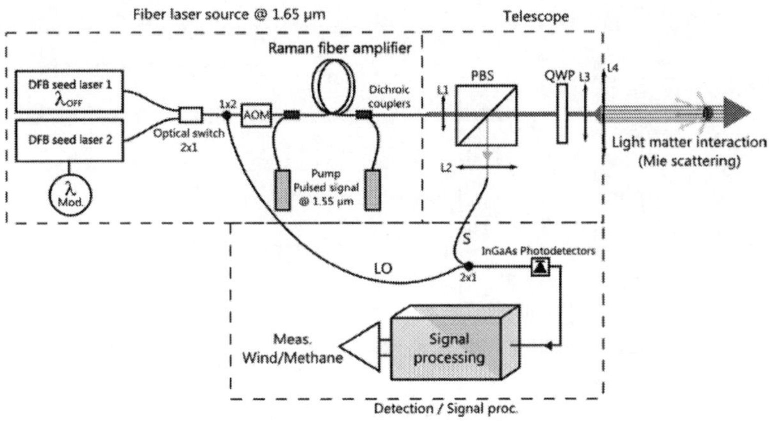

Figure 1.Combined CH4-Doppler coherent lidar design

Because EDFA cannot amplify light at 1645 nm, we use stimulated Raman scattering (SRS) amplification instead. A powerful pump laser propagating into a fiber interacts non-linearly with the silica matrix and experiences a spontaneous 13 THz down-frequency Raman shift. Using DFB seeders allows stimulating this energy transfer with a 13 THz downshift from a pump beam toward a signal beam. The overall process efficiency is weak (Raman gain is about 6.10^{-14} m/W in a standard 1550 nm PM fiber) but its bandwidth is large (about 2 THz). In our case, efficient Raman amplification at 1645nm is thus obtained using high power pump lasers at 1545 nm, which is a convenient wavelength for EDFAs. A germanium-doped highly-non-linear fiber (HNLF) is used in order to enhance the Raman gain up to 8.10^{-14} m/W. The HNLF is monomode at 1645 nm and maintains polarization.

The Raman amplifier is based on a two-stage MOFPA architecture (Master Oscillator Fiber Power amplifier), formerly described in [2] The pre-amplifier is pumped in co-propagative way at 1545 nm and delivers 100 ns pulses at 20 kHz with about 3W peak-power. Then the power amplifier, pumped at 1545 nm in contra-propagative way, amplifies the signal up to 150W peak-power (about 15 µJ). The difficulties in the setup are to prevent the signal at 1645 nm, and the pumps at 1545 nm, to experience detrimental stimulated Brillouin scattering (SBS) during their respective amplifications. Various Brillouin mitigation techniques have been employed to avoid this. Frequency-chirp patterns are applied to pump master oscillators (in both amplifier stages), while a distributed mechanical strain is applied to the power-amplifier HNLF. Indeed, frequency-chirp patterns could not be applied to the signal beam at 1645 nm, since a narrow spectral width is desired for coherent detection and also for DIAL operation. Finally, accounting for optical losses of the free-space circulator and output lenses, the lidar emits in the atmosphere linearly-polarized 10-11 µJ, 100 ns pulses at 20 kHz repetition rate out of a monomode fiber.

It was also observed that the Raman amplifier induces an additional non-linear phase at the same time that it amplifies the signal [3] The non-linear added phase ϕ_{NL} has been measured proportional to the logarithm of gain G, as shown in Figure 2 (left), following:

$$\phi_{NL}(t) = \kappa \ln\left(G(t)\right) \tag{2}$$

The coefficient κ has been measured to κ=3.2 ± 0.3 for the 1545pump wavelength. This relationship was found to be mainly attributable to a Cross-Phase Modulation process (XPM). A consequence is that a Gaussian-shape temporal pulse pattern is not the optimal shape for minimizing the signal spectral width and optimizing coherent detection performance. Instead, it was found that 100 ns square-shape temporal pulses resulted in the minimal Fourier-Transform spectral width, close from 10 MHz. Finally the spectral stability of the laser output frequency has also been measured. Figure 2(right) shows the Allan deviation of the beat note frequency barycenter between the LO and the output laser signal. It shows that for accumulation times between 20 ms and 100 seconds, the frequency uncertainty is below 100 kHz, hence the laser-induced error in wind measurements should be below 0.08 m/s. For CH_4 measurements, it also suggests that DFB lasers can remain in free-running mode.

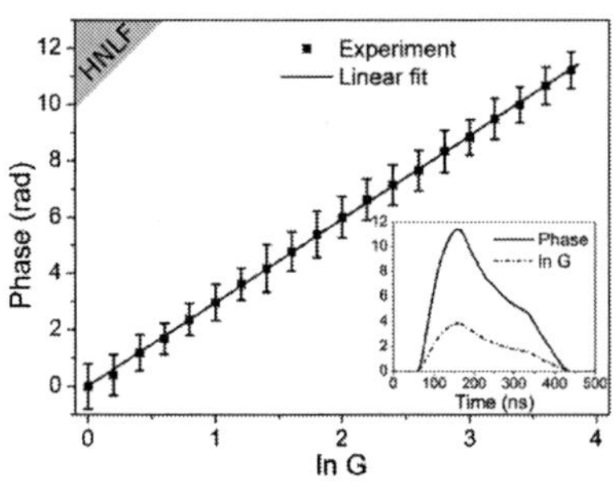

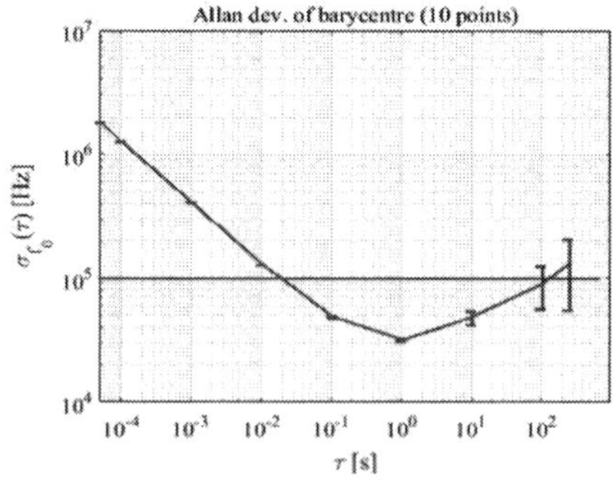

Figure 2.Left: Measured relationship between Raman amplification gain and Raman added phase. Right: Allan deviation of the fiber laser output frequency

3. First Lidar Tests

The Lidar has been placed on the laboratory rooftop, allowing conducting simultaneously range-resolved wind measurements tests, and Integrated Path Differential Absorption sensing tests (IPDA) of ambient CH_4. Atmospheric aerosols backscatter the lidar signal with a CNR that is sufficient in the first hundred meters (Figure 1Figure 3a) to derive the wind speed using a classical barycenter estimator (Figure 3b). At 2.5 km stands a communication antenna that we use as a hard target (Figure 3a) to perform IPDA CH_4 sensing. Assuming a mean CH_4concentration of 1.8 ppm in the atmosphere, the integrated methane content along the two-ways line of sight amounts to 9000 ppm.m, which is enough for detection (expected blue line in Figure 3c). A six-wavelength step-scan modulation along the absorption line centered on 1645.54 nm is applied on the ON-line laser (using current control), while the OFF-line laser is fixed. At each step, the lidar delivers ON-OFF sequences with a switch rate of 1.8 Hz during 4 seconds. Then the CNR ratio (Online accumulated CNR/Offlineaccumulated CNR) is computed. The six-point plot is shown in yellow line in Figure 3c. The baseline slope reflects the DFB power slope when tuning the current. Though this first measurement lacks details,the CH_4 absorption line is clearly visible.This is an encouraging result, calling for more accurate future experiments.

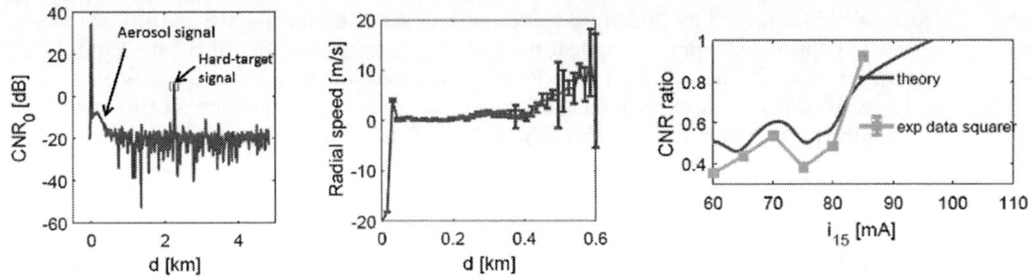

Figure 3 a) Signal CNR vs propagation distance; b) Range-resolved wind speed measurement up to 400-500 m; c) IPDA methane signature at 2.5 km range.

4. Conclusion and perspectives

We have reported a new coherent fiber lidar system at 1645 nmfor simultaneous measurements of wind speed and methane concentration. The performance of a specially-designed two-stage Raman fiber amplifier has been detailed. The laser is now integrated in a complete lidar architecture that is ready for a thorough test campaign. Very first results look encouraging and need further analysis. Future work will include more wavelength steps for IPDA measurements. Range-resolved methane measurements within an artificial gas plume shall also be tested. Another perspective, though in a different field, is to examine the potential of this technology for airborne/spaceborne applications, or for cross-checking tests of the currently developed 1645.5 nm MERLIN CH_4 satellite Lidar [4] .

5. Acknowledgments

This work was funded by Total E&P and ONERA partnership under NAOMI contract.

6. References

[1] F. Gibert et al. "Two-micrometer heterodyne differential absorption lidar measurements of the atmospheric CO_2 mixing ratio in the boundary layer", Appl. Opt. **45**, pp. 4448-4458 (2006)

[2] N.Cézard et al. "1.6 micron fiber laser source for CH_4 gas leak detection", Proceedings of ILRC 27, EPJ Web Of Conferences 119, 05010 (2016).

[3] P. Benoit et al, "Nonlinear phase added by a Raman fiber amplifier to a single-frequency seed laser," in *Conference on Lasers and Electro-Optics*, OSA Technical Digest (2016), paper SW4P.4.

[4] C. Pierangelo et al. "MERLIN (Methane remote sensing lidar mission): an overview", Proceedings of ILRC 27, EPJ Web Of Conferences 119, 26001 (2016).

Measurement of CO_2 mixing ratio profiles using ground based 3-wavelength 1.6 μm CO_2-DIAL with temperature measurement techniques in the lower-atmosphere

Yasukuni Shibata, Chikao Nagasawa and Makoto Abo
Tokyo Metropolitan University
6-6, Asahigaoka, Hino, Tokyo, Japan
sibata@tmu.ac.jp

Abstract: High-accurate vertical carbon dioxide (CO_2) profiles are highly desirable in the inverse method to improve quantification and understanding of the global sink and source of CO_2, and also global climate change. We have developed a ground based three-wavelength 1.6 μm direct detection differential absorption lidar (DIAL) to achieve simultaneously measurements of vertical CO_2 concentration and temperature profiles in the atmosphere. The high accuracy of absorption cross sections with temperature dependence is achieved by this three-wavelength DIAL. The three-wavelength DIAL system is constructed from the optical parametric generator and the direct detection receiving system that included a near-infrared photomultiplier tube operating at analog mode. In this paper, some experimental measurement results of temperature profiles and CO_2 mixing ratio profiles are reported from 0.4 to 2.5 km altitude.

Keywords: DIAL, CO_2, temperature

1. Introduction

Atmospheric gases, including carbon dioxide (CO_2), methane, ozone, chlorofluorocarbons, and water vapor, absorb some of the energy radiated by the Earth's surface, and thereby, retain and warm our planet. Before the Industrial Revolution, the atmospheric levels of CO_2 were around 280 ppm[1]. On May 9, 2013, the daily average concentration of CO2 in the atmosphere surpassed 400 ppm for the first time at the Mauna Loa Observatory in Hawaii, where the modern record of observations began back in 19582. IPCC 2013 has reported that the concentration will reach at least 440 ppm—more than 1.5 times the preindustrial level—by 2050[1]. Accurate predictions of atmospheric CO_2 concentration rely on the knowledge of its sources and sinks, transports, global distribution, and its variability with time[2]. In comparison with the ground-based monitoring network, CO_2 measurements for vertical profiles in the troposphere have been limited to campaign-style aircraft and commercial airline observations with the limited spatial and temporal coverage.

We have developed a ground based direct detection 1.6 μm differential absorption lidar (DIAL) to achieve measurements of vertical CO_2 profiles in the atmosphere[3–4]. As the spectra of absorption lines of any molecules are influenced basically by the atmospheric temperature, it is important to measure that simultaneously so that the better accuracy of the DIAL measurement is realized. The temperature dependency of the CO_2 density is 0.25 %/°C near the surface. Moreover, the CO_2 concentration is often evaluated by the mixing ratio (in units of μmol CO_2 per mol dry air, ppm). Because the air density is related by the ideal gas law, the mixing ratio is also related by the atmospheric temperature. The temperature dependency of the CO_2 mixing ratio is 1.3 ppm/°C near the surface.

So, we have developed new direct detection 1.6 μm CO_2 DIAL system for simultaneous measurements of the CO_2 concentration and temperature profiles in the lower-atmosphere. Laser beams of three wavelengths around a CO_2 absorption spectrum is transmitted alternately to the atmosphere. Moreover, the value of the retrieved CO_2 concentration will be improved remarkably by processing the iteration assignment of CO_2 concentration and temperature, which measured by these DIAL techniques.In this paper, the 3-wavelength 1.6 μm DIAL system that can measure simultaneously the temperature profiles with the CO_2 concentration profiles in the atmosphere and an experimental result of the simultaneous measurement of atmospheric temperature profiles and CO_2 concentration profiles are reported.

2. Temperature measurement by the 1.6 μm CO_2 DIAL

The mixing ratio C_x is the ratio of the air density n_a and the gas density n_x. Because atmospheric temperature T is related by the ideal gas law, the mixing ratio is also related by the atmospheric temperature. The pressure profile can be calculated quite accurately by the pressure height equation[5]. The new CO_2 DIAL technique for the atmospheric CO_2 concentration and temperature using three laser lines located on one 1.6 μm CO_2 absorption spectrum. In addition to the center absorption line λ_{on} (1572.992 nm) and the weak absorption λ_{off} (1573.137 nm) used for the conventional DIAL measurement, a third wavelength λ_T for temperature measurement is adopted to the off- center absorption line. Temperature profiles by the DIAL technique can be obtained from the ratio of receiving signals λ_T and λ_{on}.

$$\frac{S_T(z)}{S_{on}(z)} = \frac{E_T}{E_{off}} exp\left\{-2\int_0^z n_{co2}(Z)\Delta\sigma(Z,T)dZ\right\} \quad (1)$$

$$\Delta\sigma(z,T) = \sigma_{on}(z,T) - \sigma_T(z,T) \quad (2)$$

The atmospheric temperature $T(z)$ is calculated from the CO_2 density $n_{co2}(Z)$ which is obtained the DIAL measurement and the differential absorption cross section $\Delta\sigma$ with the initial temperature profile $T_i(z)$. We apply the iterative method to procedure for deducing the CO_2 concentration and the atmospheric temperature. Here, objective analysis data and U.S. Standard etc. are applied to the initial temperature profile.

3. Simultaneous measurement of CO_2 concentration and temperature

The direct detection 1.6 μm CO_2 DIAL system for simultaneous measurements of the CO_2 concentration and temperature profiles in the atmosphere has been conducted. Figure 1 shows a block diagram of the 1.6 μm CO_2 DIAL system, which was composed of an optical parametric generator (OPG) transmitter excited by a laser-diode pumped Nd:YAG laser with a high repetition rate of 500 Hz[4]. The partial power of λ_{on} and λ_T injection seeders were as split off and directed through wavelength-controlled units, respectively. The OPG output with three injection seeders was transmitted vertically into the atmosphere through a collimator with a 75 mm diameter. Three injection seeders are connected to an optical fiber switch and the output signals is switched for every shot in turn of λ_{on}, λ_{off}, λ_{on} and λ_T. The atmospheric backscatters are collected by a 25 cm diameter telescope. The optical receiver includes a near-infrared photo-multiplier tube (PMT) operating in the analog mode and a narrowband interference filter with a 1.0 nm FWHM. The system parameters are summarized in Table 1.

Table 1. DIAL system parameters

Laser wavelength	1.57 μm
Laser energy	2 mJ /pulse
Pulse repetition rate	500 Hz
Frequency stability	+-10 MHz
Telescope aperture	25 cm
Filter bandwidth	1.0 nm

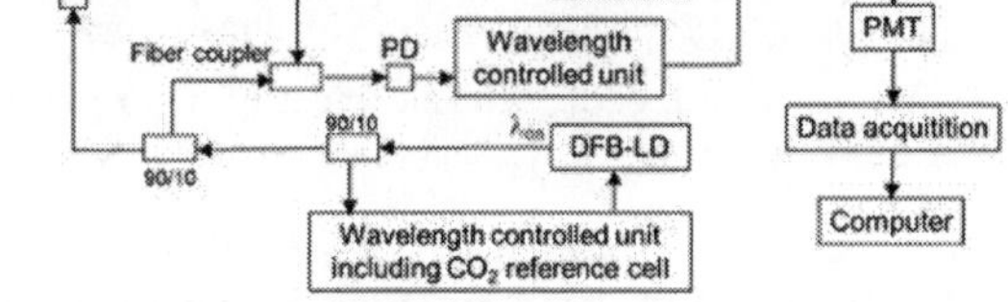

Figure 1. Block diagram of the 1.6 μm DIAL system for measurement of CO2 mixing ratio profiles.

Simultaneous CO_2 concentration and atmospheric temperature measurements were conducted from 0.4 to 2.5 km by the 3-wavelength DIAL on February 2, 2018 from 4:42 to 8:50 LT. Figure 2 shows a time-altitude color contour plot of CO_2 mixing ratio and atmospheric temperature measured by 3-wavlength DIAL on February 2, 2018. These profiles, each one based on an integration time of 30 minutes, are shown. The atmospheric temperatures below 1 km altitude are rising after sunrise (6:37LT). Temperature change from 7: 00 LT (-3.1 °C) to 9: 00 LT (1.0 °C) at 0.39 km altitude was 4.1 °C. If it is assumed that the air temperature within this time is the same, the CO_2 mixing ratio will differ by about 5.3 ppm. The simultaneous lidar observation of atmospheric temperature profiles greatly contributes to the improvement of the observation accuracy of the CO_2 mixing ratio in the lower atmosphere with large daily variation of atmospheric temperature.

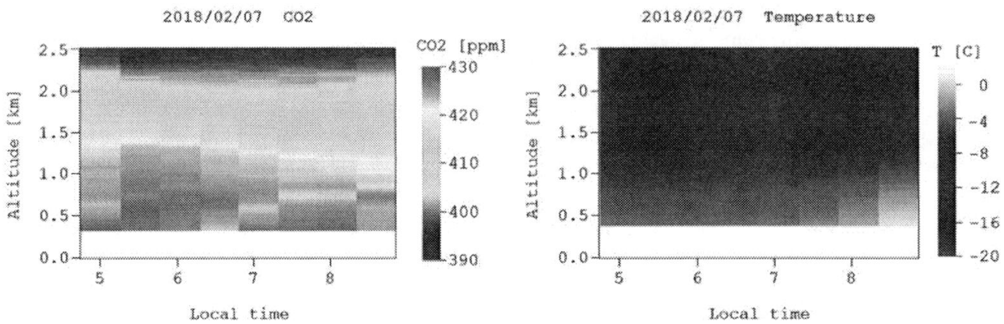

Figure 2. Continuous observations of the 3-wavlength DIAL for CO_2 mixing ratio and atmospheric temperature measurements from Tokyo Metropolitan University on February 2, 2018.

4. Conclusion

The development of the 3-wavelengths 1.6 μm range-resolved DIAL system for simultaneous measurements of the CO_2 concentration and temperature profiles in the troposphere is conducted. We found that the accuracy of the CO_2 mixing ratio profiles measured with the 3-wavelengths DIAL is improved significantly by the atmospheric temperature profile. By conducting a longer continuous observation, we plan to clarify the daily fluctuation of the vertical CO_2 concentration profile.

ACKNOWLEDGMENTS

This work was financially supported by the System Development Program for Advanced Measurement and Analysis of the Japan Science and Technology Agency.

REFERENCES

[1] F. Stocker et al., "IPCC, 2013: Climate Change 2013: The Physical Science Basis. Contribution of Working Group I to the Fifth Assessment Report of the Intergovernmental, Panel on Climate Change," Cambridge University Press, Cambridge, United Kingdom and New York, NY, USA, 1535 pp, doi:10.1017/ CBO9781107415324 (2013).

[2] B. B. Stephens, et al., "Weak Northern and Strong Tropical Land Carbon Uptake from Vertical Profiles of Atmospheric CO_2", Science, 316, 1732–1735 (2007).

[3] D. Sakaizawa, et al., "Development of a 1.6 μm differential absorption lidar with a quasi-phase-matching optical parametric oscillator and photon-counting detector for the vertical CO_2 profile", Appl. Opt. **48**, 748-757 (2009).

[4] Y. Shibata, et al., "Development of 1.6 μm DIAL using OPG/OPA transmitter for measuring atmospheric CO_2 concentration profiles", Appl. Opt.,**56**, 1194-1201 (2017).

Investigations on the performance of lidar measurements with different pulse shapes using a multi-channel Doppler lidar system

Albert Töws and Alfred Kurtz
Cologne University of Applied Sciences
Steinmüllerallee 1, 51643 Gummersbach, Germany
albert.toews@gmail.com

Abstract: Pulse distortion during pulse amplification and the consequential influence on measurement results is a problem for low and high power pulsed lidar systems. We present a multi-channel coherent Doppler lidar system with automatic feedback control of pulse shapes and durations on each wavelength-channel allowing this system to send laser pulses simultaneously with individual pulse parameters into the atmosphere. The influence of different pulse shapes and durations on the lidar performance in terms of the resulting signal-to-noise ratio, Doppler spectrum width, and velocity precision in atmospheric measurements is investigated. A comparison of two main transmitted pulse shapes will be discussed: a Gauß- and square-shaped pulse.

Keywords: Coherent Doppler lidar, multi-channel, pulse-shape control, Gauß- and square-shaped pulse

1. Introduction

Fiber based Doppler lidar systems are usually designed in master oscillator power amplifier (MOPA) architecture with erbium doped fibers (EDFA) as power amplifiers [i.a. 1, 2]. These fibers are pumped with laser diodes to achieve the population inversion which is needed for stimulated amplification. Usually, pulse lengths of 100 ns to 800 ns are applied in Doppler lidar systems as a compromise between range resolution and velocity precision. Injecting such long seed pulses into fiber amplifiers leads to decreasing population inversion during the pulse duration. Therefore, the gain decreases while the seed pulse passes the fiber amplifier. Thus, fiber amplifiers generate significant pulse distortions at typical pulse lengths and repetition frequencies in lidar systems.

In our lidar system, pulses on multiple different wavelength channels are simultaneously generated and one monostatic telescope for all channels is used. Thus, the same line-of-sight can be measured at the same time applying different pulse shapes and durations.

2. Methodology

The multi-wavelength lidar system [3] shown in Figure 1 consists of five overall units. The master oscillator unit contains multiple external cavity diode lasers. The wavelength channels are chosen from the ITU-grid near 1.55 µm. One laser light output of the master oscillator is directed to the pulse shaping unit, where the laser light is demultiplexed, so it is possible to shape the pulses for each channel with electro-optical modulators (EOM). Then the multiplexed pulses are frequency shifted by an acousto-optic modulator (AOM). The shifted pulses are amplified with a multi-stage EDFA.

The pulse-shape control circuit uses a part of the output power to measure the power and the shape of the output pulses. In order to achieve this, the wavelength-channels are demultiplexed and illuminate separate photodiodes. It is essential to pre-shape the pulses with the AOM for gain saturation compensation. The feedback control unit iteratively calculates a corrected seed pulse shape for each channel generated by the EOMs.

Via circulator the amplified and shaped pulses are directed to the collimator, where the light is decoupled. The polarization sensitive transceiver is realized by a quarter-wave plate and the circulator. In the detection unit the wavelength-channels of the local oscillator and the backscattered light are demultiplexed, thus, every channel can be coupled separately to its balanced detector. The signal processing unit extracts the information on the wind velocity and the backscattered power.

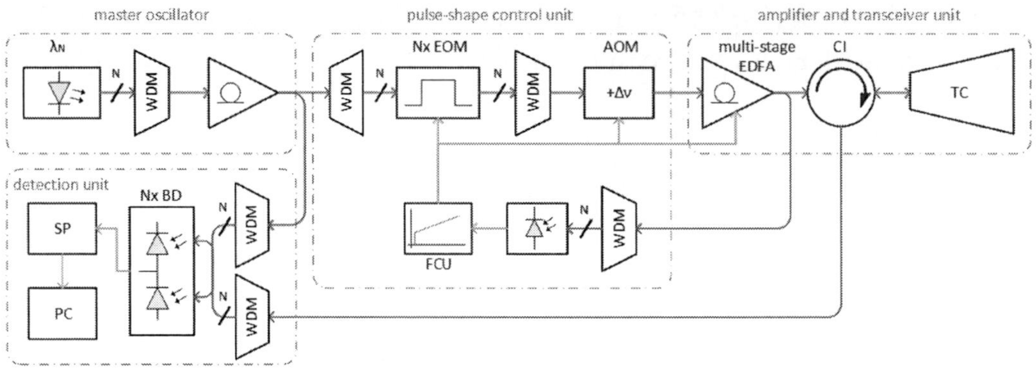

Figure 1. Overview of the multi-channel coherent Doppler lidar system.

3. Atmospheric observations with multiple pulse durations

To compare different pulse durations under atmospheric conditions, two channels of the multi-wavelength system can be utilized. Sending identically shaped pulses on two wavelength-channels, the evaluated return-signals show minor differences concerning estimated velocity and width of the Doppler-spectra. In contrast, by applying different pulse-shapes there are significant changes in the evaluated return-signals. In the first step we investigate the influence of different pulse durations of square-shaped pulses on the atmospheric return signal.

The chosen FWHM durations are 300 ns and 600 ns. The measured pulse energies at SBS threshold are 2.2 μJ and 4.3 μJ, respectively. Figure 2c shows the two regulated square-shaped pulses with different pulse durations. The resulting range gate weighting function (RWF), Figure 2b, is the convolution of the probing pulse and the FFT-window, and the effective longitudinal size of the sensing volume is defined at $\Delta z = \mathrm{RWF}(0)^{-1}$ [4]. The calculated RWFs for both measured probing pulses are $\Delta z = 46$ m for the shorter pulse and $\Delta z = 91$ m for the longer pulse. The line-of-sight velocity, Figure 2a, is estimated by applying the FFT with zero-padding and a peak finding algorithm. 30,000 spectra (3 s) were averaged for this measurement. Because of the longer sensing volume, the longer probing pulse channel (red line) shows the wind structure more smoothed along the line-of-sight than the shorter pulse. The channel with the shorter pulse determines the wind structure in more detail. This can especially be seen around 250 m and 450 m, where the blue channel estimates the velocity up to 0.4 m/s lower than the red channel due to wind gradients.

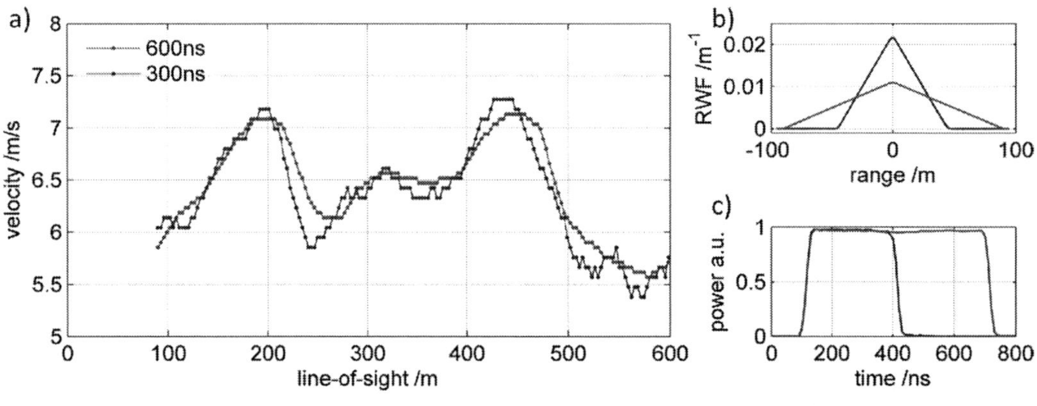

Figure 2. Atmospheric comparison measurement of line-of-sight velocity for different pulse durations.

The Doppler spectrum width over the range for both pulse durations is shown in Figure 3a. This spectral width contains information on the instrumental broadening and the broadening caused by changes of the wind velocity within the sensing volume. This might be wind shear or turbulent fluctuations. By comparing Figure 2a to Figure 3a it can be obtained that the Doppler spectrum width increases in the vicinity of 250 m and 480 m and the range dependent wind velocity shows a significant decrease of velocity. The change in bandwidth is more pronounce for longer pulses. Therefore, the increased Doppler width in Figure 3a is an indication for a stronger wind gradient. The Doppler spectrum of longer pulses experiences stronger broadening due to the larger contribution of different velocities. Therefore, the difference between those spectral widths is not constant over range and depends on the level of wind gradient.

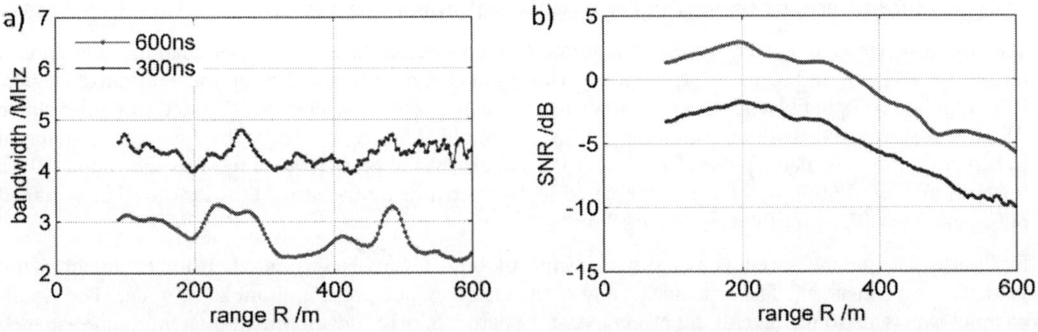

Figure 3. Measured a) FWHM bandwidth and b) narrowband SNR for different probing pulse lengths.

The measured SNR is proportional to the pulse energy of the probing pulse. Compared to the 300 ns pulse the pulse energy of a 600 ns pulse is higher by a factor of two due to SBS limitation. This results at least in a 3 dB higher SNR for the longer pulse. Additionally, the spectral width of the longer pulse is smaller by a factor of two. According to the lidar equation [5] , the higher pulse energy and the lower spectral width should result in a 6 dB higher SNR. Figure 3b shows that the SNR of the longer pulse is 4-5 dB higher. In regions where the Doppler spectrum experiences stronger broadening due to wind gradients within the sensing volume the difference in the SNR is reduced.

According to the variance summation model [6] , both, the reduced spectral width and the higher SNR, lead to a higher velocity precision for longer pulses. To measure the velocity precision, the difference between two successive registered peak-frequency values is evaluated to prevent long-term influences of wind velocity changes. The time between those measurements was around 0.05 s to ensure that the change in velocity of the particles is negligible in such a short time window. The difference of measured peak-frequencies of successive registered spectra is displayed in Figure 4. The distributions of each channel are fitted with the Gaussian function. For comparison the range gate located at a range around 350 m at high SNR-level was used. The standard deviation of the Doppler peak-frequency of the longer pulse is 0.14 MHz and 0.5 MHz for the shorter pulse for that range gate.

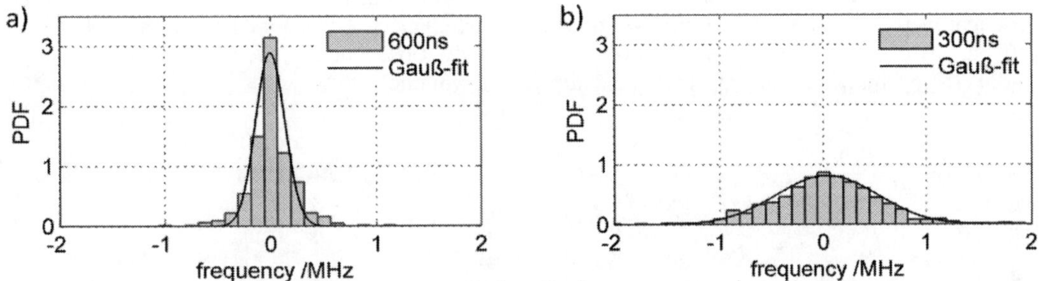

pulse duration	measured $\sigma[\bar{V}_r]$	SNR_n	bandwidth δf	model $\sigma[\bar{V}_r]$
600 ns	0.08 m/s	1.6	1.15 MHz	0.06 m/s
300 ns	0.27 m/s	0.6	1.82 MHz	0.20 m/s

Figure 4. Comparison of velocity precision of two different square-shaped pulse durations.

The calculated velocity precisions are also listed in Figure 4. The results using the variance model are well comparable to the measured velocity precision and show that the velocity precision of the longer pulse is better by more than a factor of two.

According to the velocity precision, the statistical error of the determined velocity in Figure 2a is in the order of the line thickness in the case of applied 600 ns pulses and in the order of some minor oscillations for the 300 ns curve. There are significant differences of the measured wind velocity in case of wind gradients in the region around 250 m and 480 m, which is also confirmed by the increasing linewidth of the 600 ns curve in Figure 3a.

4. Atmospheric observations with different pulse shapes

This multi-wavelength system can also regulate different pulse shapes on every channel. The chosen pulses are a Gauß- and square-shaped pulse with a FWHM duration of 300 ns for both pulse shapes. Both regulated output pulse shapes are shown in Figure 5c. Each pulse energy is limited to a pulse shape dependent stimulated Brillouin scattering (SBS) threshold. The pulse energy reached with a standard EDFA for the Gauß-shaped pulse is 3.9 μJ, for the square-shaped pulse it is 3.4 μJ. The calculated RWFs for both measured probing pulses are depicted in 5b, which result in $\Delta z = 46$ m for the square-shaped pulse and $\Delta z = 58$ m for the Gauß-shaped pulse.

In Figure 5a, the measured velocity over range of both channels is plotted. Both estimated wind velocities are consistent. The sensing volume of the Gauß-shaped pulse is about 25 % wider. The Gauß-channel shows a stronger averaging effect over the spatial velocity development than the square-shaped pulse, especially seen in the range near 200 m and 450 m.

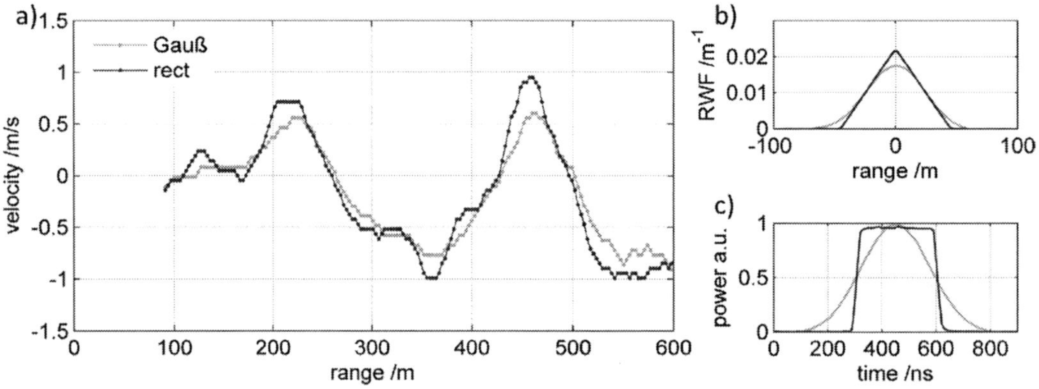

Figure 5. Atmospheric comparison measurement of line-of-sight velocity for different pulse durations.

The Doppler spectrum width (FWHM) of both pulse shapes is shown in Figure 6a. At 500 m, the spectral width of the Gauß-channel is enhanced due to stronger wind gradient. Also around 260 m, a weaker gradient merely increases the spectral width of the Gauß-channel. The squared-pulse channel was not affected by the smaller fluctuations in radial wind velocity. Consequently, the Gauß-shaped pulses are more effected by wind changes due to the wider sensing volume.

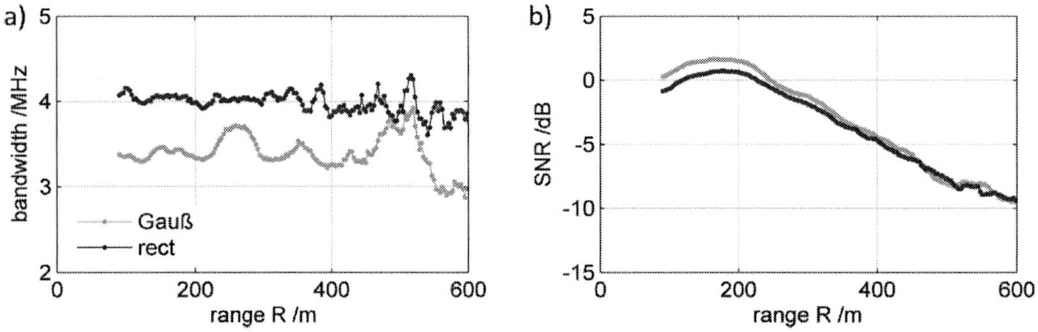

Figure 6. Measured a) FWHM bandwidth and b) narrowband SNR for different probing pulse shapes.

The Gauß-shaped pulse has a higher pulse energy and a lower spectral width. Simulations show that the spectral width at FWHM of the square-shaped pulse is about 1.3 times broader than the spectral width of the Gauß-shaped pulse. With the higher pulse energy of the Gauß-channel, this should result in a 1.8 dB higher SNR for the Gauß-channel. However, the measurement results show just a slightly higher SNR, Figure 6b. Around 500 m, the SNR of the Gauß-shaped pulse equals the SNR of the square-shaped pulse due to the spectrum broadening caused by the stronger velocity gradients.

The distribution of the line-of-sight velocity and the Gaussian function fit for both pulse shapes are shown in Figure 7. For comparison the range gate located at a range around 150 m at high SNR-level was used. The standard deviation of the Gauß-shaped pulse is 0.19 MHz and 0.31 MHz for the square-shaped pulse. The figure also shows the measured and calculated parameters for the comparison of two different pulse shapes. As expected, the velocity precision of the Gauß-shaped pulse is higher. However, measurements in the region of 500 m show a worse velocity precision of the Gauß-channel as the Doppler spectrum is strongly broadened due to wind gradients.

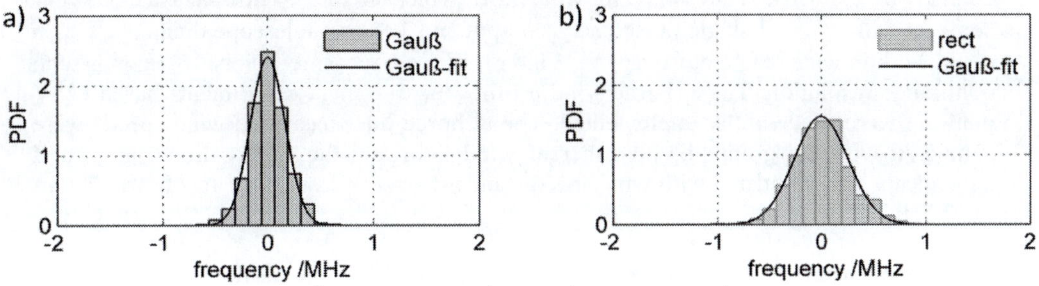

pulse shape	measured $\sigma[\bar{V}_r]$	SNR_n	bandwidth δf	model $\sigma[\bar{V}_r]$
Gauß	0.09 m/s	1.6	3.39 MHz	0.08 m/s
rect	0.14 m/s	1.4	4.16 MHz	0.11 m/s

Figure 7. Comparison of velocity precision between Gauß- and square-shaped pulse.

5. Conclusion

Different pulse shapes and durations have an impact on pulse energy, Doppler spectrum width, SNR, and velocity precision. The Gauß-shaped pulse has a slightly higher pulse energy and a better velocity precision. Due to the wider sensing volume, the Gauß-shaped pulse is more sensitive to wind gradients. The sensing volume of the square-shaped pulse is smaller. Hence, the wind gradients do not affect the performance of the square-shaped pulse as strong as they affect the Gauß-shaped pulse.

In this paper we described and discussed representative atmospheric measurement examples which show the usefulness measuring with multiple pulse shapes simultaneously.

6. References

[1] S. Kameyama, T. Ando, K. Asaka, Y. Hirano, and S. Wadaka, "Compact all-fiber pulsed coherent Doppler lidar system for wind sensing", Appl. Opt. **46**, 1953-1962 (2007).

[2] N. S. Prasad, R. Sibell, S. Vetorino, R. Higgins, and A. Tracy. "An all-fiber, modular, compact wind lidar for wind sensing and wake vortex applications," in *Laser Radar Technology and Applications XX; and Atmospheric Propagation XII,* (Society of Photo-Optical Instrumentation Engineers, Baltimore, 2015), 94650C.

[3] A. Töws and A. Kurtz, "A multi-wavelength LIDAR system based on an erbium-doped fiber MOPA-system," in Lidar Technologies, Techniques, and Measurements for Atmospheric Remote Sensing X, (Society of Photo-Optical Instrumentation Engineers, Amsterdam, 2014), 92460T.

[4] V. Banakh and I. Smalikho, Coherent Doppler wind lidars in a turbulent atmosphere (Artech House, 2013).

[5] S. Kameyama, T. Ando, K. Asaka, and Y. Hirano, "Semianalytic pulsed coherent laser radar equation for coaxial and apertured systems using nearest Gaussian approximation," Applied Optics **49**, 5169-5174 (2010).

[6] S. W. Henderson, P. Gatt, D. Rees, and R. M. Huffaker, Laser Remote Sensing, (Taylor & Francis Group, 2005), Chap. Wind Lidar

Validation of airborne 1.5 μ m all-fiber coherent lidar measurement of wind profiles

Xiaopeng Zhu (a), Jiqiao Liu (a), Xin Zhang (a), Yuan Liu (a), Xiaolei Zhu (a), Weibiao Chen (a)
(a) Key Laboratory of Space Laser Communication and Detection Technology,
Shanghai Institute of Optics and Fine Mechanics, Chinese Academy of Sciences
No.390 Qinghe Road, Jiading, Shanghai 201800, China
xp_zhu@siom.ac.cn , liujiqiao@siom.ac.cn

Abstract: A high pulse energy 1.5 μ m all-fiber coherent Doppler lidar airborne engineering prototype is developed for wind profiles measurement. The lidar system is very compact with 300 μ J single pulse energy output and 100 mm telescope diameter, which could work in wide temperature range . Wind profiles are measured every 16 seconds with 60m range resolution. The airborne wind profiles measurement experiments based UAV platform were successfully implemented. The airborne lidar measured wind speeds were compared with the ground-based coherent wind lidar and the GPS balloon, and good agreements were validated with wind speed standard error of less than 1 m/s from 1700 m to 3700 m.

Keywords: Wind measurement, Pulsed Doppler lidar, coherent detection, all-fiber .

1. Introduction

A coherent Doppler lidar (CDL) is proved to be an effective tool for wind profiles measurement[1] , windshear[2] and wake vortices detection and warning[3] in clear air with high temporal and spatial resolution and high accuracy in lower atmosphere layer. The laser source is the most important part in a CDL system. With the development of single frequency laser technology, 10 μ m CO_2 laser[4] ,1 μ m solid laser[5], 2 μ m solid laser[6], 1.6 μ m solid laser[7] and 1.5 μ m fiber laser[8] are utilized as the lidar transmitter source successively. Over the past decade, single frequency pulsed fiber lasers are studied widely. All-fiber CDL system at 1.5 μ m has attracted much attentions because of its eye safety, compact size, flexible arrangement, and mature fiber components technology from telecommunication industry[9]. It is also suitable for airborne wind profile measurement because of its stable performance in harsh environment. Several companies have developed commercial all-fiber CDL systems since 2004[10].

In this paper, an airborne all-fiber pulsed coherent Doppler lidar is presented. The campaign experiment was implemented on an unmanned aerial vehicle (UAV), and the measured wind profiles are compared with GPS sounding balloon and ground-based coherent wind lidar.

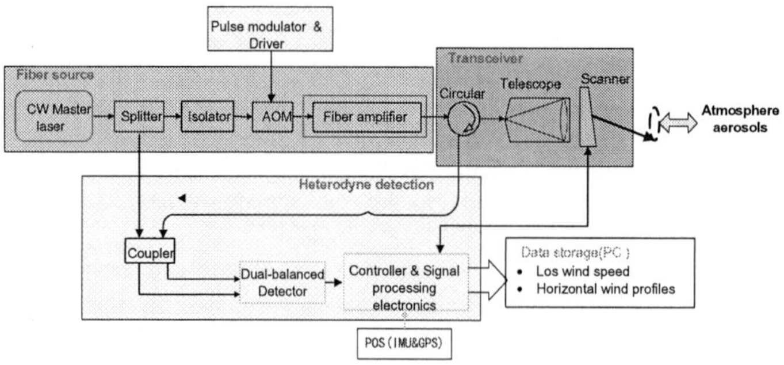

Figure 1. The block diagrams of the CDL system

2. System configuration

The block diagrams of the CDL system is shown in Fig 1. The CDL system consists of all-fiber laser source and optical transceiver and heterodyne detection unit with real-time signal processor. The CDL system is compact with size of 450mm×400mm×400mm and mass of 40 kg. The lidar could be controlled via UAV data chain and the measured data is downloaded in real time at every line of sight (LOS) wind speed. Further estimate of the 3-dimensional wind profiles could be implemented in the computer. The specifications of the system are summarized in the table 1.

Table 1. The specifications of the CDL system

Parameters	value
Wavelength	1540 nm
Pulse energy	300 μJ
Pulse width	500 ns
Pulse repetition rate	10 kHz
Power of LO	0.8 mW
Range resolution	30 m
Telescope aperture	Φ100 mm
Scanner	Φ100mm 20° deflection angle
IF	160 MHz
AD	1G/s ,10bit
Accumulated pulses	10000

An eye-safe, narrow-linewidth, polarized, pulsed, fiber laser based on master oscillator power amplifier (MOPA) configuration is acted as the lidar source[11]. A single frequency Distributed Feed Back Laser Diode (DFB-LD) with 10 kHz linewidth and 20 mw power is used as the seeder laser. The seeder is divided into two parts by the 1×2 optical coupler. One part (about 0.8 mW) is sent to the balanced detectors as the local oscillator for heterodyne detection. The other part is pulsed modulated and frequency shifted by the AOM at repetition rate of 10 kHz, pulse duration of 400ns. The IF (Intermediate frequency - IF) frequency is 160MHz. Through multi-stages fiber amplifiers, fiber laser with pulse energy of 300 μ J, linewidth of 1.1MHz at wavelength of 1540nm is obtained.

The optical transceiver combines a circulator and a telescope and an optical scanner. The high power circulator is used to separate transmitting fiber laser and signal backscattered by aerosols. The optical diameter of the telescope is 100mm. The output beam from telescope is conically scanned by a compact lens scanner with 20° deflection angle. VAD scanning could be implemented in less than 20 seconds with 8 different azimuth angles.

In the heterodyne detection unit the local oscillator from seeder is mixed with signal backscattered by aerosols from transceiver, then it is received by the InGaAs balanced detector. The heterodyne signal from the balanced receiver is sampled by the high speed AD converter with 1G/s rate. The sampling time is 40 μ s and 30 m range resolution is realized with 200 ns range gate. Fast Fourier Transform (FFT) of 1024 points and power spectrum accumulation with 10000 pulses are implemented in every range gate. Accumulated power spectra in every azimuth are saved and downloaded.

3. Experiments

Airborne CDL validation experiment is carried out in Shanxi province, China on Mar 29, 2017. The CDL system is installed in an UAV. For airborne CDL, the speed and attitude angles should be corrected to calculate accurate wind profiles. So the POS system composed of Inertial Navigation System (INS) and global positioning system (GPS) is installed in the aircraft with the lidar together to record aircraft speeds and attitude angles (yaw ,pitch and roll). Wind profiles measured from airborne CDL are compared with ground-based CDL and GPS sounding balloon in the campaign simultaneously to assess the airborne CDL performance. The distance between the aircraft ground track and the ground lidar is about 4 km, and for the balloon is about 7.7 km. The whole flight time is lasted about 1 hour. The horizontal wind profiles are shown in Figure 2.

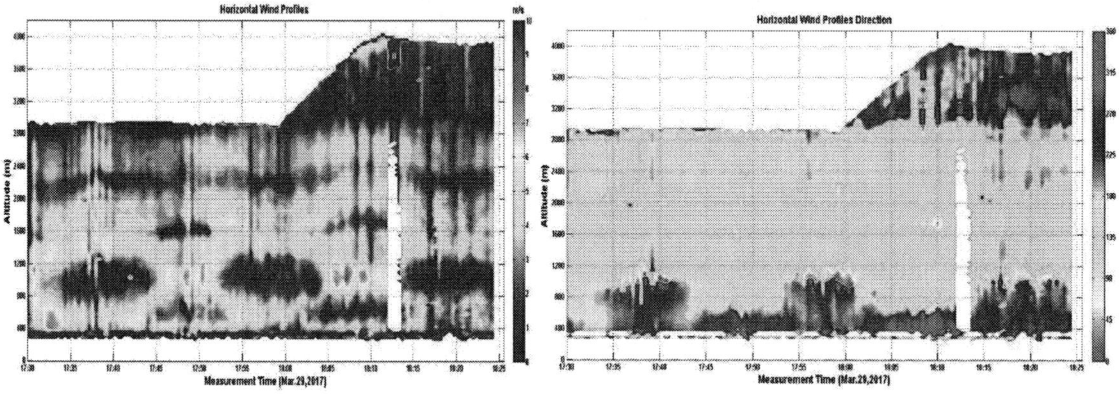

Figure 2. The horizontal wind speeds (left) and directions (right) profiles.

For one case the airborne lidar measured wind speeds were compared with the ground-based coherent wind lidar and the GPS balloon. Good agreements were validated with wind speed standard errors of 1 m/s and 0.91 m/s respectively from 1700 m to 3700 m which are shown in Figure 3. But in lower altitude the deviations are larger. The results is caused by the distances between airborne lidar, ground-based lidar and balloon, and the mountains at the site of south and east. The height of the mountains are about 1.6 km-1.7 km.

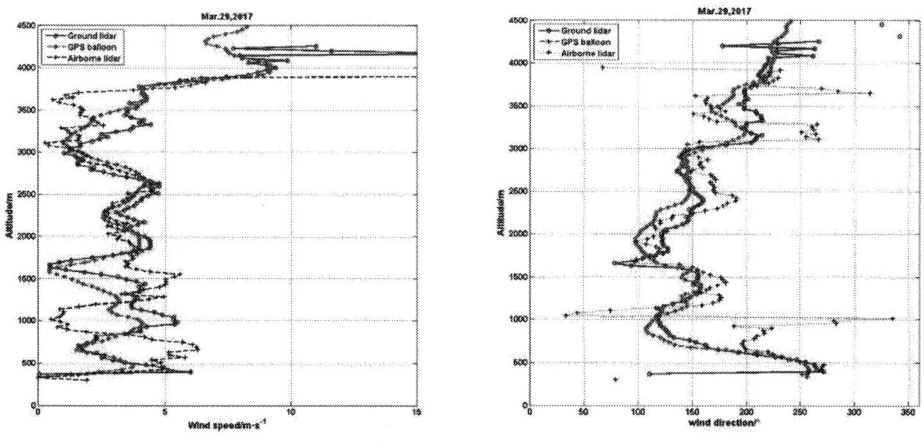

Figure 3.The wind speed profiles and direction comparisons between the three systems.

4. CONCLUSIONS

A high pulse energy 1.5 μm all-fiber coherent Doppler lidar airborne engineering prototype is developed for wind profiles measurement. The lidar system is very compact with 300 μJ single pulse energy output and 100 mm telescope diameter, which could work in wide temperature range. Wind profiles are measured every 16 seconds with 60m range resolution. The airborne wind profiles measurement experiments based UAV platform were successfully implemented. The airborne lidar measured wind speeds were compared with the ground-based coherent wind lidar and the GPS balloon, and good agreements were validated with wind speed standard error of less than 1 m/s from 1700 m to 3700 m.

5. References

[1] Asaka, K., T. Yanagisawa, and Y. Hirano, 2001:1.5-μm eye-safe coherent lidar system for wind velocity measurement. SPIE. 4153: p. 321-328.

[2] Frehlich, R., 1994:Coherent Doppler lidar signal covariance including wind shear and wind turbulence. Appl. Opt. 33(27): p. 6472-6481.

[3] Dolfi-Bouteyre, A., et al., 2007:1.5 μm all fiber pulsed lidar for wake vortex monitoring, in 14th Coherent Laser Radar Conference.

[4] Huffaker, R.M. and R.M. Hardesty, 1996:Remote sensing of atmospheric wind velocities using solid-state and CO2 coherent laser systems. Proceedings of the IEEE. 84(2): p. 181-204.

[5] Kane, T.J., et al., 1987:Coherent laser radar at 1.06 um using Nd:YAG lasers. Opt. Lett. 12(4): p. 239-241.

[6] Jirong, Y., et al. An all solid-state 2um laser system for space coherent wind lidar. in Aerospace Conference Proceedings, 2000 IEEE. 2000.

[7] Hannon, S.M., S.R. Vetorino, and J.V. Pelk, 2007:Next generation Doppler lidar sensor at 1.6μm, in 14th Coherent Laser Radar Conference: Colorado.

[8] Ando, T., et al., 2008:All-fiber Coherent Doppler LIDAR for wind sensing. Materials Research Society. 1076: p. 1076-K04-05.

[9] Kameyama, S., et al., 2007:Compact all-fiber pulsed coherent Doppler lidar system for wind sensing. Applied Optics. 46(11): p. 1953-1962.

[10] Ando, T., S. Kameyama, and Y. Hirano, 2008:All-fiber Coherent Doppler LIDAR technologies at Mitsubishi Electric Corporation, in 14th International Symposium for the Advancement of Boundary Layer Remote Sensing IOP.

[11] Zhang, X., et al., 2014:Single-frequency polarized eye-safe all-fiber laser with peak power over kilowatt. Applied Physics B. 115(1): p. 123-127.

From Windcube#0001 to Windcube#1000 : Doppler Lidar as a mature technology.

Cariou Jean-Pierre, Thobois Ludovic, Spencer Peter
LEOSPHERE
14-16 rue Jean Rostand, 91400 Orsay, France

Abstract: 10 years after the Windcube #0001, Leosphere has delivered its 1000th Windcube Doppler Lidar. Still based on pulsed fibered technology at 1.5µm, the Windcube is now an industrial product widely used for Wind Resource Assessment and Weather Monitoring. Tracing the path of technological progress from laboratory to industry in hardware and software, but also the progress in metrology, the presentation expresses the various and demanding application requirements, the continuous improvements of coherent lidar solutions and the slow but inexorable acceptance of remote sensors in wind energy and meteorological applications.

Keywords: Coherent Laser Radar, Windcube, Pulsed Fibered Amplifier, Wind Energy.

1. Introduction

Modern operational wind lidars were born during the nineties, with first commercial lidars developed by Coherent Technologies in the US. Technology was based on solid state Ho,Tm :Yag lasers at 2µm. In early 2000's, just after the burst of the Internet bubble, first fiber laser based lidars were developed in Europe (Qinetiq, Onera) leading to private company foundation (Natural Power/ZephIR 2003, Leosphere 2004, Halo Photonics 2005). Similar developments appeared in US (Catch the Wind 2008) and Japan (Mitsubishi 2006). This new fiber lidar technology encountered the raising wind energy market and progressively took market shares in wind measurement sensors.

Now in 2018, more than 1700 fiber laser based lidars are operating among the word mainly for wind energy applications, proving than coherent Doppler lidars have reached a good level of maturity and have definitively left the laboratory to field operation. This paper focuses on Leosphere Windcube technology and evolution, starting from serial number #001 in 2007 to #1000 in 2018.

2. Windcube Technology

Windcube lidar technology was born at ONERA (French Aerospace Lab) and transferred to Leosphere in 2006. It is based on fibered architecture and fibered amplification. A pigtailed narrow linewidth DFB laser at 1543nm is used in a Master Oscillator Power Amplifier (MOPA configuration). The major power is modulated into pulses with an acousto-optic modulator (AOM) and amplified in a multistage EDFA (Erbium Doped Fiber Amplifier) operating around 1550nm (Telecom C band). A fraction of the MO is used as Local oscillator (LO).

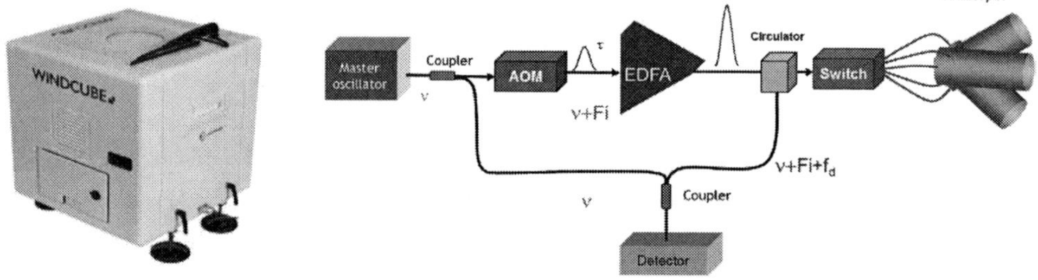

Figure 1: Pulsed Windcube lidar architecture.

In short range Windcubes, the laser pulses are sequentially transmitted to 4 to 5 different telescopes by a non-mechanical fibered switch. In long range lidars, a post telescope hemispherical scanner head is used. Different Windcubes differ by the EDFA pulse energy and telescope size.

The 1550nm wavelength is eye-safe in most conditions, benefits of a low molecular extinction and a good Doppler sensitivity. Moreover, most components are available on the shelf at reasonable prices. Figure 1 illustrates the general architecture of a Windcube pulsed lidar.

Figure 2 illustrates the evolution of the Windcube family. Scanning lidars were released in 2010, mostly for airport hazard monitoring and severe weather meteorology.

Year	2005	2007-2010	2010-2018	2010-2018
Model	Prototype	Windcube v1	Windcube v2	Windcube 100S/200S/400S
Qty	0	90	800	130

Figure 2 : Windcube evolution timeframe

3. Assessing the reliability

Coherent lidars are interferometers and performance requires sub-wavelength alignment and beam quality. Since Leosphere has the ambition of manufacturing lidars for demanding operational applications, mainly in wind energy, reliability is a critical point.

Fibered technology is intrinsically reliable. Light is guided and beam alignment is easy, with low loss connectors or splices. So fibered architectures are not so sensitive to vibrations and temperature change, compared to bulky free air lidar architectures. Fiber components benefit also from high reliability standards, as Telcordia. Critical components are active components, such as pump diodes. But different grades exist and similar diodes are used in submarine long haul fibered networks, with lifetimes over 15 years. In lidars however, vibration conditions and temperature range can be more challenging. For wind turbine applications for example, mechanical and thermal designs must take into account vibrations (IEC 60068) and external temperature range from -20°C to 60°C.

From 2010 to now, Windcube lidars designs have been modified to improve the reliability and stability over external conditions, while keeping the same MOPA architecture. Rotating prism used in version 1 has been replaced by optical switch in version 2 (2010). Metallic mounts have been replaced by athermal telescopes with monolithic alignment (Patent WO2012045976A1). Variation of EDFA gain over temperature has been improved by design.

Software stability and reliability is also very important. When analyzing the main failures, hardware no longer accounts for the majority of root cause. Software problems make up over 50% of the current output of systematic root cause analysis.

4. Assessing the metrology

The main purpose of a wind lidar is to give to the user an available and accurate windspeed measurement. The second purpose is to provide data with reduced sensitivity to various atmospheric environments (vertical shear, turbulence, complex terrain, ...). Assessing the Windcube metrology has been a concern at Leosphere and reference laboratories (DTU, Deutsche Windguard, DNV GL ...) since 2010. Before lidars, met masts were the reference for wind energy consultants and developers. Lidars were then compared to masts and hundred calibration campaigns were performed. First in flat terrain, and progressively in complex terrain.

Figure 3 gives the results of standard deviation of the speed difference between Windcube and reference along the years (10 minutes horizontal speed average). Measurements from 2006 to 2009 are from DTU with a 110m mast reference [1], 2012-2017 data come from Leosphere, after the Lidar to Lidar calibration has been validated by DTU. Terrain is flat, turbulence intensity and shear coefficients are low.

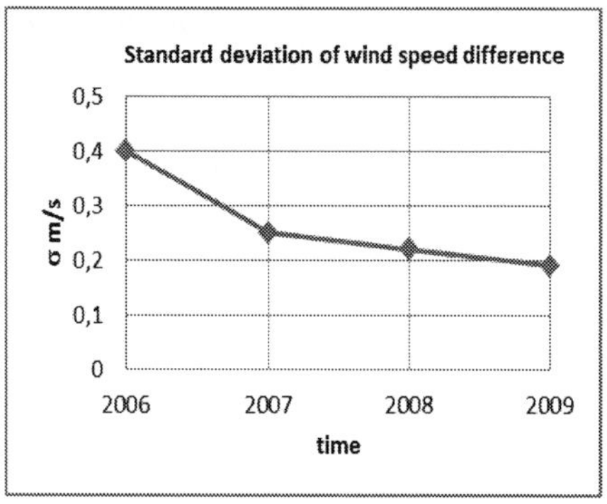

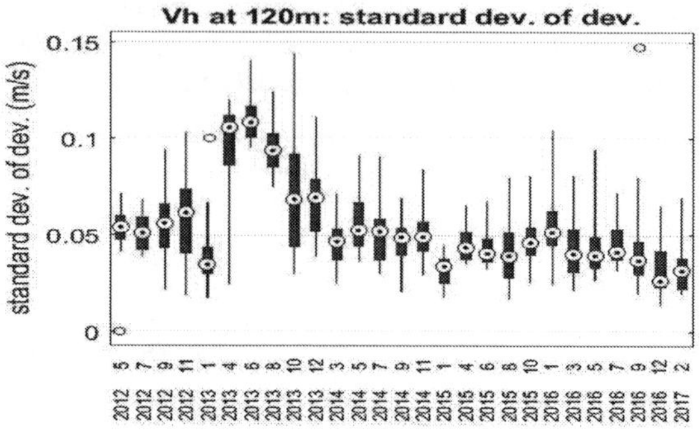

Figure 3: Windcube accuracy measurements along the time

We observe a fast accuracy improvement in the early years, followed by a stability period. In flat terrains, the wind speed accuracy is indeed very good, and differences with reference are negligible.

In complex terrain, the lidar metrology is affected by the key assumption of horizontal flow homogeneity. To improve this limitation, research activities conducted since 2010 mostly have demonstrated the feasibility to combine the lidar wind measurements with CFD models. This combination reduces the impact of the orography on the averaged wind flow for light to moderate complex terrain. Without correction, the lidar mean error can reach 10% at low wind speed. With correction, the mean error is reduced to 2%.

The uncertainty studies described above lead to a better understanding of the lidar metrology but also the determination of the quantitative sensibility of each environmental parameters (EP) (shear, turbulence intensity, rain,..) on lidar mean error. It has been demonstrated that among all the classifications, the sensibilities of the Windcube to EPs are the same. Given these results obtained progressively year after year, the Windcube lidar started to be used more and more for the different applications especially wind resource assessment. The expert and standardizing groups helped to share the advancements of the understanding of the metrology of coherent lidars. For example, the IEC standard n°61400-12-1 ed.2 Annex L, published in 2017 and related to Power curve measurements, prescribes a process to make sure that uncertainty is consistent across sites and devices based on well-defined device calibration method, requirements on device classification (2 devices/2 sites, observed sensitivity within 50% of the mean sensitivity) and requirements on on-site verification of RSD (deviations must be lower than combined sensitivity and calibration uncertainty).

Many on-going groups are continuing for introducing the use of remote sensors in the standards like the IEA task 32 and the IEC standard 61400-50-3 for wind resource assessment. IEA Task 32 was set up in 2011 to support the deployment of lidar for wind energy applications and identify the barriers to the deployment of lidar for wind energy. In 2015, participants identified four main application areas where an international collaboration between researchers, device manufacturers, end users and other stakeholders could mitigate barriers. These application areas include site assessment, wind turbine power performance, wind turbine controls, and the use of lidar to measure complex flow. The application areas are currently at different technology readiness levels: lidar systems are already used by industry for site assessment and power performance, and so the barriers are more related to implementation. For loads and control as well as for use in complex flow applications, there are still open research questions to solve [2]. IEC standard 61400-50 should be released before 2019.

To sum-up, if the worldwide experts in wind measurements agree with the use of Lidar remote sensor as a stand-alone unit for power curve measurement and wind resource assessment in flat terrain, the operational people are on the way mostly to adopt the technology in the next 5 to 10 years. This is also important to highlight that there are some regional discrepancies in the LIDAR adoption. The consulting companies in Europe have already adopted the technology for years where as in USA and in China for example, the adoption is much slower but the interest is raising.

Finally, a recent jointed ISO/WMO standards has been released in July 2017 related to the use of Coherent Doppler lidars (ISO 28902-2:2017) in order to develop the use of Coherent Doppler Lidar for all applications related to atmospheric measurements.

5. Lidars in Wind Energy

From almost zero units in 2008 and approximately half of the total fleet in 2008, the wind energy industry now accounts for more than 84% of the total cumulative system production of Leosphere. In that same period, the total fleet in the wind energy industry has grown by 25 times (Figure 4). Within the overall growth, we have seen the penetration of lidar as a remote sensing technology follow the waves of maturity in the wind industry with the Americas holding overall market share but with a transition from North America to Latin America. Europe has been displaced as the growth engine by Asia as a market for the Windcube and now accounts for more of the market than the Americas.

While past performance is no guarantee of future success, we do not see these established trends changing in the near future.

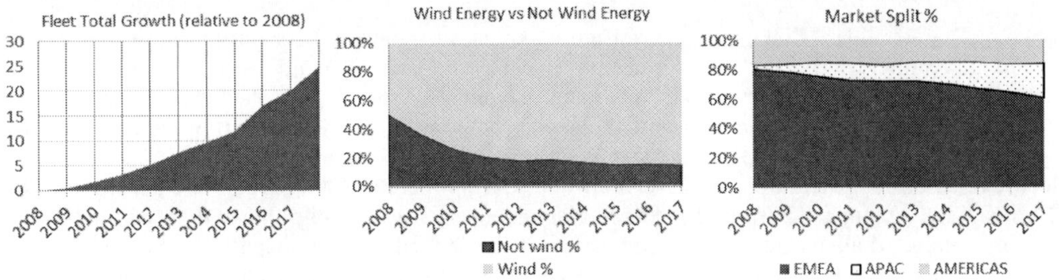

Figure 4 : Evolution of the Windcube market

6. Conclusion and perspectives

The Windcube lidar, together with other lidars based on similar fiber technology, opened 10 years ago the industrial market to coherent lidars, thanks to the wind energy industry growth. Year after year, assessing the metrology acknowledged by international standards, in flat terrain and then in more demanding conditions, the Windcube wind profiler has been deployed in operation widely around the world, for wind resource assessment and power curve. Meanwhile, scanning Windcubes are more and more adopted for wind hazard monitoring at airports. Next challenges concern the lidar metrology in complex terrain, and the use of this reliable and efficient fiber technology to turbine mounted lidars, for real time turbine control.

[1] Lindelöw, P. J. P., & Courtney, M. (2009). From Prototype to Standardization: Five years of lidar Anemometry in the Wind Energy Industry. In Conference Proceedings (pp. 255-258)

[2] Andrew Clifton, Peter Clive , Julia Gottschall , David Schlipf, Eric Simley, Luke Simmons, Detlef Stein, Davide Trabucchi, Nikola Vasiljevic and Ines Würth, IEA Wind Task 32: Wind Lidar Identifying and Mitigating Barriers to the Adoption of Wind Lidar. Project Report , Remote Sensing 2018, 10(3), 406; doi:10.3390/rs10030406 . http://www.mdpi.com/2072-4292/10/3/406/htm

Mobile coherent doppler lidar prototype v2 for wind sensing

Toshiyuki Ando(a), Eisuke Haraguchi(a), Hitomi Ono(a) and Akihiro Fijie(a)
(a)Mitsubishi Electric Corporation, Information Technology R&D Center
5-1-1 Ofuna, Kamakura, Kanagawa 247-8501, Japan
Ando.Toshiyuki@ap.MitsubishiElectric.co.jp

Abstract: We have developed ultra-compact coherent doppler lidar prototype version 2 for wind sensing. This new mobile lidar has its dimension of 39x29x16 cm, its weight of 2.9kg, and power consumption of 31.4W. The new mobile lidar can be operated by a lithium battery and can be continuously measured wind profile over 2.5 hours, thanks to the low consumption optical transceiver / signal processor sub system, called coherent doppler lidar engine. Preliminary experiments have been performed by this mobile lidar prototype 2 for line of sight wind velocities with the maximum horizontal range more than 1km.

Keywords: Coherent Doppler Lidar, Serrodyne Modulation, Pulsed Modulation.

1. Introduction

Coherent Doppler lidar (CDL) systems are attractive sensors for wind sensing because they offer a method of remote wind speed measurements in clear atmospheric conditions. An all-fiber CDL system using a 1.5-micron wavelength has many advantages, such as compactness, eye-safety, and reliability, due to using commercial-off-the-shelf components from telecom products [1]. The deployment of this CDL is expanding steeply for many applications of wind energy [2]. Mitsubishi Electric released 1st generation commercial all-fiber CDL system in 2006[3], and 2nd generation CDL, DIABREZZA™ compact' for wind resource assessment in 2014. While the CDL deployments at a steep mountain, small islands, or disaster areas, it needed not only transportable but also battery drivable. The mobile CDL proto type 1 has also demonstrated two-hour successive operation without exchanging batteries [4]. This CDL, however, was not enough small to be transported by one person. This paper presents a new ultra-compact CDL updated by newly developed sub-systems [4].

2. System Configuration

Figure 1 shows a schematic block diagram and outer view of the mobile coherent Doppler Lidar prototype V2. This new lidar has an all-in-one design consisting of an optical transceiver board directly connected with a signal processor, an optical high power amplifier(OHPA), an optical antenna via a fiber circulator, and a lithium polymer battery. The dimension is 39(W)x29 (H)x16(D)cm, and its weight is 2.9kg.

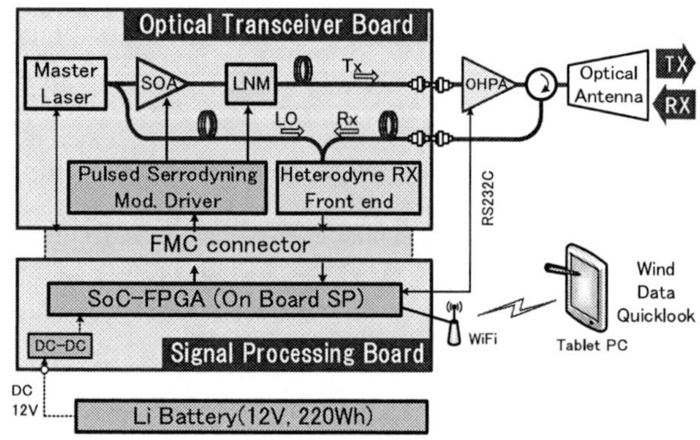

Fig.1 Schematic block diagram and outer view of the mobile Doppler lidar proto type V2.

The optical transceiver is combined a conventional fiber-optic heterodyning receiver with a newly developed coherent optical pulse seeder based on pulsed-serrodyning modulation. In the signal processing board a system-on-a-chip (SoC) solutions have been adopted by using a large Field Programmable Gate Array (FPGA) with an internal processor core. Measured wind data can be displayed on a tablet PC via the WiFi after on-board signal processing for wind speed estimation.

3. Optical Transceiver Unit

Fig.2 shows the block diagram and outer view of the optical transceiver unit as a coherent pulse seeder combined with a heterodyne receiver. All fiber-optic components are commercial off the shelf components in the optical communication which have enough high reliability with Telcodia GR468-core compliant. An Integrable Tunable Laser Assembly (ITLA) was used as a master laser with line width of 200 kHz and wavelength of 1550 nm. Then its output was split into a local oscillator (LO) and a seed light to a pulsed serrodyning modulator which is consisting of a semiconductor optical amplifier (SOA), a lithium niobate optical phase modulator (LNM) and their drivers with digital-to analog converters (DAC). The pulsed serrodyning modulation is our newly developed technique to realize both pulse modulation and frequency shift without a double-pass AO (Acousto optic) modulator currently used in our CDL system [3]. In a receiver section optical return signals (RX) are phonically mixed with a LO signal, then detected as intermediate frequency (IF) signals at balanced two photo diodes (PD). In order to accommodate a bandwidth of 100MHz corresponding to a range of Doppler frequency of +/- 35m/s, the heterodyning signal is sampled at rate of 216MSps with analog-to digital converter (ADC) and transferred to the signal processing board.

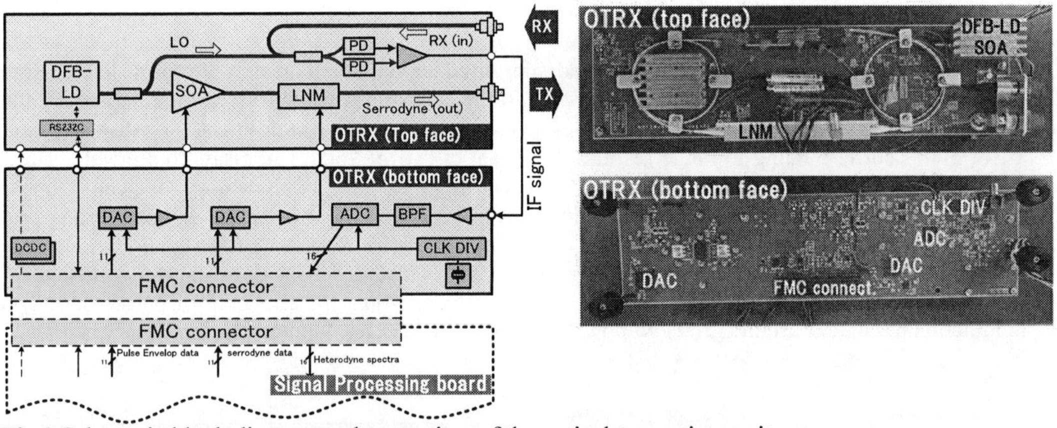

Fig.2 Schematic block diagram and outer view of the optical transceiver unit.

4. Pulsed Serrodyning Modulation

Serrodyine frequency shifting is realized by applying a saw-tooth phase modulation to the optical signals. Recently frequency shift of a few hundred megahertz as required in CDL can be obtained with a good quality by technological advances in high speed driver electronics. While, SOA is the promising to optical pulse modulator with small foot print if the issue s of frequency chirp is solved. We have newly developed a pulsed serrodyning modulation which realizes both frequency shift and compensation of frequency chirp in SOA[5].

Figure 3 shows the temporal signals of pulsed serrodyning modulation for an optical intensity of a SOA and a optical phase of a LNM. The saw-tooth phase modulation is applied within only the pulse-on period, T1 over pulse repetition interval (PRI). If the complete saw-tooth phase modulation with cycle of Tm is applied, the output optical frequency is shifted with an offset frequency of 1/Tm.

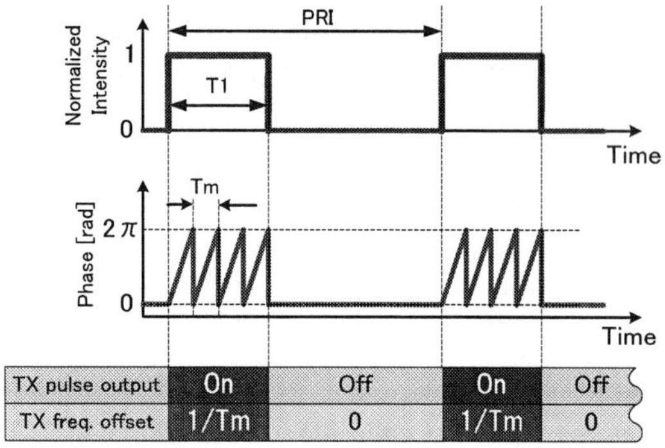

Fig.3 Temporal signals of pulsed serrodyning modulation for an optical intensity of a SOA and a optical phase of a LNM.

It is worthy to note that no frequency shift is occurred within pulse-off periods because of not applying the saw-tooth signal. This leads effective rejection performance of unwanted beat noise between optically internal reflections within pulse-off periods and the LO signal in the heterodyne receiver without such a special pulsed modulator as a double-pass AOM with a high extinction ratio.

5. Experimental Results

In order to confirm whether the pulsed-serrodyning modulation does correctly perform, line of sight (LOS) wind Doppler spectra have been evaluated in the case of pulsed-serrodyne modulation and that of single-pass AO modulation as a reference. The measuring condition as follows: the pulse-on period, T1 of 500ns corresponding to range resolution of 75m, a PRI of 250 μ s, the saw-tooth cycle, Tm of 6.17 ns corresponding to IF of 162MHz as shown in fig.3. Figure 4 shows the LOS wind Doppler spectra at distance of 500m. In the case of single-path AO modulation the spectrum has a strong peak around zero Doppler velocity which may be caused by beat noise between a LO signal and internally reflection of leaky lights within pulse-off periods. This unwanted beat noise makes it difficult to measure wind signals near the zero-Doppler velocity. Meanwhile, the spectrum of pulsed-serrodyne modulation has a peak around -1m/s without any noise peak at zero-Doppler velocity.

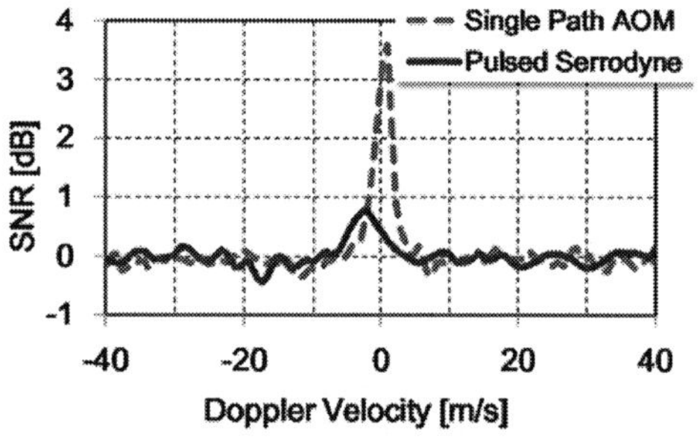

Fig.4 LOS wind Doppler spectra at distance of 500m

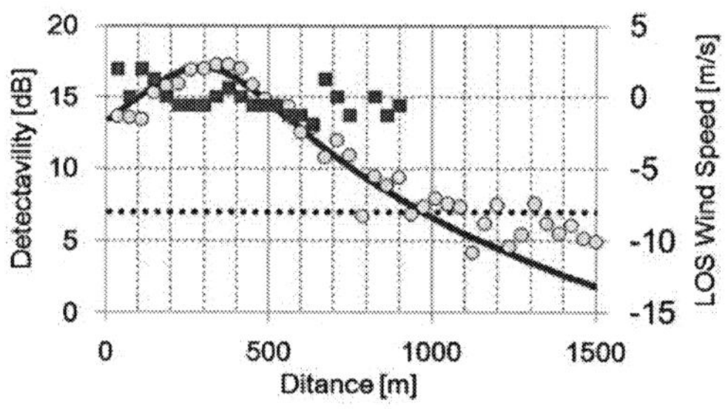

Fig.5 LOS wind velocity and detectability with respect to distance

Figure 5 shows the measurement data of the LOS wind velocity with respect to distance and that of their detectability. Theoretical calculation for detectability is also shown as a function of distance. The setting parameters are as follows: the optical peak power of 30W at a fiber end of an OHPA, the aperture diameter of 50mm, the focusing distance of 500m and the integration number of 4000. The back scattered coefficient is assumed of 8.3e-8 $m^{-1}sr^{-1}$ taking into account of measured number of aerosol using a particle counter. The measured detectability has considerably agreed with theoretical curve, which indicated the LOS wind velocities are measured up to 900m because of larger detectability than the detection limit of 7dB.

Figure 6 shows that the temporal variation of supplying power from fully charged batteries for continuously wind sensing under the same measurement condition as in fig. 6. In this figure the present mobile CDL ran the wind measurement over 5 hour 20 minutes without exchanging battery. The average supplying power was 31.4W.

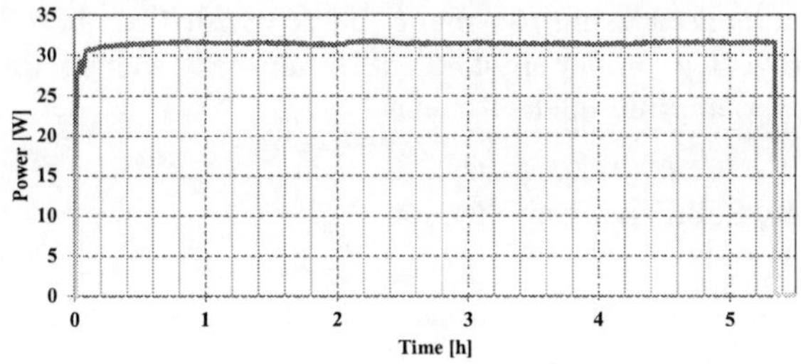

Fig.6 Battery supplying power under continuously wind sensing

6. Summary

We have developed a new mobile coherent Doppler lidar prototype for wind sensing with its dimension of 39x29x16 cm. The pulsed-serrodyning modulation makes it possible to realize both frequency shift and compensation of frequency chirp within pulse-on period of the SOA as pulse modulator. The new mobile lidar can be operated by a lithium battery and can be continuously measured wind profile over 5 hours, thanks to the low consumption optical transceiver / signal processor sub system. Preliminary experiments have been performed by this mobile lidar prototype 2 for line of sight wind velocities with the maximum horizontal range more than 1km.

7. References

[1] S. Kameyama et al, "Compact all-fiber pulsed coherent Doppler lidar system for wind sensing", Applied Optics, vol. 46, 1953 (2007).

[2] T. Mikkelsen, N. Angelou, K. Hansen, M. Sjöholm, M. Harris, C. Slinger, P. Hadley, R. Scullion, G. Ellis, and G. Vives, "A spinner-integrated wind lidar for enhanced wind turbine control", Wind Energy, 16, 625-643 (2012).

[3] T. Ando et al., "All-fiber coherent Doppler LIDAR technologies at Mitsubishi Electric Corporation", IOP Conference Series: Earth Environmental Science, 1, 012011(2008).

[4] T. Ando et al., "Mobile coherent Doppler LIDAR system for wind sensing", Proc. of 17th CLRC, (2013)

[5] T. Ando et al., "New coherent Doppler lidar engine integrating optical transceiver with FPGA signal processor", Proc of. 18th CLRC, 145 (2016).

[6] E. Haraguchi, et al, "Pulsed serrodyning technique as applied the wind sensing coherent doppler lidar", Proc of. 18th CLRC, 141 (2016).

Comparison of FMCW-LiDAR system with optical- and electrical-domain swept light sources toward self-driving mobility application

Napat J.Jitcharoenchai, Nobuhiko Nishiyama, Tomohiro Amemiya, and Shigehisa Arai
Dept. of Electrical and Electronic Engineering, Tokyo Institute of Technology
E-mail: j.n.aa@m.titech.ac.jp

Abstract: The suitable configuration of FMCW-LiDAR to realize compact LiDAR system on Si-photonics platform for self-driving mobility application is investigated. Especially, signal comparison with two types of wavelength swept light sources is discussed. One is an optical-domain thermally wavelength-tuned laser by sweeping injection current. The other is an electrical-domain wavelength-tuned source by the combination of CW laser with an external modulator which superimposes electrical frequency shift. Signals and 3-dimentional images were taken and compared using two light sources with fiber-based coherent detection setup. The results revealed similar signal-to-noise ratio and images for two configurations if applying k-clock resampling interferometer.

Keywords: FMCW-LiDAR, Swept light source, Self-driving application

1. Introduction

Self-mobility applications got high attentions and developments of technologies for such applications are going in worldwide. Light detection and ranging (LiDAR) sensor is one of the crucial components for these technology as an environment mapping and sensing device. The commercial grade LiDAR usually comes with the bulky size and consisted of many parts of mechanical movement results in high physical disturbance and high production cost. To find an alternative to a bulky LiDAR system, the compact size LiDAR system was expected to be developed [1-3]. We are also proposing to realize the one-chip size Frequency-modulated Continuous-wave (FMCW) LiDAR with the slow light and Si-photonics technology with hybrid III-V laser integration to minimize the unit size and making more stable light sweeping while lower the fabrication cost [4-6].

The sweeping FMCW system employ the different in frequency or the beat frequency of transmitted and received light signal to measure the distance value. Thus, the system required a linear relationship of time and frequency in order to recover back the correct and accurate distance value. To realize high performance FMCW-LiDAR, wavelength swept light source is a key device. Although the required optical power level of FMCW-LiDAR is much lower than that for time of flight (ToF)-LiDAR, narrow line width, low-noise and linear wavelength sweeping by time are required. To tune the wavelength, several methods can be considered. One of the method is direct thermal tuning of lasers [7]. Another method is using an external modulator to superimpose swept frequency. In this paper, we compared acquired images from FMCW-LiDAR setup using two types of swept light sources.

2. System setup in Optical-Domain Method

In order to design the system, we have to verify the performance with the large components setup firsts before making one-chip LiDAR. Therefore, all experiments in this paper used optical fiber based system. The scanning of light was realized by Galvano scanner. Figure 1. shows the setup with thermal modulated swept laser diode (we called "optical-domain" light source). To compensate k-clock linearity of the wavelength sweeping, a Mach-Zehnder interferometer (MZI) was installed. Figure 2. shows the time average spectrum of lasers in the setup. We used distributed feedback (DFB) lasers with total output power of 10 dBm. Injection current was modulated with the repetition rate of 100 kHz. By changing the injection current, refractive index was changed, resulting wavelength change. Tuning bandwidth was 0.03 nm (3.7 GHz). However, in this thermal tuning, the relationship between the time and frequency of the sweeping is distorted. This is the reason for the k-clock MZI in the system. It works as a calibration system to recover the non-linear relationship back to the linear one.

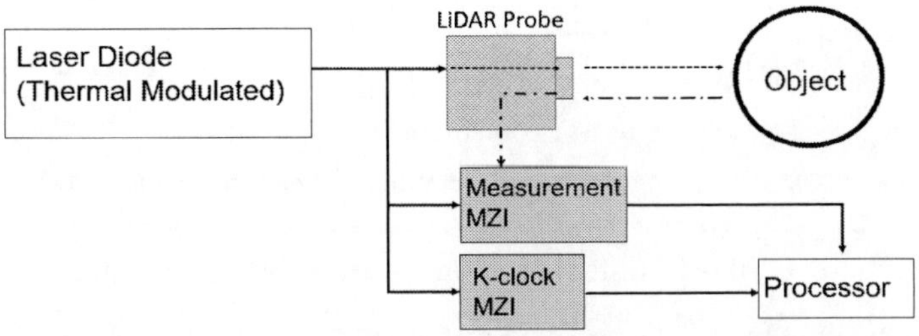

Figure 1. LiDAR system setup in optical-domain method.

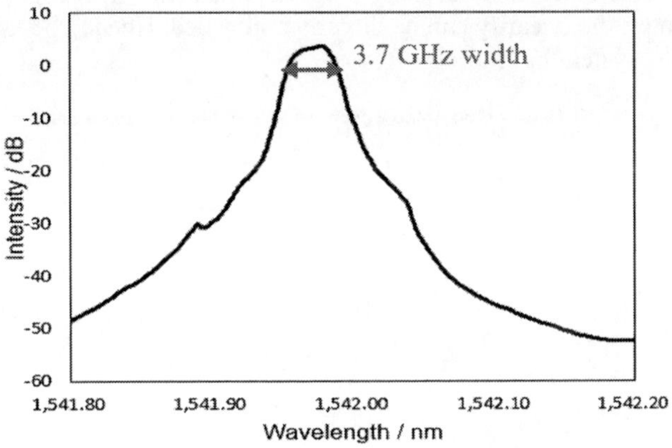

Figure 2. Direct thermal modulated spectrum of the Laser Diode (time averaged).

Since the wavelength (or frequency) is single in this case, the result of beat signal is simple. Let this modulation light signal be $A\exp(j\omega_a t)$ and the reflected signal be $B\exp\{j(\omega_a + \Delta\omega)t\}$, the intensity of the signal after interference will be

$$\begin{aligned} I &= |A\exp(j\omega_a t) + B\exp\{j(\omega_a + \Delta\omega)t\}|^2 \\ &= A^2 + B^2 + 2AB\cos(\Delta\omega) \end{aligned} \quad (1)$$

The beat frequency $\Delta\omega$ is yielded and some DC portion is remained. Therefore, this modulation method has a simple setup; however, it also comes with unwanted problems. The modulation speed is limited by the thermal distribution in the modulation process. Also, k-clock calibration is needed as mentioned above.

3. System setup in Electrical-Domain Method

To solve the heat distribution and non-linearity problem in modulation, the electrical-domain method was introduced. The system was set based on the former setup; however, the laser was not getting modulated in this setup, but the external modulation method was employed. An arbitrary wave generator (AWG) and a Lithium-Niobate (LN) modulator were introduced in the system to sweep the wavelength. The system setup is shown in Fig. 3.

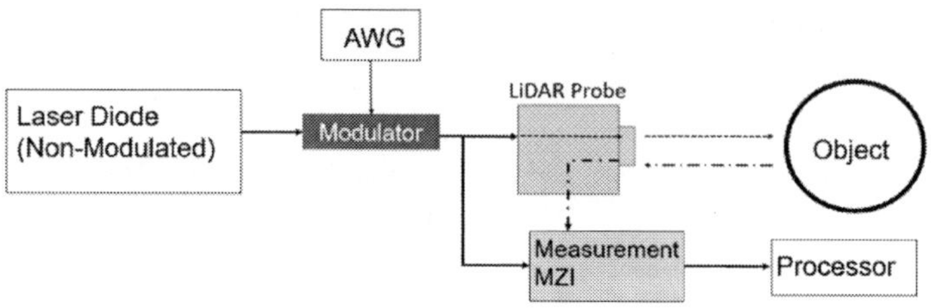

Figure 3. LiDAR system setup in electrical-domain method.

The AWG generated the electrical sweep frequency starting from 8 GHz to 13 GHz with the repetition rate of 200 kHz. The laser wavelength as carrier frequency was set to be 1550 nm. The electrical signal was applied to the LN modulator, resulting the electrical frequency was superimposed to the optical frequency. In this case, we have double sideband wavelength spectrum as shown in Fig. 4. Since the modulation is conducted in the modulator, the linearity can be directly controlled. Hence, the k-clock MZI can be disable (physically connect to the system but turned-off).

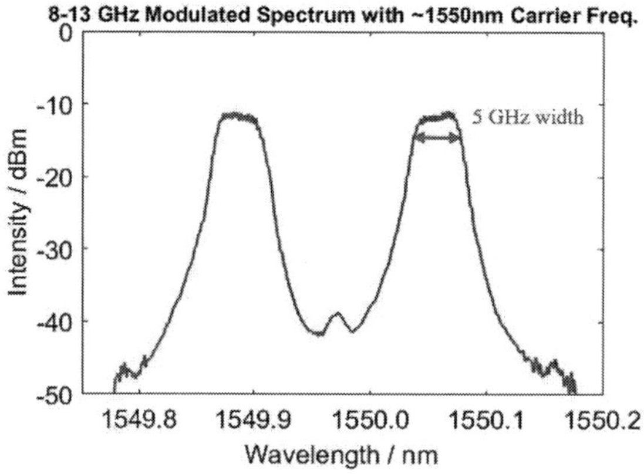

Figure 4. Spectrum of signal after LN modulation (time averaged).

Similar to the Optical-Domain Method, the beat frequency can be derived. Let the modulation light signal be $A\exp(j(\omega_c + \Delta\omega_1)t) + A\exp(j(\omega_c - \Delta\omega_1)t)$ and the reflected signal be $B\exp(j(\omega_c + \Delta\omega_1 + \Delta\omega_2)t) + B\exp(j(\omega_c - \Delta\omega_1 - \Delta\omega_2)t)$, the intensity after the interference will be

$$\mathrm{I} = |\,\mathrm{A}\exp\{j(\omega_c + \Delta\omega_1)t\} + \mathrm{A}\exp\{j(\omega_c - \Delta\omega_1)t\} + \mathrm{B}\exp\{j(\omega_c + \Delta\omega_1 + \Delta\omega_2)t\} + B\exp\{j(\omega_c - \Delta\omega_1 - \Delta\omega_2)t\}|^2 \quad (2)$$

After conjugate squared and term distribution, 5 different frequency terms of sinusoidal function is written below

$$\mathrm{I} = \{4\mathrm{AB}\cos(\Delta\omega_2)\} + \{4\mathrm{A}^2 + 4\mathrm{B}^2\} + \{2\mathrm{A}^2\cos(\Delta\omega_1)\} + \{4\mathrm{AB}\cos(2\Delta\omega_1 + \Delta\omega_2)\} + \{2\mathrm{B}^2\cos(2\Delta\omega_1 + 2\Delta\omega_2)\} \quad (3)$$

Using the band-pass filter to omit the unwanted frequency components, left the beat frequency term.

$$\mathrm{I} = 4\mathrm{AB}\cos(\Delta\omega_2) \quad (4)$$

Hence, the beat frequency $\Delta\omega_2 = \omega_b - \omega_a$ is yielded (When ω_a and ω_b is the start and stop sweeping frequency)

Although the beat frequency $\omega_a - \omega_b$ for distance measurement is yielded, other beat frequencies appear unlike optical-domain one. If these frequencies are close to main beat frequency, they can be noise or ghost.

So, enough separation to filter out may be needed. In our measurement, the frequency separation is 8 GHz and it should not affect to the image.

4. LiDAR Image Results

First, we measured the object 1-m away from the LiDAR probe. The LiDAR system can capture the image in the resolution of 512x512 pixels. The 2-demesional image comparison between Optical- and Electrical-domain setup can be seen in Fig. 5.

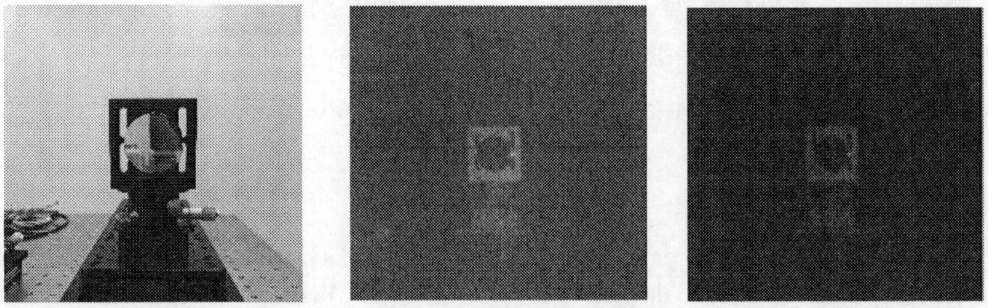

Figure 5. Image comparison of 1. Image from camera (left), 2. Image from optical-domain system (middle), 3. Image from electrical-domain System (right).

These result shows the electrical-domain method can also achieved the similar image compared with that of the optical-domain method. The only difference is the image brightness due to the lower output power of the electrical-domain method. This is due to higher total optical insertion loss due to the LN modulator. The 1-dimensional intensity at a certain length of LiDAR scanning are shown in Fig. 6. Peak-to-noise ratio are similar for both methods.

Next, the effect of k-clock compensation was investigated. The comparison can be seen in Fig. 7. This comparison shows that the image by the electrical-domain method can produce a non-distorted image even though the k-clock system is disable. Thus, k-clock MZI can be removed in from the chip fabrication design.

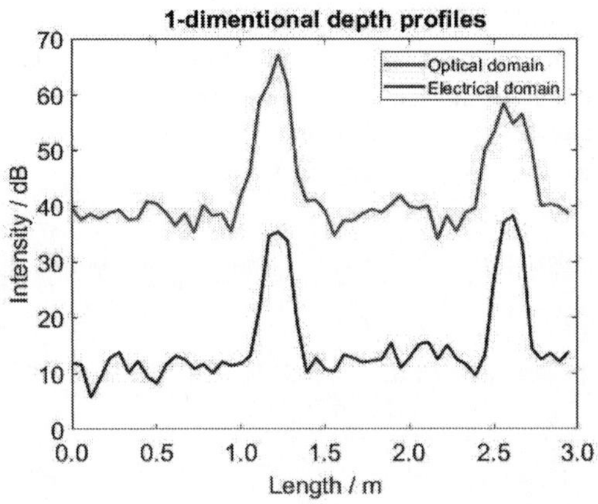

Figure 2. 1-dimensional depth profile of the object.

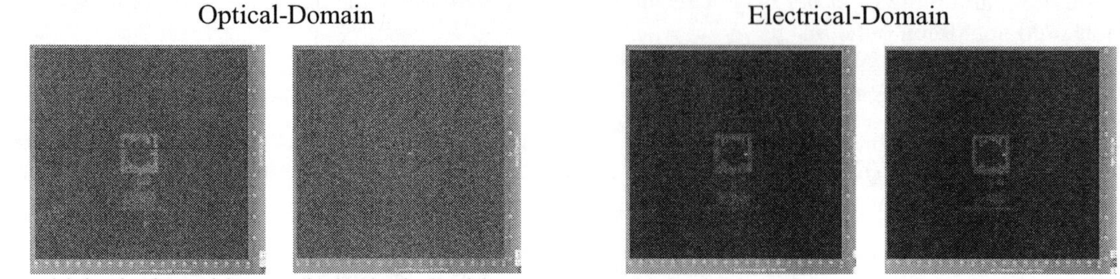

(a) k-clock enable (b) k-clock disable (c) k-clock enable (d) k-clock disable

Figure 7. Images of optical- and electrical-domain methods with/without k-clock.

5. Conclusion

The image performance comparison is carried out for two-types of swept light sources, thermally modulated optical-domain method and electrical-domain method in FMCW LiDAR. The qualities of images were compared and the result shows that the both methods can achieved similar quality of image with same peak-to-noise ratio. The electrical-domain one showed linear frequency sweep to eliminate k-clock compensation.

6. References

[1] C.V. Poulton et al., "Coherent solid-state LiDAR with silicon photonic optical phased arrays," Optics Letters, Vol. 42, No. 20 (2017)

[2] P. Adany et al., "Chirped Lidar Using Simplified Homodyne Detection," Journal of Lightwave Technology, Vol. 27 No. 16 (2009)

[3] S. Arafin, "Advance InP Photonic Integrated Circuits for Communication and Sensing", IEEE Journal of Selected Topics in Quantum Electronics, Vol. 24, No. 1 (2018)

[4] T. Baba et al., "Line Beam Scanner using Slow-Light Waveguides in Si Photonics," Solid State Devices and Materials International Conference, H-8-04 (2017)

[5] T. Baba et al., "Investigation of Si-Photonics Slow Light LiDAR," JSAP Autumn Meeting, 14p-B4-10, (2016)

[6] Y. Hayashi et al., "GaInAsP/silicon-on-insulator hybrid laser with ring-resonator-type reflector fabricated by N2 plasma-activated bonding," Japanese Journal of Applied Physics, Vol. 55, pp. 082701-1-7., (2016)

[7] N. Nishiyama et al., "Investigation on Application of OCT system for FMCW-LiDAR System," IEICE Society Conference, C-3-22 (2017)

Synthetic Aperture Imaging Ladar Ground Test and Flight Demonstration

Yu Zhou (a), Zhiyong Lu (a), Jianfeng Sun (a), Qian Xu (a), Zu Luan (a), Lijuan Wang (a), Guangyuan Li (a), and Guo Zhang (a)

(a) Shanghai Institute of Optics and Fine Mechanics, Chinese Academy of Sciences

Abstract: This report describes the working principle, hardware and software structure and the first-round flight test results of the auto-heterodyne synthetic aperture laser imaging radar(ASAIL) demonstration system. At the same time, the results of the ground test with 10km distance of the modified synthetic aperture laser imaging radar system and the preliminary results of the inverse synthetic aperture laser imaging (ISAIL) of the UAV target are introduced. Finally, the improve proposals of auto-heterodyne synthetic aperture aperture laser imaging radar system and the future application directions of synthetic aperture laser imaging technology are discussed.

A quantitative evaluation of skin phantom by skin TD-OCT

Dilimulati Adili (a), and Tatsuo Shiina (a)
(a) Graduate School of Science and Engineering, Chiba University
Chiba University 1-33 Yayoi-cho, Inage-ku, Chiba-shi, Chiba 263-8522 JAPAN
dakaro@chiba-u.jp

Abstract: Optical Coherence Tomography (OCT) has been practically used in ophthalmology, however, its quantitativeness is more valued for the evaluation of human skin. In this study, Time-Domain typed Optical Coherence Tomography (TD – OCT) was specialized to evaluate human skins quantitatively, and skin phantom measurements were assessed. At first, skin phantoms at different concentrations of titanium dioxide were prepared and measured according to its representation of dermis on human skin. Same increment of the extinction coefficient was obtained as the concentration increases. This OCT measurement result was compared with the result from the transmittance experiment. The difference between two results was considered as the effect of multiple scattering in the strong scattering medium. Next, the TD-OCT measurement was performed on the human skin. Extinction coefficients were evaluated with criterions of accurateness in medical field. Based on the linearity of the specialized TD-OCT, the extinction coefficient obtained by this method could accurately represent optical property of dermis and epidermis of human skin.

Keywords: Optical Coherence Tomography, Extinction coefficient, Human skin, Skin phantoms.

1. Introduction

Generally, humans have the same skin structure. However, the quality of skin varies depending on genetics and lifestyle. Furthermore, the character of the skin changes day by day with age, environment, and hormone's work. In recent years, as the concern for the skin shifted from diseases to skincare and beauty, the demand for skin examination devices is increasing. Current skin diagnoses are performed by specialists after a physical examination, a patch test or a microscopic observation that are targeted only for the skin surface [1]. Thus, the deviation occurs when it comes to the evaluation of the actual optical properties of the skin. A quantitative evaluation of the optical characteristics of human skin is probable, due to the existing Optical Coherence Tomography (OCT) used in ophthalmology.

OCT is a non-invasive ophthalmologic diagnostic technique that renders in vivo cross-sectional view of the retina. OCT utilizes a concept known as interferometry to create a cross-sectional map of the retina with the accuracy of 10 -15μm. OCT was first introduced in 1991 [2] and has found many uses outside of ophthalmology, where it has been used to image certain non-transparent tissues. Acquisition of the OCT signal is based on splitting the incident low coherent light beam into two paths: (1) a sample beam and (2) a reference beam, which follow two different paths but with the same length. When the reflected light beams (derived from each of these two paths) simultaneously reach the detector, they induce an interference signal. The amplitude of this interference signal is measured and used to acquire the image.

Swept-Source OCT (SS-OCT) and Frequency-Domain OCT (FD-OCT) are the current standard of OCT techniques. In recent years, as technology advances, the OCT also developed in fields other than ophthalmology, i.e. skin diagnosis. Although, the FD-OCT and SS-OCT have been popular with their high resolution and high acquisition speed [3-6], it is difficult to accurately judge the optical characteristics of the skin due to the nonlinearity of the backscattered light signals derived from indirect signal processing. On the other hand, TD-OCT can measure the human tissue with high linearity because of simple magnification of return intensity. Therefore, TD-OCT is the efficient way to measure and evaluate the human skin quantitatively.

In this study, for quantitative evaluation of the human skin, skin phantoms made with titanium dioxides were utilized and their extinction coefficients were evaluated by the specialized TD-OCT. These results were then compared with other results from transmittance measurement. Then, actual human skin in section of the forearm and the upper arms were measured and evaluated with the same procedure that in the medical field. This method is expected to increase the accuracy of skin measurement values in diagnosis and contribute to early detection and prediction of skin diseases [7-9]. Furthermore, it could also be expected to provide guidelines for the formulation of cosmetic products.

2. Experimental setup and Method

The schematic diagram of the skin OCT used in this research is shown in Figure 1. The basic configuration is the Michelson interferometer. Super Luminescent Diode (SLD) is used as a light source for this TD-OCT. The SLD is in the near-infrared light with a central wavelength of 1.31μm, and since the measurement depth is less than 1.0 mm, the absorption was neglected [10-11]. In the present apparatus, a variable optical path mechanism is used as a reference optical path. This variable optical path consists of a rotating corner reflector and a fixed mirror to create an approximately linear motion. In this setup, the rotation angle was $\pm 20^{\circ}$ and the deviation of the linear motion was about 1%. Therefore, measurement with good reproducibility was possible, and a stable scanning can be performed. Skin phantoms (model paint with titanium dioxide) with volume 5%, 10%, 20%, 30% and 50%, were measured. Figure 2 shows the result of plotting the skin phantoms placed on a light box. It can be seen that the transmitted light intensity decreases with the increase of concentration from (5%) to (50%).The skin phantoms were made by an applicator, and the thickness of the phantom were fixed as 200μm. In addition to the titanium dioxide(TiO_2), the phantom was also included with dispersion resin and other solutions.

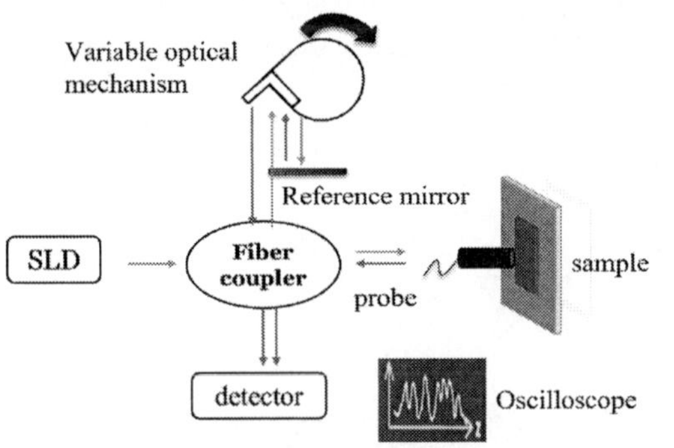

Figure 1. Basic schematics of the experiment setup.

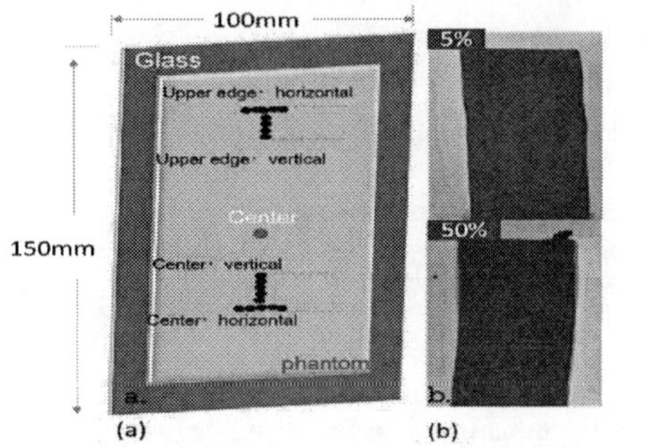

Figure 2. Appearance of the phantom (a) Measured spots (b) Appearance on the light box.

3. Result and discussion

Figure 3 shows the measurement results. Here, the time difference caused by rotation of the reflector was converted to the media depth (L). The graph is indicated in log scale and the slope represents the extinction coefficient (σ) that was derived from the equation (1):

$$\mathrm{Log}\frac{I}{I_0} = (-2\sigma L) \tag{1}$$

Where, I and I_0 are incident and reflected light intensities. Back scattered light from the depth of 1 mm was obtained. The smoothly declined linear trend of the slope indicates the homogeneous characteristics of the material of the phantom.

Figure 4 shows the derived extinction coefficients, it can be observed that, as the increment of the concentration, the extinction coefficient of the backscattered light decreases slowly. Furthermore, it can also be inferred that even if the concentration doubles, the change in the extinction coefficient remains almost constant.

Figure 5 indicates the transmittance measurement result. The extinction coefficient gradually increases as the media concentrations increases. The increment of the extinction coefficient was smooth as the growth of concentrations. However, the trend of the increment was different with that of the OCT result.

Figure 6 presents the adjusted and corrected results of OCT measurement showed in Figure 4. The maximum optical path length (1 mm) was the consequence of original thickness that based on the fixed standard of 200 μm of the provided phantom. The original thickness of the phantom was fixed to the 200 μm when it was made. However, the derived optical thickness of the phantom was turned out to be 1mm according to the OCT measurement result. The actual thickness of the phantom measured as 16.8 μm, 14.1 μm, 29.4 μm, 23.0 μm and 50.0 μm for 5%, 10%, 20%, 30%, and 50%, respectively by confocal microscopy. The difference of phantom thickness was dependent on the particle amount and vapor process of dispersion resin. Therefore, the thickness was normalized to 50 μm based on 50% concentration and other phantoms were also normalized the same way. Due to the different thickness of the phantom, as the concentration increases, digits of the extinction coefficient changes. Therefore, the correction was carried on the extinction coefficient according to the different thickness of the phantom. As a result, the increment tendency of modified results showed the same trend with the result from transmittance measurement.

The adjusted and corrected OCT results in Figure 6 have a large fluctuation when it is compared with result from the transmittance measurement. This variation was considered as follows: as for the OCT measurement, the material texture and density evenness will be strongly influenced due to small spots that carried on by the microscopic measurement. For the transmittance measurement, however, the observation size was large 6 mm. Therefore, the result appears as a smooth increment.

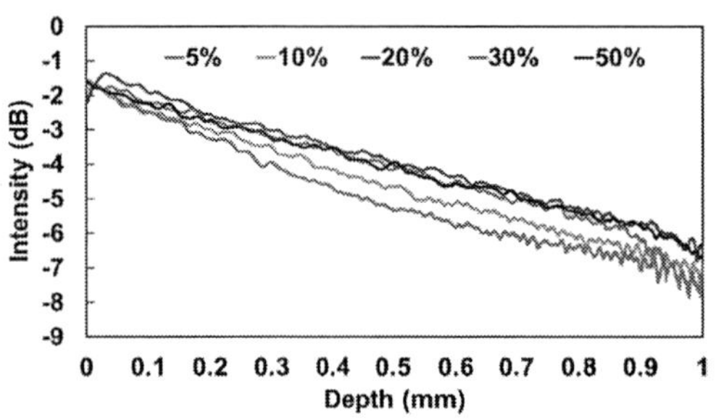

Figure 3. Measurement results of phantom under skin OCT devices.

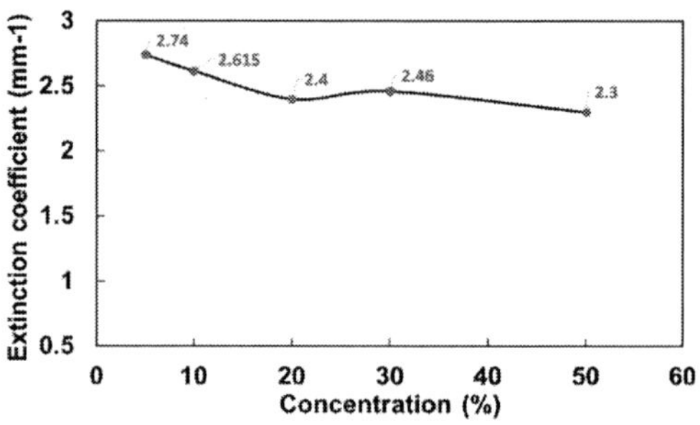

Figure 4. Relationships of extinction coefficients and concentrations.

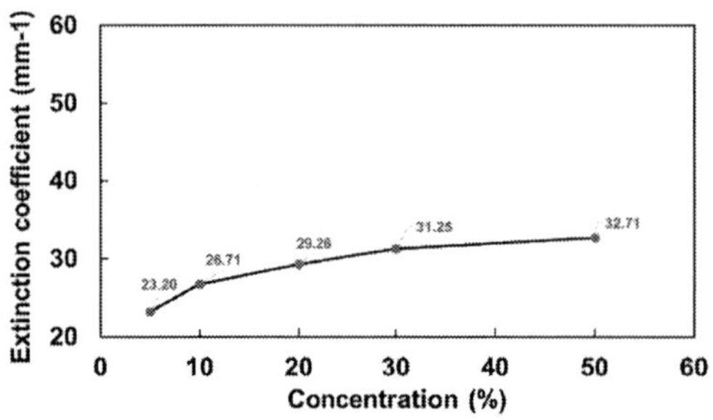

Figure 5.Extinction coefficient of Transmittance Measurement.

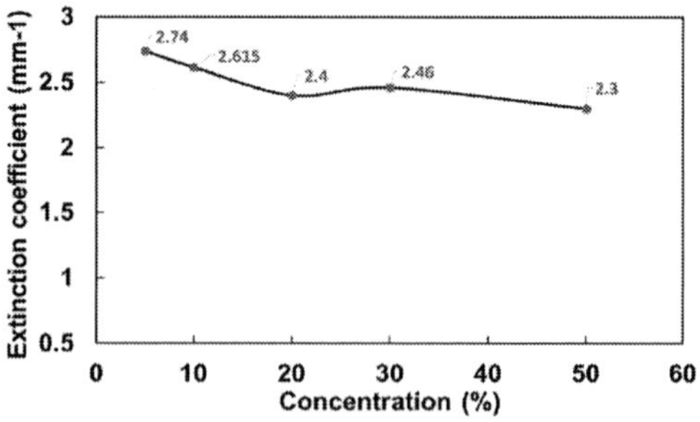

Figure 6. Relationships of adjusted and corrected extinction coefficients and concentrations.

4. Summary

In this paper, our work was mainly conducted by following two subjects: at first, due to the complexity of the human skin, we purposed to prepare, measure and evaluate 5 phantoms with different concentrations 5%, 10%, 20%, 30% and 50%. Optical property of these titanium dioxide phantoms was resemblance to the human skin. Next, we implemented same OCT measurements for human skins applying the demonstrated evaluation method. To achieve the objective of this paper, we applied the TD-OCT system to obtain the extinction coefficient from both phantom and human skin. Thickness of the phantoms was measured by confocal microscopy to confirm the actual thickness of the phantom. Transmittance measurement was also applied to compare the extinction coefficient with that of TD-OCT measurement results. As a result, attenuation tendency of backscattered light was obtained from the phantoms in different concentration. Almost constant extinction coefficients were observed as the increment of concentrations. And the difference between OCT and transmittance measurement results were considered as effects of (1) consideration of actual thickness of phantoms and (2) variation of the extinction coefficient as the change of concentration that was led by the distinction of the measurement size of objected area. At last, the extinction coefficients of the human skin were measured and evaluated as 7.3 for forearm and 7.0 for upper arm, which are reasonable on the medical criterion.

Since the OCT measurement can be considered as a type of microscopic analysis, it can be applied for the quantitative determination of micro structures (such as skin roughness) and larger structures (such as wrinkles) on skin diseases. We also quantified the internally reflected light from the human skin. This study measured the optical characteristics (such as extinction coefficient) in human skin for dermis and epidermis. For a cosmetic product to achieve a brilliance effect, the determination of backscattered light is important. These measurements and results provide guidelines for formulation of cosmetic products.

5. References

[1] T. Gambichler, R. Matip, G. Moussa, P. Altmeyer, K. Hoffmann "In vivo data of epidermal thickness evaluated by optical coherence tomography: Effects of age, gender, skin type, and anatomic site," Journal of Dermatological Science 44(3) 145–152(2006).

[2] D. Huang, E. A. Swanson, CP Lin, et al, "Optical coherence tomography," Science. 254(5035), 1178-1181(1991)

[3] J. G. Fujimoto, C. Pitris, S. A. Boppart, and M. E. Brezinski, "Optical Coherence Tomography: An Emerging Technology for Biomedical Imaging and Optical Biopsy," Neoplasia. 2(1-2), 9–25(2000).

[4] S. H. Yun, G. J. Tearney, B. E. Bouma, B. H. Park, and J. F. de Boer, "High-speed spectral-domain optical coherence tomography at 1.3 μm wavelength," Opt. Express 11, 3598-3604 (2003).

[5] Robert J. Zawadzki, Steven M. Jones, Scot S. Olivier, Mingtao Zhao, Bradley A. Bower, Joseph A. Izatt, Stacey Choi, Sophie Laut, and John S. Werner, "Adaptive-optics optical coherence tomography for high-resolution and high-speed 3D retinal in vivo imaging," Opt. Express 13, 8532-8546 (2005)

[6] M. E. Brezinski and J.G. Fujimoto, "Optical Coherence Tomography: High-Resolution Imaging in Nontransparent Tissue," IEEE J. Quant. Electron. 5(4), 1185-1192(1999).

[7] A. J. Coleman, T. J. Richardson, G. Orchard, A. Uddin, M. J. Choi, K. E. Lacy "Histological correlates of optical coherence tomography in non-melanoma skin cancer," Skin Res Technol 19(1) 10-9(2013).

[8] T. Shiina, Y. Oritani, M. Ito, and Y. Okamura, "Long Optical-Path Scanning Mechanism for Optical Coherence Tomography," Applied Optics 42(19), 2795-3799(2003).

[9] T. Shiina· F. Tetsuo, "paint evaluation by portable OCT," Proceedings of the Electronics, Information and Systems Conference. Electronics, Information and Systems Society, I.E.E. of Japan. OS 3-5, 700-705(2013), (written in Japanese).

[10] T. Shiina. S. Miyazaki. T. Honda. "Factory Built-in Type Simplified OCT System for Industrial Application," IEEJ Transactions on Electronics, Information and Systems 129(7) (2009), (written in Japanese).

[11] S. Tabuchi, T. Shiina, "Skin diagnosis and quantitative evaluation by portable OCT," Papers of Technical Meeting on Instrumentation and Measurement, IEE Japan, IM-14-060, 15-20(2014), (written in Japanese).

FIB-SEM-EDS Analysis of Aerosols: Implications to Particle Optics

Vikas Goel[a,b], Sumit K. Mishra[a,b], Ajit Singh Ahlawat[a,b], Chhemendra Shrama[a], Vijay Narain Ojha[a,b]

(a) CSIR-National Physical Laboratory, New Delhi, India-110012

(b) Academy of Scientific and Innovative Research, Kamla Nehru Nagar, Ghaziabad, Uttar Pradesh, India- 201002

Abstract: In retrieval techniques, aerosol generally considered as chemically homogeneous spheres which results in erroneous results. There is a severe need for multi-dimensional analysis of aerosol. Therefore, mineral dust particles were analyzed using focused ion beam (FIB) coupled with energy dispersive X-ray spectroscopy (EDS) and scanning electron microscopy (SEM). In present study, particle was observed in core-shell type structure, where shell was rich in Fe, Ca, C, Al, and Mg and core were rich in Si. In conventional SEM-EDS, the particle was observed as homogeneous mixture of Ca, Si, Cu, S, K, Mg, C, Fe and Al. Single scattering albedo (SSA) was calculated using FIB-SEM-EDS results and conventional SEM-EDS, difference between two results was observed to increase with increasing wavelength and the maximum difference was observed at 860 nm i.e. 19.38%. This shows that homogeneous spheres assumption is putting 19.38% uncertainty in results, which could have a tremendous effect on radiative forcing estimations.

Keywords: Radiation budget, FIB-SEM-EDS, Core-shell, SSA.

1. Introduction

Aerosol particles are major constituent of atmosphere, they affect earths' radiation budget. This effect can be better understood with assessment of their optical property (Single Scattering Albedo; SSA). The optical property of aerosols depends upon their chemical composition [1], shape [2, 3], size [4] and internal structure [5]. In satellite observations the shape of aerosols are generally assumed as spherical and their composition is assumed to be homogeneous throughout the particle [6]. The conventional techniques like scanning electron microscopy [7] (SEM), Energy dispersive X-ray spectroscopy (EDS) and transmission electron microscopy [8] (TEM) used for the analysis are not capable to reveal the internal structure of the aerosol particle. Therefore, there is a severe need for detailed multi-dimensional analysis of aerosol particles which is capable of imaging embedded phases and voids hidden from the conventional SEM or TEM-EDS.

In the present work, dust particles were analyzed using Focused Ion Beam (FIB) coupled with SEM and EDS. The gallium (Ga) was used to cut the particle which was followed by taking EDS spectra of exposed the surface. This procedure was repeated may time (2~3) on a single particle, that reveals its internal structure.

2. Experimental Section

In the current study, the aerosol particles were collected from Jaisalmer (26.9157° N, 70.9083° E), Rajasthan, India using Envirotech® APM801 sampler. The particles were collected on tin substrate. After collection the particles were stored in desiccator and brought to the CSIR-National Physical Laboratory, New Delhi for individual particle analysis using Auriga Cross beam FIB-SEM-EDS.

3. Result and Discussion

Mineral dust dominated the overall loading of aerosol in atmosphere [9]. Conventional SEM-EDS and FIB-SEM-EDS analysis of dust particle is shown in fig (1). The results of conventional SEM-EDS show that the particle is rich in Ca, Si, Cu, S, K, Mg, C, Fe, O and Al (fig 1a). After taking the conventional SEM-EDS spectra the particle was milled with high energy focused Ga gun. The milling cuts the particle and exposes its internal structure. EDS spectra of exposed internal surface shows the particle is rich in Si and O only (fig 1b). The atomic percentage of Si and O confirms the presence of SiO_2 inside the particle. The same particle was milled second time at the same position to check the elemental composition of the core of the particle. The EDS spectra recorded after second milling shows the presence of SiO_2 rich core. The advanced FIB-SEM-EDS analysis shows that the particle is having core-shell type structure, where the core is rich in SiO_2 and shell is rich in Fe, Ca, C, Al, and Mg.

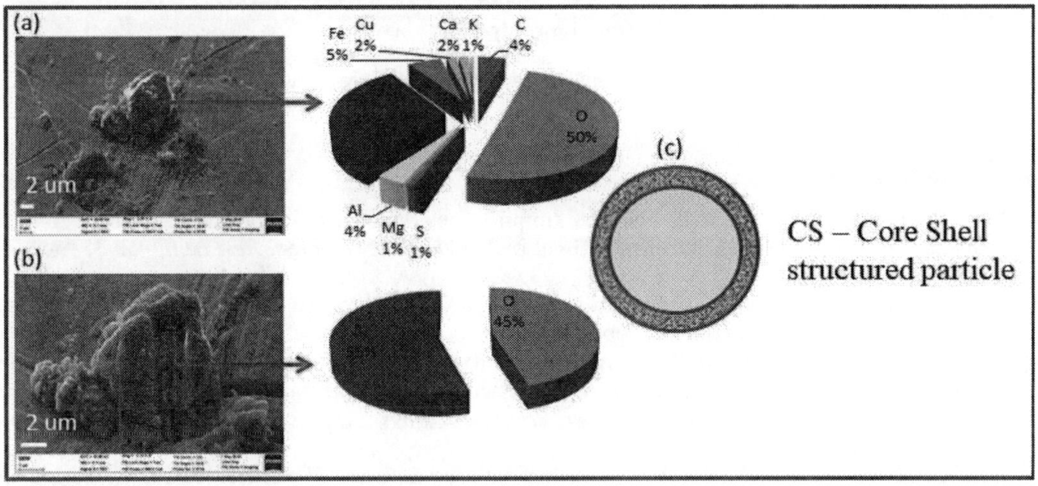

Figure 1. FIB-SEM-EDS analysis of dust particle

Optical property (SSA) of dust particle shown in fig (1) was simulated using T-matrix and core-shell-shell code. The spectral variation of SSA is shown in fig (2). Here, SSA was simulated for five different structures of particles i.e. Quarts homogeneous sphere (S1), Quartz homogeneous spheroid (SPH1), multiple species homogeneously mixed sphere (S2), multiple species homogeneously mixed spheroid (SPH2) and core-shell structured particle (CS). The SSA of S1 and SPH1 was observed to be near to 1 due to completely scattering characteristic of quartz. Chemical composition of S2 and SPH2 was determined by conventional SEM-EDS, SSA for S2 and SPH2 was observed to be very low near to 0.56. The low value of SSA shows the highly absorbing character which was attributed to the presence of hematite and carbon in particle. The effect of shape between SSA of S1, SPH1 and S2, SPH2 was observed to be negligible. Chemical composition of CS particle was determined by advanced FIB-SEM-EDS, SSA for CS was observed to be ranging from 0.54-0.66. The increasing character if SSA in CS was observed due to spectral variation of imaginary part of refractive index of hematite. The SSA of CS was found to be between S1, SPH1 and S2, SPH2. The S1 and SPH1 calculation overestimating the real value, whereas, the S2 and SPH2 were underestimating the real value. The difference between the SSA calculated using conventional SEM-EDS results (S2, SPH2) and advanced FIB-SEM-EDS results (CS) was observed to increase with wavelength and maximum difference was observed at 860 nm. The difference between S1 and CS at 860nm was observed to be 19.38%. Therefore, the conventional results are putting high degree of error in optical property simulations.

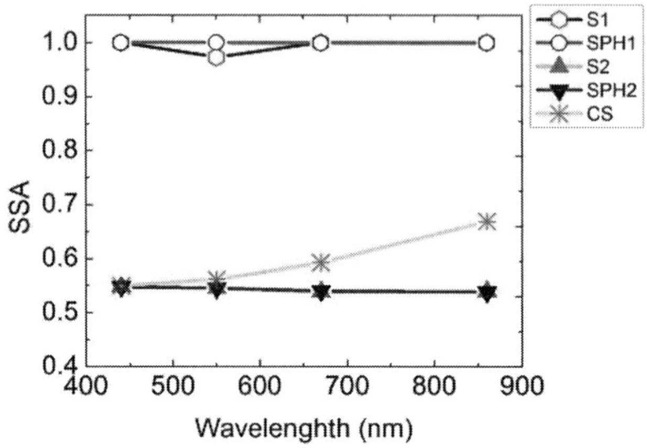

Figure 2. Single Scattering Albedo of dust particle shown in fig.1.

4. References

[1] S. K. Mishra, S. Dey, and S. N. Tripathi, "Implications of particle composition and shape to dust radiative effect: A case study from the Great Indian Desert," *Geophys. Res. Lett.*, vol. 35, no. 23, Dec. 2008.

[2] S. K. Mishra, "Morphology of Atmospheric Particles over Semi-Arid Region (Jaipur, Rajasthan) of India: Implications for Optical Properties," *Aerosol Air Qual. Res.*, vol. 2015, no. 3, 2015.

[3] S. K. Mishra, S. N. Tripathi, S. G. Aggarwal, and A. Arola, "Effects of particle shape, hematite content and semi-external mixing with carbonaceous components on the optical properties of accumulation mode mineral dust," *Atmospheric Chem. Phys. Discuss.*, vol. 10, no. 12, pp. 31253–31300, Dec. 2010.

[4] S. K. Mishra and S. N. Tripathi, "Modeling optical properties of mineral dust over the Indian Desert," *J. Geophys. Res.*, vol. 113, no. D23, Dec. 2008.

[5] J. M. Conny, "Internal Composition of Atmospheric Dust Particles from Focused Ion-Beam Scanning Electron Microscopy," *Environ. Sci. Technol.*, p. 130709074405007, Jul. 2013.

[6] D. Ganguly, H. Gadhavi, A. Jayaraman, T. A. Rajesh, and A. Misra, "Single scattering albedo of aerosols over the central India: Implications for the regional aerosol radiative forcing: SINGLE SCATTERING ALBEDO CENTRAL INDIA," *Geophys. Res. Lett.*, vol. 32, no. 18, p. n/a-n/a, Sep. 2005.

[7] J. Arndt *et al.*, "Scanning electron microscopy-energy dispersive X-ray spectrometry (SEM-EDX) and aerosol time-of-flight mass spectrometry (ATOFMS) single particle analysis of metallurgy plant emissions," *Environ. Pollut.*, vol. 210, pp. 9–17, Mar. 2016.

[8] W. Li, L. Shao, Z. Wang, R. Shen, S. Yang, and U. Tang, "Size, composition, and mixing state of individual aerosol particles in a South China coastal city," *J. Environ. Sci.*, vol. 22, no. 4, pp. 561–569, Jan. 2010.

[9] M. O. Andreae, R. J. Charlson, F. Bruynseels, H. Storms, R. Van Grieken, and W. Maenhaut, "Internal Mixture of Sea Salt, Silicates, and Excess Sulfate in Marine Aerosols," *Science*, vol. 232, no. 4758, pp. 1620–1623, Jun. 1986.

Development of the monostatic fiber-based laser Doppler vibrometer using double-pass frequency shifter reducing the impact of internal reflection light and improving demodulation performance

Yuichi NISHINO, Ryota KOJIMA, Masaharu IMAKI and Shumpei KAMEYAMA
Mitsubishi Electric Corporation, Information Technology R&D Center,
5-1-1 Ofuna, Kamakura, Kanagawa, 247-8501

Abstract: The fiber-based laser Doppler vibrometer is useful for many applications including the infrastructure health/condition monitoring, owing to its compactness, reliable components, eye-safety, and so on. However, the fiber-based configuration suffers from its internal reflection light which impacts the demodulation performance of vibration waveform, if the configuration is coaxial. Therefore, the past vibrometer must be bistatic to prevent this impact, and the sensitive alignment of the transmitter and receiver was needed. Here, we show the monostatic fiber-based vibrometer using the double-pass frequency shifter. This configuration realizes the discrimination of the internal reflection light and the received echo from the target without the above mentioned alignment. We developed the prototype and demonstrated its performance in sensing the road surface vibrations which were caused by passing vehicle. The demonstration implies the future application of this vibrometer to the road condition (sliding friction) monitoring for the traffic safety.

Keywords: Coherent Laser Radar, Doppler vibrometer, Acousto-Optic Modulator

1. Introduction

The laser Doppler vibrometer measures the frequency of the target from the time variation of the Doppler shift. It has the spatial light circuit type and the fiber-based type. The former is composed monostatic and the latter is composed bistatic. If the configuration is coaxial, the fiber-based configuration suffers from its internal reflection light which impacts the demodulation performance of vibration waveform. Therefore, the past vibrometer must be bistatic to prevent this impact, and the sensitive alignment of the transmitter and receiver was needed. Here, we show the monostatic fiber-based vibrometer using the double-pass frequency shifter. This configuration realizes the discrimination of the internal reflection light and the received echo from the target without the above mentioned alignment.

We noted that the vibration sound of the road surface varies depending on the condition of the road surface, we developed the prototype and demonstrated its performance in sensing the road surface vibrations which were caused by passing vehicle. In the following sections, we explain the system configuration and the experimental results.

2. System Configuration

The configuration of laser Doppler vibrometer is shown in Fig. 1. And the photo of transmitting and receiving apparatus is shown in Fig. 2. The system consists of the transmitting and receiving apparatus, signal analyzer and personal computer (PC). The transmitting and receiving apparatus consists of the laser, the collimator, the Acousto-Optic Modulator (AOM), the telescope and the balanced receiver. Each component is connected with polarization maintaining optical fiber. The 1.5 μm laser is divided local light and transmitted light by the coupler 1. The transmitted beam is irradiated to the target through the AOM and the telescope. The received light is converted to electric signal by the balanced receiver. The received signal is processed in the signal analyzer, where Fourier transform is executed. The vibration spectrum is calculated from the fluctuation amount of the peak frequency. If AOM is placed between the coupler 1 and the coupler 2, when the amount of the Doppler shift is very small, the received echo from the target overlaps the internal reflection light. And in case that the received echo from the non-moving target is detected, the Doppler shift becomes zero and overlaps with the internal reflection light.

Here, we placed AOM between the collimator and the telescope. By shifting frequency only to the received echo from the target, it is possible to discriminate it from the internal reflection light. The received echo from the target with small Doppler shift amount and the received echo from non-moving target can also be detected. In addition, tilting the telescope diagonally avoided the influence of the internal reflection light. The internal reflection light component was cut by the filter, and only the received echo component from the target was passed.

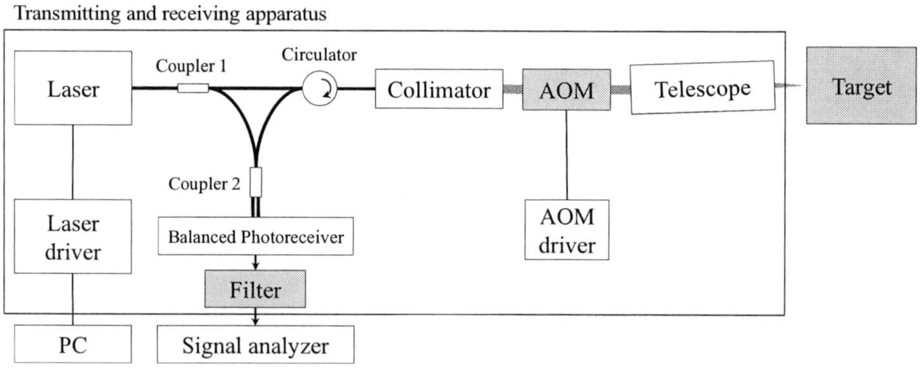

Fig. 1. Configuration of laser Doppler vibrometer.

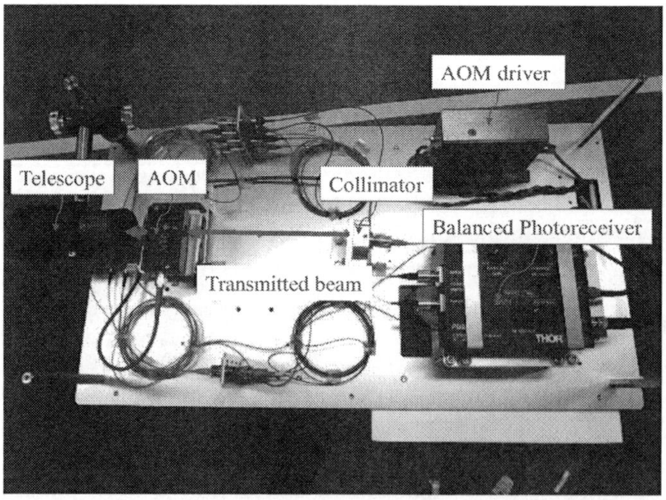

Fig. 2. Photo of the transmitting and receiving apparatus.

3. Experimental Results

The appearance of the experiment is shown in Fig. 3. The prototype was set on the road with a tripod. The measurement was performed on the dry condition and the wet condition. The sufficiently dried asphalt road surface was used as the dry condition. On the wet condition, the water sprayed so that a puddle was always formed near the laser irradiation position.

The experimental result of the vibration spectrum is shown in Fig. 4. It was the waveform acquired by performing Fourier transform on the demodulated signal output from the signal analyzer. In the case of the dry condition, it was confirmed that the peak occurred at 14 to 15 Hz. On the other hand, in the case of the wet condition, it was found that its spectral line width was broadened.

The relationship between the friction coefficient and the frequency width is shown in Fig. 5. The friction coefficient was measured by the sliding resistor. The frequency width of the vibration spectrum was calculated by determining the secondary moment in the range 50 Hz before and after the peak frequency. It was confirmed that there was a correlation between the coefficient of friction of the road surface and the vibration spectrum of the road surface which were caused by passing vehicle.

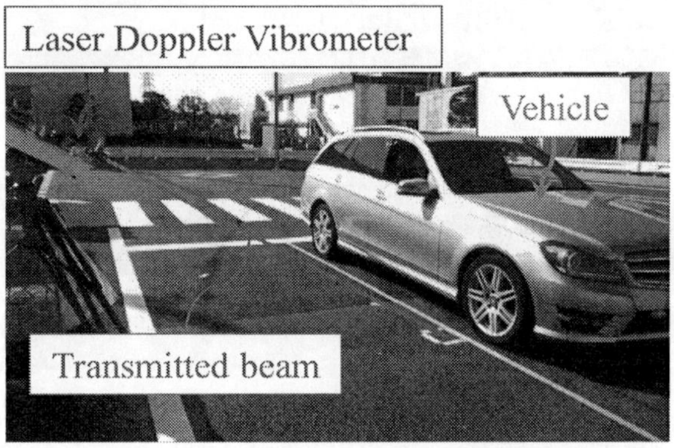

Fig. 3. Appearance of the experiment.

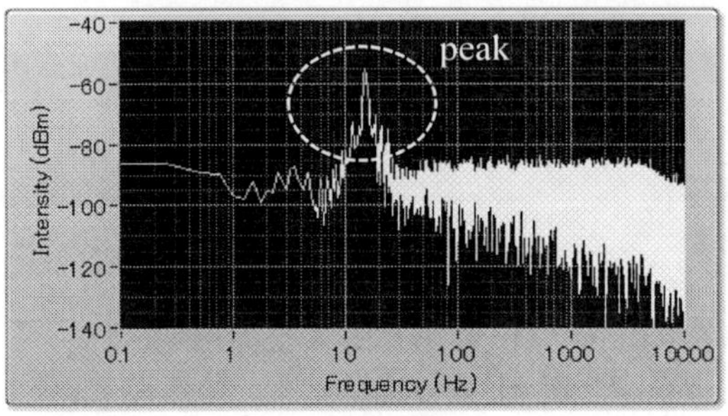

(a) Dry condition.

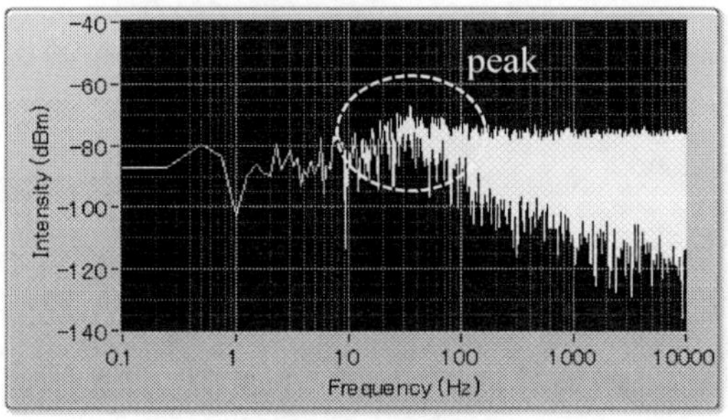

(b) Wet condition.

Fig. 4. Experimental result of the vibration spectrum.

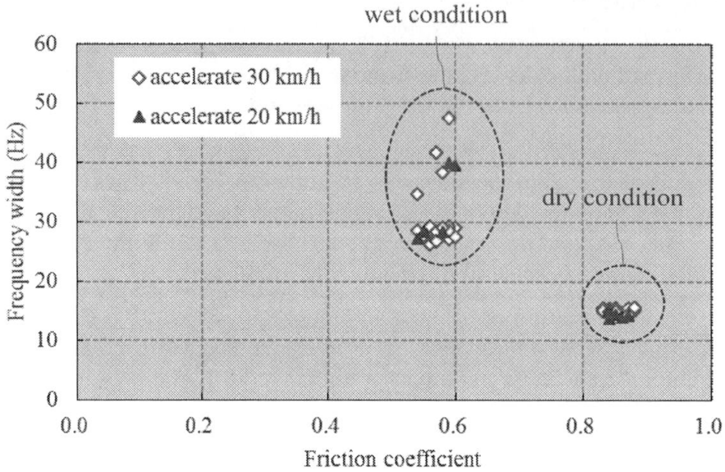

Fig. 5. Relationship between the friction coefficient and the frequency width.

4. Summary

We developed the prototype and demonstrated its performance in sensing the road surface vibrations which were caused by passing vehicle. As a result, it was confirmed that there was a correlation between the coefficient of friction of the road surface and the vibration spectrum of the road surface. The demonstration implies the future application of this vibrometer to the road condition (sliding friction) monitoring for the traffic safety.

5. References

[1] M. Mizuma *et al.*, "Pulsed all-fiber laser Doppler vibrometer", 24th Laser Sensing Symposium, p59-60 (2005).

[2] S. Kameyama *et al.*, Proc. Of Coherent Laser Radar Conference, 43-46, (2003).

[3] S. Kameyama *et al.*, Proc. Of Coherent Laser Radar Conference, 174-177, (2005).

Mid Infrared Superconducting Hot Electron Bolometer Mixers with nano-antennas

A. KAWAKAMI(a), H. SHIMAKAGE(b), J. HORIKAWA(c), M. HYODO(d), S. SAITO(a), S. TANAKA(a), and Y. UZAWA(e)

(a) National Institute of Information and Communications Technology, Kobe, Hyogo 651-2492, Japan
(b) Graduate School of Science and Engineering, Ibaraki University, Hitachi, Ibaraki 316-8511, Japan
(c) National Institute of technology, Fukui College, Sabae, Fukui 916-8507, Japan
(d) Faculty of Mechanical Engineering, Kanazawa University, Ishikawa 920-1192, Japan
(e) Advanced Technology Center, National Astronomical Observatory of Japan, 181-8588, Japan

kawakami@nict.go.jp

Abstract: To improve the response performance of mid-infrared (MIR) detectors, we designed and fabricated a superconducting hot electron bolometer mixer (HEBM) by using MIR-antenna technology. For designing antennas and circuits, the surface impedance of gold (Au) thin film at cryogenic temperatures was evaluated. The fabricated HEBM was formed with a superconducting strip, twin-slot-antenna, and distributed circuits. The design frequency was set at 61 THz. Under MIR-light irradiation, dark pulses were observed when the HEBM was biased near the critical current. From the pulse width, we considered that the MIR HEBM have 0.9 GHz IF-bandwidth. For evaluating the HEBM properties, measurement setup without beam splitter was constructed. IF output power characteristics of the MIR HEBM was observed at 61.3 THz and it was slightly changed due to the difference in the thermal load temperature between 1100 K and 300 K. The receiver noise temperature was estimated at about 17,000 K (DSB).

Keywords: antenna, HEBM, mid-infrared, NbN

1. Introduction

Electromagnetic waves generally exhibit both particle and wave properties. Many conventional MIR detectors are designed with structures and mechanisms based on the particle nature of light. However, such detectors usually suffer from a tradeoff between sensitivity and response speed because sensitivity depends on the detection area, and the detector volume and the electrostatic capacitance of a detector tends to increase with increasing area. Therefore, there were several reports of IR detection utilizing the wave nature of light [1], [2]. We also propose the use of nano-antennas for infrared detectors [3], [4].

Significant research and development in HEBM has led to the emergence of low noise electromagnetic wave receivers in the terahertz frequency range [5], [6], and we also contributed to this research and development effort [7]. Such mixer consists of a planar antenna and a small detector located at the feeding point of the antenna, and the upper limit of the operating frequency is considered to be determined by the physical size of the device. From these background, we started researching MIR HEBMs to clarify the usefulness of MIR-antenna technology. In this paper, we report on the evaluation of the noise temperature of the MIR HEBM at 61 THz.

2. DESIGN AND FABRICATION OF MIR HEBMS

For the simulation of MIR circuits, the dielectric constant of the magnesium oxide (MgO) substrate [8] and the complex surface impedances of the Au films were needed. The impedances could be derived using the measured complex refractive indices, however, these derivations were insufficient for designing superconducting MIR devices operating at cryogenic temperature. Therefore, we corrected the surface impedance at cryogenic temperatures by using a FTIR with sample cooling system. For the evaluation, resonator arrays constructed of thin film strips of Au were fabricated. The resonant frequency was observed as absorption properties at around 61 THz, and when the temperature changed from 300 K to 10 K, it was found that the resonant frequency shifted to the low frequency side. By using resonant frequency under cryogenic, we corrected the surface reactance by fitting to the simulated results [4].

The HEBMs were formed from a superconducting strip, nano-antenna, and distributed circuits. By using the corrected surface impedances, the MIR circuits were designed using Sonnet EM simulator for operation at 61.3 THz. Figure 1 shows the design and the calculated antenna impedance. The antenna was a twin-slot antenna, and the slot length and width were decided by the simulated results and set at 2200 and 200 nm, respectively. The size of the choke filter was also decided. At the designed frequency, the antenna impedance was expected to be $Z_{Ant.} = 250 - j6\ \Omega$.

For impedance matching between the antenna and the superconducting strip, we used a niobium (Nb, 2.5-δ nm)/ niobium nitride (NbN, 5 nm) bilayer as the superconducting strip to reduce sheet resistance. Here, for electric contact, the surface of the bilayer was etched by argon ion beam and the etched depth δ was estimated at less than 1 nm. Fabricating the MIR HEBMs with an antenna requires the building of fine structures on nano-scales. We developed a new fabrication process using electron beam lithography for all our lithography processes [3], [4]. Figure 2 shows a SEM image of the fabricated MIR HEBM, the inset is a magnified view of the detector. The detector width and length were approximately 0.18 and 0.26 μm, respectively. The normal resistance of the HEBM was approximately 200 Ω.

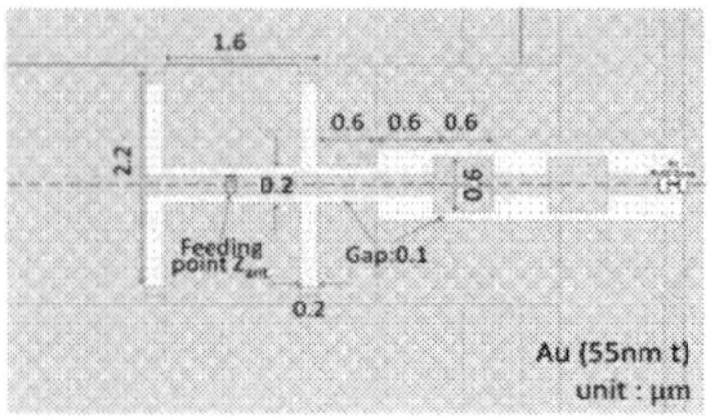

(a) Design of the twin-slot antenna

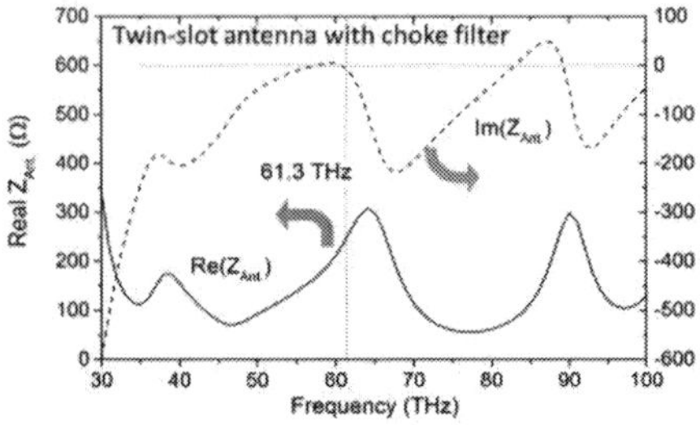

(b) Simulated impedance of the twin-slot antenna at feeding point

Figure 1. Design and impedance of twin-slot antenna.

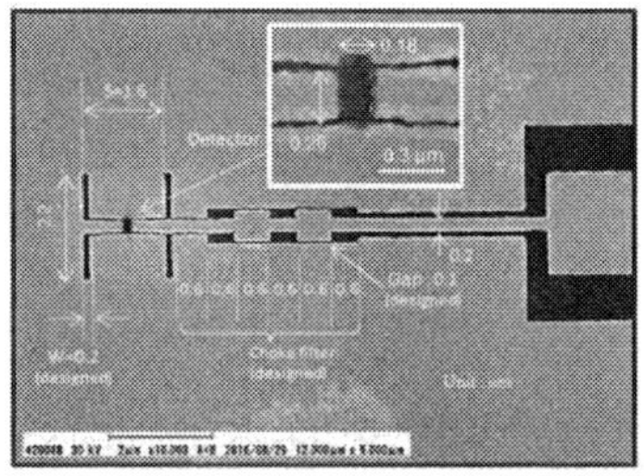

Figure 2. SEM image of the MIR HEBM.

3. EVALUATION OF MIR HEBM

The IF bandwidth of the HEBM is defined by the frequency at which the IF output is reduced by 3 dB. We tried estimating the bandwidth by evaluating the relaxation time of HEBM. When the bias current was set near the critical current, dark pulses were observed under MIR irradiation. Figure 3 shows the pulse which was averaged by 4096 pulses. The obtained the full width at half maximum of the pulse was about 0.21 ns. Assuming that the pulse width of the dark pulse is caused by the relaxation time of HEBM with Low-pass-filter model, the IF bandwidth of HEBM was estimated to be about 0.9 GHz.

For evaluating the HEBM properties, measurement setup without beam splitter was constructed. Figure 4 shows the setup. The HEBM was cooled by a GM refrigerator. A mid-infrared quantum cascade laser (QCL) with a wavelength of 4.89 μm was used as the local oscillator (LO). In this case, the LO power was irradiated from the surface of the HEBM. In order to concentrate the LO power, an offset parabola miller was

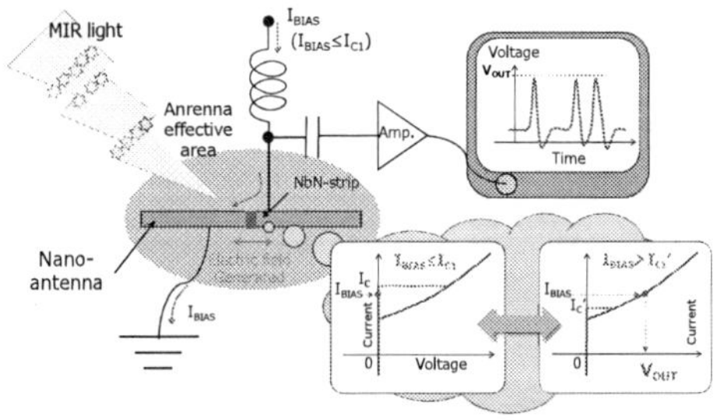

(a) Observation of HEBM dark pulse

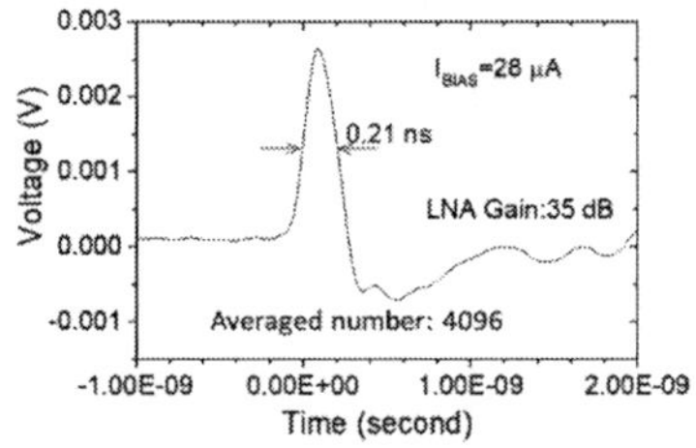

(b) Dark pulse waveform

Figure 3. Waveform of the dark pulse.

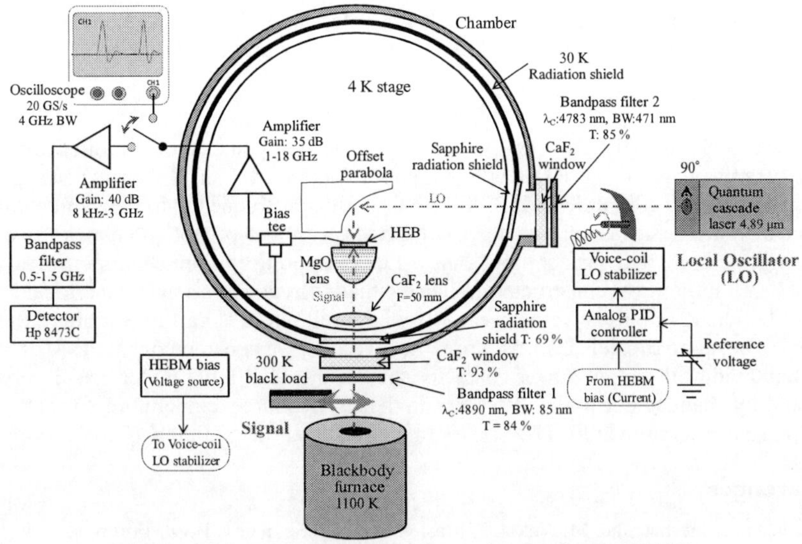

Figure 4. Measurement setup for evaluating MIR HEBMs.

used. However, we set the miller at defocus position to reduce the influence of the vibration of the GM refrigerator. The signal from the thermal load was irradiated from the back of the HEBM through an anti-reflection coated MgO hyper-hemisphere lens and an CaF_2 lens. A black body furnace set at 1100 K and 300 K black load were used as reference signals. Bandpass filter 1, CaF_2 window and sapphire radiation shield were used to exclude thermal radiation, except for the signal around LO-wavelength from the furnace. The total transmittance in the signal input optical path was estimated to be about 50%. The IF signal was amplified by a 1–18 GHz cooled low-noise amplifier and an 8 kHz to 3 GHz room temperature amplifier. The signal was then filtered by a bandpass filter centered at 1 GHz with bandwidth of 1 GHz and the IF power was monitored using a Schottky diode detector.

To suppress the effect of periodic changing of the coupling efficiency between the HEBM and QCL caused by the vibration of the GM refrigerator, we used stabilizing of the LO irradiation power by the voice coil. HEBM was biased by DC voltage source. Since the flowing current depends on the irradiation power, the PID output was acquired by comparing the amount of current flowing with the setting value, and stabilization of the LO irradiation power was performed by physically limiting the LO permeation amount by the voice coil.

Fig. 5 shows the pumped and unpumped I-V characteristics and IF output power characteristics of the MIR HEBM. The IF output showed a maximum at around 0.15 mV. We observed the change in the IF

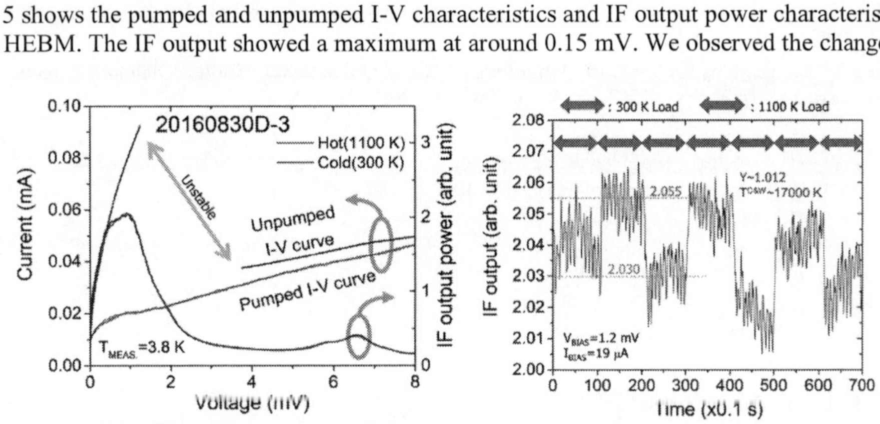

I-V and IF output power characteristics of the MIR HEB as a function of bias voltage.

IF output power – Hot/Cold Loads

Figure 5. I-V and IF output power characteristics of the MIR HEB as a function of bias voltage.

output due to the difference in the signal source temperature slightly, and estimated the receiver noise temperature at 61 THz as about 17,000 K (DSB) [9] from the IF output difference.

4. CONCLUSION

To design superconducting MIR devices, the complex surface impedance of Au thin films at cryogenic temperatures was evaluated using FTIR with a sample cooling system. Resonator arrays constructed with Au thin film strips were fabricated. By fitting the resonant frequency to the simulated results, the surface impedance at 10 K was corrected. MIR HEBMs with a twin-slot antenna were designed and fabricated using the corrected surface impedance. For evaluating the mixer properties, measurement setup without beam splitter was constructed. When the bias current was set near the critical current under MIR light irradiation, dark pulses were observed, and the full width at half maximum of the dark pulse was evaluated to be approximately 0.21 ns. From these results, we considered that the MIR HEBM have 0.9 GHz IF-bandwidth. IF output power characteristics of the MIR HEB mixer was observed, and the output was slightly changed due to the difference in the signal source temperature. From the difference, the receiver noise temperature at 61 THz was estimated to be about 17,000 K (DSB).

5. References

[1] Y. Lobanov, M. Shcherbatenko, M. Finkel, S. Maslennikov, A. Semenov, B. M. Voronov, A. V. Rodin, T. M. Klapwijk, and G. N. Gol'tsman, "NbN Hot-Electron-Bolometer Mixer for Operation in the Near-IR Frequency Range," *IEEE Trans. Appl. Supercond.*, vol. 25, p. 2300704, 2015

[2] Y. Yao, R. Shankar, P. Rauter, Y. Song, J. Kong, M. Loncar, and F. Capasso, "High-Responsivity Mid-Infrared Graphene Detectors with Antenna-Enhanced Photocarrier Generation and Collection," *Nano Lett.*, 14(7), pp 3749-3754, 2014

[3] A. Kawakami, S. Saito, M. Hyodo, "Fabrication of Nano-Antennas for Superconducting Infrared Detectors," *IEEE Trans. Appl. Supercond.*, vol. 17, pp. 632–635, 2011

[4] A. Kawakami, H. Shimakage, J. Horikawa, M. Hyodo, S. Saito, S. Tanaka, Y. Uzawa, "Design and Fabrication for the Construction of MIR HEB Mixers," *IEEE Trans. Appl. Supercond.*, vol. 27, p. 2300105, 2017

[5] S. Krause, V. Mityashkin, S. Antipov, G. Gol'tsman, D. Meledin, V. Desmaris, and M. Rudzinski, "Reduction of Phonon Escape Time for NbN Hot Electron Bolometers by Using GaN Buffer Layers," *IEEE Trans. THz Sci. Technol.*, vol. 7, no. 1, pp. 53–59, 2017

[6] D. Cunnane, J. H. Kawamura, M. A. Wolak, N. Acharya, T. Tan, X. X. Xi, and B. S. Karasik, "Characterization of MgB2 Superconducting Hot Electron Bolometers," *IEEE Trans. Appl. Supercond.*, vol. 25, no. 3, 2300206, 2015

[7] Y. Irimajiri, A. Kawakami, I. Morohashi, M. Kumagai, N. Sekine, S. Nagano, S. Ochiai, S. Tanaka, Y. Hanado, Y. Uzawa, and I. Hosako, "Development of a Superconducting Low-Noise 3.1-THz Hot Electron Bolometer Receiver," *IEEE Trans. THz Sci. Technol.*, vol. 5, no. 6, pp. 1154–1158, 2015

[8] R. E. Stephens and I. H. Malitson, "Index of Refraction of Magnesium Oxide," *J. Res. Natl. Bur. Stand.*, 49 pp. 249-252, 1952

[9] A. R. Karr, "Suggestions for Revised Definitions of Noise Quantities, Including Quantum Effects," *IEEE Trans. Microwave Theory Tech.*, vol. 47, no. 3, pp.325-329, 1999

Demonstration of Athermal and Light-weight Optical Telescopes For Airborne LIDAR System

Akihiro Fujie, Yoshichika Miwa, Hitomi Ono, Jiro Suzuki, and Toshiyuki Ando
Information Technology R&D Center
Mitsubishi Electric Corporation
5-1-1 Ofuna, Kamakura, Kanagawa, 247-8501 Japan
Fujie.Akihiro@cw.MitsubishiElectric.co.jp

Abstract: In airborne LIDAR system, optical telescope is required to be small, lightweight, and insensitive to temperature and pressure change in flight. In this paper, we designed athermal light-weight optical telescopes with CFRP (Carbon Fiber Reinforced Plastic) barrel and frame structures. We also evaluated optical performance with Shack-Hartmann wavefront Sensor under flight environment.

Keywords: Airborne LIDAR System, Optical System, Wavefront Sensor

1. Introduction

Development of the system which can foresee outbreak of the turbulence by measuring the wind velocity in front of an aircraft is desired for many years, because primary cause of aircraft accident is the unexpected turbulence. On this subject, we are developing airborne LIDAR (Laser Imaging Detection And Ranging) system, which can detect CAT (Clear Air Turbulence) [1-3]. This system enables us to measure wind velocity in fine weather, compared to radar detection system that are already installed in aircraft. However, in this airborne LIDAR system, the optical telescope, which has the function of transmitting laser beam into the atmosphere and receiving aerosol backscattered beam, is required to be small, light-weight, and to be robust against environmental conditions in flight. In particular, it is necessary to maintain optical performance against environmental temperature changes, that is, to minimize the change in focal length of the optical telescope against environmental temperature changes. This is because, the change in focal length of the optical telescope causes the reception efficiency of the backscattered beam to be lowered, and also causes deterioration of the wind speed measurement accuracy and measurement distance. It is also important to maintain optical performance against high loads caused by acceleration and deceleration of the airplane.

In this paper, we designed athermal optical telescope with CFRP (Carbon Fiber Reinforced Plastic) structures taking into consideration of CTE of CFRP as shown in Figure 1. Furthermore, we evaluated optical performance with Shack-Hartmann wavefront Sensor under flight environment.

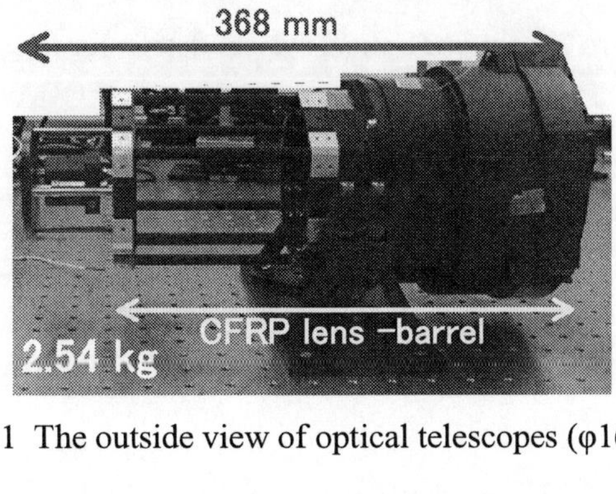

Figure 1 The outside view of optical telescopes (φ162 mm)

2. Design Concept of Athermal Optical Telescope

As shown in Chapter 1, in designing optical telescope of airborne LIDAR, it is required that the optical antenna is required to be small, robustness, light-weight, and insensitive to temperature and pressure change in flight. In this study, we adopted CFRP (Carbon Fiber Reinforced Plastic), which has low density, high rigidity ratio and low thermal expansion, for the optical telescope lens barrel. Furthermore, we adopted a passive athermal design method that realizes temperature compensation by appropriately setting the optical and mechanism design parameters based on aberration theory. In this method, as shown in Equation 1, under the condition that the lens barrel of the telescope constructed with CFRP, we designed lens barrel, lens material, and lens shape to compensate the thermal aberration in each lens with the aberration which caused by changing of the distance between lenses by the linear expansion of the CFRP lens barrel under the environment temperature change.

$$\Delta f = f\left(\alpha_g - \frac{1}{n-1}\frac{dn}{dT} - \alpha_n\right)\Delta T = 0 \qquad (1)$$

f: Focal-length, α_g: Coefficient of thermal expansion of the lens, α_n: Coefficient of Linear expansion of CFPR lens barrel, n: Refractive index of the glass material constituting the lens, T: Temperature

3. Result of system design for telescope

(i) Optical system design

Figure 2 shows the result of optical system design. We also evaluated wavefront aberration of optical telescope with simulation, based on optical system design as shown Fig 2. Figure 3 shows temperature dependence of wavefront error with (a) CFRP lens barrel and (b) Aluminum lens barrel. In this study, the optical telescope with Aluminum barrel should not show athermal performance, because we applied the passive athermal design method for CFRP lens barrel. As Fig. 3 shows, for the case of telescopes with CFRP lens barrel, the wavefront error doesn't change from 15°C to 25°C. On the other hand, in optical system with Aluminum lens barrel, large thermal aberration occurs at 15°C and 25°C. The nominal wavefront error is 0.0086 λ, RMS at 20°C. The wavefront error with Aluminum lens barrel is over 0.05 λ, RMS at 25°C, and it shows significant dependence of the temperature. On the other hand, in the case of CFRP, the wavefront error is suppressed under 0.01 λ, RMS in this temperature range.

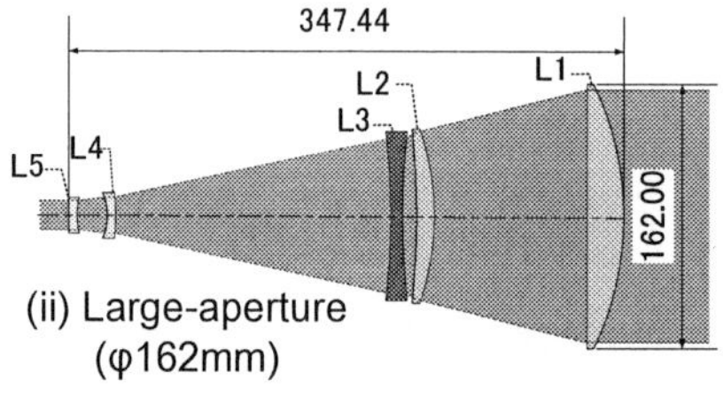

Figure 2 Result of optical system design for telescope

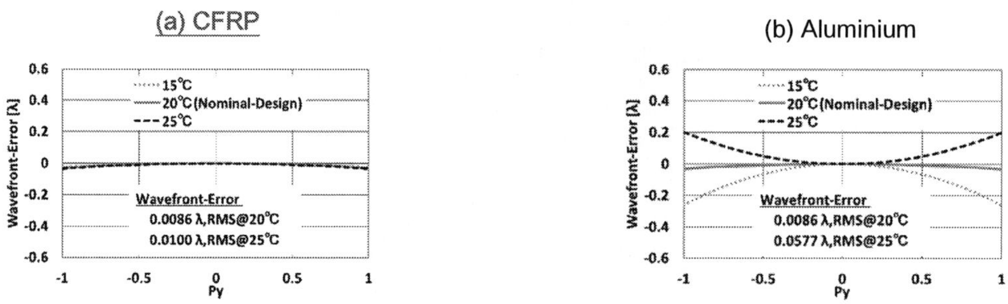

Figure 3 Temperature dependence of wavefront error of the our telescope

(ii) Mechanical system design

Figure 4-(a) shows appearance of the optical telescope. In this study, the telescope includes three tolerance compensators in order to compensate the wavefront aberration which caused by manufacturing tolerances of lens, lens barrel, and alignment errors during assembly of these components. The Compensator C1 is a part for adjusting the distance L3-L4, the Compensator C2 is a part for adjusting the eccentricity of L1, the compensator C3 is a part for adjusting the distance L2-L3. Figure 4-(b) shows the effect of tolerance compensators on wavefront aberration. From this result, it can be seen that wavefront aberration due to manufacturing tolerance and alignment errors can be effectively reduced by mounting all the compensators C1 to C3. It was expected that the wavefront aberration 0.07 λ, RMS could be achieved with a probability of 90% or more by simulation. This wavefront aberration result of 0.07 λ RMS corresponds to the reception efficiency loss of backscattered beam of 1.0 dB on wind measurement.

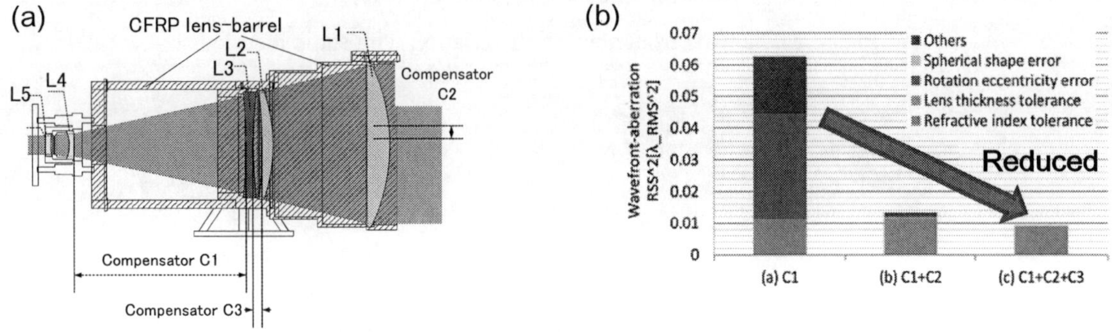

Figure 4 Result of mechanical system design

4. Results of optical performance evaluation of telescope

As shown in Fig. 1, we prototyped optical telescope based on result of optical and mechanical system design. As a result of applying CFRP to the lens barrel and optimizing the structure, the weight of the telescope become 2.54 kg, 70 % lighter than the conventional one which weight is 7.1kg.

Furthermore, we evaluated its optical performance by a Shack-Hartmann wavefront sensor [4]. The sensor is a wavefront measuring method using Hartmangram image, which consists of multiple focusing beam spot spatially divided by the micro lens array. The tilt of wavefront in front of each lens of microlens causes displacement of focusing beam spot. In other words, we can get tilt of each place of light flux so that of the wavefront of measurement object can be calculated by differential value of wavefront which is equal to tilt.

In this study, we evaluated the change of the wavefront aberration under the situation where the static load was applied to the prototype telescope. The wavefront aberrations with the static load of 1 G to 6 G are shown in Figure 6. From this result, compared to the wavefront aberration without the load, the wavefront aberration fluctuation is about 0.002 λ, RMS with the static load of 1 G to 6 G is applied. This result indicates that the structure of the optical telescope designed in this report has appropriate robustness. Furthermore, the transmitted wavefront shape of optical telescope shown in Figure 7, where the wavefront shape was maintained against application of static load. In conclusion, we confirmed that the optical telescope designed in this report was expected to be applicable to airborne LIDAR.

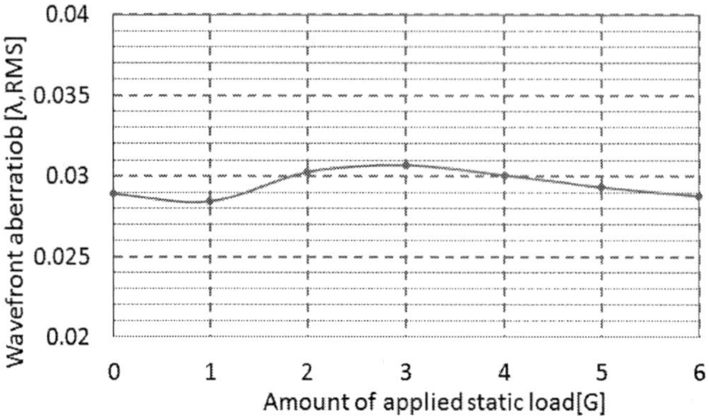

Figure 5 The wavefront aberration variation with static load

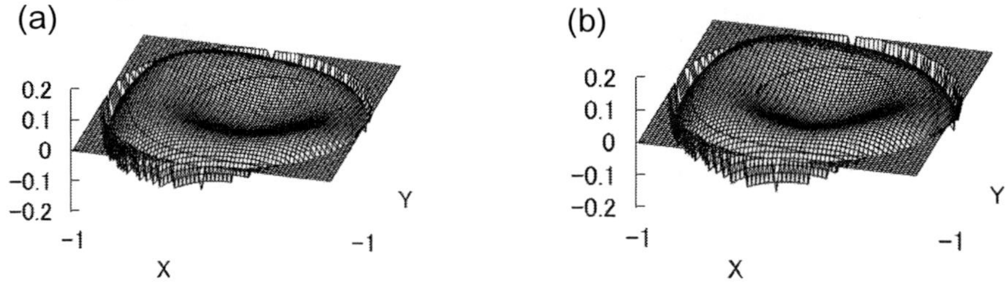

Figure 6 The transmitted wavefront shape of optical telescope

[(a): Not static load applied (b): 3G static load applied]

5. Summary

In airborne LIDAR system, the optical telescope is required to be small, light-weight, and to be robust against environmental conditions in flight. We developed an engineering model of lightweight athermal telescope with aperture diameter of 162 mm, length of 368 mm and weight of 2.54kg. Performance of athermal properties was experimentally verified as variation of wavefront error of less than 0.01 λ under environmental temperature at 15°C to 25°C. Furthermore, wavefront errors were also evaluated as their variation of 0.002 λ under static load was externally applied from 1G to 6G. In conclusion these results showed that the present athermal telescope has the primary performances for airborne LIDAR.

6. References

[1] H. Inokuchi, H. Tanaka, and T. Ando, "Development of an onboard doppler LIDAR for flight safety," Journal of Aircraft, Vol. 46, No. 4, pp. 1411-1415, 2009.

[2] H. Inokuchi, T. Akiyama, and S. Machida, "Development of an onboard safety avionics system using a doppler LIDAR," Proceedings of 18th Coherent Laser Radar Conference, Colorado, U.S.A., 2016.

[3] H. Ono, Akihiro Fujie, Yoshichika Miwa, Jiro Suzuki, Toshiyuki Ando, "Designing and Evaluating Athermal Optical Antenna for Airborne LIDAR System" International Conference on Space Optical Systems and Applications 2017, Okinawa, JAPAN, 2017.

[4] Y. Miwa, T. Endo, J. Suzuki, and T. Ando, "Linearity test of Shack-Hartmann wave front sensor," The JSAP 61st Spring Meeting, Tokyo, Japan, 17p-E4-1, 2014.

Development of conductively cooled, Q-switched Tm,Ho:YLF MOPA system

Makoto Aoki (a), Atsushi Sato (b, a), Shoken Ishii (a), Kohei Mizutani (a), and Katsuhiro Nakagawa(a)
(a) National Institute of Information and Communications Technology
4-2-1 Nukui-kitamachi, Koganei, Tokyo, 184-8795 Japan
(b) Tohoku Institute of Technology
35-1 Yagiyama-Kasumi, Taihakuku, Sendai, 982-8577 Japan
E-mail address: maoki@nict.go.jp

Abstract: A conductively cooled, diode-pumped, Q-switched Tm,Ho:YLF master oscillator and power amplifier (MOPA) was developed for coherent lidar applications. By optimizing laser rod parameters and oscillator configurations, the master oscillator output energy of 74 mJ with a pulse width of 120 ns was obtained at 30 Hz and a crystal temperature of -40 °C. As a single-pass power amplifier, the pulses were amplified into 129 mJ pulses without the appearance of amplified spontaneous emission. Injection-seeded, single-frequency operation was achieved without any loss of pulse energy owing to the adoption of a ring resonator for the oscillator.

Keywords: Conductive cooling, Q-switch, Tm,Ho:YLF laser, MOPA, Coherent Doppler wind lidar

1. Introduction

The National Institute of Information and Communications Technology (NICT) is conducting research and development on lidar technologies related to the future space-based coherent Doppler wind lidar (CDWL) [1-3]. Recently, we demonstrated a Q-switched Tm,Ho:YLF laser oscillator conductively cooled to -80 °C with an average output power of 7.28 W, corresponding to a Q-switched pulse energy of 104 mJ and a pulse repetition frequency (PRF) of 70 Hz [4]. Although the pulse energy and PRF satisfy the requirements for a laser transmitter used for a super low altitude test satellite (SLATS) developed by the Japan Aerospace Exploration Agency (JAXA) [5,6], the realization of cooling to the laser operating temperature of -80 °C for the space-based system has many difficulties in terms of limitations such as system complexity, lack of coolant, and large power consumption. A higher operating temperature is required to improve the feasibility of space-based lidars.In 2 μm holmium (Ho) lasers operating on the $^5I_7$ to $^5I^8$ transition and defined as a quasi-four-level laser, the thermal population at a low laser level introduces reabsorption loss, which leads to an increase in laser threshold and a reduction in slope efficiency. By utilizing a moderately short laser rod, which leads to a high pump density, with lower laser threshold, the slope efficiency is higher, and it is effective in high-temperature and high-pulse-energy operation. However, some issues are raised when using a short laser rod under an excessively high pump density. It yields the ground-state depletion (GSD) of Ho ions, which leads to a decrease in the Tm-Ho energy transfer efficiency. Therefore, to enable the high-pulse-energy operation of the Tm,Ho:YLF laser at a high temperature, it is important to optimize the Ho doping concentration in conjunction with the high pump density.

In this paper, we demonstrate a conductively cooled, diode-pumped, Q-switched Tm,Ho:YLF master oscillator and power amplifier (MOPA). On the basis of the results of numerical simulations using a laser rate-equation model [7], we selected three Tm,Ho:YLF laser rods with different Ho doping concentrations and examined their lasing characteristics at 30 Hz and -40 °C to evaluate their applicability as the master oscillator. As a result, a 100-mJ-class Q-switched pulse energy was demonstrated in the conductively cooled, single-pass MOPA.

2. Instruments

In our previous study, a Tm,Ho:YLF laser rod of 33 mm length and 4 mm diameter with 4% thulium (Tm) and 0.4% Ho doping concentrations was selected on the basis of the result of numerical simulations and the aspect of heat removal from the laser rod. A lasing performance meeting the requirements for the space-based CDWL was demonstrated at a rod temperature of -80 °C. However, the pump density was still relatively low, resulting in high thresholds due to reabsorption losses at the higher operating temperature. In addition, a roll-off of the slope efficiency occurred at high-pump-density regions owing to the GSD of the Ho ions. The use of the shorter and additional Ho-doped laser rod is effective for the lasing at a higher rod temperature.

In this study, we selected three laser rods, all 22 mm in length and 4 mm in diameter, with 4% Tm doping concentration but different Ho doping concentrations of 0.4, 0.7, and 1.0%. The rod is conductively cooled from three directions by three copper heat sinks controlled at a temperature of -40 °C with an accuracy of ±0.2 °C by a refrigerator system. The rod covered with indium film was strongly clamped by the heat sinks using setscrews from a housing of the pump module and polyamide rings attached to both sides of the heat sinks. The laser rod was side-pumped through a gap in the heat sinks by three pump modules placed 120° apart around the laser rod. The wavelength of emitted light from the pump modules was set to 792 nm by controlling the temperature of the pump modules. The pump module was put in a vacuum chamber.

Figure 1 shows the configuration of a Q-switched Tm,Ho:YLF MOPA. A 3.86-m-long ring resonator is used for the master oscillator because the single-frequency, long-pulse Q-switched operation is required for CDWL applications. The laser resonator consists of a 75% reflecting output coupler at 2.05 μm, three flat high-reflection (HR) mirrors, a crystal quartz acousto-optic Q-switch module antireflection-coated at 2.05 μm, and two pairs of prism-type and concave HR mirrors. One of the flat HR mirrors is mounted at a piezoelectric actuator and used for the resonator length control for the injection seeding. Pairs of prism-type and concave mirrors are used for miniaturizing the oscillator with the resonator length. The negative thermal lensing effect of the YLF laser rod is compensated by the concave HR mirrors. The curvature of the concave mirrors is calculated by ray transfer matrix analysis. The single-frequency and unidirectional operation is accomplished with a seed laser, which is injected into the resonator through the back of the output coupler. When the injection-seeded operation is not carried out, one output beam from the oscillator is retroreflected into the resonator by using a HR mirror for the unidirectional operation. The output from the oscillator is amplified by a Tm,Ho:YLF amplifier pump module with a structure similar to that of the oscillator. A plano-convex lens is used for matching between the output beam from the oscillator and the pump distribution in the laser rod for the amplifier. A single-pass amplifier was adopted in terms of the energy extraction efficiency and beam pattern.

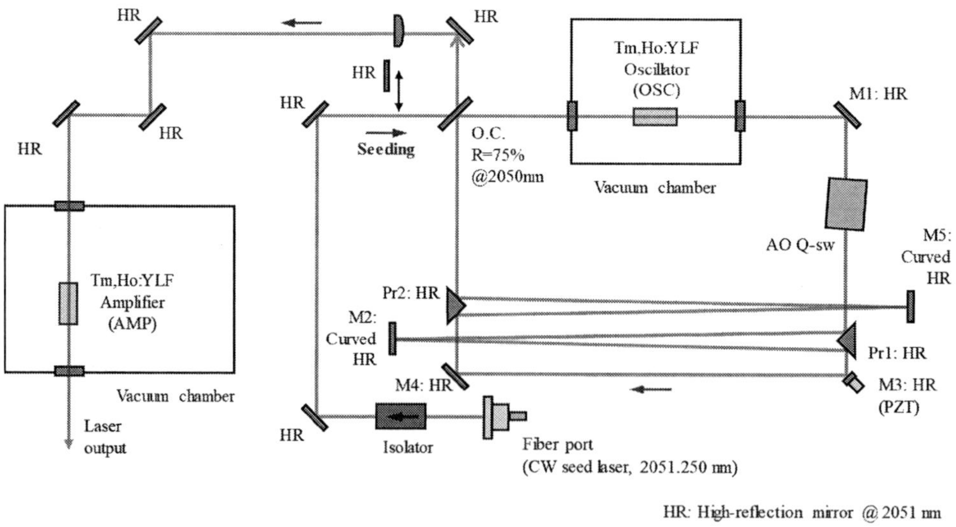

Figure 1. Configuration of the Q-switched Tm,Ho:YLF MOPA.

3. Results

Figure 2 shows the pump-energy-dependent output pulse energies of the 22-mm-long Tm,Ho:YLF laser rods at 30 Hz and -40°C in normal and Q-switched mode operations. The experimental results obtained using the previous pump module with the 33-mm-long rod are depicted in the same figure. The lasing threshold with the Ho doping concentrations of 0.4 and 0.7% decreases owing to the high pump density compared with that of the 33-mm-long rod. Despite the higher pump density, the threshold with the Ho doping concentration of 1.0% is larger than that of the 33-mm-long rod because the excess Ho doping leads to a large reabsorption loss by the thermal population at the lower laser level. While the output energies showed linear increases with pump energy in the normal operations, the roll-off of slope efficiency occurred at the high-pump-density region owing to GSD of the Ho ions in the Q-switched operations. As can be seen from Fig. 2(b), the roll-off was improved by the increase in the Ho doping concentration. The use of the 22-mm-long and 0.7%-Ho-doped laser rod leads to better results than the others at the Q-switched operation. An oscillator output energy of 74 mJ with a pulse width of 120 ns and a near-diffraction-limited beam quality was obtained. One of the prominent concerns using the shorter 22-mm-long rods was the decrease in heat removal efficiency from the lateral surface of the laser

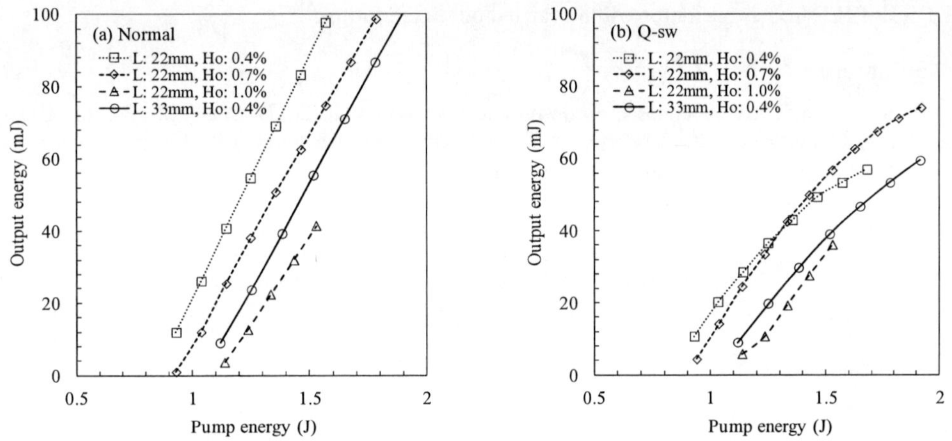

Figure 2. Output energies of the Tm,Ho:YLF laser oscillators at 30 Hz and -40 °C in (a) normal and (b) Q-switched mode operations as a function of pump energy.

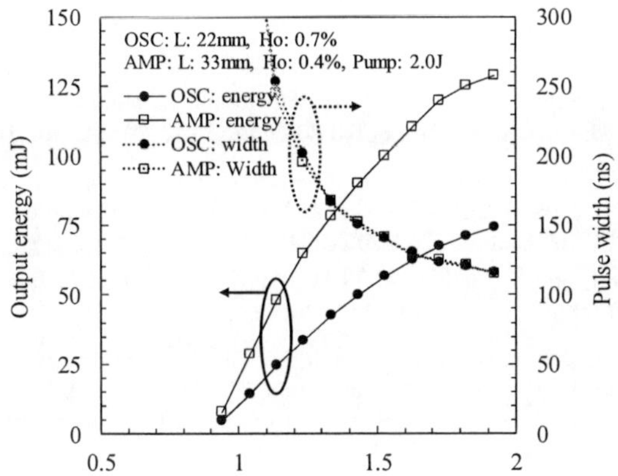

Figure 3. Output energies and pulse widths of the Q-switched Tm,Ho:YLF MOPA at 30 Hz and -40 °C as a function of pump energy.

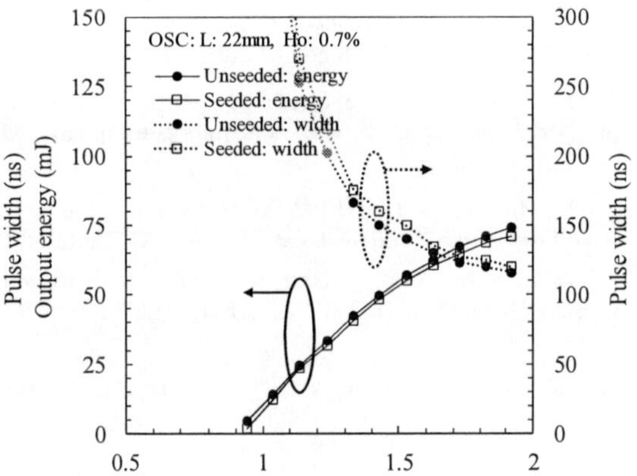

Figure 4. Output energies and pulse widths of the injection-seeded Tm,Ho:YLF oscillator at 30 Hz and -40 °C as a function of pump energy.

rod. The results revealed that the pump modules with the 22-mm-long rod possess sufficient heat removal efficiency under the experimental conditions.

Figure 3 shows the output energies and pulse widths of the Q-switched Tm,Ho:YLF MOPA at 30 Hz and -40 °C as a function of pump energy. The pump modules with the 22-mm-long and 0.7%-Ho-doped rod were used for the master oscillator. Although it is better to use a similar module for the power amplifier, we used the existing module with the 33-mm-long rod owing to the lack of an extra pump module for the 22-mm-long rod. A single-pass amplified pulse energy of 129 mJ with a pulse width of 115 ns was obtained at pump energies of 1.9 and 2.0 J for the oscillator and the amplifier. The parasitic oscillation did not occur under the maximum pumping condition. The single-pass gain and calculated energy extraction efficiency were 1.74 and 0.36, respectively.

Figure 4 shows the output energies and pulse widths of the injection-seeded Tm,Ho:YLF oscillator at 30 Hz and -40 °C as a function of pump energy. We used a solid-state laser with an emission wavelength of 2,051.250 nm for the seed laser. The injection-seeded single-frequency operation for the oscillator was achieved by a ramp-and-fire technique with a negligibly small loss of the pulse energy owing to the adoption of the ring resonator. A single-frequency output energy of 71 mJ with a pulse width of 125 ns and a near-diffraction-limited beam quality was obtained. The amplifier experiment of the injection-seeded Tm,Ho:YLF master oscillator will be carried out in the future.

4. Summary

We developed a conductively cooled, diode-pumped, Q-switched Tm,Ho:YLF MOPA for CDWL applications. A Q-switched output energy of 74 mJ with the pulse width of 120 ns for the master oscillator was obtained at 30 Hz and -40 °C by using the newly developed pump module. The single-pass gain and energy extraction efficiency for the power amplifier without injection seeding were 1.74 and 0.36, respectively. The laser pulse from the master oscillator was amplified to 129 mJ with the temporal properties and no parasitic oscillations or damage to the element of the amplifier. Further work in development of the Tm,Ho:YLF laser is on the way to satisfy the specification of the SLATS-based CDWL transmitter.

5. Acknowledgement

The authors would like to thank Mr. H. Fukuoka of Hamamatsu Photonics K.K. for his advice and fruitful discussions.

6. References

[1] S. Ishii, K. Mizutani, H. Fukuoka, T. Ishikawa, B. Philippe, H. Iwai, T. Aoki, T. Itabe, A. Sato, and K. Asai, "Coherent 2 μm differential absorption and wind lidar with conductively cooled laser and two axis scanning device," Appl. Opt. **49**, 1809–1817 (2010).

[2] K. Mizutani, T. Itabe, S. Ishii, M. Aoki, K. Asai, A. Sato, H. Fukuoka, T. Ishikawa, and K. Noda, "Diode-pumped 2-μm pulse laser with noncomposite Tm,Ho:YLF rod conduction-cooled down to -80°C," Appl. Opt. **54**, 7865-7869 (2015).

[3] K. Mizutani, S. Ishii, M. Aoki, H. Iwai, R. Otsuka, H. Hukuoka, T. Ishikawa, and A. Sato, "2 μm Doppler wind lidar with a Tm:fiber-laser-pumped Ho:YLF laser," Opt. Lett. **43**, 202-205 (2018).

[4] A. Sato, M. Aoki, S. Ishii, R. Otsuka, K. Mizutani, and S. Ochiai, "7.28-W, High-Energy, Conductively Cooled, Q-switched Tm,Ho:YLF Laser," IEEE Photo. Techno. Lett. **29**, 134-137 (2017).

[5] S. Ishii, P. Baron, M. Aoki, K. Mizutani, M. Yasui, S. Ochiai, A. Sato, Y. Satoh, T. Kubota, D. Sakaizawa, R. Oki, K. Okamoto, T. Ishibashi, T. Y. Tanaka, T. T. Sekiyama, T. Maki, K. Yamashita, T. Nishizawa, M. Satoh, and T. Iwasaki, " Feasibility Study for Future Space-Borne Coherent Doppler Wind Lidar, Part 1: Instrumental Overview for Global Wind Profile Observation," J. Meteorol. Soc. Jpn. **95**, 301-317 (2017).

[6] A. Noda, M. Homma, and M. Utashima, "The Study of a Super Low Altitude Satellite," Trans. Japan Soc. Aero. Space Sci. **7**, Pf17-Pf22 (2009).

[7] A. Sato, M. Aoki, S. Ishii, R. Otsuka, K. Mizutani, and S. Ochiai, "High-average-power operation of a 100-mJ-class, conductively cooled, Q-switched Tm,Ho:YLF laser," 12th Conference on Lasers and Electro-Optics Pacific Rim (CLEO-PR 2017), (Marina Bay Sands Expo and Convention Center, Singapore, 2017), 1-4H-4.

Measurements of Water Vapor Profiles with Compact DIAL for Local Heavy Rainfall Prediction in the Tokyo Metropolitan Area

Makoto Abo(a), Tetsu Sakai(b), Phong Pham Le Hoai(a), Yoshinori Shoji(b)
(a) Tokyo Metropolitan University
6-6 Asahigaoka, Hino, Tokyo 191-0065, Japan
(b) Meteorological Research Institute
1-1 Nagamine, Tsukuba, Ibaraki 305-0052, Japan
abo@tmu.ac.jp

Abstract: In recent years, the frequency of occurrence of locally heavy rainfall that can cause extensive damages, has been increasing in Japan. For early prediction of the heavy rainfall, numerical weather model is employed using the conventional meteorological station data. However, the lead time and accuracy of the prediction are limited because of the coarse spatial and temporal resolutions of the data. To improve them, it is useful to measure the water vapor distribution upwind cumulus convection beforehand and assimilate the data into the model. For that purpose, we have been developing compact NIR water vapor differential absorption lidar (DIAL). We show the results of the measurements with lidar in summer when the local heavy rainfall frequently occurs in Japan.

Keywords: differential absorption lidar, water vapor, tapered semiconductor optical amplifier

1. Introduction

Locally heavy rainfall is a severe weather hazards to affect urban areas in Japan. For early prediction of the heavy rainfall, numerical weather model is employed using the conventional meteorological station data. However, the lead time and accuracy of the prediction are limited because of the coarse spatial and temporal resolutions of the data. To improve them, it is useful to measure the water vapor distribution upwind of cumulus convection beforehand and assimilate the data into the model. Figure 1 shows schematic of this study.

For that purpose, we have developed compact and mobile diode-laser-based differential absorption lidar (DIAL) that can measure the vertical distribution of water vapor in the lower troposphere [1-3].

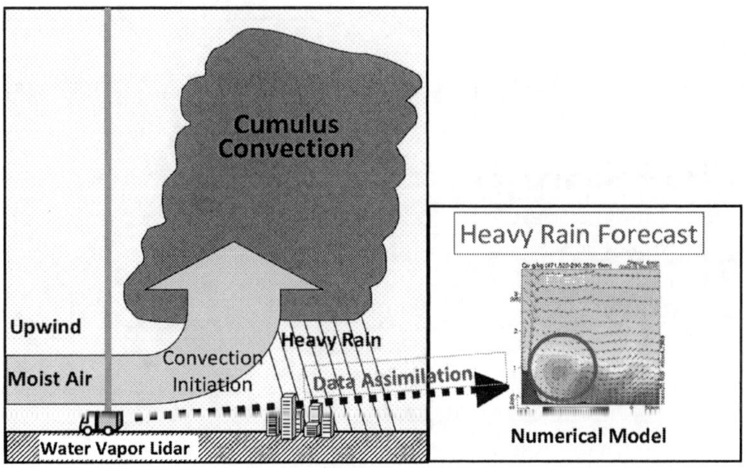

Figure 1. Schematic of this study.

2. DIAL System

The DIAL employs two distributed Bragg reflector (DBR) lasers operating at 829.054 nm for the online wavelength and 829.124 nm for the offline wavelength with tapered semiconductor optical amplifier (TSOA) in a master oscillator power amplifier (MOPA) configuration. The lidar measure the vertical distribution of the water vapor in the lower troposphere with a vertical resolution of a few tenths of a meter and temporal resolution of a few tenths of a minute.

Table 1. Specifications of DIAL System

Transmitter	
Laser	DBR diode laser x2 (seed laser) +TSOA
Wavelength	829.124 nm (Off-line), 829.054 nm (On-line)
Pulse Energy	4 uJ
Repetition	10 kHz
Receiver	
Telescope	Schmidt-Cassegrain
Diameter	35 cm
Field of View	0.3～1 mrad (adjustable)
Detector	PMT (Hamamatsu H9422P-50)
Operation	Photon Counting
Interference Filter	
Center Wavelength	829.11 nm
FWHM	0.3 nm
Transmission	65%

3. Measurement Results

Figure 2 shows the results of comparison with JMA mesoscale analysis (MSM) and radiosonde at Tsukuba, 80 km northeast from the lidar observation site Hino.

We show the results of the 24H continuous measurements with lidar in summer (26 August 2016) when the local heavy rainfall frequently occurs in Japan in Figure 3.

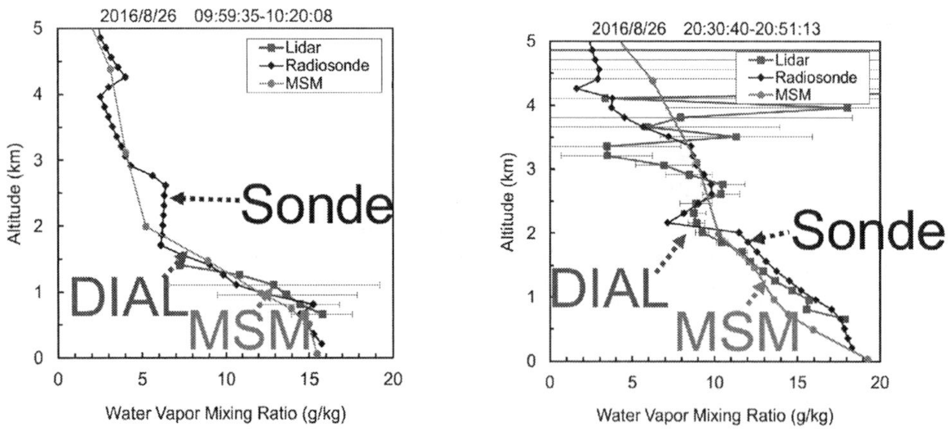

Figure 2. The results of comparison with JMA mesoscale analysis (MSM) and radiosonde at Tsukuba (80 km northeast).

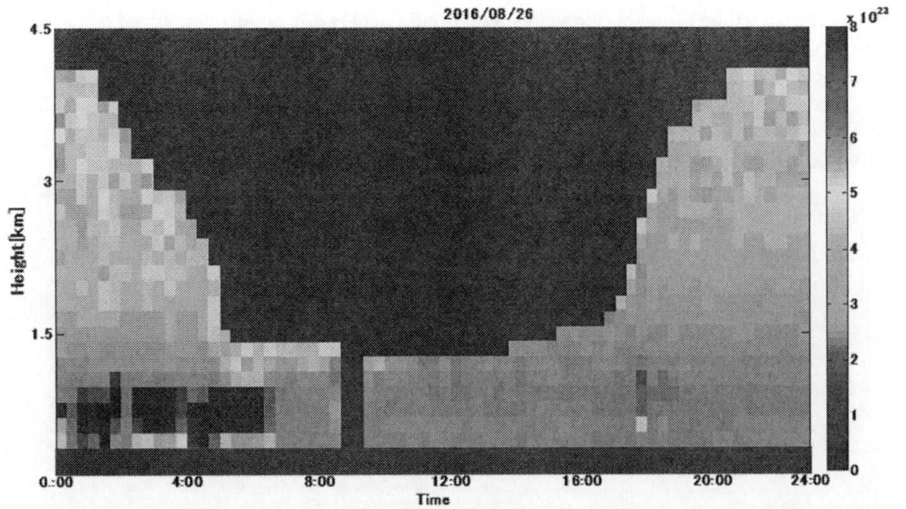

Figure 3. Temporal and height cross section of water vapor mixing ratio obtained with compact DIAL over Hino (35.7°N, 139.4°E) on 26 August 2016.

4. Conclusion

We have been developing water vapor lidar for local heavy rainfalls forecast. Now measurement range is 0.5–3 km in night time and 1 km in daytime with 150 m vertical and 20 min time resolution with uncertainty within 30%.

We are challenging to increase daytime measurement range to 2 km for DIAL using etalon. We are planning to conduct field experiment in Tokyo in the summer of 2018 and to assimilate the lidar data to mesoscale model and study the impact on it.

Figure 4. Strategy of early prediction of the heavy rainfall with numerical weather model and water vapor profiles observed by lidar.

Acknowledgements

This work was supported by JSPS KAKENHI Grant Number JP26282115.

References

[1] Le Hoai, P.P., M. Abo, T. Sakai, "Development of field-deployable Diode-laser-based water vapor DIAL", EPJ Web of Conferences 119 05011, doi: 10.1051/epjconf/20161190501 (2016).

[2] K. S. Repasky, D. Moen, S. Spuler, A. R. Nehrir, and J. L. Carlsten, "Progress towards an autonomous field deployable diode laser-based differential absorption lidar (DIAL) for profiling water vapor in the lower troposphere", Remote Sens., **5**, 6241–6259, doi:10.3390/rs5126241 (2013).

[3] S. M. Spuler, K. S. Repasky, B. Morley, D. Moen, M. Hayman, and A. R. Nehri, "Field-deployable diode-laser-based differential absorption lidar (DIAL) for profiling water vapor", Atmos. Meas. Tech., **8**, 1073-1087, doi:10.5194/amt-8-1073-2015 (2015).

HIGH PEAK POWER DOPPLER LIDAR BASED ON A 1.5μM COMPRESSIVE-STRAINED-SINGLEMODE FIBER AMPLIFIER

L. Lombard, B. Augère, A. Durécu, D. Goular, F. Gustave, M. Valla, A. Dolfi-Bouteyre

ONERA – The French Aerospace Lab, BP 80100, F-91123 Palaiseau, France

laurent.lombard@onera.fr

RÉSUMÉ

Wind Doppler coherent lidars based on pulsed single frequency fiber sources are usually limited by SBS (stimulated Brillouin Scattering) in the final fiber booster. This nonlinear effect limits the peak power achievable to around 50-100W in singlemode fibers, 200-300W in commercial LMA (large mode area) fibers and up to 2kW in specialty fibers. Onera has developed a patented SBS mitigation technique using compressive strain on fibers. This technique has been applied to a singlemode fiber amplifier and a peak power of up to 300W has been generated before the arising of SBS. This peak power is comparable to LMA fiber booster performances with additional advantages: the singlemode geometry allows using standard telecom fiber components and optimizes coherent mixing with the singlemode local oscillator. We present the details of this technique, the achieved laser performances and tests results of a lidar using this compressive-strained-source.

MOTS-CLEFS : *fiber amplifier, lidar, nonlinear optics*

1. INTRODUCTION

Coherent wind Lidars are increasingly used for wind measurement and turbulence assessment with applications in wind farm projects optimization or aircraft security during take-off and landing. Laser pulses are emitted through the atmosphere and wind speed is measured using Doppler-induced frequency shift on the backscattered laser light. [1] Laser sources with excellent spatial beam quality, narrow linewidth and typical pulse duration ranging from ~100 ns to 1 μs are required. Pulsed master oscillator power fiber amplifiers (MOPFA) at 1.5 μm are well adapted, versatile sources but with peak power limited in typical fibers to a few 100s W by nonlinear effects especially Stimulated Brillouin Scattering (SBS).
SBS limits the peak power achievable to around 50-100W in singlemode fibers, 200-300W in commercial LMA (large mode area) fibers and up to 2kW in specialty fibers. Onera has developed a several patented SBS mitigation techniques using compressive strain on fibers.

2. SBS MITIGATION IN SINGLEMODE FIBERS BY COMPRESSIVE STRAIN GRADIENT

The first technique consists of a mechanical structure applying tensile strain distribution along the fibre length [1]. The second patented technique we developed is based on a compressive strain distributed along the fibre [2]. This technique reduces considerably the fibre aging.
The principle of the compression technique is presented in Fig.1 left. We place the fibre in a plastic tube and elongate the tube of a certain percentage of its initial length. The next step is to glue the fibre in the tube and release the tube in order to recover its initial length. By doing so, we apply a longitudinal compressive constraint on the fibre.

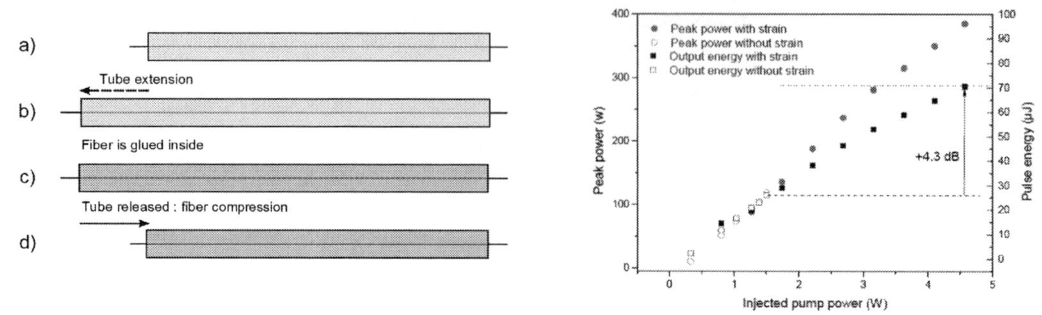

Figure 1 : (left) Principle scheme of the compression strain procedure patented by ONERA; (right) Output peak power and Pulse energy as a function of the pump power with and without applying the compressive strain gradient on the amplifier.

Fig. 1 right presents the peak power and pulse energy that we obtained with this amplifier. The compressive strain gradient allowed us to reach 385 W peak power instead of the 120 W limit corresponding to P_{sbs} without strain. We then achieved an increase of 5.1 dB on the SBS threshold resulting in a 385 W single mode fibre amplifier whereas commercial ones are limited to powers below 100 W [3[1][2]. The amplifier is spliced to a 15 cm long PM1550 single mode pigtail to allow transport of the optical power at the end output.
Thanks to this technique, we built a high peak power, 1.5µm compressive-strained-singlemode fiber amplifier with the performances of a LMA fibre amplifier and the advantages of a singlemode fiber amplifier: beam quality and direct splice to further standard singlemode components.
To perform Lidar tests with this amplifier, we replaced the output single mode pigtail by a fiber polarization beam splitter. In this configuration and for 200-ns pulses, the output peak power is 160W, currently limited by SBS in fiber cords of polarization beam splitter and the pulse energy is 35 µJ.

3. Doppler Lidar Tests

In order to compare the compressive-strained-singlemode fiber (CSSF) amplifier and the LMA fiber amplifier performances in a lidar configuration, we have replaced the LMA fiber amplifier (fig. 2) by the CSSF amplifier (fig. 3) in a standard coherent wind lidar configuration.
Aerosols present in the atmosphere will backscatter a very small part of the laser beam. Thanks to a quarter waveplate, the backscattered beam is separated from the output beam on a polarizing beam splitter and coupled into a singlemode fiber. The coupled light is mixed with the local oscillator and the resulting light is detected for further signal processing. In the case of CSSF amplifier, a fiber circulator can advantageously replace the polarizing beam splitter, associated collimators and quarter waveplate thanks to the fact that the EDFA output fiber is singlemode and compatible with standard components.

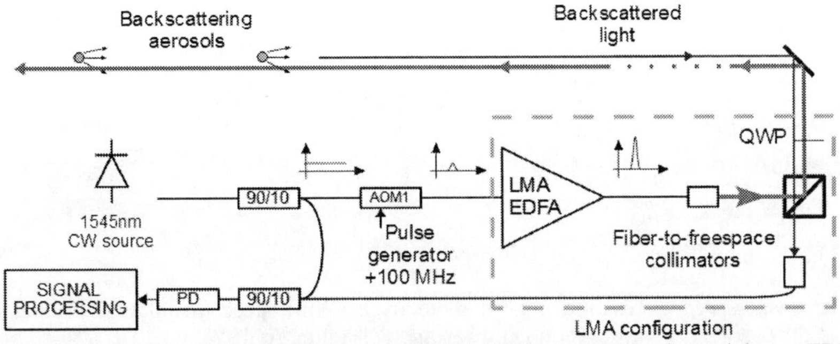

Fig.2 Principle of a wind lidar using a LMA fiber amplifier. AOM : acousto-optic modulator ; EDFA : erbium doped fiber amplifier ; QWP quarter waveplate ; PD photodiode. The quarter wave allows coupling of back-scattered light to collection fiber optics. The dashed rectangle delimits the amplifier and optics to be replaced by a CSSF amplifier and a fiber optical circulator. Fiber paths are represented in black, free-space emission in red and free-space backscaterred light in blue.

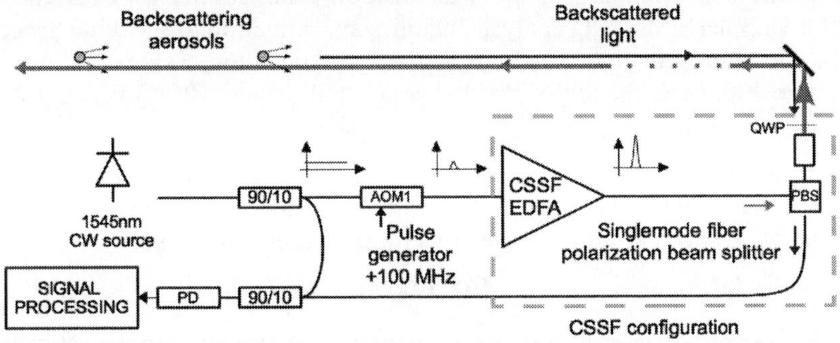

Fig. 3 LIDAR setup with CSSF laser source and a fiber polarizing beam splitter. Fiber paths are represented in black, free-space emission in red and free-space backscaterred light in blue.

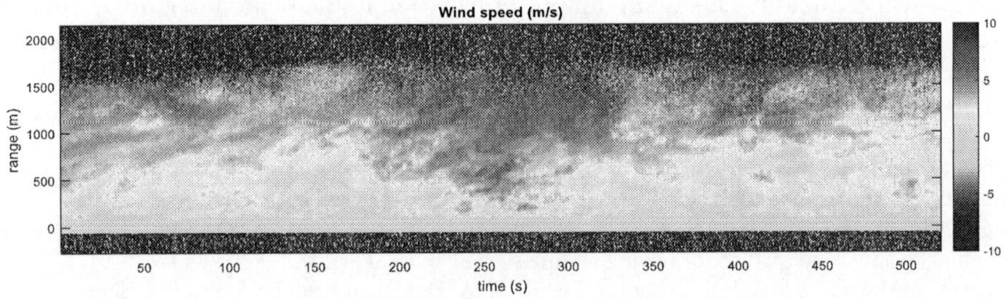

Fig. 4: wind speed obtained with polyfit estimator (ref CLRC2016) versus time and range.

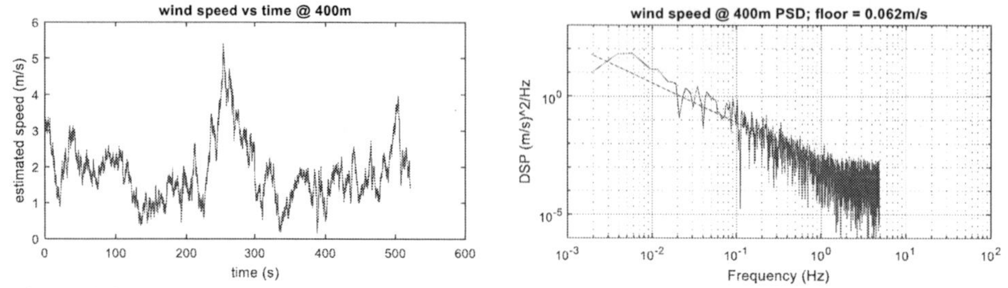

Fig. 5: (left) wind speed versus time at range 400m. (right) Wind speed @ 400m PSD (blue), -5/3 slope Kolmogorov fit and noise floor (red).

The backscattered signal is acquired for each pulse, and its frequency shift is analyzed for various distances, ranging from 100 to 2000 m, using the following procedure. A power spectral density (PSD) is computed for each distance and laser shot. Then PSDs are averaged for 1,000 shots, corresponding to 0.1 s acquisition time, resulting in PSDs, ready to be analyzed for Doppler shift. PSD analysis includes an estimation of the wind speed through a polyfit [4]. The estimated wind speeds are reported on fig. 4. Fig. 5 shows the estimated wind speed at range 400m and the log of the power spectral density (PSD) versus frequency. On the PSD log–log graph we can visualize the $-\frac{5}{3}$ slope of the wind turbulence, and the flat level of the estimator noise [5]. The vertical axis value is expressed in terms of equivalent wind speed variance (the PSD value has been integrated over the band). For this distance, the experimental wind speed accuracy is 0.06m/s for 0.1s acquisition time.

CONCLUSION

We have developed a CSS amplifier with a peak power comparable to LMA fiber booster performances with additional advantages: the singlemode geometry allows using standard telecom fiber components such as circulators and optimizes coherent mixing with the singlemode local oscillator. This CSSF amplifier has been implemented in a lidar. First tests show that compressive-strained-singlemode fiber based amplifiers are compatible with coherent lidars.

RÉFÉFENCES

[1] Canat, G. and Lombard, L. and Gouët, J.L. and Fleury, D. (2016) Optic fiber amplifier having a high Brillouin threshold and method for making such an amplifier. Patent US9236704 B2

[2] Canat, G. and Renard, W. and Lombard, L. and Fleury, D. Device comprising an optical fiber having a high Brillouin threshold and method for manufacturing such a device. Patent US9653875 B2

[3] G. Canat, S. Makon-Makon, L. Lombard, J. Le Gouët, A. Durécu, « High peak power Erbium-Ytterbium single frequency pulsed fiber amplifier using a compressive strain gradient", SM4Q.6, CLEO 2016

[4] L. Lombard, A. Dolfi-Bouteyre, M. Valla, and G. Canat, "Performance of Frequency Estimators for real time display of long range Lidar wind map," in *Coherent Laser Radar Conference Proceedings* (2016).

[5] R. Frehlich, S. M. Hannon, and S. W. Henderson, Bound.-Lay. Meteorol. 86, 233 (1998).

Improvement of SNR at near range using pulsed serrodyne technique on coherent laser transmitter for wind sensing lidar

Eisuke Haraguchi(a), Hitomi Ono(a), Akihiro Fujie(a), Toshiyuki Ando(a),
(a) Mitsubishi Electric Corporation, Information Technology R&D Center,
5-1-1 Ofuna, Kamakura, Kanagawa, Japan, 247-8501
Haraguchi.Eisuke@cw.MitsubishiElectric.co.jp

Abstract: In the Coherent Doppler LIDAR (CDL), the minimum measurable range is restricted by the internal scattered light. We have developed Coherent Doppler LIDAR (CDL) system for wind sensing using a semiconductor optical amplifier (SOA) and an optical phase modulator operated at pulsed serrodyne modulation. In the pulsed serrodyne modulation, the amount of frequency shift within a single pulse can be arbitrarily changed. Using this technique, heterodyne beat signals from internal scattered light at near range are efficiently rejected by electrical bandpass filter. In this paper, we demonstrated the measurement of the scattered light from hard target at 40m distance within the internal scattered light by controlling the amount of frequency shift within a single pulse duration which pulse width is 500 ns.

Keywords: Coherent Laser Radar, Internal scattered light, Pulsed serrodyne modulation.

1. Introduction

All-fiber coherent Doppler LIDAR (CDL) system has many advantages such as its compactness, eye-safety and reliability thanks to using commercial off the shelf components from telecom products[1]. The CDL can be used for aviation safety, leisure, global environment, city & traffic safety disaster prevention, and wind resource assessment[2][3]. In the CDL system for wind resource assessment, the CDL is required compactness, portable, high resolution and long measurable distance. In the pulsed CDL system capable of long distance measurement, the measurement of near range is limited by the internal scattered light. We have developed the ultra compact CDL system for wind sensing using a semiconductor optical amplifier and an optical phase modulator operated at pulsed serrodyne modulation[4][5]. In the pulsed serrodyne modulation, the amount of frequency shift within a single pulse can be arbitrarily changed.

In this paper, we demonstrated the measurement of the scattered light from hard target within the internal scattered light by controlling the amount of frequency shift within a single pulse duration.

2. The principle of improving SNR within the internal scattered light

In the pulsed serrodyne technique, we use a pulsed serrodyne modulator that consists of a semiconductor optical amplifier (SOA) and a phase modulator (PM). The SOA is operated by direct modulation with a pulse width of T1, and the PM is operated using pulsed saw-tooth phase modulation with a linear slope of the phase change and a peak amplitude that equals 2π radians. The pulse width of the saw-tooth phase modulation is T2. The optical signal of pulsed serrodyne phase modulation is shown in Eq. (1), where E_S, f_0 and T_m represent the field amplitude, the optical carrier frequency and the cycle of sawtooth waveform, respectively.

$$E_S(t) = E_S \exp[i2\pi(f_0 + 1/T_m)t] \quad (1)$$

In the pulsed serrodyne modulation, the SOA and the optical phase modulator are modulated independently. Using this technique, in the frequency domain the internal scattered light is separated light

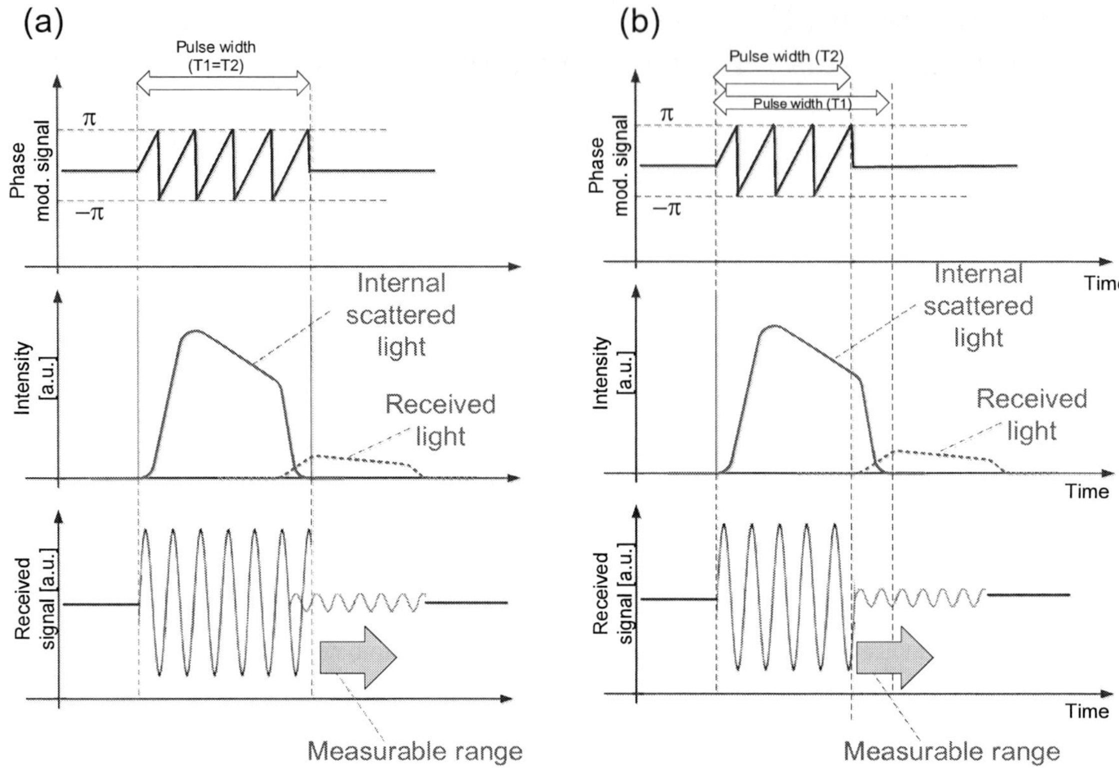

Figure 1.Schematic diagram of loopback experimental setup.

from the scattered light from the short range. Figure 1 shows the image of the phase modulation signal, received optical pulse shape, and the received beat signal, respectively. Figure 1 (a) is shown the case of the pulse width of the SOA (T1) is the same as the pulse width of the serrodyne modulation. In this case, the measurable range is limited by the internal scattered light. Figure 1 (b) shows the case of the pulse width of the PM (T2) is shorter than the pulsed width of the SOA (T1). In this case, since the beat signal frequency between the local light and the internal scattered light in the pulse falling region (T1-T2) is a DC component, we can measure within the internal scattered light.

3. The experimental setup and experimental results

Figure 2 shows the schematic diagram of loopback experimental setup. Distributed Feed Back Laser Diode (DFB-LD) was used as a master laser with line width of 200 kHz and wavelength of 1550 nm. The output from DFB-LD was split into a local and a signal light. The signal light was coupled to the PM and the SOA. The SOA was directly modulated to obtain the pulsed signal which pulses width was 500 ns and pulse repetition frequency was 4 kHz. The PM was operated at serrodyne modulation, which pulse width is 500 ns or 380ns. This signal pulses was amplified by an Erbium Doped Fiber Amplifier (EDFA) which output power was 20 W. The signal laser from EDFA was attenuated by the optical attenuator about 40 dB. Attenuated signal laser was propagated through the delay fiber which length is 40 m. This fiber length corresponds to a delay time of 400ns. The reflected light from fiber end were separated by the circulator and combined with a local laser. Then, the signal light and the local light are received by the balanced receiver. The received signal is subjected to band limitation by a band pass filter, and then the received signal is evaluated by using Digital Sampling Oscilloscope (DSO) with sampling rate of 2G samples/s.

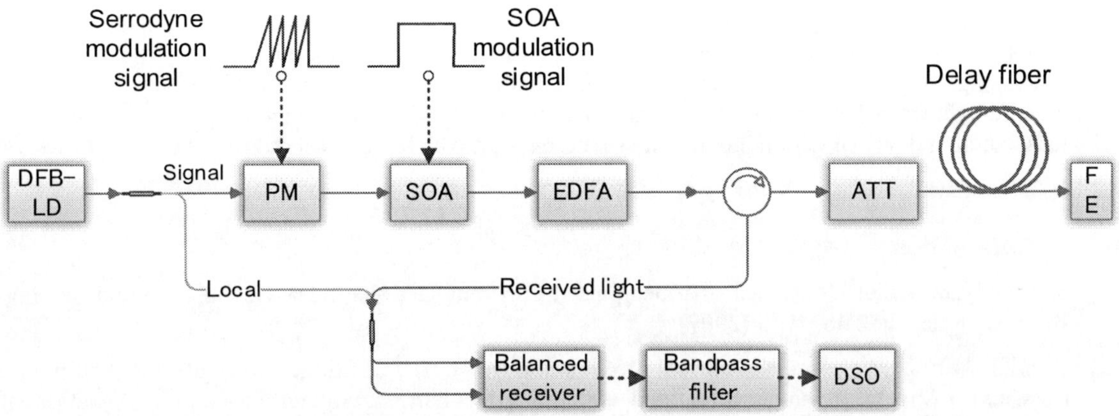

Figure 2.Schematic diagram of loopback experimental setup.

Figure 3 shows the experimental results in time domain of the pulse shape, the serrodyne modulation signal and the received signals. Figure 3 (a) is in the case of the pulse width of the serrodyne modulation signal is 500 ns, Figure 3 (b) is in the case of the pulse width is 380 ns. In Figure 3 (a), it can be seen that the beat signals of internal scattered light and reflected light are overlapped and can't be separated. On the other hand, in the case of the pulse width of serrodyne modulation signal is shorter than the pulse width of the SOA, it is clearly demonstrated the internal scattered light and the reflected light can be separated.

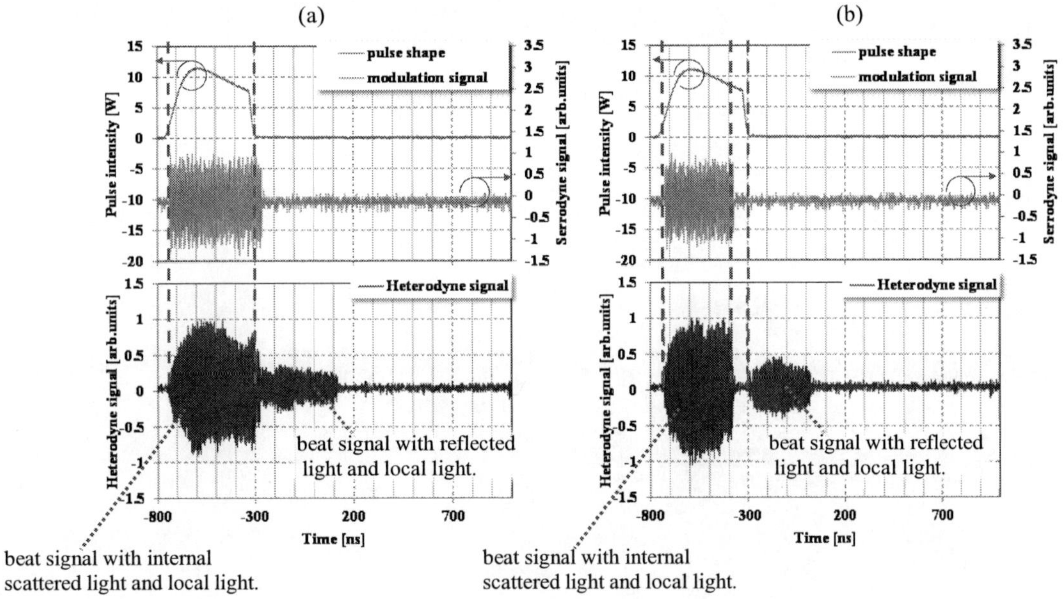

Figure 3. The heterodyne signals between local and back – reflected light. (a) Same as the intensity pulse width and the serrodyne modulation width, (b) The serrodyne modulation width is shorter than the intensity pulse width.

4. Summary

In this paper, we proposed the technique of measuring the nearest range using pulsed serrodyne modulation. In this system, the SOA and the optical phase modulator are modulated independently in time domain. Using this technique, in the frequency domain the internal scattered light is separated light from the scattered light from the short range. We have demonstrated the measurement of the scattered light from hard target at 40m distance within the internal scattered light which pulse width is 500 ns.

5. References

[1] S. Kameyama et al, "Compact all-fiber pulsed coherent Doppler lidar system for wind sensing", Applied Optics, vol. 46, 1953 (2007).

[2] T. Mikkelsen, N. Angelou, K. Hansen, M. Sjöholm, M. Harris, C. Slinger, P. Hadley, R. Scullion, G. Ellis, and G. Vives, "A spinner-integrated wind lidar for enhanced wind turbine control", Wind Energy, 16, 625-643 (2012).

[3] S. Wu, B. Liu, J. Liu, X. Zhai, C. Feng, G. Wang, H. Zhang, J. Yin, X. Wang, R. Li, and D. Gallacher, "Wind turbine wake visualization and characteristics analysis by Doppler lidar", Opt. Exp. 24, A762-A780 (2017).

[4] T. Ando, et al, "New coherent Doppler lidar engine integrating optical transceiver with FPGA signal processor", Proc of. 18th CLRC, 145 (2016).

[5] E. Haraguchi, et al, "Pulsed serrodyning technique as applied the wind sensing coherent doppler lidar", Proc of. 18th CLRC, 141 (2016).

Development of the low-cost and high-performance coherent Doppler lidar system

Kuniaki Higashi (a, b), Jun-ichi Furumoto (a)

(a) Research Institute for Sustainable Humanosphere, Kyoto University
Gokasho, Uji, Kyoto, Japan
(b) Metro Weather Co. Ltd., Japan
20-21, Nishiura, Gokasho, Uji, Kyoto, Japan
higashi@metroweather.jp

Abstract: Metroweather Co., Ltd. and Kyoto University collaborates to develop the high-performance and low-cost coherent Doppler Lidar (CDL). This new lidar system enable us to detect the detailed structure of wake turbulence with the temporal resolution of < 10 seconds. The new CDL can also be applied to the dual observation to determine the horizontal wind components from several co-operated CDL data. The result of performance test and the potential of the new CDL will be presented in the session.

Keywords: High-performance and low-cost, Coherent Doppler Lidar

1. Introduction

Our project aims to make an innovation by developing the high performance CDL detectable within the range of around 20 km. The cost of our lidar system is estimated to 1/10 of the conventional instruments. This research focuses on the development of the high-performance and low-cost CDL , including the design and assembly of a prototype CDL. This system will enable each to monitor and assess the wake turbulence risk autonomously. To detect wake turbulence generated below an airplane, the spatial and temporal resolution of CDL should be improved to ten meters and ten seconds, respectively. The development activity last year was devoted to establish the method to extract detailed atmospheric parameters. The feasibility study of a new unique beam steering system for air-borne CDL was also researched.

2. Approach

To downsize CDL, small coherent telescopes are adopted, and the coherency with them, weatherproof, and waterproof properties were kept at the proper level for the CDL cases which should have enough durability as meteorological instruments (Fig.1). Fiber laser, significantly developed in recent studies, is adopted in this approach, and combined with precision optical instrument technology and Digital Signal Processing developed in Kyoto university over many years.

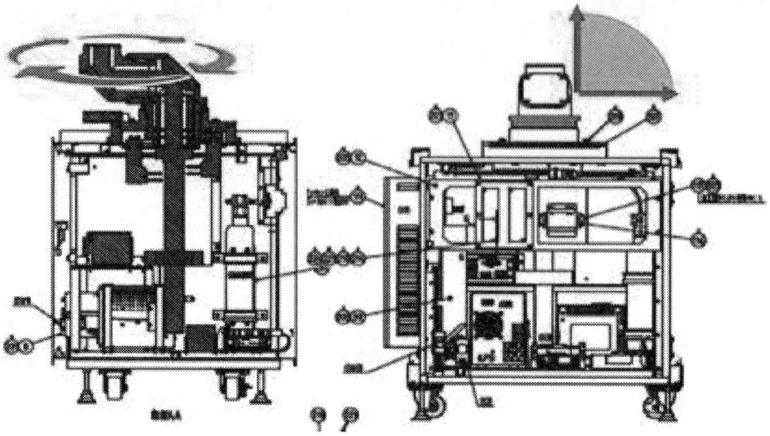

Figure 1. Design of prototype CDL.

3. Work completed

The development activity last year was devoted to establish the method to extract detailed atmospheric parameters. The feasibility study of a new unique beam steering system for air-borne CDL was also researched. It was possible to downsize to one fourth of CDL prototype 01 (height 2m, length 1m, width 1m) in the study for a year. The diameter is 20cm and the length is 2m for the prototype 01, and the diameter is downsized to 10cm, and 0.5m for the length with keeping the coherency on downsized CDL. The power supply system and optical components were also optimized thorough past experiments for prototype 01, and components were reviewed to downsize without affecting the performance (Fig. 2 and Fig.3).

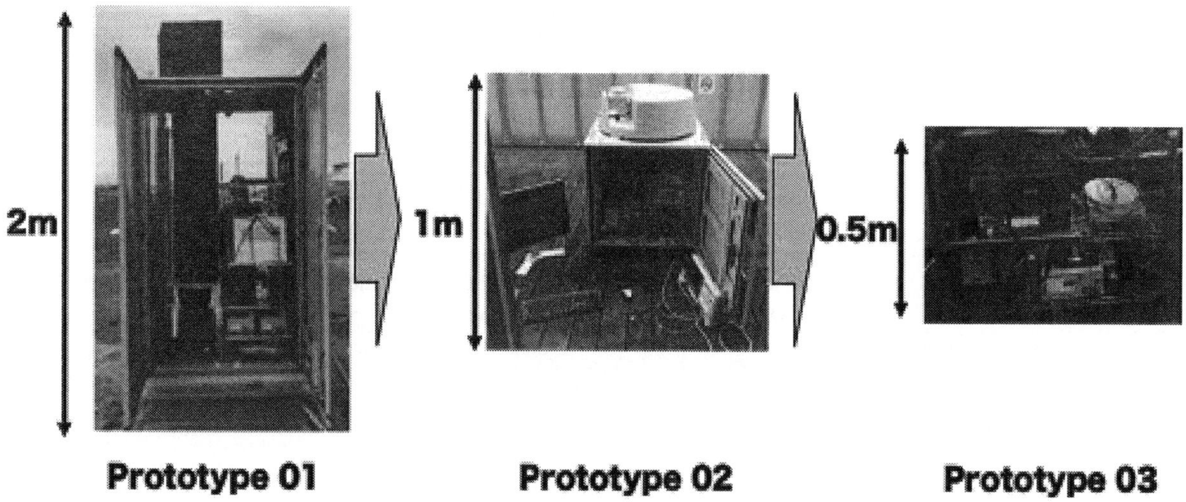

Figure 2. Miniaturization of CDL.

Figure 3. Recently manufactured CDLs on March 2018.

4. Results

The performance of the new CDL was investigated with comparing the data by anemometers established in the past and CDL developed around airport so far. Figure 4 shows the data by anemometers and CDL. The black line indicates wind speed by CDL, and × represents it by anemometers. Comparing the data by anemometers to the data by new CDL, they are obviously enough close for the evidence to insist that the new CDL has the same performance as existing anemometers. The CDL has been downsized with keeping the performance in this study so far, and we will further pursue technology to downsize optical components and keep progressing to implement the airborne CDL on airplanes.

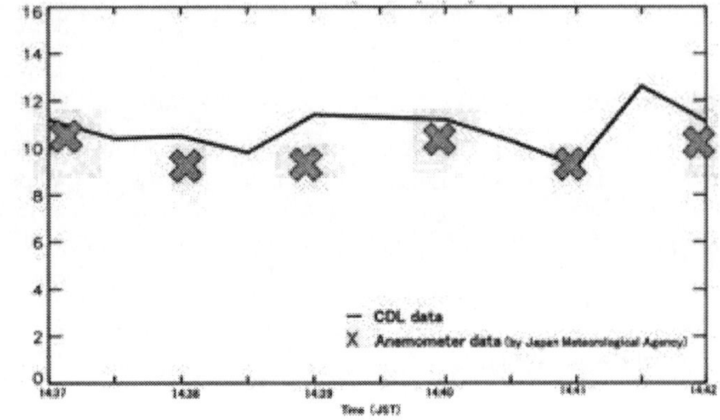

Figure 4. The trends in the wind speed observed by anemometers and new CDL at Sendai air port.

First measurement results of water vapor profile using 1.53 micron coherent differential absorption LIDAR

Masaharu IMAKI, Kenichi HIROSAWA, Takayuki YANAGISAWA, and Shumpei KAMEYAMA
Mitsubishi Electric Corporation, Information Technology R&D Center,
5-1-1 Ofuna, Kamakura, Kanagawa, Japan, 247-8501
Lead Author e-mail address: Imaki.Masaharu@dc.MitsubishiElectric.co.jp

Abstract: We have been developing a ground based coherent differential absorption LIDAR for water vapor profile using 1.53 micron laser wavelength. A coherent lidar has an advantage in daytime measurement compared with incoherent lidar because the influence of background light is greatly low. In addition, the lidar can simultaneously measure wind speed and water vapor density. We demonstrated the water vapor and wind velocity measurements in vertical range, and the height profile of water vapor density and wind velocity were observed up to 2500 m.

Keywords: water vapor, differential absorption, coherent

1. Introduction

Water vapor in the atmosphere affects growing cumulonimbus which causes localized heavy rainfall, therefore the measurement of water vapor density profile is required for the early prediction of this rainfall.

Height profile of water vapor measurement using incoherent differential absorption lidar (DIAL) technique was reported in Ref. 1 and 2. Using this technique, the continuous height profile measurement was demonstrated, but the measurement accuracy was influenced by solar background light.

On the other hand, coherent DIAL technique can overcome this issue and was reported in Ref. 3 and 4. The coherent lidar is based on heterodyne detection by using local laser light, the affect of measurement accuracy by the solar background light is relatively low. The good agreement result on water vapor density in comparison with that of radiosonde soundings was shown.

Since late 1990s, we have been developing the wind sensing coherent Doppler LIDARs with 1.5 μm wavelength. The LIDAR with 1.5 μm wavelength has advantages such as its compactness and reliability by using telecom products. Here, we consider to apply this technology to DIAL, and perform preliminary study on a ground based coherent DIAL for vertical profiling of water vapor density using 1.53 μm laser wavelength [5,6]. In the followings, we discuss the concept, configuration, and first measurement results of the lidar.

2. System configuration

The system configuration is shown in Fig. 1. A CW laser lights of 1.5 μm wavelength from two laser diodes are locked to the ON and OFF wavelength for water vapor. Either ON or OFF wavelength lights are selected by an optical switch. Selected laser light is modulated with a pulse waveform, is amplified, and is transmitted to the atmosphere. A part of the CW light is tapped and used for a local light for heterodyne detection. The received lights are converted into electrical signals by a photodetector. The received signal is analog-to-digital converted and spectral analysis and accumulation is performed in the signal processor. A coherent lidar has an advantage in daytime measurement compared with incoherent lidar because the influence of background light is greatly low.

In addition, the lidar can simultaneously measure wind speed and water vapor density. The wind velocity is measured from Doppler shift frequency, and the water vapor density is measured from amplitude of received spectrum. Therefore, this lidar system has the potential to observe the changes in water vapor density due to wind velocity.

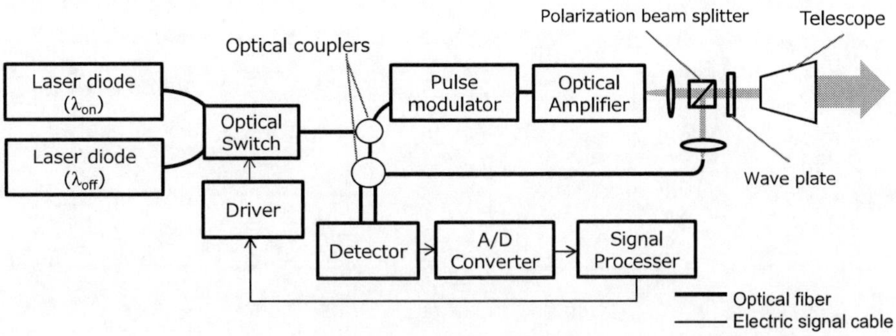

Figure 1. Schematic diagram of coherent differential absorption lidar.

A differential absorption optical depth DAOD and water vapor density n are derived from a ratio of the normalized measured intensities of the two wavelengths, and are written by

$$DAOD(z_i) = \ln\left(\frac{P_{off}(z_{i+1})}{P_{on}(z_{i+1})} \cdot \frac{P_{on}(z_i)}{P_{off}(z_i)}\right), \qquad \cdots(1)$$

$$n(z_i) = \frac{DAPD(z_i)}{2 \cdot \Delta z \cdot (k_{on} - k_{off})}, \qquad \cdots(2)$$

where $P_{off}(z)$ is received power of OFF wavelength in electric domain and $P_{on}(z)$ is received power of ON wavelength in electric domain at range z, Δz is range resolution of the lidar, k_{on} and k_{off} are absorption coefficient of ON wavelength and OFF wavelength, respectively.

3. Measurement results of water vapor density

The high profile of water vapor and wind velocity were measured at March 30, 2018. The lidar system parameters are described as follows. The output laser energy is 0.3 mJ, pulse width is 1.4 μs, repetition frequency is 16 kHz, and the receiver diameter is 110 mm. The number of laser shot is 30,000 for each wavelength, which corresponds to the 10 minutes observation time.

The measurement results are shown in Fig. 2. The laser beam is transmitted to the atmosphere at zenith angle of laser beam is 35.4 deg. The height profile of relative signal value which corresponds to aerosol density and clouds is shown Fig. 2(a), water vapor is shown Fig. 2 (b), and wind velocity of line of sight is shown Fig. 2 (c). The water vapor density was measured up to 2500 m height, and the wind velocity was measured, simultaneously. The high aerosol density layer was detected at 2700 m height, and the high water vapor density was observed under the aerosol layer compared with height profile up to 2000 m.

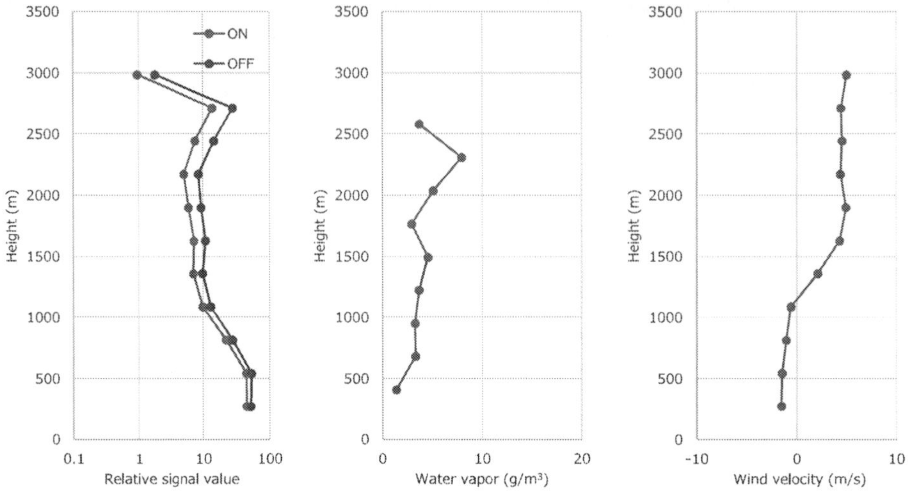

Figure 2. Height profile of relative signal value which corresponds to aerosol density and clouds (a), water vapor (b), and wind velocity of line of sight (c).

4. Summary

We have been developing a ground based coherent differential absorption LIDAR for water vapor profile using 1.53 micron laser wavelength, and we demonstrated the water vapor measurements in vertical and horizontal range.

We demonstrated the water vapor and wind velocity measurements in vertical range, and the height profile of water vapor density and wind velocity were observed up to 2500 m.

5. Rererences

[1] A. R. Nehrir, *et al.*," Eye-Safe Diode-Laser-Based Micropulse Differential Absorption Lidar (DIAL) for Water Vapor Profiling in the Lower Troposphere," Journal of Atmospheric and Oceanic Technology **28** (2011) 131-147.

[2] S. M. Spuler, *et al.*, "Field-deployable diode-laser-based differential absorption lidar (DIAL) for profiling water vapor," Atmospheric Measurement Techniques **8** (2015) 1073-1087.

[3] M. J. Kavaya, *et al.*, "Monte Carlo computer simulations of ground-based and space-based coherent DIAL water vapor profiling," Appl. Opt. **28** (1989) 840-851.

[4] R. M. Hardesty, "Coherent DIAL measurement of range-resolved water vapor concentration," Appl. Opt. **23** (1984) 2545-2553.

[5] M. Imaki, *et al.*, "Preliminary study on ground based coherent differential absorption LIDAR for vertical profiling of water vapor density using 1.53 μm wavelength," *Proceedings of CLRC 18* (2016).

[6] M. Imaki, *et al.*, "Development of wavelength locking circuit for 1.53 micron water vapor monitoring coherent differential absorption lidar," *Proceedings of ILRC 28* (2017).

Coherent Doppler LIDAR system switching the LOS direction by wavelength of laser

Nobuki Kotake(a), Yojiro Watanabe(a), Eisuke Haraguchi(a), Shumpei Kameyama(a), Toshiyuki Ando(a) and Takayuki Yanagisawa(a)
(a) Mitsubishi Electric Corporation, Information Technology R&D Center
5-1-1 Ofuna, Kamakura, Kanagawa, Japan, 247-8501

Abstract: The expectation for coherent Doppler LIDAR (CDL) systems applied for wind energy and airport safety is rapidly increased and high reliability and the mass productivity are strongly required. To satisfy the requirements, we have developed a new coherent Doppler LIDAR (CDL) system that switch line-of-sight directions by switching the wavelength without a mechanical scanner. The CDL consists of μITLA, WDM and multi-beam scanner. We demonstrated wind velocity and wind direction measurement at more than 250-m distance using this system.

Keywords: Coherent Laser Radar, Non-mechanical system, multi-beam and scanner

1. Introduction

The role of coherent Doppler LIDAR (CDL) systems for wind energy and airport safety is rapidly increased. High reliability and the mass productivity are also demanded because (i) the setting environment is severe (heavy rain, salt damage, etc.) and (ii) the difficulty of replacement, in case used for turbine control. The conventional CDL uses the mechanical switch (wedge prism and motor, etc.) for switching the laser direction. We want to apply the passive switch to control the direction without mechanical devices for improving the reliability. In this paper, we introduce the new CDL system concept that switch line-of-sight directions by switching the wavelength without a mechanical scanner and the demonstration result.

2. System configuration

Figure 1 shows a non-mechanical CDL system using μITLA for the source light, wavelength-division-multiplexing (WDM) for switching the optical path which are used for the optical communication, and a multi-beam direction telescope with several fiber adaptor. This telescope can switch the LOS direction by position of that adaptor. By decreasing the number of the telescope, this concept contributes to the mass productivity. Moreover, we use LN modulator and the Semiconductor Optical Amplifier (SOA) which are also distributed for optical communication.

The new components (gray squares) are changed for realizing our concept to the general fiber-based CDL system[1]. The operation flow is as follows. (1) μITLA omit the seed laser with wavelength λ_1, (2)that laser is modulated by LN modulator, (3) modulated laser is pulsed by SOA, (4) the pulsed laser amplified by Fiber-Amplifier (FA), (5) WDM switch the optical path depending on the wavelength (6) the laser is omitted to the air through telescope which can change the laser direction depending on the position of connecting adaptor (described later). (7) The received signal is detected with heterodyne detection through circulator, 4port copular, and Balanced Receiver (B.R.). (8) line-of-sight (LOS) wind velocity is calculated with FFT in the signal processing as like conventional CDL. (9)Return to (1) with changing the wavelength λ_2. The omitting laser direction can be changed by changing the wavelength.

We developed the CDL system as shown in fig.2. The new device “multi-beam scanner” as shown in fig.3 is installed, moreover, the integrated TRX board which includes μITLA, LNM, SOA, B.R. and signal processor (see fig.1) is also installed.

The telescope has 5 fiber adaptor whose positon related to the laser direction is different. The laser direction is vertical, 0-deg(north) , 90-deg(east), 180-deg(south) and 270-deg(west) in azimuth angle, respectively. The cone angle is 11deg except for vertical beam. The wave front aberration is set less than 1/14 λ (λ=1550nm) switch is well known as martial criteria. The photo of developed multi-beam scanner is shown in fig. 3. In case of using optical switch for changing the laser direction, same number of telescopes as number of direction is needed. To decrease the system size, and increase the mass productivity, we realize the function with only one telescope.

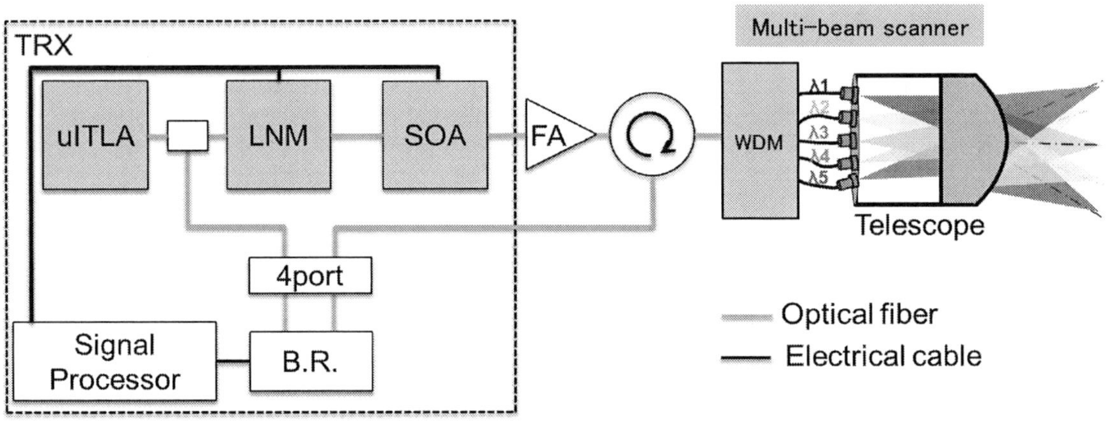

Figure 1. System configuration.

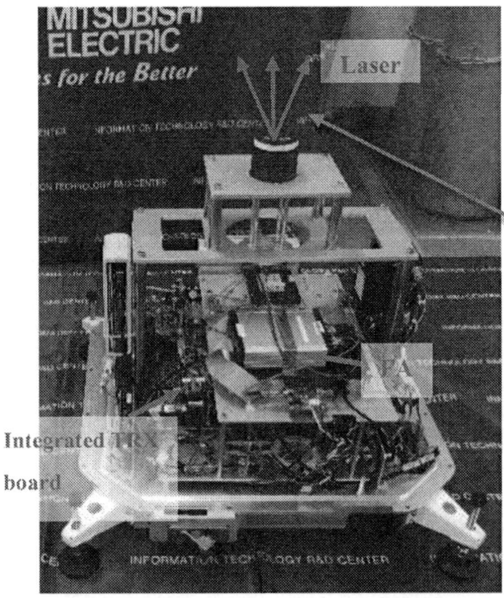

Figure 2. Photo of developed system

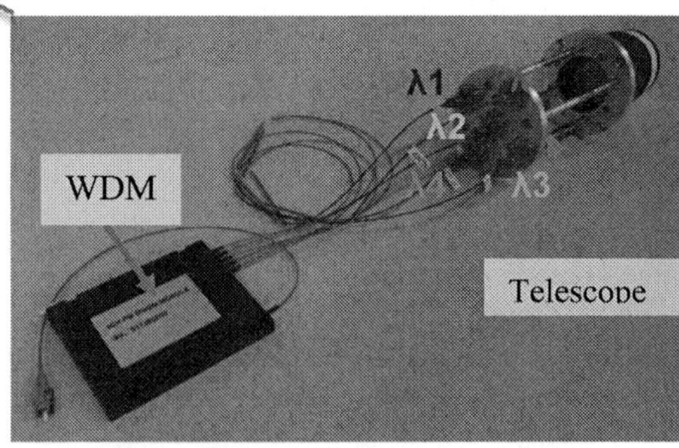

Figure 3. Photo of multi-beam scanner

3. 1st Trial and measurement results

Using developed system, we test the measurement of wind velocity as shown in fig.4. In this time, we test using 3 kinds of wavelength (3 directions) for the demonstration. As the result, we can measure the wind velocity at more than 250-m distance.

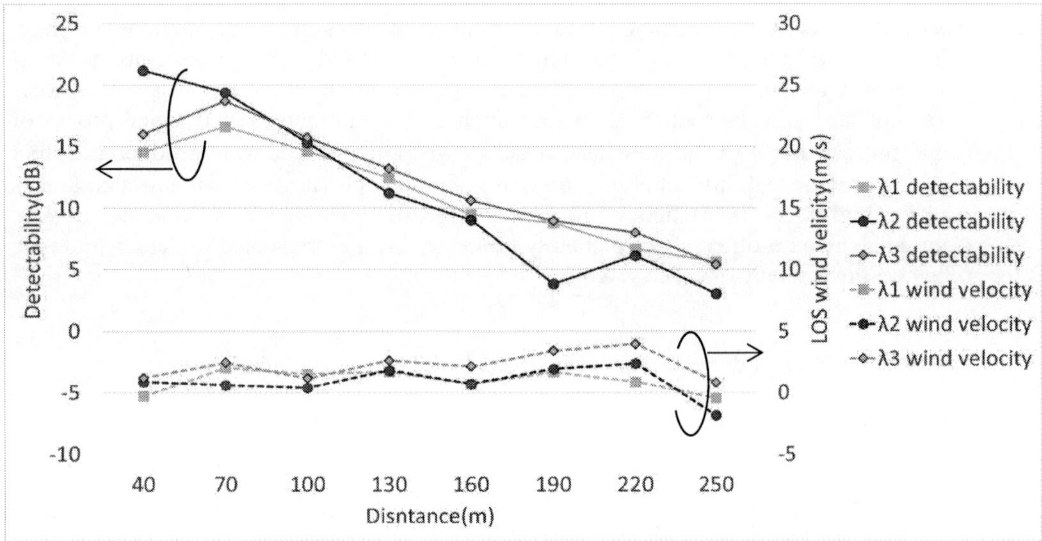

Figure 4. Wind velocity measurement demonstration result

3. Summary

We introduced a new concept of wind lidar system for realizing high reliability and high mass productivity using μ ITLA, WDM and multi-beam scanner. Using this system, we demonstrate wind velocity measurement at more than 250-m distance. This system is easy to extend the number of LOS directions and to realize the simultaneous measurement.

4. References

[1] S. Kameyama T. Ando, K. Asaka, Y. Hirano, and S. Wadaka,"Compact all-fiber pulsed coherent Doppler lidar system for wind sensing," Applied optics, **46**, No.11, pp.1953-1962, 2007.

High efficiency all fiber coherent lidar for wind measurement

Anran Zhou, Yuli Han, Dongsong Sun, Fei Han, Jun Zheng, Nannan Zhang
CAS Key Laboratory of Geospace Environment, School of Earth and Space Sciences, University of Science and Technology of China, 230026
No.96, JinZhai Road Baohe District,Hefei,Anhui
anranzh@mail.ustc.edu.cn

Abstract: Recently a high efficiency all-fiber coherent lidar system at 1.55 μm wavelength is developed to achieve real-time measurements of wind fields. The system employs 80 μJ pulse energy at 10 kHz pulse repetition rate. In addition, the system consists of a fiber-based optical transceiver unit, a 2-axis scanner, and a multicore digital signal processor (DSP) for real-time signal processing. The lidar with compact and waterproof architecture is able to provide mobility, reliability, and ease of field deployment for wind measurements. And the high receiver efficiency provides for measurement range of wind up to 5km. Continuous wind profiles of the planetary boundary layer is presented to demonstrate the robustness and stability of the system.

Keywords: coherent doppler lidar, wind measure, planetary boundary layer.

1. Introduction

The high accuracy and high resolution wind profiles throughout the lower atmosphere layer are critical for improving forecasting, global pollution tracing, hurricane tracking, identifying possible hazardous weather conditions for aviation, and so on. Coherent Doppler lidar (CDL) systems are have been proven as accurate, reliable sensors for the remote sensing of atmospheric wind velocities under clear to moderate weather conditions [1]. Thus they have been developed for a long time by many researchers. The first wind-sensing CDL system with a 10.6 μm continue wave (CW) CO_2 laser was developed for the remote measurement of atmospheric wind velocity in the 1970s [2]. NOAA measured wind fields of the atmosphere boundary layer using a pulsed CO_2 Doppler lidar and revealed good accuracy [3]. Since the late 1980s, the development of high efficiency, high power 2 μm solid-state coherent Doppler lidar has produced high velocity resolution measurements of wind fields. Russell Targ et al. developed the first airborne 2 μm lidar for real-time wind field measurements [4]. Kavaya et al. completed the 2 μm Solid-state CDL system using a 2 μm Tm:YAG solid-state lasers, which realized the measurable range of 20 km [5]. And this was used for both ground-based platform and airborne platform [6]. Recently, with commercialized telecom fibers and fiber-based components can be used, all-fiber CDL system develop rapidly.

In this paper we demonstrate a high efficiency all-fiber pulsed CDL system for wind profiles sensing. The system has been improved in the pulsed laser with output energy of 80 μJ and the telescope with effective aperture of 50 mm, and two-dimensional scanning mode. Field experiments are implemented in Shenzhen, China and the results have been shown to be closely correlated with the results measured by ultrasonic anemometers. In addition, the observation results of boundary layer wind speed and wind direction are present in this paper.

2. System configuration

The system consists of following components (Figure 1): i) seed laser, (ii) AOM, (iii) fiber amplifier, (iv) circulator, (v) detector, (vi) signal processor and (vii) telescope. The seed laser has two outputs: a 1.55 μm cw, which is used as a local oscillator (LO), and pulsed laser, which is modulated and frequency shifted by an acousto-optic modulator (AOM).The modulated pulses are amplified to 80 μJ by a high power Erbium-Doped Fiber Amplifier (EDFA) then transmitted in to the atmosphere by a telescope. The backscattering of atmosphere aerosol is received by the same telescope and pass through a circulator. Both LO and backscattered signals are mixed by a 2×2 optical coupler which connects to a balanced detector. Finally, the digitalized heterodyne signals are processed in the signal processor card to obtain and visualize the line-of-sight (LOS) velocity of the wind. The temporal and range resolution of the system is 1 s and 30 m, respectively.

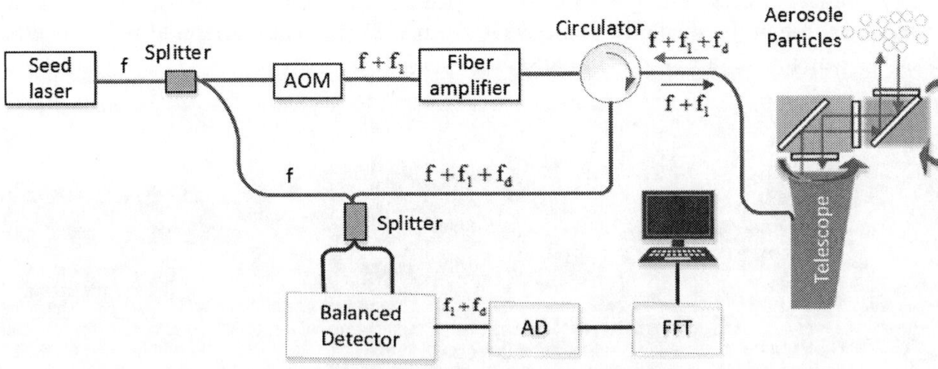

Figure 1. The CDL system setup

Since the detection range and accuracy of measure winds is initially determined by whether there is sufficient carrier to noise ratio (CNR) from the measurement data, high receiver efficiency, which mainly depend on wave front quality and optics transmission efficiency, is extremely important. However, due to the very small core diameter of single-mode fiber, the backscattered light is difficult to fully couple into the fiber. So a new telescope and the alignment method for it is developed to achieve high receiver efficiency. The structure of the off-axis 50 mm telescope is shown in Figure. 2.

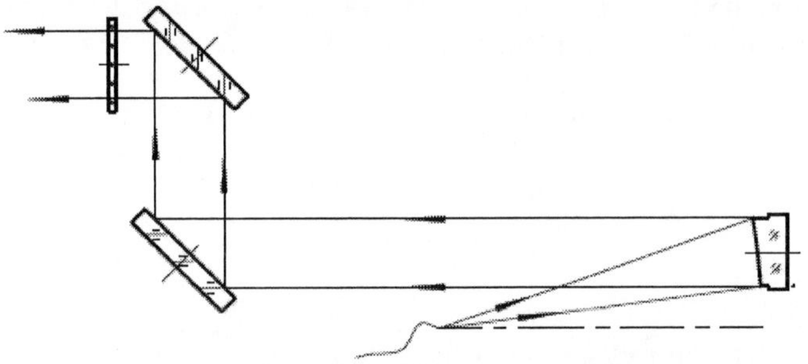

Figure 2. Schematic diagram of the off-axis telescope

For real-time signal processing, a multicore digital signal processor (DSP) is programed to perform the FFT of the heterodyne signal from the analog to digital converter card and export the power spectrum to PC. After getting the Doppler shift, the LOS wind velocity would be retrieved by the PC through the method of maximum likelihood discrete spectral peak (MLDSP). The specifications of the system are summarized in the table 1.

Table 1. The specifications of the system

Parameter	Value	Parameter	Value
Wavelength	1.55 μm	Telescope aperture	50 mm
Pulse energy	80 μJ	ADC sample rate	500 Msps
PRF	10 KHz	Detector bandwidth	200 MHz
Pulse width	300 ns	Scan speed	0.5°/s to 20 °/s

3. Experiments and results

The wind field of planetary boundary layer had been observed by using the CDL system at Shenzhen, south of China, in January 2018. To get the wind velocity and direction, velocity azimuth display (VAD) scan mode was applied. For the VAD scanning, the elevation of the laser beam is constant and the azimuth angle is continuously changed. After getting 8 LOS wind speed of different azimuth angle, the wind profiles would be retrieved through the method of sine wave fitting (SWF). The horizontal wind speed and wind direction data of 72 hours are presented in Figure 3.

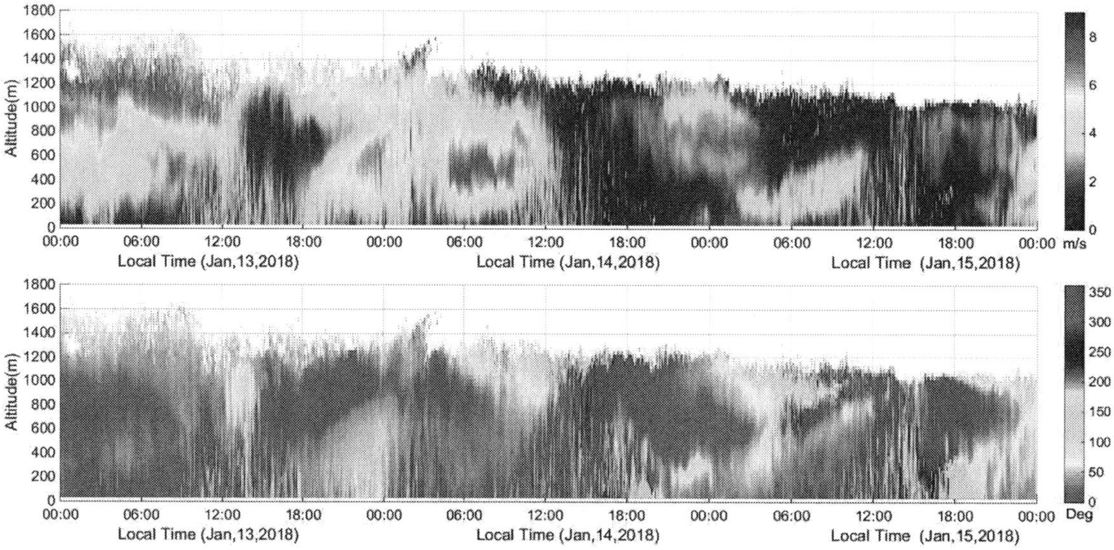

Figure 3. Wind profiles of planetary boundary layer in Jan 13-15

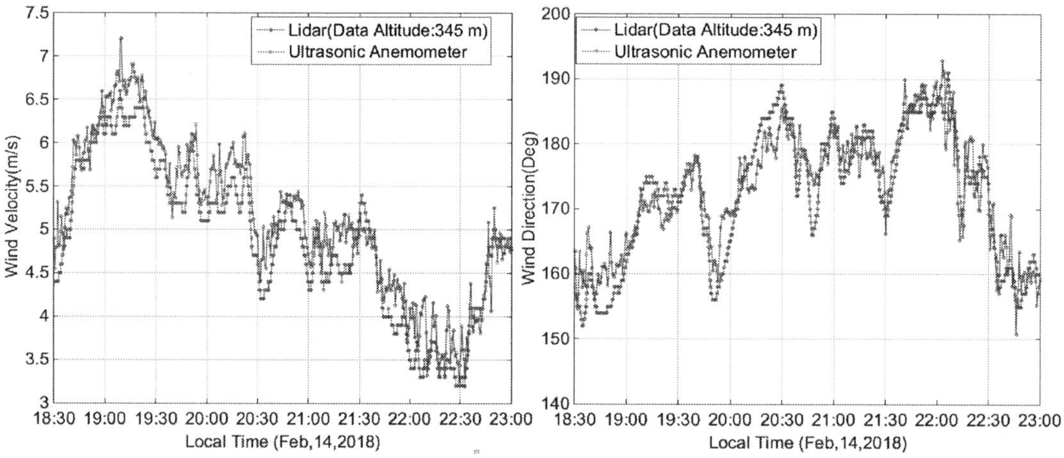

Figure 4. Time series of the wind velocity and wind direction from measurements by the sonic anemometer and the lidar at altitude of 345 m

To validate the measurement result, a comparison with the measurements from the ultrasonic anemometer was taken out. The ultrasonic anemometer which is located in Shenzhen meteorological tower, about 350 m above mean sea level (MSL) and 220 m from the lidar. The 10-s average wind data from the ultrasonic anemometer and the wind profiles measured by the system are found to be correlated (Figure. 4). The Histograms of the velocity difference and direction difference is shown in the figure 5.

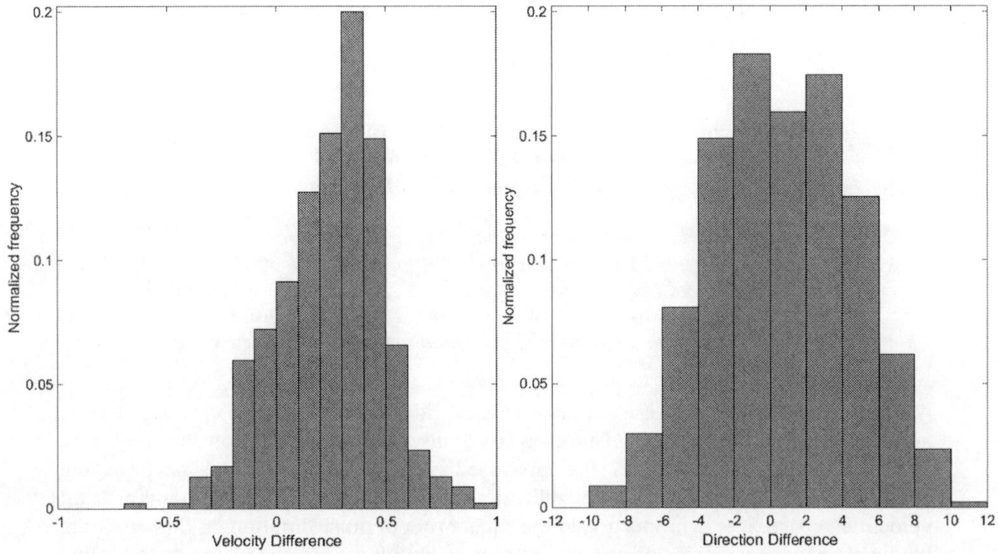

Figure 5. Schematic diagram of the velocity and direction difference between lidar and anemometer

4. Conclusion

A high efficiency all fiber coherent lidar system with all-fiber structure and newly optimized telescope is developed in University of Science and Technology of China. The 2-axis scanner and multicore digital signal processor enable the system to monitor specifically area with high time and space resolution, and provide high-precision LOS wind speed. To verify the performance of the system, several experiments were carried out. One verification test was taken out during Jan-Feb 2018, and over 72 hours continuous data is presented to demonstrate the stability of the system. Comparison experiments was carried out in Feb 2018, which showed good agreement with sonic anemometer.

5. References

[1] R. M. Huffaker and R. M. Hardesty, "Remote sensing of atmospheric wind velocities using solid-state and CO2 coherent laser systems," Proc. IEEE 84, 181–204 (1996).

[2] R. M. Huffaker, A. V. Jelalian, and J. A. L. Thompson, "Laser-Doppler system for detection of aircraft trailing vortices," Proc. IEEE 58, 322–326 (1970).

[3] B.F. Weber , F.F. Hall, M.E. Jackson, "Wind measurement accuracy of the NOAA pulsed infrared Doppler lidar", Applied Optics , 23 (15) :2503 (1984).

[4] Targ R, M.J. Kavaya, R.M. Huffaker, "Coherent lidar airborne windshear sensor: performance evaluation," Applied Optics,30(15):2013-2026 (1991).

[5] M. J. Kavaya, S. W. Henderson, J. R. Magee, C. P. Hale, and R. M. Huffaker, "Remote wind profiling with a solid-state Nd:YAG coherent lidar system," Opt. Lett. 14, 776–778 (1989).

[6] R. Targ, M. J. Kavaya, R. M. Huffaker, and R. L. Bowles, "Coherent lidar airborne windshear sensor: performance evaluation,"Appl. Opt. 30, 2013–2026 (1991).

Experimental Verification of Wind Lidar with Long-Duration Frequency-Modulated Pulse

Hiroshi Yamasuge(a), Eiichi Yoshikawa(b), Tomoo Ushio(a), Shoken Ishii(b), Makoto Aoki(c), and Kazuki Hashimoto(b)
(a) Tokyo Metropolitan University
6-6 Asahigaoka, Hino, Tokyo, Japan 191-0065
(b) Japan Aerospace Exploration Agency
6-13-1 Osawa, Mitaka, Tokyo, Japan 181-0015
(c) National Institute of Information and Communications Technology
4-2-1 Nukui-Kitamachi, Koganei, Tokyo,Japan 184-8795
Lead Author e-mail address: yamasuge-hiroshi@ed.tmu.ac.jp

Abstract: In our previous research, a new wind lidar which accomplishes high power and high range resolution by transmitting laser pulses with long duration and frequency modulation was proposed. In the proposed wind lidar, analytical performance evaluation under ideal conditions indicated that accuracies of wind ranging and velocimetry increases in proportion to the square root of pulse duration. In this research, we investigated accuracies of modulated pulse shaping that is one of issues to realize the proposed wind lidar. We developed an experimental system which directly returns generated laser pulses to its receiver, and digitally measured the returned waveform. Although, in the returned signal, the returned waveform is distorted compared to its ideal waveform, the result of matched filter processing showed sufficient performances that the mainlobe width was almost maintained, and peak power was reduced very little by –0.3 dB as compared with the ideal waveform.

Keywords: Coherent Laser Radar, Doppler, Frequency Modulation.

1. Introduction

In recent years, demand for wind lidars has increased greatly because of their technological progress and lowered price. Through wind lidars are used for various practical purposes, wind lidars with higher output and higher resolution have been desired. So far, researches and developments have made much efforts for high power output by improving the performance of laser sources and optical amplifiers.

As a new approach of a wind lidar for higher power and higher resolution, a wind lidar transmitting laser pulses with long duration and frequency modulation was proposed. A method that achieves high power and high resolution by transmitting long duration pulse and applying modulation and demodulation techniques has been utilized in radar. In lidar, however, large Doppler shifts due to very high carrier frequencies compared with radar becomes problematic. Since a frequency of the received signal is largely shifted, a correlation between a reference signal (which is normally a replica of the transmitting signal) and the received signal is lost. Therefore, it is impossible for lidars to apply pulse compression by matched filter as for radars. The proposed approach prepares multiple reference signals which are frequency-shifted corresponding to pre-supposed relative velocities, and matched filter processing is performed with each of them. Results of matched filter processing derive a range-velocity profile of received signal. According to analytical performance evaluation under ideal conditions, accuracies of wind ranging and velocimetry increase in proportion to the square root of pulse duration while maintaining the range resolution [1].

For demonstrating the proposed wind lidar, several issues have to be solved. One of issues is accuracy of pulse shaping. In the proposed wind lidar, since chirp-like frequency modulations derives large errors in velocimetry (because of characteristics in its ambiguity function [2],) a random phase modulation is used. Although waveform of phase modulations has discontinuous changes, the discontinuous waveform cannot be shaped accurately due to response times of optical instruments. In this research, we studied accuracy of waveform shaping, that is, actual waveforms were obtained by an experimental model, and signal processing performances given by the actual waveform were examined.

2. Experimental Model

In order to verify accuracy of waveform shaping, we developed an experimental model that directly detect modulated signal. Figure 1 shows the optical setup of the experimental model. In the experimental model, a laser souse with wavelength of 1550 nm is used, and the optical signals modulated by Acousto-optic Modulator (AOM) is directly returned to a detector. The laser transmitted from the laser source is divided via the coupler, one is to the AOM and the other is to heterodyne detection. The modulation signal generated by arbitrary waveform generator (AWG) is input to AOM, and the laser after AOM is modulated. In future observation, modulated signals will be emitted to free space and scattered signals form aerosol will be received. In this experiment, however, they are not released to free space but are directly returned to its receiver. The returned signal is converted to an intermediate frequency by the heterodyne detection. Table 1 shows the parameters of the signal used in this experiment. The carrier frequency and the intermediate frequency are 194 THz (1550 nm), 80 MHz, and a pulse length is 10 μsec.

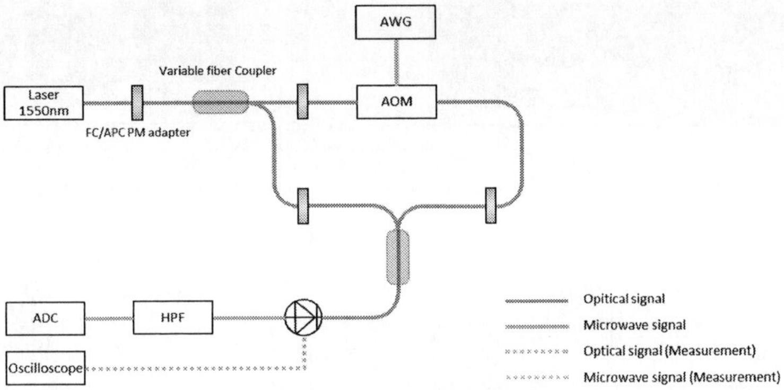

Figure 1. Experimental setup

Table 1. Parameter of signal in experiment

Parameter	Value
Carrier frequency	194 THz (1550 nm)
Intermediate frequency	80 MHz
Modulation	8-PSK
Bandwidth	2 MHz
Pulse duration	10 μsec

3. Results

In Figure 2, an experimental waveform (right) and a corresponding ideal waveform (left) are compared. Major disagreements between the two are seen at times switching phases. The experimental waveform indicates that the time required for a phase switch is about 0.12 μsec. Meanwhile, in specifications of the AOM, its rise time is 0.1 μsec, which means that the waveform distortion is mainly caused by the AOM's rise time.

Figure 3 shows spectra of the ideal waveform (left) and the experimental waveform (right). Sidelobes are lower in the experimental waveform than the ideal waveform. Looking at the experimental waveform again (Figure 2), the experimental waveform is shaped as multiplying a window function to the ideal waveform. Thereby, it is speculated that the reduction of sidelobes is caused by a windowing due to the system's response time.

Figure 4 shows range-velocity profiles which are resulted from the matched filter processing, whose left and right plots are ones calculated from the ideal and the experimental waveform, respectively. Since the experimental model directly returns the transmitting signal, it can be considered as its received signal is from a single particle at zero meter in distance and zero meter per second in velocity. Comparison of the two plots showed that the mainlobe width was almost maintained, and peak power was reduced by –0.3 dB.

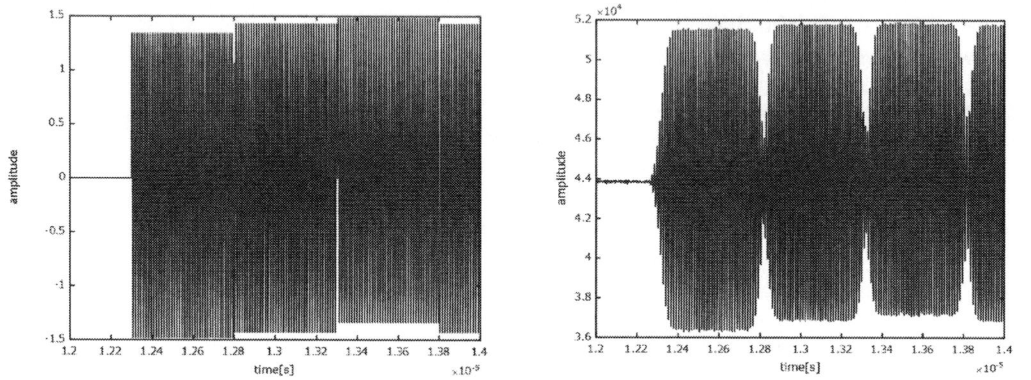

Figure 2. Ideal waveform (left) and experimental waveform (right)

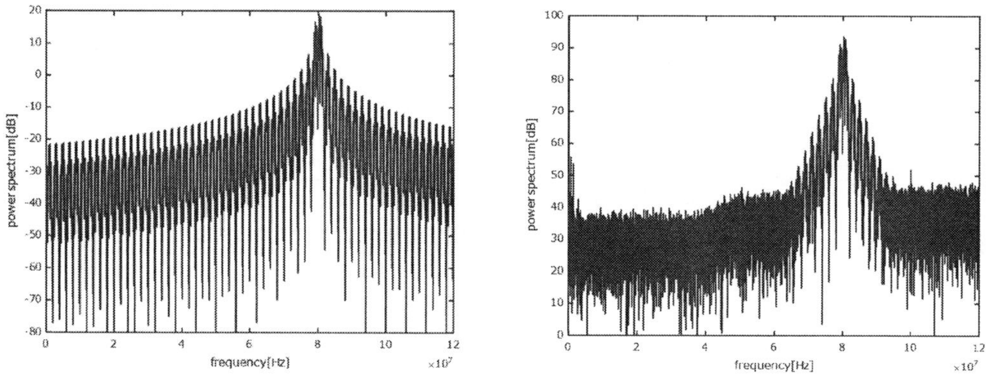

Figure 3. Spectrum of ideal waveform (left) and experimental waveform (right)

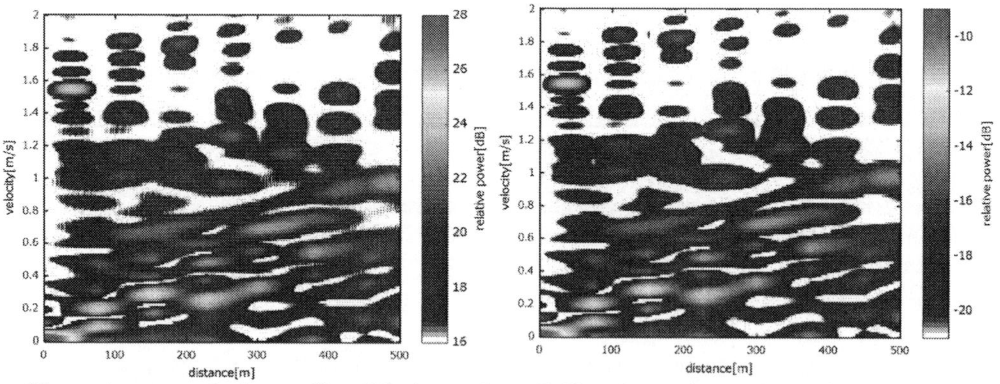

Figure 4. range-velocity profile of ideal waveform (left) and experimental waveform (right)

4. Conclusion

In this research, we experimentally verified accuracies of the waveform shaping which is one of issues for realizing our proposed wind lidar transmitting laser pulses with long duration and frequency modulation. In the proposed method, since phase modulation is used, there are discontinuous points in the waveform of the modulated signal. And, discontinuous waveform cannot be accurately generated due to a response time of the optical instruments. In order to verify the accuracy of pulse shaping, an experimental model which directly returns generated laser pulses to its receiver was developed. As a result of experiments carried out, the returned waveform was distorted around the discontinuous points compared with the ideal waveform. The distortion almost corresponded to a rise time of its optical device (AOM). By comparing range-velocity profiles resulted from the returned and the ideal waveforms, moreover, it is confirmed that the mainlobe width was almost maintained, and peak power was reduced by –0.3 dB. Therefore, it is concluded that the influence of pulse shaping is negligible for realizing the proposed wind lidar.

In the future, we will demonstrate the proposed wind lidar by further development of the experimental model to emit lasers in free space and observe scattered waves.

5. References

[1] Eiichi Yoshikawa, Tomoo Ushio, "Wind ranging and velocimetry with low peak power and long-duration modulated laser," Optics Express 25 8, 8845-8859 (1997).

[2] Merrill I Skolnik, *Radar Handbook Third Edition* (McGraw-Hill Education, 2008), Chap. 8.

All semiconductor pulsed Doppler lidar for wind sensing

Cariou Jean-Pierre, Pureur Vincent, Hwryniack Eric
LEOSPHERE
14-16 rue Jean Rostand, 91400 Orsay, France

Abstract: Coherent Doppler lidars are usually powered by optical fiber amplifiers or solid state bulk amplifiers. Thanks to recent improvements of semiconductor optical amplifiers (SOA), we have demonstrated the first 1.5 µm all semiconductor pulsed Doppler lidar for wind sensing. Low pulse energy typically available in SOAs is balanced by high pulse repetition frequency, allowing range resolved measurement of wind from 40 m to about 200 m with 20 m range resolution. While amplifying, the SOA shapes the light pulse leading to a quite simple, compact and cost effective setup. First results are shown and early atmospheric wind measurements are compared with a reference Windcube fibered lidar.

Keywords: Coherent Laser Radar, Semiconductor Optical Amplifier.

1. Introduction

For 10 years, Leosphere and other lidar manufacturers have designed and developed pulsed Doppler wind lidars based on Erbium Doped Fiber Amplification (EDFA) technology, coming from telecom industry but modified to operate in pulsed mode. Beyond proven reliability and accuracy, one main advantage is the good tradeoff between pulse energy and pulse repetition frequency, allowing middle or long range measurements with low average power.

Fiber amplification technology is however relatively complex and expensive, requiring different high quality components, such as high contrast modulator, multi-stage fiber amplifiers with polarization maintaining (PM) active fibers, and high power circulators. So, if fiber amplification positively competes with solid state or planar amplifiers for high end applications, other technologies can be considered when compactness, electrical consumption and cost are the main requirements.

Hansen and Pedersen [1] demonstrated continuous wave (CW) all semiconductor Doppler lidar in recent years. This paper is a first attempt to design and test an all semiconductor pulsed Doppler lidar based on Semiconductor Optical Amplifier, able to measure wind speed at short ranges (≤ 200 m) for wind energy applications.

2. Semiconductor lidar architecture

In the experimental setup, the pulse modulator and the fiber amplifier are replaced by a SOA as shown on Figure 1. The SOA is driven by a current pulse, so that no other modulator is required to shape the pulses.

The master oscillator is a DFB semiconductor diode, operating at 1543 nm. The power is divided between the Local Oscillator (LO) (1 mW on the detector after the mixing coupler) and the main signal (20 mW at the SOA input).

Following the SOA, the amplified pulse is slightly focused by a telescope in the atmosphere. Using a quarter wave plate (QWP), the backscattered light is mainly crosspolarized. A polarization beam splitter (PBS) is used as a beam circulator to direct the signal onto the mixing coupler. A single detector generates the electric signal, with a 125 MHz bandwidth.

The electric signal is then amplified and sampled at 250 MHz and is spectrally analysed with a real time FPGA processor.

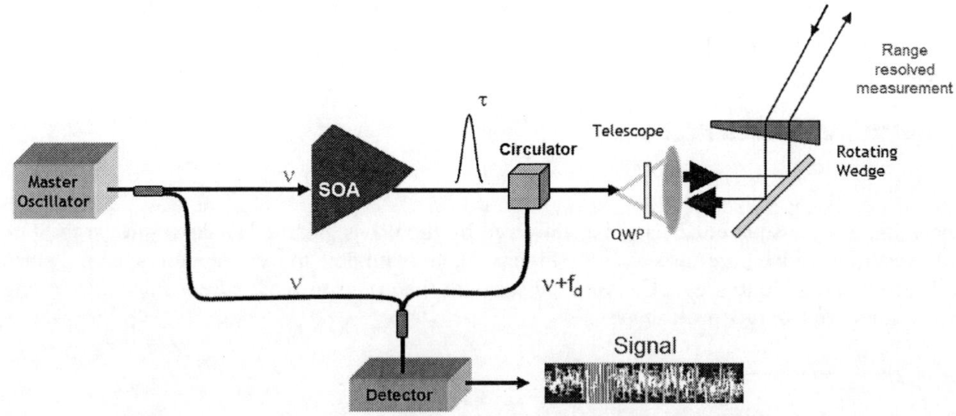

Figure 1: Experimental setup

The tested SOA is usually operating in C band (1530-1565 nm) for telecoms. It is built in InGaAsP active material on a InP substrate. The optical wave is guided along the component, like in a Fabry Perot laser, but the antireflective coating of the device facets avoids the laser effect. Optical gain is spectrally modulated by residual facet reflections, but tilted facets limit modulations.

Modulation bandwidth is high, compatible with muti GHz telecom bandwidths. Pulses can be very short, or can present short rise and fall times (some ns). Moreover, pulse repetition frequency can be high, allowing efficient modulation form factors.

The saturation power is a key parameter, corresponding to the maximum continuous output power, with saturated input. In pulsed mode, the peak power can exceed the saturation power, so that the average power is close to the saturation power. Optical damage on facets is however an associated risk limiting the SOA lifetime.

The Amplified Spontaneous Emission (ASE) is the other important parameter. Photons created by spontaneous recombination are amplified along the SOA, creating a wide bandwidth noise and limiting the maximum gain. The noise factor represents the SNR loss between output and input optical signals.

The next table summarizes the SOA characteristics. The tested unit is the BOA1004XP manufactured by Covega, and integrated in a butterfly casing.

Table 1: SOA parameters

Parameter	Value	Unit
Typical drive current	600	mA
Center wavelength	1550	nm
Spectral bandwidth	85	nm
Saturation power	15	dBm
Gain ripple	0.05	dB
Noise factor	0.05	dB

The driver board controls both the SOA temperature (Peltier cooler) and the drive current. It is configured to deliver ~ 0.2µs pulses, to be compared to the Windcube V2 lidar. The peak power and PRF are determined to respect SOA characteristics and lidar processing unit as well, while optimizing the figure of merit :

$$FOM = E\, t_{\mathrm{p}} \sqrt{PRF.t_{\mathrm{i}}}$$

Compared with the Windcube pulsed fibered lidar, pulse energy is here much lower. To increase the FOM of the pulsed semiconductor lidar, the PRF is increased up to 300 kHz and integration time is increased up to 10 s. The resulting ambiguous range $R_{\mathrm{amb}} = \frac{c}{2\,PRF} = 500$ m is enough for short range measurements.

3. Laboratory characterization

3.1. Pulse shape

The SOA is driven by a square 0.2 µs pulse delivered by the driver board. Thanks to the large SOA bandwidth, resulting pulses are also square (Figure 2), maximizing the energy for a given pulse duration. It is also possible to create Gaussian pulses with the driver to reduce frequency chirp along the pulse and improve Doppler peak shape.

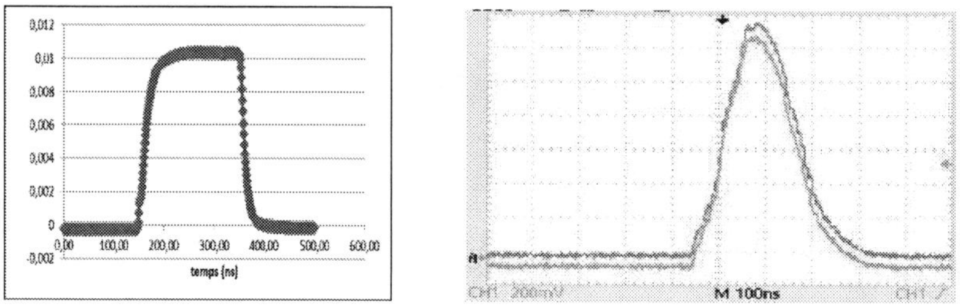

Figure 2: Optical output pulses – a) square pulse (0.2 µs) and b) Gaussian pulse (0.165 µs)

3.2. Power and Energy

The mean power is measured by a power meter and converted to peak power, knowing PRF=300 kHz and pulse duration.

Table 2: SOA driver characteristics

Parameter	**Name**	**Value a)**	**Value b)**	**Unit**
Pulse shape		Square	Gaussian	
Average power	P_{m}	15	12.9	mW
Peak power	P_{p}	250	261	mW
Pulse duration	t_{p}	0.2	0.165	µs
Pulse energy	E	0.05	0.043	µJ
Pulse repetition frequency	PRF	300	300	kHz
Integration time	t_{i}	10	10	s
Figure of Merit	FOM	17.3	12.3	µJ.µs

The Windcube FOM is around 200µJ.µs. With the semiconductor lidar, we then forerecast around 10dB drop in SNR.

4. First field experiments

The semiconductor amplifier is integrated in the same optical setup than the Windcube: same master oscillator, same telescope, same signal processing.

The Figure 3 a) shows the detector output signal, sampled by the acquisition board.

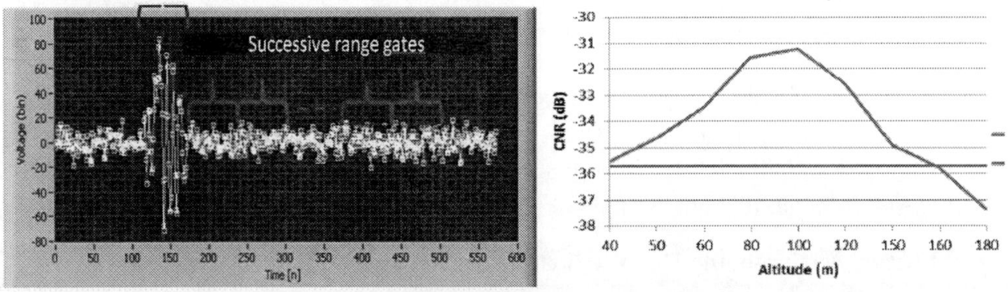

Figure 3: a) Time series - b) CNR plot (blue) – CNR Threshold (red)

The Narcisse reflected pulse is used as the zero distance while the pulse mean frequency is used as the zero velocity frequency. The sampling frequency is 250 MHz, so that each bin is corresponding to 0.60 m distance interval. Successive range gates are selected along the backscattered signal, corresponding to 40, 60 … 200 m altitude.

The CNR, defined as the Carrier to Noise Ratio in the total acquisition bandwidth, is plotted on Figure 3 b). The CNR threshold is adjusted to limit the False Alarm Rate to 5%. The CNR Lorentzian shape corresponds to a 40 mm telescope aperture and a focus around 100 m.

Using a 4 beam scanning configuration, and sequential measurements every 10 s in North, East, South and West directions, horizontal wind is measured and averaged over 10 minutes. Results are compared to a reference Windcube, measuring at the same time and the same place. Figure 4 highlights the correlation plot between both lidars.

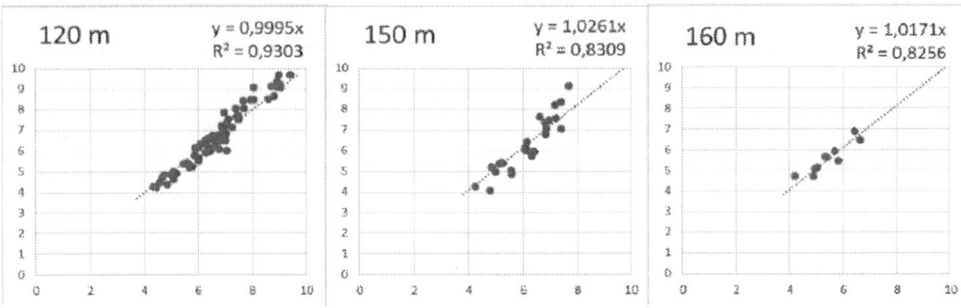

Figure 4: Correlation plots between Pulsed Semiconductor Lidar and Windcube

Even if pulse energy is low, we demonstrate that the SNR is good enough to perform an accurate wind speed measurement up to 160m with precision less than 0.5 m/s.

5. Conclusion

An all semiconductor pulsed lidar has been developed and tested in the atmosphere. Using a SOA as laser amplifier and pulse shaper, the pulse energy is 0.05 µJ at 300 kHz. Even with this low energy, it is possible to measure the wind speed in the atmosphere up to 160 m in 10 s integration time with a precision better than 0.5 m/s. These early results show that a very compact all semiconductor pulsed Doppler lidar can be designed for atmospheric sensing. Constant improvement in SOA technology would improve integration time and maximum range in the next future [2].

[1] All semiconductor laser Doppler anemometer at 1.55 µm, René Skov Hansen and Christian Pedersen2 2008 / Vol. 16, No. 22 / OPTICS EXPRESS 18288

[2] Monolithic master oscillator power amplifier at 1.58 µm for lidar measurements, M. Faugeron; M. Krakowski; Y. Robert; E. Vinet; P. Primiani; J. P. Le Goëc; O. Parillaud; F. van Dijk; M. Vilera; A. Consoli; J. M. G. Tijero; I. Esquivias; Proceedings Volume 10563, International Conference on Space Optics — ICSO 2014; 105630U (2017)

Airship-borne Doppler lidar wind measurement ability simulation

Decang BI[1] Weibiao Chen[1] Xiaopeng Zhu[1] Cunfeng zhao[1] Guyu Hu[1] Jie Meng[1]
zhengyang Jiang[1] Liyuan Jiang[1] Song dan[1] Jiqiao Liu[1] xiaolei Zhu[1]

([1]*Shanghai Institute of Optics and Fine Mechanics ,shanghai,jiading, 201800, China*)

Abstract Airship-borne Doppler lidar is an effective method to obtain stratosphere wind. Through number of simulation when airship fly at 25km altitude with 30°nadir looking, the Doppler lidar measurement ability with 4 scanning direction is simulated: ≤1m/s wind speed error within 2km detection distance, ≤1.65m/s speed error and 5°direction error within 4km distance,≤2m/s wind speed error and 10°direction error within 5km detection distance.

Keywords Doppler wind lidar; wind field; airship-borne

1. Introduction

Because of the thin air density on the tropopause and bottom of the mesosphere ,the molecular backscattering signal is dominant position. As a result the dierct Doppler lidar is more effective for wind vector detection. Wind distribution can be identified and be used for aircraft safety. Up-to-date there is only one research work published : NOAA reported one airship-borne principium wind lidar model with singe 45° nadir looking in 2006, It given capability of 2m/s Line-Of-Sight speed error and 2km detection range.

2. System parameters design and principle

Tabel.1 showed the direct detection lidar system parameters. The lidar adopted double edge Fabry-Perot interferometer as the frequency discriminator to resolute molecular backscattering Doppler frequency variety and in order to calculate wind vector.

Table 1 Doppler lidar system parameters

Parameters	Value	Remark
wavelength	355nm	
Transmitter Power	1Watt	
The effective Receive Aperture	100mm	
Receiving field angle	0.21mrad	
Scan directions	4~8	

1 ***Corresponding author:** Decang BI, E-mail: bidecang@siom.ac.cn

Filter bandwidth	FWHM=0.5nm&Trans m=0.8	
Detector QE	~40%	
Sun radiation	1.3W/cm²•nm	
Nadir looking Angle	30°	
Altitude resolution	300m	Range resolution=600m
Supposed fly altitude	25000m	

The atmosphere parameters included the transmission and backscattering coefficient were from Hitrans 2008. About the back noise calculation, planetary albedo of sea land and cloudy were 0.07,0.04 and 0.55 respectively. About the molecular spectrum used Cabannes scattering line-shape [1]to simulate the measurement ability.

The airborne Doppler lidar measurement model was showed in fig.1. f_1, f_3 and f_2, f_4 are in the orthogonal plane according to the airborne coordination system. When lidar worked once, four measurement directions divided the cone angle evenly. Although the four direction were not orthogonal in the coordination system WGS84, the inversion algorithm could makeup them by accurate attitude from Inertial Measurement Unit(IMU).

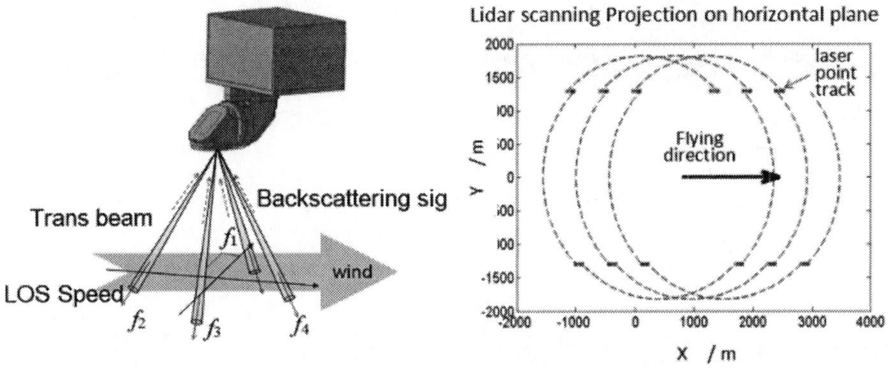

Fig. 1 airship work mode example: four directions scanning measurement

When campaign experiment, every direction integration time could be 30s or increasing to 300s in order to achieve higher SNR. According to the vector decompose relationship, the four Line-Of-Sight observation data can form four equations（1）。

$$\begin{cases} V_{LOSV-f_1} = V_{wind}(V,\theta_x,\theta_y,\theta_z) \times \cos(\theta_1) \times \cos(\psi_1) \\ V_{LOSV-f_2} = V_{wind}(V,\theta_x,\theta_y,\theta_z) \times \cos(\theta_2) \times \cos(\psi_2) \\ V_{LOSV-f_3} = V_{wind}(V,\theta_x,\theta_y,\theta_z) \times \cos(\theta_3) \times \cos(\psi_3) \\ V_{LOSV-f_4} = V_{wind}(V,\theta_x,\theta_y,\theta_z) \times \cos(\theta_4) \times \cos(\psi_4) \end{cases} \quad (1)$$

Where the $V_{wind}(V,\theta_x,\theta_y,\theta_z)$ included the speed value V and the direction angle $\theta_x,\theta_y,\theta_z$ which were the angle between the wind vector and the X,Y,Z coordination system in

WGS84 respectively. For the equation(1),there are four unknow wind parameters $V_{wind}(V,\theta_x,\theta_y,\theta_z)$ and four independent equations thus we can obtain the analytic solution results.

3 simulation results and discussion

The total detection error includes: random error, environment parameters error(temperature non-consistency error, Pressure variety error)and the attitude error.

(1) random measurement error

The simulated backscattering signal as fig.4 showed. there is only one photon came with 600m range bin in the range of 7km. The SNR of single shot with 30s time integration was showed in fig.5. The backscattering SNR is equal to 1 at 19km altitude in clear day or at 20km in cloudy. About the random error can be expressed[2] ΔV_{rand_err}

$$\Delta V_{rand\_err} = \frac{1}{S \times SNR_{\text{int}}} \tag{2}$$

Where the SNR_{int} could be calculated from four direction detection data as follow:

$$\frac{1}{SNR_{\text{int}}} = \frac{1}{SNR_{f_1}} + \frac{1}{SNR_{f_2}} + \frac{1}{SNR_{f_3}} + \frac{1}{SNR_{f_4}} \tag{3}$$

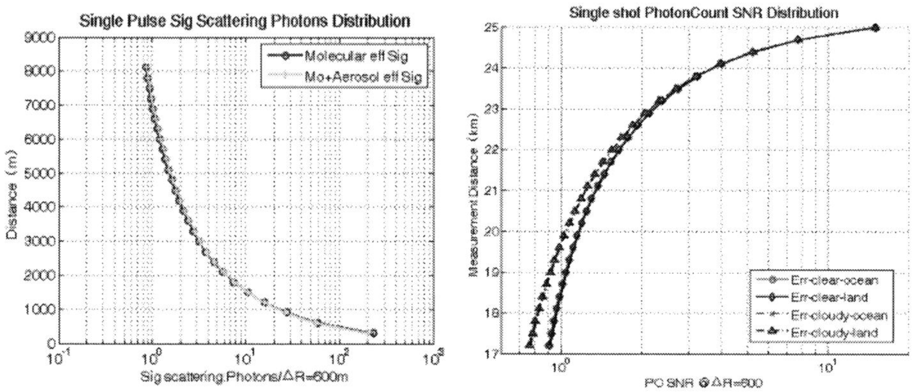

Fig.2 single pulse simulating backscattering photons and backscattering signal-noise-ratio

(2) wind error from temperature environment parameter

Because of the temperature is non-uniform in measurement process so the measurement is not consistent. The temperature error ±5℃ (range from 205K to 215K) will give rise to ±0.0145m/s, or temperature caused the wind error is 2.9‰ as fig.4 showed.

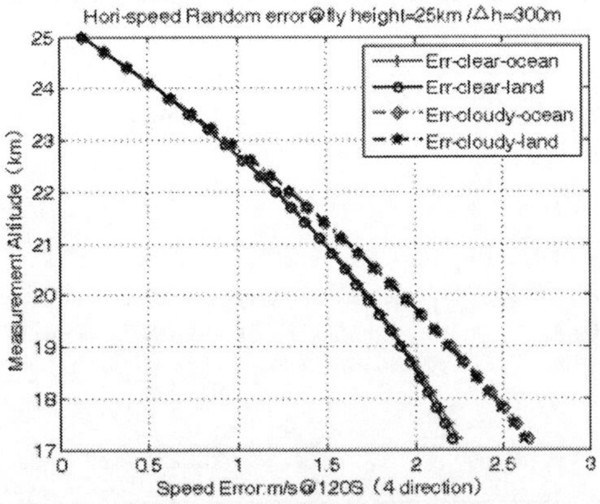

Fig.3 Random wind measurement simulation error

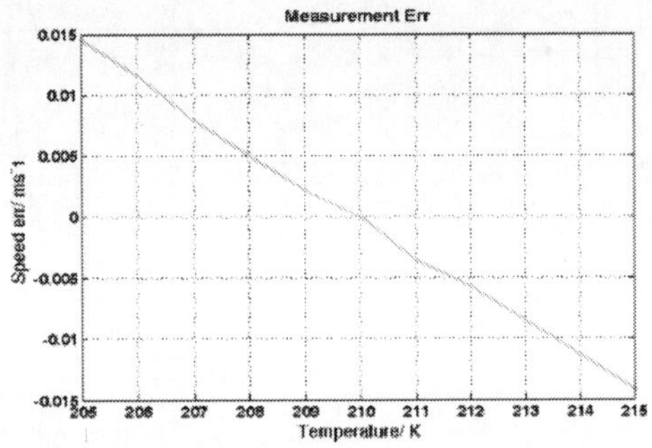

Fig. 4 wind simulation error under different temperature error with 1m/s input condition

Because the air pressure just caused little error(5hPa~10^{-4}m/s)， thus we do not discuss them.

The transmitter laser frequency stability is more important, 3MHz stability will cause 0.531m/s speed error.

(3) The attitude error .supposed the flying speed 10m/s the IMU attitude error is 0.2° for three axis. then the speed error will be :

$$\mathrm{V} = \sqrt{[\frac{V_{LOSV\perp}}{\cos(\Psi)}]^2 + [\frac{V_{LOSV\parallel}}{\cos(\Psi)}]^2} \leq 0.15\mathrm{m/s} \qquad (4)$$

(4) The total error

According to the independent speed variety, the total error can be expressed as follow:

$$V_{\mathrm{erf}} = \sqrt{V_{random}^2 + V_{PT}^2 + V_{freq}^2 + V_{atitu}^2} \qquad (5)$$

So the simulation total error distribution was listed as table 2

Table 4 airship-borne lidar wind simulation error

Error items	Technical error	Speed error	Total error
Attitude	0.2°	0.15m/s	
Frequency stability	3 MHz	0.531m/s	
temp	1°C	0.145m/s	
2km range random err		0.82m/s	1m/s
4kmrange random err		1.5	1.65m/s
Wind direction err			10°

The total wind speed error and direction error was showed in fig5.

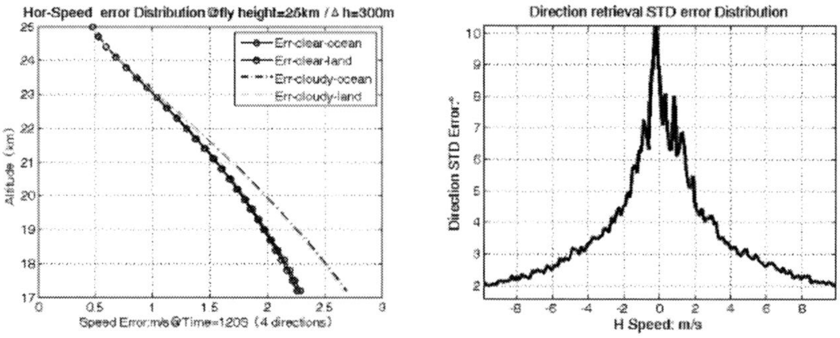

Fig. 5 horizontal wind error simulation distribution and wind direction error @1.6m/s

4. Results

From this work we obtained the airship-borne Doppler measurement ability when fly 25000m altitude with 30° nadir looking with 30s time integration: 1.65m/s speed error and 10° direction error within 4km range. This ability may be brought into operation.

Reference:

[1] Tenti G, Boley C D, Desai R C. On the Kinetic Model Description of Rayleigh–Brillouin Scattering from. [J]. Canadian Journal of Physics, 1974, 52(4):285-290.

[2] direct-detection Doppler wind measurements with a Cabanne-Mie lidar:A comparsion between iodine vapor filter and Fabry-Perot interferometer method. Chiao-Yao She, Jia Yue,Zhao-Ai Yan, John W. Hair, Jin-Jia Guo, Song-Hua Wu, and Zhi-Shen Liu ., applied optics,2007,46(20)

A Study on the Convective Initiation by Combining Coherent Doppler Lidar and Large Eddy Simulation Model in the Urban Area

Naohiro Iwamoto (a), Jun-ichi Furumoto (a), Kuniaki Higashi (b, a)
(a) Research Institute for Sustainable Humanosphere, Kyoto University
Gokasho, Uji, Kyoto, Japan
(b) Metro Weather Co. Inc., Japan
20-2, Nishiura, Gokasho, Kyoto, Japan
naohiro_iwamoto@rish.kyoto-u.ac.jp

Abstract: This study focused on the characteristics of convective initiation by combining Coherent Doppler Lidar (CDL) and Large Eddy Simulation (LES), aiming to explicate the developing mechanism of the active convective cell, which often causes localized heavy rainstorms. The convergence near the ground is one of the most important triggers of the convective activity. The data observed by CDL is useful to detect in detail convective field of the complicated surface region such as urban area. The CDL was installed at the top of 200 m-height building at central of Tokyo and observed the radial wind velocity with the range resolution of 100 m. The observed data were assimilated into the WRF (Weather Research and Forecasting) -LES model by using the three-dimensional variational assimilation method. The anomaly component of CDL-derived wind velocity recognized streaky structure prior to the of precipitation, and the model result showed the signature similar to the CDL observation.

Keywords: Convective Initiation, Coherent Doppler Lidar, Large Eddy Simulation, Assimilation.

1. Introduction

This study focused on the characteristics of convective initiation, aiming to explicate the developing mechanism of the active convective cell, which often causes localized heavy rainstorms. The convergence near the ground is one of the most important triggers of the convective activity. The data observed by Coherent Doppler Lidar (hereafter, referred as CDL) is useful to detect in detail convective field in the urban area, where various surface structures such as tall buildings, rivers, and green effect the atmospheric behavior in the boundary layer. Large Eddy Simulation (hereafter, described as LES), which explicitly represents turbulence structure in the boundary layer, is very useful for assimilating radial wind velocities with CDL.

In this study, we research on the atmospheric structure of the boundary layer by combining LES model which was high resolution and observation with CDL installed in the urban area.

2. COHERENT DOPPLER LIDAR MEASUREMENT

A CDL installed at the top of 200 m-height building in the central of Tokyo metropolitan area was used to obtain radial wind velocity field during May and October 2015. The location and observation area were illustrated in Figure 1. The "WINDCUBE 200S" made by Leosphere, Co., was employed for this observation. The observational configuration of the CDL measurement were listed in Table 1.

Table 1. Specification and configurations of the CDL observation

Scan mode	PPI
Elevation angle (deg.)	0
Azimuth angle (deg.)	0 – 360
Range Resolution (m)	100
Temporal Resolution (min.)	6
Observation Range	200-4,000 m

The horizontal wind velocity was calculated by applying Velocity Azimuth Display (VAD) method to the PPI images obtained every six minutes. The typical results of observed radial Doppler velocity and anomaly component after removed background wind are shown in Figure 2.

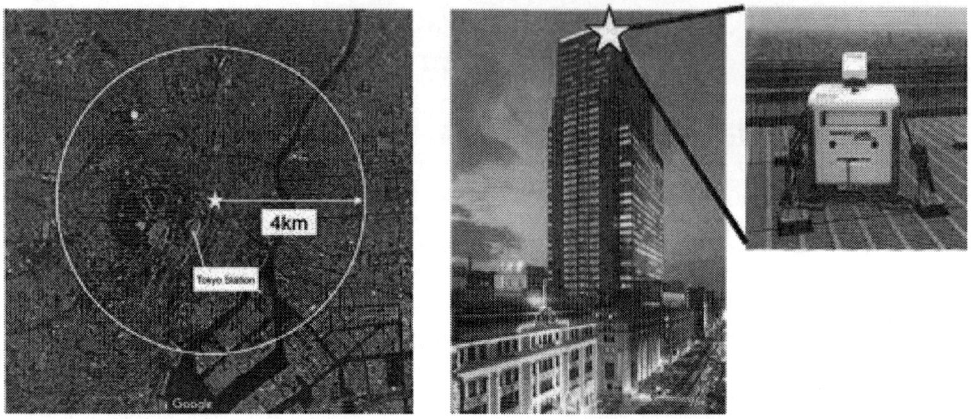

Figure 1. The observation area of the CDL measurement is shown by the white circle in the left figure. The overview of CDL system and

In order to analyze the signature of convective initiation prior to active convective activity, we, here, selected the CDL data during 0245 and 0257 UTC on August 10, 2015. Note that, in the observation area, precipitation of 30 mm/h was detected by the operational weather radar measurement.

In Fig. 2, the clear streak structures with the spacing of several hundred meters are observed prior to the precipitation events. To examine the relationship between this structure and the generation of precipitation event, the numerical forecast model of Large Eddy Simulation (LES) was employed.

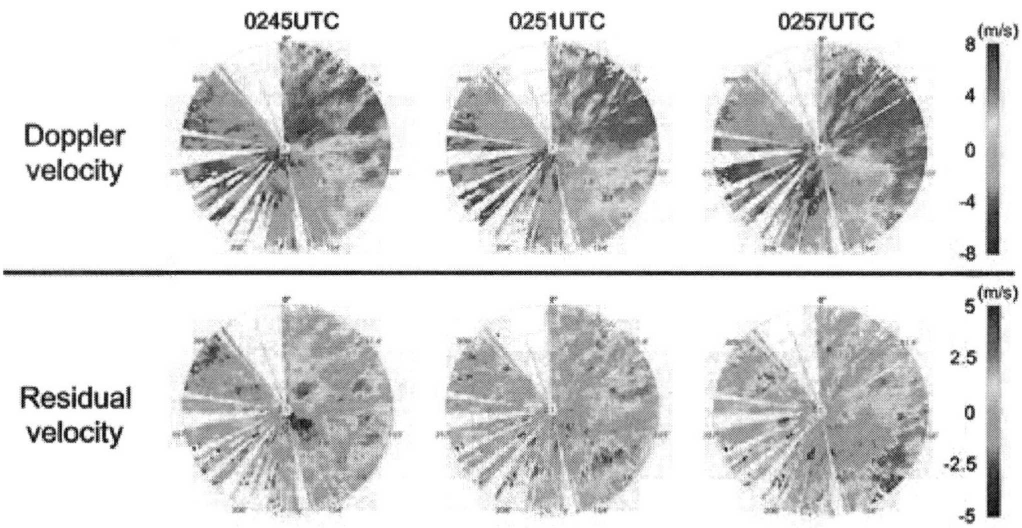

Figure 2. Top panel shows the radial velocity observed with CDL. The anomaly component of each top panels after removed background wind velocity are shown in the bottom panels.

Table 2. Model configurations of the numerical and LES model.

Model dynamics	RANS (domain 1-3), LES (domain 4)
Vertical levels	60
PBL/turbulence parameterization	MYNN Level3 (domain 1-3). Not used in domain 4.
Micro physical parameterization	WSM6
Cumulus parameterization	Kain-Fritsch (domain 1), not applied (domain 2-4)
Radiation parameterization	Dudhia, RRTM
Initial and boundary condition	JMA-MSM (domain1)
Sea surface temperature	NCEP RTG SST HR
Ground surface temperature	NCEP NFL

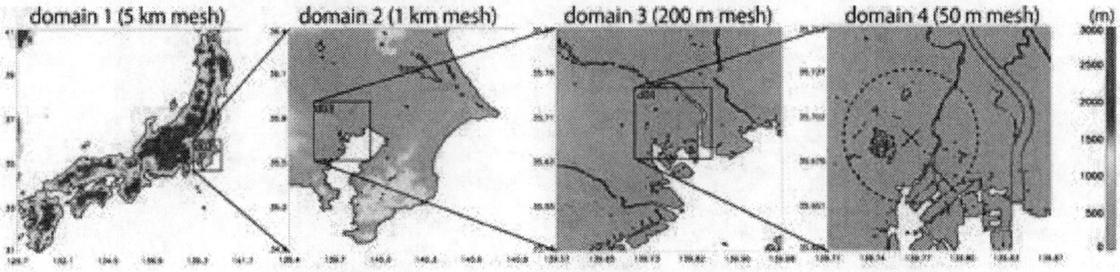

Figure 3. The calculation region of numerical simulation is shown. The × and dashed circle in the domain 4 shows the installation location and observation area of CDL, respectively.

3. Large Eddy Simulation

In this study, LES model embedded on the Weather Research and Forecasting Model Version 3.8.1 (hereafter, referred as WRF-LES) was used. Figure 4 illustrates domain configuration of the numerical experiments. In the finest domain, grid-size turbulence is explicitly calculated with LES scheme, on the other hands, in the larger three domains the turbulence effect was implicitly regarded by the parameterization method of MYNN Level 3[1]. The radial velocity observed by the CDL at 0303UTC was assimilated to the WRF-LES model by using the three- dimensional variational assimilation method.

Figure 3 shows the horizontal distribution of horizontal wind convergence near the surface at 0400 UTC on August 10, 2015. To analyze the detailed structure in the convergence line, Figure 4 shows the horizontal distribution of the anomaly component of wind speed of 260 m-height at 0340, 0345, 0350, 0355, and 0400 UTC on August 10, 2015. The similar streaky structure with the CDL results are indicated in the model forecast. This streak structure appears in parallel with the meso-β scale convergence line in Figure 3.

4. References

[1] M. Nakanish, and H. Niino, "Development of an Improved Turbulence Closure Model for the Atmospheric Boundary Layer", Jounal of the Meteorological Society of Japan, Vol.87 No.5, pp.895-912, (2009)

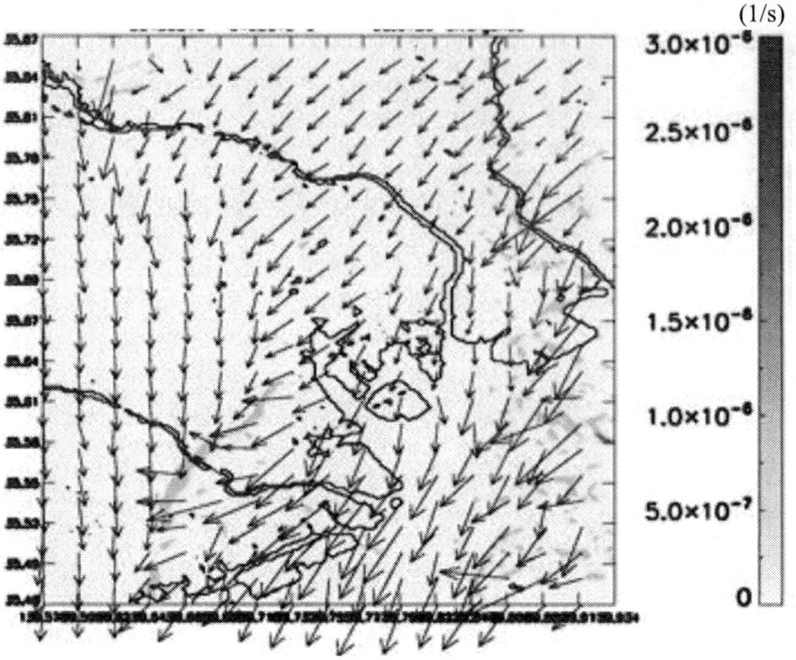

Figure 3. The horizontal distribution of horizontal wind convergence near the surface at 0400 UTC on August 10, 2015.

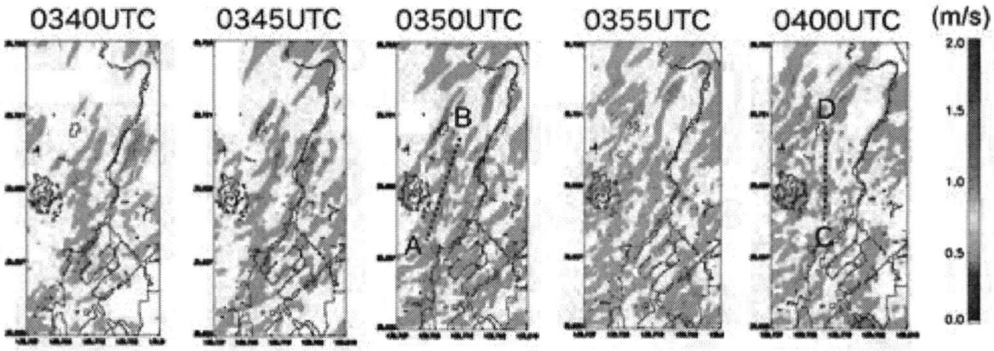

Figure 4. The horizontal distribution of horizontal wind-speed of 260 m-height at 0340, 0345, 0350, 0355, and 0400 UTC on August 10, 2015.

Feasibility study of data assimilation using a mobile water vapor

S. Yoshida(a), T. Sakai(a), T. Nagai(a), S. Yokota(a), H. Seko(a), and Y. Shoji(a)
(a) Meteorological Research Institute
Nagamine 1-1, Tsukuba, Ibaraki, 305-0052, Japan
syoshida@mri-jma.go.jp

Abstract: Vertical profiles of water vapor mixing ratio (q_v) obtained with a Raman lidar (RL) are assimilated into the Japan Meteorological Agency non-hydrostatic model using three-dimensional local ensemble transform Kalman filter as a feasibility study. Comparison of forecast and analysis of humidity field with precipitable water vapor (PWV) observed by Global Navigation Satellite System (GNSS) on surface indicates that data assimilation of vertical profiles of q_v reduces errors in forecast and analysis of humidity field. However, data assimilation of vertical profiles of q_v degrades estimation of humidity field in the late of the data assimilation experiment. The assimilation of q_v produces heavy precipitation three hours prior to the actual precipitation, causing cold outflow prevailing on surface in the model domain. The cold outflow accompanied by the early precipitation decreases PWV around thunderstorms, resulting in increasing errors in forecast and analysis of humidity field.

Keywords: Raman lidar, water vapor, data assimilation

1. Introduction

Water vapor in atmosphere exerts strong influence on initiations and development of thunderstorms. Small change on water vapor horizontal distribution sometimes causes strong impact on weather phenomena [e.g. 1]. Knowledge of water vapor distribution in atmosphere is very important for precipitation prediction with a use of data assimilation. In a couple of decades, data assimilation of water vapor obtained with space-borne imager [e.g. 2], Global Navigation Satellite System (GNSS) [e.g. 3], and water vapor lidar on airplane [4] and ground [5,6] are assimilated to numerical models and improvement of analysis and forecast of humidity field was presented.

A water vapor Raman lidar (RL), which emits laser pulses into air and receives Raman backscattered light form air molecules, provides vertical profile of water mixing ratio (q_v) by emitting lidar pulses vertically. One important and major advantage of water vapor lidar observation is providing vertical profiles of q_v that are not observed by GNSS. Therefore, development of the data assimilation of q_v obtained with lidars has been paid attention by a lot of researchers. This study examine the assimilation of vertical profiles of q_v obtained with a mobile RL into Japan Meteorological Agency non-hydrostatic model (JMA-NHM) [7] using three-dimensional local ensemble transform Kalman filter (LETKF). The quality of the data assimilation experiment is assessed by comparison of model-derived precipitable water vapor (PWV) with independent GNSS-observed PWV on ground. The impact of the data assimilation of the vertical profile on forecast and analysis is evaluated.

2. Data and Method

In the feasibility study, we focused on a heavy rainfall occurred on Kanto plane in Japan on 17 August 2016. At 1800 UTC on 17 August 2016, humid southerly winds prevailed near south part of the Kanto plane because a stationary front, accompanying an extratropical cyclone at Sea of Okhotsk, was located northern part of Kanto plane. The stationary front caused a group of cumulonimbus clouds around low level convergence between 0900 and 2100 UTC. Local heavy rainfall, exceeding 100 mm/2h, was recognized near the low level convergence around 1800 UTC.

We conducted RL observation to obtain vertical profile of q_v in Tsukuba, Japan, from 2 August 2016 to 6 December 2016. The RL was consisted of a Nd:YAG laser, and a beam expander, a telescope, and photomultiplier tubes (PMTs), and a transient recorder. The laser emitted light pulses at 355 nm with

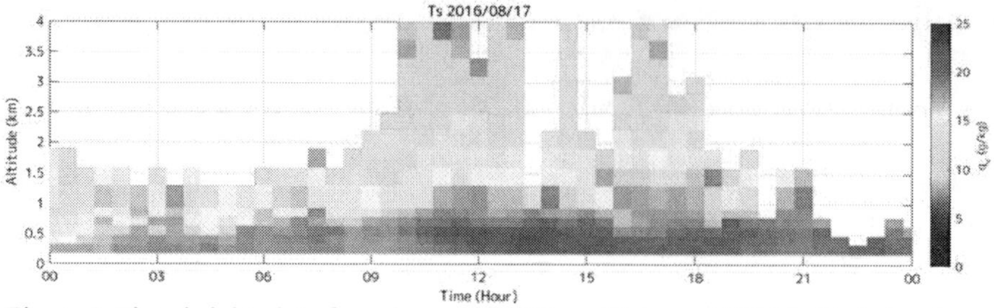

Figure 1. Time-height plot of qv observed by RL on 17 August 2016 in Tsukuba, Japan

pulse energy of 200 mJ and 10 Hz. The emitted beams were expanded to a diameter of about 5 cm by the beam expander. The light backscattered by atmospheric molecules and particles were received by the telescope and divided into three wavelengths, that are, 355 nm for elastic backscatter light, 386.7 nm for Raman nitrogen, and 407.5 nm for Raman water vapor with PMTs. Each light signals through the PMTs were digitized in the transient recorder. Figure 1 shows a time-height plot of q_v obtained with RL on 17 August 2016. It appears that q_v at low level (generally less than 1 km) increased by 2 g/kg between 0900 UTC and 1200 UTC.

To check the observational bias of RL, we performed comparison of q_v observed by RL and that estimated in high-resolution local analysis (LA) data, which provides meteorological parameters based on calculation of JMA-NHM. The comparison was performed using data observed between 2 August 2016 and 30 September 2016. Figure 2a shows a scatter plot of q_v obtained with the RL versus q_v estimated in LA. Figure 2a shows that most q_v obtained by RL agrees with q_v estimated in LA. To discuss the relationship in details, Figures 2b and 2c, respectively, show vertical profiles of bias (RL q_v minus LA q_v) and root mean square (RMS) between RL q_v and LA q_v. As seen in Figure 2b, the biases are positive below altitudes of 2.5 km, indicating that RL q_v has wet bias in the low level compared to LA.

In this study, we employed nested LETKF as data assimilation system. In the outer LETKF, the interval of horizontal grid was 15 km. The mesoscale analysis and global forecast were employed every 6 hours as boundary conditions. The JMANHM was employed to perform an ensemble forecast. In the outer LETKF, no data are assimilated. In the inner LETKF, we performed three assimilation experiments to explore the impact of RL data assimilation. The first assimilation experiment is termed as "TEST1", in which RL-observed q_v is directory assimilated. The second assimilation experiment is termed as "TEST2", in which RL-observed q_v is assimilated after the bias value shown in Figure 2b is subtracted from RL-observed q_v. The observation errors employed in TEST2 are set to RMS values shown in Figure 1c divided by square root of 2. The last assimilation experiment is termed as "CNTL", in which no data is assimilated. In TEST1 and TEST2 experiments, RL data are assimilated in a 3-h assimilation window and no other data are assimilated.

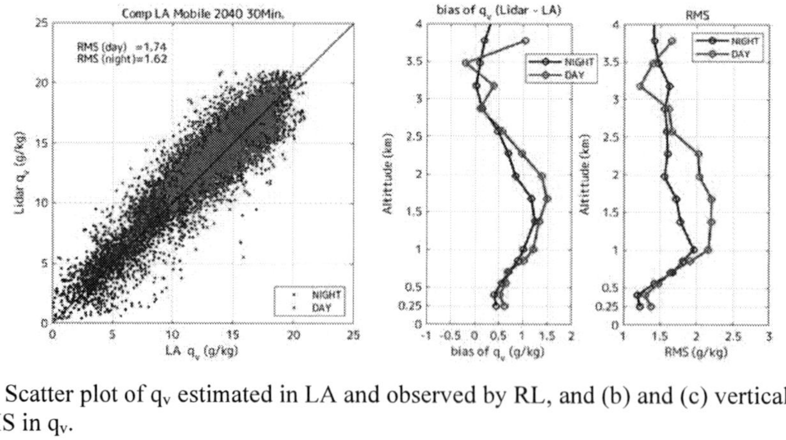

Figure 2. (a) Scatter plot of q_v estimated in LA and observed by RL, and (b) and (c) vertical profiles of bias and RMS in q_v.

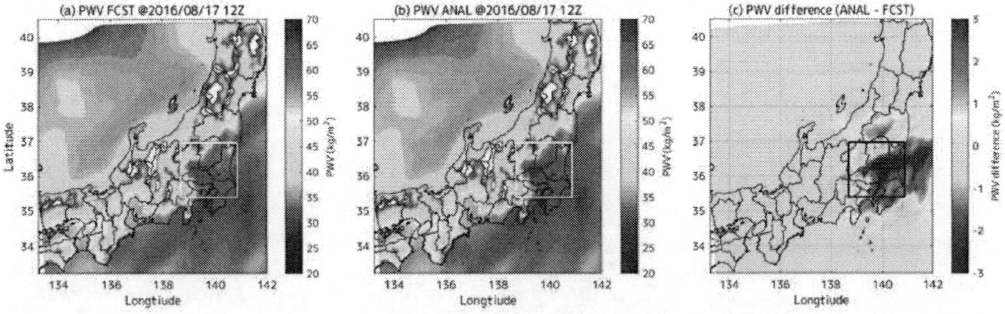

Figure 3. Comparison of ensemble means of PWV between TEST1 and GNSS observation. (a) PWV forecast and (b) analysis. (c) Difference of ensemble means of PWV (analysis - forecast).

3. Results

To explore the impact of assimilation of RL data, Figure 3 shows ensemble means of PWV in TEST1 at 1200 UTC. Figure 3 indicates that the RL data assimilation contributes to increasing PWV. The assimilation of RL has influence on the areas within about 200 km of the RL observation site as seen in Figure 3c. We perform comparison of GNSS-observed PWV with ensemble means of PWV in forecast and analysis to discuss the bias and root mean square (RMS). We performed two quality controls (QC) before the comparison to eliminate GNSS-observed PWV that might have large error. Firstly, we eliminate GNSS PWV when altitude differences between model surface and PWV site elevation are larger than 40 m in this comparison. Secondly, we eliminate GNSS PWV when precipitation more than 5 mm/h is recognized within 25 km of GNSS observation site.

The bias and RMS between forecast and GNSS observation, respectively, are -0.82 kg/m^2 and 2.75 kg/m^2. While, the bias and RMS between analysis and GNSS observation, respectively, are -0.27 kg/m^2 and 2.45 kg/m^2. Here, the bias indicates ensemble means of PWV minus GNSS-observed PWV. Since RMS and absolute value of bias in analysis is smaller than those in forecast, the assimilation of RL data contributes to improvement of humidity field at 1200 UTC. It should be noted that similar results are obtained in TEST2 experiment. Overall, the assimilation of RL seems to improve estimation of humidity field at 1200 UTC.

To examine the impact of the data assimilation, Figure 4 shows the bias and RMS between ensemble means of PWV and GNSS-observed PWV every 20 minutes. We applied the same QC scheme employed in Figure 3 before the comparison. As shown in Figure 4a, the absolute values of bias in TEST1 and TEST2 assimilation experiments are closer to zero compared to those in CNTL almost all the time. Moreover, Figure 4b shows that RMS in TEST1 and TEST2 is lower than the CNTL before 21 UTC. These results suggest that data assimilations of RL data contribute to improvement of humid filed estimation up to the 21 UTC. However, after 21 UTC, the data assimilations degrades forecast and analysis of humidity field.

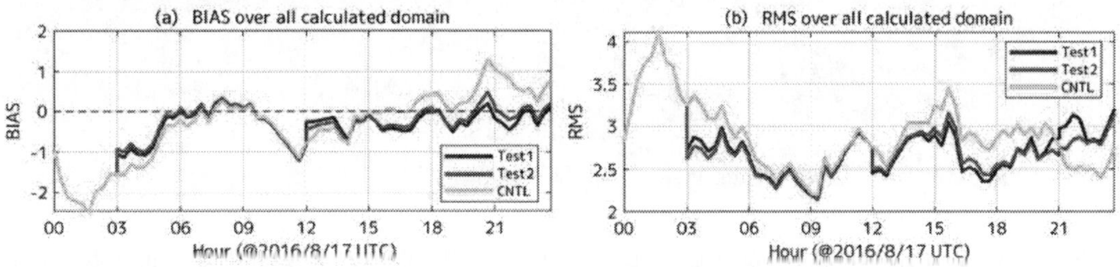

Figure 4. Time series of bias and RMS between ensemble means of PWV in TEST1 and GNSS-observed PWV.

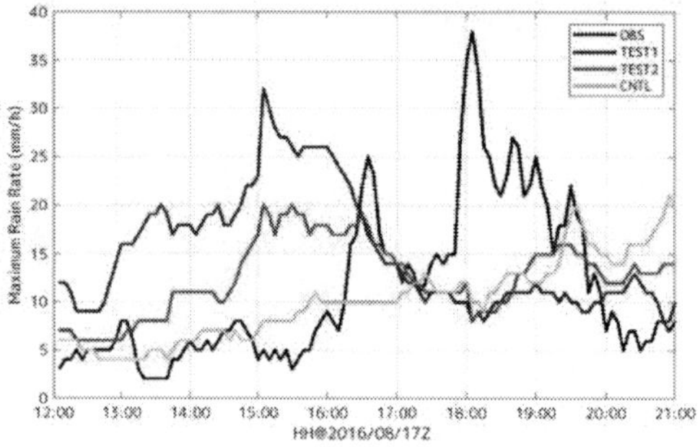

Figure 5. Maximum rain rate for RAM, TEST1, TEST2, and CNTL, in the area where heavy precipitation is recognized.

4. Discussion and summary

In the previous section we showed that RMS associated with TEST1 and TEST2 after 21 UTC becomes larger than those of CNTL. When we limit the area for calculating bias and RMS from all model domain, which corresponds to Figure 4, to an area indicated by white rectangle in Figure 3a, degradation of bias and RMS is recognized after 1800 UTC (not shown). These results indicate that data assimilation of RL degrades forecast and analysis of humidity field after 18 UTC. Figure 5 shows time series of the maximum rain rate every 5 minute for TEST1, TEST2, CNTL, and Radar-AMeDAS (RAM) around area where heavy rain fall occurred. Maximum rain rate observed by RAM has two peaks at 1630 UTC and 1805 UTC. On the contrary, peaks of rain rate in TEST1 and TEST2 are recognized only around 1500 UTC. It seems that rain rate peak in RAM at 1805UTC might correspond to the rain rate peak at 1505 UT in TEST1 and TEST2 data assimilation experiments because location of heavy precipitation of RAM at 1805 UTC are close to location of heavy precipitation of TEST1 at 1500 UTC. This suggests that data assimilation experiment of TEST1 and TEST2 produced heavy rainfall 3 hour prior to the actual heavy rainfall. The heavy precipitation produced in TEST1 and TEST2 might cause cold outflows on surface from thunderstorms. These cold and dry outflows prevailed around Tochigi prefecture and Gunma prefecture, where actual heavy precipitation occurred. Eventually, the sequence of procedures following heavy precipitation produced three hour earlier to the actual heavy rain in TEST1 and TEST2 lowers PWV in forecast and analysis compared to GNSS-observed PWV.

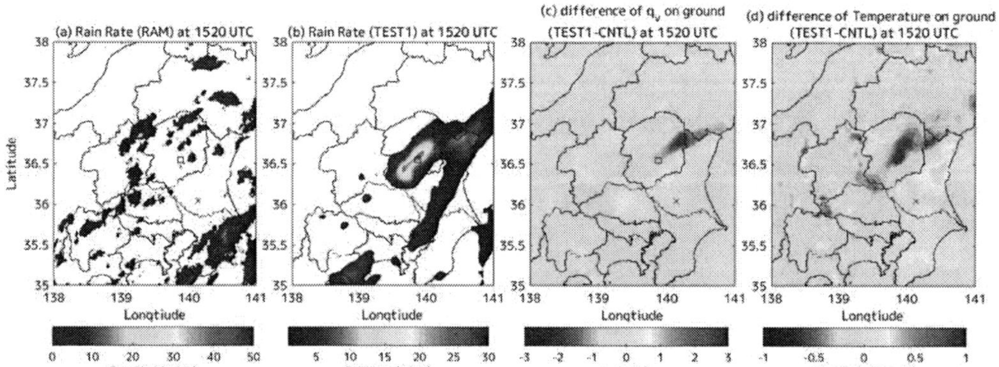

Figure 6. Observation and model results at 1520 UTC. (a) Observed rain rate, (b) ensemble mean of precipitation in TEST1, (c) difference of ensemble means of qv on ground between TEST1 and CNTL, and (d) difference of ensemble means of temperature on ground between TEST1 and CNTL.

Figure 6 shows rain rate, q_v, and temperature on surface comparison of 1520 UTC. At 1520 UTC, heavy precipitation is presented in TEST1 (Figure 6b), while no heavy rain was observed in actual (Figure 6a). Figures 6c and 6d, respectively, show difference of q_v on surface between TEST1 and CNTL, and difference of temperature on surface between TEST1 and CNTL, showing noticeable decreases of q_v and temperature on surface in TEST1 compared to CNTL. It appears that cold flows might have begun to propagate from thunderstorms accompanied by heavy precipitation around 1500 UTC in TEST1, resulting in decreasing q_v and temperature on surface. In TEST2 experiment, similar results to TEST1 experiment are obtained (not shown).

One possible reason why heavy precipitation has started three hours prior to the actual precipitation is overestimation of q_v and PWV in data assimilation process at 1200 UTC. At that time, overestimation of PWV in TEST1 and TEST2 compared to GNSS observations are recognized around area where PWV are high (exceeds 65 km/m^2) (not shown). Thunderstorms in TEST1 and TEST2 are triggered around the PWV overestimated area and developed northeastwards. Finally, they cause precipitation around south part of Tochigi prefecture, where heavy rainfall is presented in TEST1 and TEST2 around 1500 UTC. Therefore, it appears that overestimated PWV field "mis-triggered" thunderstorms so that they cause heavy precipitations three hours prior to the actual precipitation.

In this presentation, we performed data assimilation of RL data and showed that data assimilation of vertical profiles of q_v contributes to improvement of forecast and analysis of humidity field. However, forecast and analysis of humidity field were degraded after heavy precipitation three hours prior to actual are produced in the model. For further improvement of RL data assimilation, development of assimilation scheme and modeling is needed.

5. Acknowledgement

This work was partially supported by JSPS KAKENHI grant 17H00852.

6. References

[1] V. Wulfmeyer, R. M. Hardesty, D. D. Turner, A. Behrendt, M. P. Cadeddu, P. Di Girolamo, P. Schlüssel, J. Van Baelen, and F. Zus, "A review of the remote sensing of lower tropospheric thermodynamic profiles and its indispensable role for the understanding and the simulation of water and energy cycles," *Rev. Geophys.*, **53**, 819–895 (2015).

[2] Zapotocny, T. H., W. P. Menzel, J. Jung, and J. P. Nelson, "A four-season impact study of rawinsonde, GOES, and POES data in the Eta Data Assimilation System. Part I: The total contribution," *Wea. Forecasting*, **20**, 161-177 (2005).

[3] Seko, H., T. Miyoshi, Y. Shoji and K. Saito, "Data Assimilation Experiments of Precipitable Water Vapor using the LETKF system -An intense rainfall event over Japan 28 July 2008," *Tellus*, *63A*, 402-412 (2011)

[4] F. Harnisch M. Weissmann C. Cardinali and M. Wirth, "Experimental assimilation of DIAL water vapour observations in the ECMWF global model," *Quart. J. Roy. Meteor. Soc.*, **137**, 1532–1546 (2011)

[5] M. Grzeschik, H.-S. Bauer, V. Wulfmeyer, D. Engelbart, U. Wandinger, I. Mattis, D. Althausen, R. Engelmann, M. Tesche, and A. Riede, "Four-dimensional variational data analysis of water vapor Raman lidar data and their impact on mesoscale forecasts," *J. Atmos. Ocean. Technol.*, **25**(8), 1437–1453 (2008).

[6] T. Somekawa, Y. Sato, M. Takahashi, N. Takada, and M. Fujita, "Data assimilation of water vapor for a numerical weather prediction model using the Raman lidar system, " *Journal of the Remote Sensing Society of Japan*, **33**, 5, 360-366 (2013)

[7] K. Saito et al., "The operational JMA nonhydrostatic mesoscale model," *Mon. Wea. Rev.*, **134**, 1266–1298 (2006)

Retrieval of $PM_{2.5}$ profile using the Doppler Lidar across New York State Mesonet

Bhupal Shrestha, Dr. Everette Joseph
Atmospheric Sciences Research Center, University at Albany, Albany, NY, USA
bshrestha@albany.edu

Abstract: The University at Albany has developed a Mesonet, a network of 17 enhanced sites across the NY state, comprised of Doppler Lidar, Microwave Radiometer and Sun Photometer. The goal of the network is to enhance information content by combining all sensors for Numerical Weather Prediction and air quality forecasting.

The vertical profile of $PM_{2.5}$ is very important for the study of air quality and its forecasting. The commonly used method only estimates surface $PM_{2.5}$, which is based on its relationship with Aerosol Optical Depth (AOD) from satellite and Sun Photometer. In this study, the relationship of aerosol extinction coefficient retrieved from the Doppler Lidar inversion, with surface $PM_{2.5}$ is presented. Then, the $PM_{2.5}$ profile is retrieved using the extinction coefficient profile. Three cases of high, low and medium $PM_{2.5}$ are presented and compared with surface $PM_{2.5}$ measurements. The experiment results show that Doppler Lidar can be used to continuously monitor $PM_{2.5}$ profile and hence, the retrieval technique when applied throughout the network is a good sign for real time 3D monitoring of $PM_{2.5}$.

Keywords: Doppler Lidar, aerosol extinction coefficient, AOD, $PM_{2.5}$.

1. Introduction

$PM_{2.5}$ are the fine aerosol particles that are less than 2.5 µm in diameter, expressed as mass concentration, µg/m^3. Since these particles are so small and light, they are serious pollutant that are easily inhalable and are the cause of increased morbidity and mortality [1], particularly cardiopulmonary disease and lung cancer [2]. Similarly, $PM_{2.5}$ can scatter and absorb solar radiation, thus, can reduce the atmospheric visibility by its extinction effect. So, it is very important to monitor $PM_{2.5}$. There are various ground stations for the surface PM2.5 measurements only. In order to retrieve surface $PM_{2.5}$, the statistical approach based on its relation with AOD from satellite and Sun Photometer measurement have been studied [3-6]. There is hardly any effective retrieval technique for $PM_{2.5}$ profile in existence. In order to quantify the air quality and its variability as well as provide initial data for air quality models for air quality forecasting, the $PM_{2.5}$ mass concentration profile is very important.

$PM_{2.5}$ can be estimated through remote sensing of the optical properties of the aerosols from the space and ground [7]. Doppler Lidar is an active remote sensing instrument that is primarily used for wind measurement but can be used to detect aerosol profile and monitor atmosphere and boundary layer structure through attenuated backscatter signal. Since AOD is a column integrated aerosol extinction up to the top of the atmosphere and surface $PM_{2.5}$ is only related to aerosol loading near surface, the higher uncertainty in the relation between PM2.5 and AOD always exist. The relationship between surface $PM_{2.5}$ and AOD is found to be associated with vertical and size distributions of aerosols [7]. Thus, in this study, the aerosol extinction coefficient profile retrieved from the inversion of attenuated backscattering coefficient measured by well-calibrated Doppler Lidar is used to retrieve $PM_{2.5}$ profile.

2. Instruments

Doppler Lidar is an active remote sensing instrument that measures the radial velocity and the Carrier-to-Noise Ratio (CNR) along the line of sight. The CNR is the integral of the signal spectra normalized by the noise spectra. When the instrumental effect is removed from the CNR, the attenuated relative backscattering is retrieved.

(a)

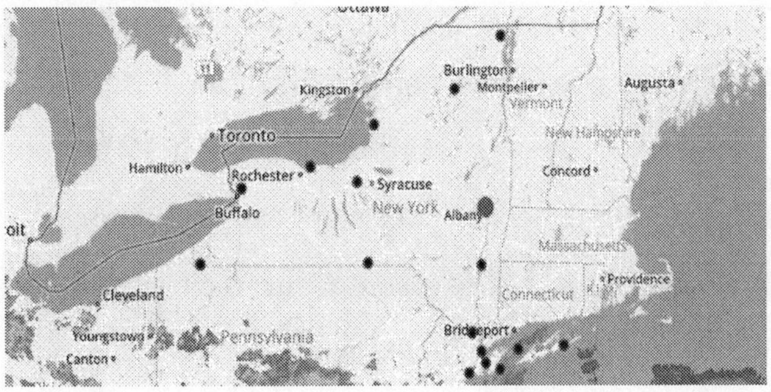

(b)

Figure 1. (a) Leosphere WindCube Doppler Lidar and (b) New York Mesonet 17 Enhanced Sites

Table 1. The specifications of Doppler Lidar in NY Mesonet

Parameter	Value
Wavelength	1.54 μm
Pulse Repetition Frequency	20 KHz
Physical Resolution	50m
Range	100 to 3000 m
Gates	117

3. Methodology

The Doppler Lidar provides radial wind and Carrier to Noise Ratio (CNR) after the interaction of laser beam with the aerosols along the line of sight. The obtained CNR values need the correction for the effect of instrumental function due to theory of focus. After the calibration of the instrument with low elevation scan and the assumption of atmospheric horizontal homogeneity, the relative backscattering coefficient profile is obtained and expressed as:

$$\beta_{rel}(z) = K.\beta(z).\exp(-2\int_0^z \sigma(r)dr) \quad (1)$$

Using the lidar inversion solution by Klett [8], and the assumption that:

$$\beta = B_o \, \sigma^k \quad (2)$$

The extinction coefficient profile is retrieved as:

$$\boldsymbol{\sigma(z)} = \frac{e^{[\frac{S(z)-S(m)}{k}]}}{\frac{1}{\sigma(m)} + \frac{2}{k}\int_z^{z(m)} e^{[\frac{S(z)-S(m)}{k}]} dz'} \quad \text{for } z < z(m),\ S(z) = \log(\beta_{rel}(z)) \quad (3)$$

The value of 'k' is considered to be 1.

The extinction coefficient profile is integrated from the ground to the certain height (z_{max}) to calculate aerosol optical depth, in order to compare with Sun Photometer retrieved AOD.

$$\text{AOD} = \int_0^{\text{Zmax}} \sigma(z)dz \quad (4)$$

If n(r) is the aerosol size distribution such that,

$$n(r) = dN/dr \quad (5)$$

where, dN is the aerosol number concentration in radius (r→r + dr), then

PM2.5 mass concentration can be written as:

$$PM_{2.5} = \int_0^{2.5\mu m} \rho(4/3)\pi r^3 n(r)\, dr \quad (5)$$

And, the aerosol extinction coefficient can be written as:

$$\sigma = \int_0^{\infty} \pi r^2 Qext\, n(r)\, dr \quad (6)$$

where, Qext is the extinction efficiency.

Combining Eq. (5) and (6), we can obtain the following relationship:

$$\sigma(z) = (3Q_{ext}\, \eta \,/\, 4r\, \boldsymbol{\rho})\, PM_{2.5}$$
$$= K(z) * PM_{2.5}(z) \quad (7)$$

Using the above relationship, PM2.5 can be retrieved using the extinction coefficient profile obtained from Doppler Lidar, where 'K' is the specific coefficient which depends on refractive index, relative humidity and aerosol size distribution. 'η' is ratio of total mass concentration to $PM_{2.5}$ concentration. Using the selected pairs of known 'σ' and PM2.5, 'K' is determined.

4. Results

Three clear sky cases are considered for the study where the average PM2.5 over 24 hour period on 25th Sep, 2017 is 14.96 µg/m^3, on 31st Mar, 2018 is 1.51 µg/m^3 and on 22nd Apr, 2018 is 5.62 µg/m^3. $PM_{2.5}$ measurement from two sites, one 3 miles north and other 4 miles south-west of lidar site, are considered for comparison with lidar retrieved $PM_{2.5}$ data.

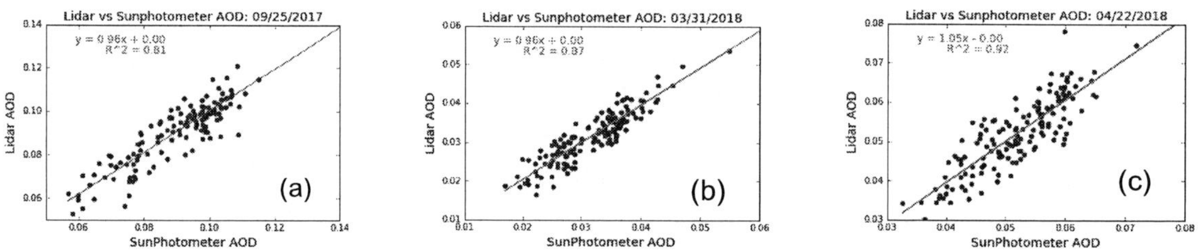

Figure 2. Comparison between Lidar and Sun Photometer AOD at Staten Island, NY (8 am to 6pm) on (a) Sep 25th, 2017, (b) Mar 31st, 2018 and (c) Apr 22nd, 2018

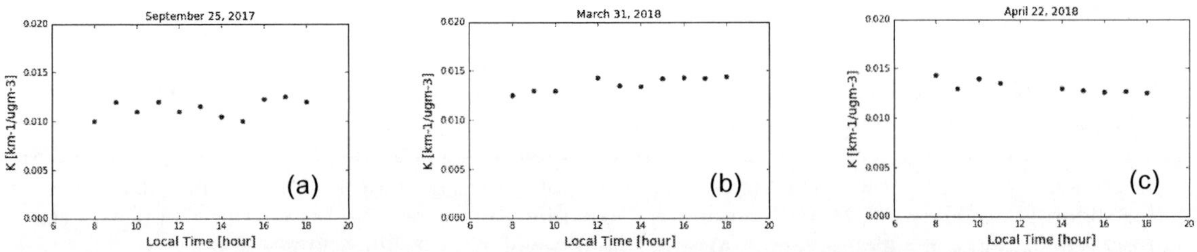

Figure 3. Specific Coefficient (K) for each hour at Staten Island, NY (8 am to 6pm) on (a) Sep25th, 2017, (b) Mar 31st, 2018 and (c) Apr 22nd, 2018

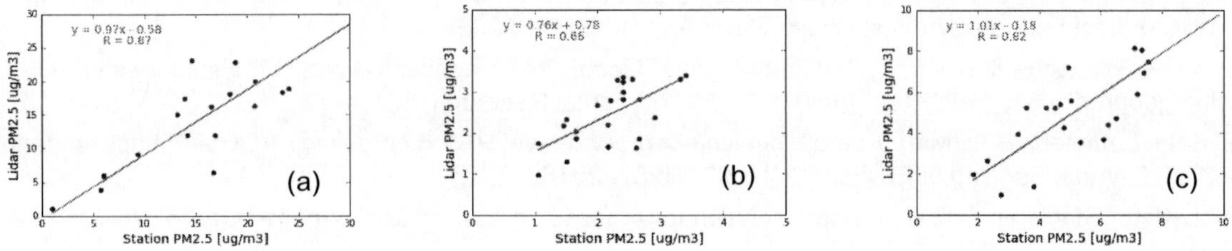

Figure 4. Comparison of surface $PM_{2.5}$ with Doppler Lidar retrieved $PM_{2.5}$ at Staten Island, NY on (a) Sep25th, 2017, (b) Mar 31st, 2018 and (c) Apr 22nd, 2018

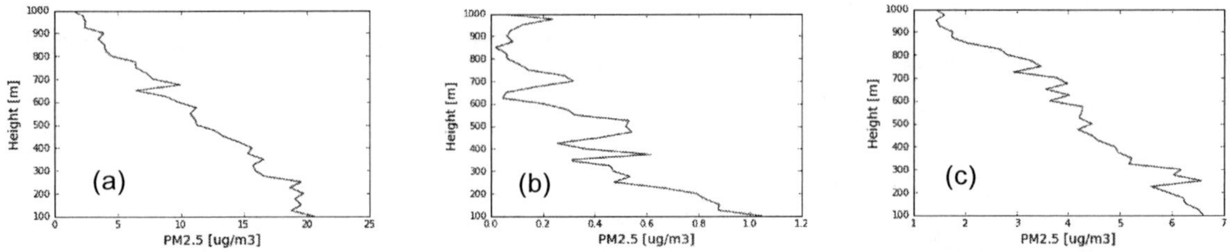

Figure 5. Vertical profile of $PM_{2.5}$ retrieved from Doppler Lidar at Staten Island, NY on (a) Sep25th, 2017 at 8am, (b) Mar 31st, 2018 at 12pm and (c) Apr 22nd, 2018 at 6pm

5. Conclusions

The validation of Doppler Lidar retrieved extinction coefficient is compared with the Sun Photometer in terms of AOD and the R^2 for three cases are found to be 0.81, 0.87 and 0.92 respectively. The specific coefficient 'K' is found to varying from 0.010 to 0.015 $km^{-1}/(\mu g/m^3)$ for all three cases, which approximately indicate that it could be considered constant. The extrapolated surface $PM_{2.5}$ from the Doppler Lidar correlates with measured surface $PM_{2.5}$ with R value of 0.87, 0.66 and 0.82 respectively. The R-value is found to be better when PM2.5 values were relatively higher. Thus, the Doppler Lidar can be used to monitor real time $PM_{2.5}$. In the near future, this technique will be applied across the NY State Mesonet for 3D characterization of $PM_{2.5}$, which will be very helpful for understanding the air quality and its forecasting.

References

[1] M. Bell, L. Ebisu & K. Belanger, "Ambient air pollution and low birth weight in Connecticut and Massachusettts", Environmental Health Perspectives, 115, 1118-1124, 2007.
[2] K. Miller, D. Siscovick, L. Sheppard, J. Sullivan & J. Kaufan, "Long term exposure to air pollution and incidence of cardiovascular events in women", New England Journal of Medicine, 356, 447-458, 2007.
[3] M. Schaap, A. Apituley & R.M.A Timmermans, "Exploring the relation between aerosol optical depth and PM2.5 at Cabauw, the Netherlands", Atmos. Chem. Phys, 9, 909-925, 2009.

[4] M. Green, S. Kondragunta, P. Ciren & C. Xu, "Comparison of GOES and MODIS Aerosol Optical Depth to AERONET AOD and improve PM2.5 mass at Bondville, Illinois", Journal of the Air and Waste Management Association, ISSN: 1096-2247, 2012.

[5] P. Gupta & S.A. Christopher, "Seven year particulate matter air quality assessment from surface and satellite measurements", Atmos. Chem. Phys, 8, 3311-3324, 2008.

[6] X. Hu, W. Lance & Y. Wang, "Estimating ground level PM2.5 concentrations in the southeastern US, using geographically weighted regression", Environmental Research, 121, 1-10.

[7] S. Li, E. Joseph & Q. Min, "Remote sensing of ground-level PM2.5 combining AOD and backscattering profile", Remote Sensing of Environment, 0034-4257, 2016.

[8] J. Klett, "Stable analytical inversion solution for processing lidar returns", Applied Optics, Vol 2, No. 2, 1981.

Data Processing Technique For the All-fiber Wind Profiler

G. A. Petrov[a,b], **N. A. Baranov**[a,c]
(a) JSC «BANS»
15, 12, Bolshaya Novodmitrovskaya st., Moscow, 127015 Russia
(b) Saint Petersburg Electrotechnical University "LETI"
ul. Professora Popova 5, 197376 St. Petersburg, Russia
(c) Dorodnicyn Computing Centre, FRC CSC RAS
Vavilov st. 40, 119333 Moscow, Russia
g.petrov@ians.aero

Abstract: Wind profiler based on continuous laser source is considers in this work. This lidar allows to solve the task of wind profile restoration up to 300 m altitude. Hardware level signal processing technic is developed by JSC «BANS». Increasing accuracy and speed of wind parameters calculation signal processing technics have been studied in this research.

Keywords: Lidar, Wind field, Wind parameters, Wind speed

1. Introduction

This paper considers the new technique of data processing for a wind lidar profilometer based on a continuous laser source. This lidar allows to solve the task of wind profile restoration up to 300 m altitude.

An effective means for determining the wind profile of the surface layer of the atmosphere is a Doppler wind lidar. Currently, the most widely spread lidars are operating with infrared laser beam at a wavelength of 1.54 microns. The lidar beam is scattered by aerosol particles which are present in the atmosphere.

VAD is the atmosphere scanning modes by the lidar laser beam. VAD scanning technic is used in wind profiling lidar. This mode provides aggregated data reception on the values of the radial projection of the wind speed in the measurement direction. This data is further converted into the value of the full vector of wind speed for all heights of scanning [1].

The lidar receives a signal backscattered from the aerosol particles. During this process the Doppler shift of receiving signal occurs. Since the aerosol particles are moved under the effect of an air flow their velocity corresponds to the wind parameters.

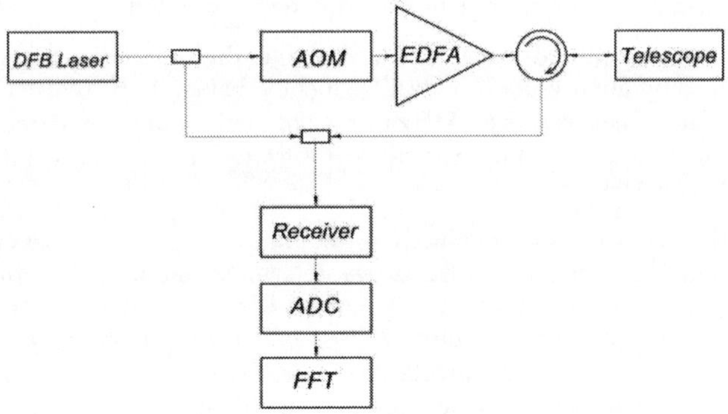

Figurc 1. Optical scheme with the local oscillator

The problem is the frequency of analyzed signal doesn't depend on the wind velocity direction. No matter if the wind velocity vector directs to or from the laser emission source the frequency stays the same. So, to solve such uncertainty in the lidar optical schemes local oscillator is used to change initial frequency of the signal. Example of such lidar optical scheme is shown on figure 1.

2. Methodology

In [2], the lidar architecture based on the quadrature reception of signals is described. This scheme makes it possible to build an optical scheme without a local oscillator. In this case, an element of the phase shift for the reference optical signal is needed in the optical scheme. A fiber optic element is used to form a phase shift of a quarter wavelength.

Fiber-optic scheme of the lidar is shown on figure 2.

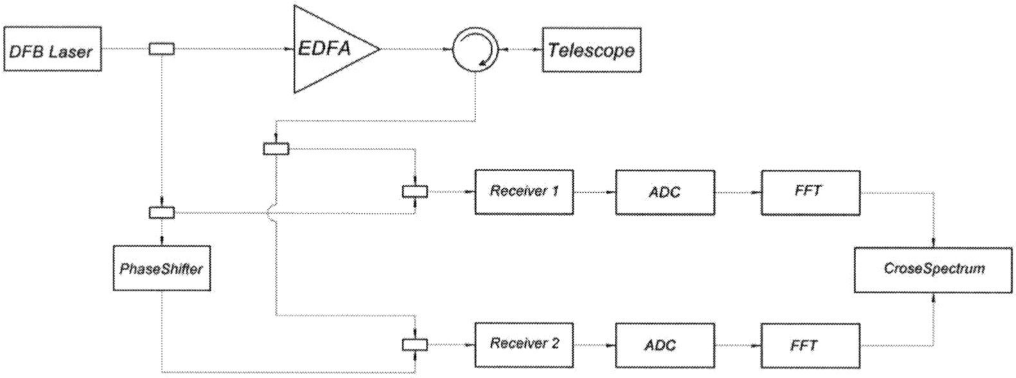

Figure 2. Scheme based on optical hybrid receiver

The backscattered radiation enters the optical hybrid receiver. At the output there are two pairs of optical signals for connection to two balanced photodetectors. Let's introduce the following notations for optical signals: S_b – reference signal without phase shift; S_{bs} – reference signal with a phase shift of a quarter wavelength; D – backscattered signal.

Photodetectors receive two mixtures of signals: $S_b + D$, $S_{bs} + D$. At the outputs, electrical signals are formed: E_b and E_{bs}:

$$E_b = S_b + D; \qquad E_{bs} = S_{bs} + D. \tag{1}$$

Then, mixture of this signals goes to the photodetector where the interference occurs with light intensity alternation. The Doppler shift determines the frequency of the new mixed signal that is transformed to the low-frequency envelope by the photodetector output.

Optical receiver redirects amplified signal to the analogue-digital converter (ADC). ADC sampling frequency is determined by utmost Doppler shift frequency that could be registered with lidar measure system. The Doppler shift frequency is 64 MHz in case of determining a maximum wind speed 50 m/s (if the direction of measurements and the direction of the wind are similar). In this case, the sampling rate should be at least 128 MHz.

The task of determining the wind speed reduces to finding the Doppler shift frequency of the received backscattered radiation. The Doppler shift frequency determines the frequency of the time beats of the signal at the output of the photodetector. The digitized and sampled signal from the photodetector output enters the digital data processing circuit. First of all, the spectrum of the received signal is calculated, after which the successively calculated spectra are summed to increase the signal-to-noise ratio. The addition of signal spectra allows statistical accumulation of the measurement information.

The signal spectrum E_b is calculated by the equations:

$$F_b = FFT(E_b); \qquad F_{bs} = FFT(E_{bs}), \tag{2}$$

where the operator $FFT()$ – Fast Fourier Transform.

To determine the Doppler frequency, two cross-spectra are preliminarily calculated. These two cross spectra are transformed into one total cross-spectrum. The calculation uses real and imaginary parts of the spectra of interference signals. Cross-spectrums C_1 and C_2 are calculated by the equations:

$$C_1 = \mathrm{Im}(F_b) \cdot \mathrm{Re}(F_{bs}); \quad C_2 = \mathrm{Im}(F_{bs}) \cdot \mathrm{Re}(F_b). \tag{3}$$

Signs of the calculated cross-spectra are uniquely determined by the sign of the frequency shift, i.e. wind direction. In this case, the signs of the cross-spectra themselves are opposite. Therefore, the total cross-spectrum is calculated by the equation:

$$C_d = C_1 - C_2. \tag{4}$$

The total cross-spectrum allows for amplitude summation of the two components. The main dependence of the sign of the cross-spectrum on the sign of the frequency shift is retained. Figure 3 shows the results of the calculation of the total cross-spectrum and its components for multidirectional wind speeds.

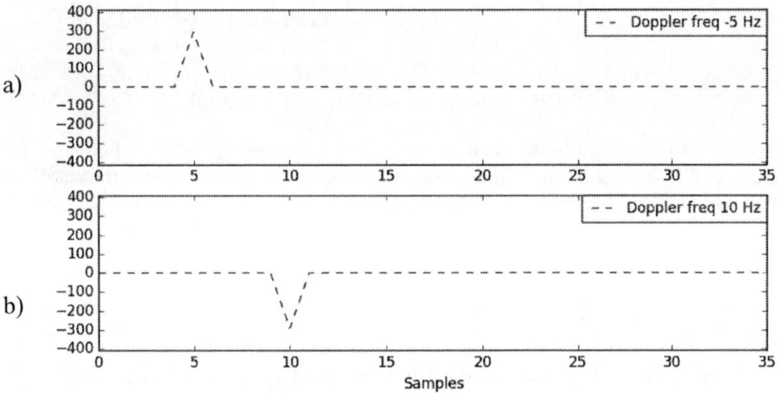

Figure 3. a) Doppler frequency ∂f= -5 Hz, b) ∂f=10 Hz

The simulation results showed that the amplitude characteristics of the total cross-spectrum and the power spectrum coincide (Fig. 4). However, the analysis of the power spectrum allows us to infer only the magnitude of the Doppler frequency shift, while the total cross-spectrum preserves information about the sign of the frequency shift, i.e. wind direction.

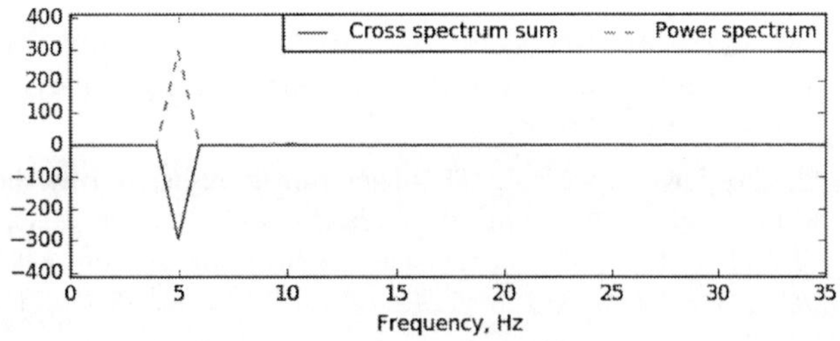

Figure 4 Comparison of the total cross-spectrum and power spectrum

The disadvantage of the described above model without a local oscillator is the need for two separate computational modules for simultaneous calculation of cross-spectra C_1 and C_2.

Combining optical envelopes without shifting E_b and with an offset E_{bs} in a complex field makes it possible to use only one FFT module for calculating a single spectrum:

$$F_c = \mathrm{FFT}(E_b + i \cdot E_{bs}) \tag{5}$$

The fiber-optic scheme of the lidar, which realizes this approach, is shown in Fig. 5.

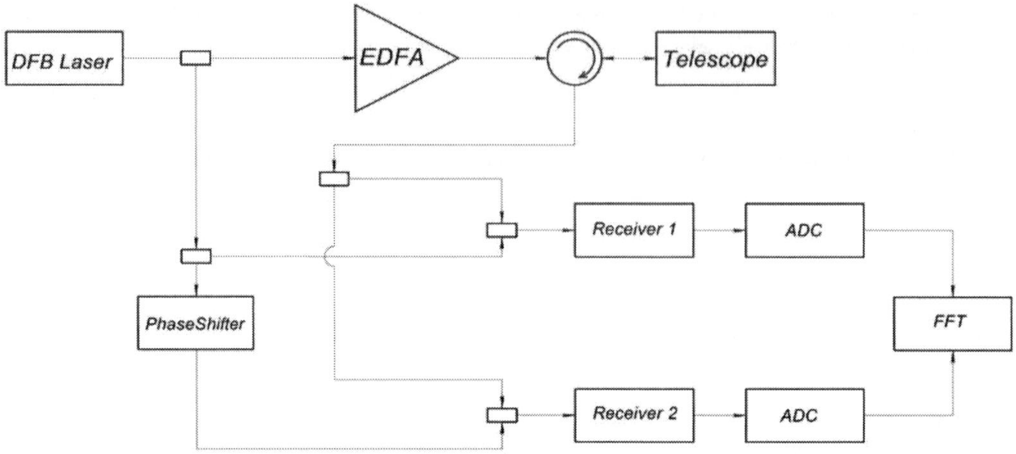

Figure 5. Optical scheme with one FFT module

A characteristic feature of the resultant spectrum F_c is the asymmetry with respect to the maximum detected frequency. Such a spectrum, unlike the classical power spectrum, makes it possible to determine both the positive and negative Doppler shifts.

3. Results

To test the efficiency of the described data processing techniques, a laboratory bench was assembled.

Laser beam is back scattered by the air flow. The wind speed and direction can be adjusted. The scheme as in fig. 2 was mounted. Received light is registered by two balanced photodetectors. The outputs of the photodetectors are connected to the oscilloscope.

On the oscilloscope we can see the electrical signals from the receivers of optical radiation. Two signals correspond to two channels of the oscilloscope. In Fig. 6 shows two screen-shots from the oscilloscope screen.

It is necessary to make sure that the direction of wind can be uniquely determined from the data recorded by two photodetectors.

The sign of the total cross-spectrum is uniquely determined by the sign of the phase difference, and hence the direction of the frequency shift. The above experiment clearly proves the prospects of using the built on an optical hybrid fiber optic scheme.

Also in fig.6 we can see the power spectrum. This spectrum is registered by the scheme, shown in fig.1. It can be seen that power spectrum doesn't depends on wind direction. Also, it is also significant that the magnitude of the pick on the power spectrum is the same as the magnitude of the cross-spectrum.

Using the proposed optical scheme based on one FFT unit allows to simplify the path of digital data processing, to perform joint processing of signals and to restore information about the sign of the frequency shift of optical radiation.

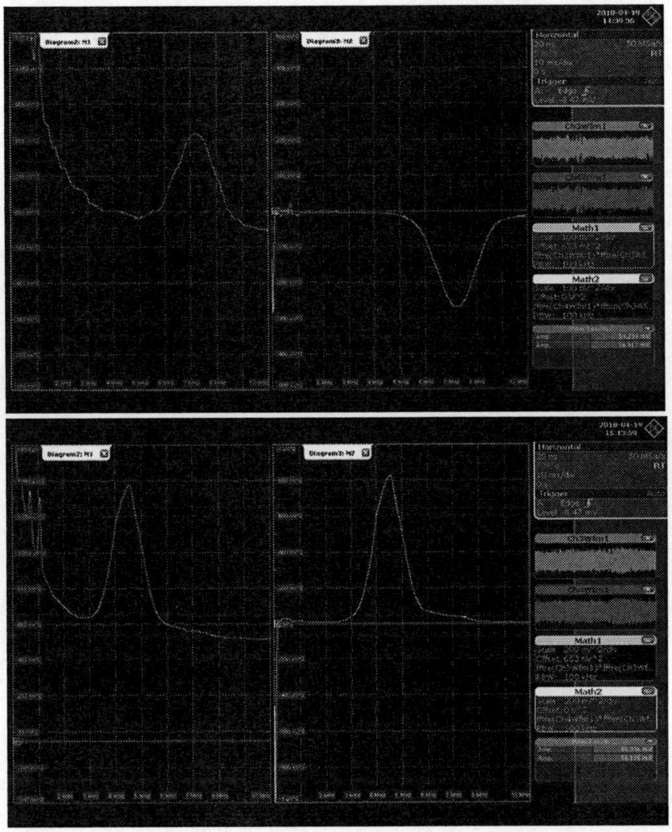

Figure 6. Wind lidar signals—power spectrum (left) and cross-spectrum (right).
The wind direction was opposite (up and down).

4. Conclusions

This study shows the joint signal processing in one FFT module, which reduces the computational complexity and improves the energy efficiency due to joint processing of correlated signals.

This spectrum analyzing makes it possible to determine both the positive and negative Doppler shift. Described technique:

- simplifies the scheme of signal processing;
- facilitates the search for Doppler frequency;
- reduces computing requirements.

Acknowledgements

The work was partially supported by the Russian Foundation for Basic Research, project 16-07-01072.

5. References

[1] Christian Werner, Dr. Claus Weitkamp, 2005: Lidar. Range-Resolved Optical Remote Sensing of the Atmosphere, Springer New York, pp 325-354.

[2] Cyrus F. Abari, Anders T. Pedersen, and Jakob Mann "An all-fiber image-reject homodyne coherent Doppler wind lidar" Optics Express Vol. 22, Issue 21, pp. 25880-25894 (2014)

Solar background noise for spaceborne lidar on specific orbits

Chuanliang Zhang (a), Xuejin Sun (a), Shijun Zhao (a), and Yanwen Liu (a)
(a) College of Meteorology and Oceanography, National University of Defense Technology

Abstract: Solar background noise is one of the main noise for the measurement of spaceborne lidar in daytime. For spaceborne Doppler wind lidar (DWL), the solar background radiation would introduce random error. In this paper, we quantified the global distribution of solar background noise on 355 nm and 532 nm using radiative transfer model (RTM) libRadtran for spaceborne DWL on specific orbit, and the orbit of Aeolus and International Space Station (ISS) was taken as an example. The bidirectional reflectance-distribution function (BRDF) was used to define the anisotropy surface reflectance on land, and the modified Cox and Muck model was used on ocean. The solar background noise varies with seasons was also taken into consideration. The results shows the solar background noise have a big difference on different orbits.

Discussion under generation range of non-diffractive beam transformed by annular beam in random media

Xiafukaiti Alifu (a), Ziqi Peng(a), Tatsuo Shiina (a) and Hiroaki Kuze (b)

(a) Graduate School of Science and Engineering, Chiba University
(b) Center for Environmental Remote Sensing, Chiba University
Chiba University 1-33 Yayoi-cho, Inage-ku, Chiba-shi, Chiba 263-8522 JAPAN
xapkat_alip@chiba-u.jp

Abstract: When an annular beam is used as a lidar transmitting beam in the atmosphere, it can make the influence of the atmosphere fluctuation smaller than a Gaussian beam, because the annular beam can self-transform to a non-diffractive beam. However, in general, light is hard to propagate at long distance in random media, such as cloud or haze, due to strong scattering of particles. In this paper, the annular beam (diameter: 40 mm) was propagated in random media (diluted processed milk, milk fat: 1.8%) and the propagated waveform was detected by a narrow view angle of 5.5 mrad. The non-diffractive beam was successfully generated at the short distance of 3 cm ~ 5 cm in random media with high media concentration of 2% ~ 23%. It was generated in wider concentration range at shorter distance in random media with higher concentration. Furthermore, the relation between concentrations and transmittance value under the generation range of non-diffractive beam was discussed.

Keywords: Lidar transmitted beam, Annular beam, Random media, Non-diffractive beam.

1. Introduction

Cloud droplets are comprised of aerosols and trace gases in the atmosphere. To explain the interaction between the aerosols and cloud droplets is one of the important challenges in aerosol and cloud research fields [1]. Light detection and ranging (LIDAR) system is one kind of optical sensing which uses light to measure the target [2]. However, surface layer of cloud was only obtained by lidar measurement because the transmitted beam could not penetrate over the 200 m depth in cloud [3]. In general, transmitted beam will gradually diminish during its propagation in random media such as cloud or dense fog, because of the influence from the fluctuation, scattering and absorption [4]. These effects destroy the propagation stability of light and decrease the transmission distance in the media. Therefore, it is hard to obtain more information from the deeper area of the target itself. Improving the propagation stability to increase the transmission distance of light in random media is important technique for achieving a long-range measurement in remote sensing field.

The non-diffractive beam was first studied by Durnin et al [5]. An ideal non-diffractive beam which has an intensity distribution like the Bessel function. The important optical property of the non-diffractive beam is self-recovery which the intensity distribution can keep while this beam propagates at a long distance even if it encounters an obstruction [6]. Most energy is included in its main lobe. Thus, transforming the waveform of light into a non-diffractive beam is a useful method for increasing the propagation stability of light in random media. Several methods were presented for generation of the non-diffractive beam [7]. One method for generating the non-diffractive beam is to propagate an annular beam in free space [8]. The annular beam self-transforms into a non-diffractive beam when it propagates at long distance in air. For this reason, the annular beam was applied for our inline type micro-pulse lidar system as a transmitted beam [9]. The result showed that the influence from the atmosphere fluctuation became smaller than that of a Gaussian beam case. However, for measurement of cloud, transmitted beam (annular beam) was still strongly scattered. Then, we focused on how to generate the non-diffractive beam during the propagation of annular beam in random media. We conducted the lab-base experiment to propagate the controlled light into the random media. In our previous work, the non-diffractive beam was successfully generated by the propagation of annular beam (40 mm diameter) at a few tens centimeters' distance (10 cm ~ 30 cm) in random media (0.3% ~ 1.2% concentrations of diluted processed milk) [10]. This non-diffractive beam was constructed by forward scattering lights. As a result, we found that when the annular beam propagated at shorter distance in higher concentration random media, the higher center peak of non-diffractive beam could be generated [11].

The aim of this paper is threefold importantly. First, we focus on the generation of the non-diffractive beam in random media with higher concentration which is close to nimbostratus and human tissue. Thus, the propagation of the annular beam was tested at shorter distances (3 cm ~ 5 cm) in random media. Second, we examined the characteristic change in generation of non-diffractive beam. Therefore, the relationship between the center intensity ratio and media concentration was analyzed by obtained experiment results (3 cm ~ 30 cm). Lastly, by analyzing the relation between media concentrations and transmittance of beam, the generation range of the non-diffractive beam was discussed.

2. Experimental system

Figure 1 shows the schematic diagram of the experimental system. A diode pumped solid state (DPSS) pulse laser with 4.6 kW peak power, 532 nm wavelength, 2 ns pulse width and 10 kHz repetition frequency was utilized. A neutral density filter was used to adjust the optical intensity. Axicon prisms with large zenith angle were applied. It maintains the polarization property of the incident beam. The annular beam was converted by a Gaussian beam with this Axicon prisms. It has an external diameter of 40 mm with a ring width of 3 mm. As shown in Figure 2, the media tanks of 3 cm and 5 cm distance were used for generation of the non-diffractive beam in higher media concentration. A diluted processed milk (milk fat: 1.8%, casein: 2.4%) with 0% ~ 30% concentration was used as a random media. The main scattering particles are a milk fat and a casein in this random media. The scattering property of the random media was decided by the milk fat because the casein is much smaller than the milk fat. The absorption was disregarded due to the small absorption coefficient (0.0066 cm-1) in this random media of 30% concentration [12]. To measure the particle size of milk fat in this random media, we implemented the Brownian motion experiment. As a result, the average particle size of milk fat was around 1.1μm in this random media.

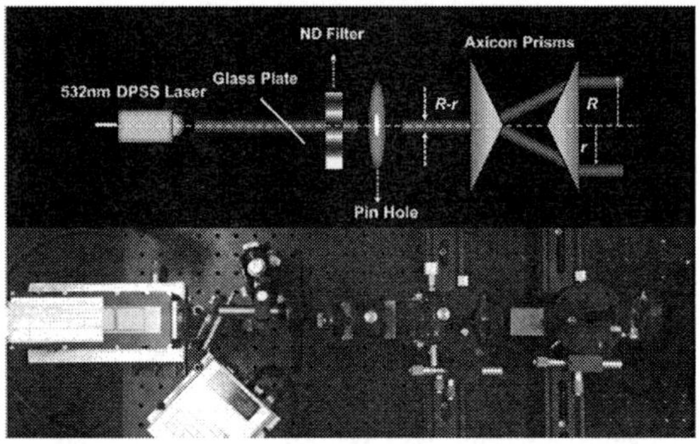

Figure 1. Schematic diagram of the experimental system

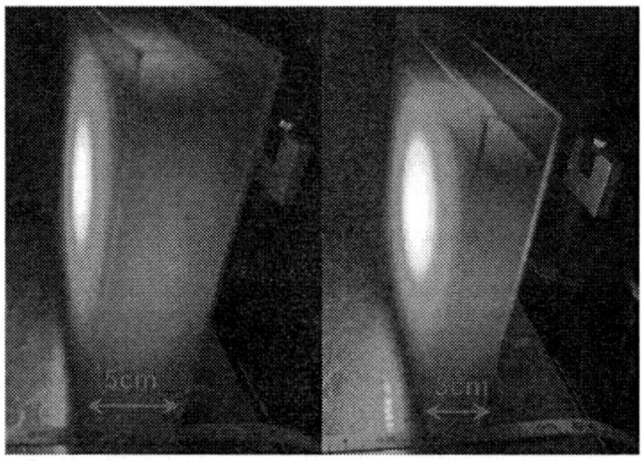

Figure 2. Media tank of 5 cm (left) and 3 cm (right) in width

The optical receiver unit consists of a collimated lens and a multiple mode optical fiber. The forward-scattered light in random media was received by its narrow view angle of 5.5 mrad selectively. The receiving unit was set on an auto mechanical stage with the small step of 0.1 mm on a horizontal-axis. A photo multiplier tube (PMT) was used as a detector to obtain the weak optical signal through the random media, and a sampling oscilloscope was used to monitor the pulsed electric signal from PMT.

3. Results and Discussion

Figure 3 shows the received scattering waveforms (the intensity was normalized at the position of 20 mm) under the media concentration that appeared the highest peak of the non-diffractive beam. At both scattering waveforms, the non-diffractive beam (6 mm in diameter) was generated at the center of the waveform. At the distance of 5 cm and 3 cm, the generation of this non-diffractive beam was achieved by annular beam propagated in random media with the concentration of 5.0% and 22.0%, respectively. These concentrations are close to the density of nimbostratus and human tissue. By analyzing the variation of the concentration at different propagation distances (3 cm ~ 30 cm), the concentration to generate the non-diffractive beam was small change in the 10 cm ~ 30 cm propagation distance, while it became large change in the 3 cm ~ 5 cm propagation distance. The non-diffractive beam on the waveform through random media was derived by forward scattering lights. The surrounding curve of the waveform was the result of multiple scattering lights. The center intensity ratio was calculated by the ratio of the maximum intensities between non-diffractive beam and multiple scattering lights.

Figure 4 shows the variation between the center intensity ratio and concentrations at different propagation distances. The highest center intensity ratio was 0.84%, 1.22% and 1.85% at media concentration of 0.4%, 0.6% and 1.0% in their propagation distance of 30cm, 20cm and 10cm, respectively. The highest center intensity ratio was 1.94% and 2.41% at concentrations of 5.0% and 22.0% in the media tank width of 5 cm and 3 cm, respectively. Thereinto, the highest peak of non-diffractive beam appeared at the propagation distance of 3 cm was almost 2.5 times higher than that at 30 cm propagation distance. The center part was subjected to stronger forward scattering at higher concentration in a short-distance random media. The change of the highest center intensity ratio at 10 cm ~ 30 cm propagation distances was increased strongly. It became smooth at 3 cm ~ 5 cm propagation distances. At the propagation distance of 3 cm, 5 cm, 10 cm, 20 cm, 30 cm, the non-diffractive beams were generated in concentration ranges of 16.0% ~ 23.0%, 2.0% ~ 6.0%, 0.6% ~ 1.2%, 0.3% ~ 0.8%, 0.2% ~ 0.5%, respectively. For comparison, the concentration range of non-diffractive beam appeared at propagation distance of 3 cm was 23 times wider than that at propagation distance of 30 cm. That is, the generation range of the non-diffractive beam at short-distance became wider than the that of longer propagation distance. We discussed this condition result below.

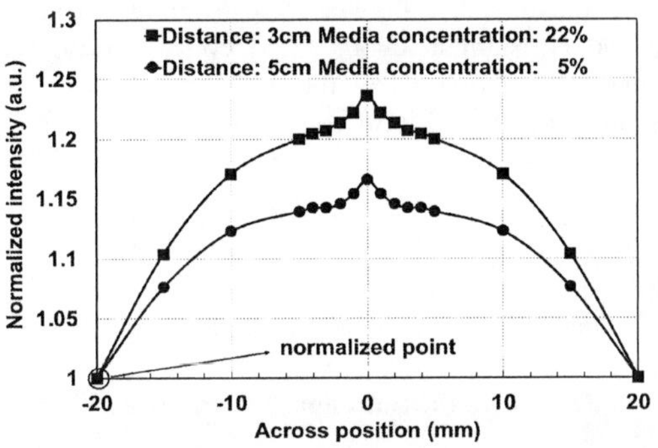

Figure 3. Scattering waveforms of annular beam at propagation distance of 3 cm and 5 cm in random media.

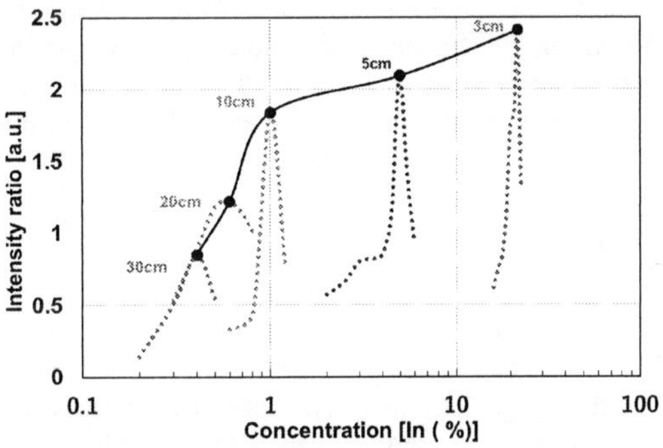

Figure 4. Analysis of center intensity ratios against media concentration at each propagation distance.

The relation between the media concentration range and beam transmittance was analyzed. Figure 5 shows the experimental results of beam transmittance against media concentrations. Given the green line (start of non-diffraction beam generation) and blue line (end of non-diffraction beam generation) in Fig 5, the feature characteristic of the concentration range became clear. The red line indicates the non-diffractive beam with highest intensity ratio and it was near the end blue line. The generation range of non-diffractive beams in longer distances (i.e. >10 cm) appeared to have a large change on the transmittance curve. On the other hand, the generation of non-diffractive beams in shorter distances (i.e. <10 cm) appeared to have a small change of the transmittance curve. It is the reason that the shorter propagation distance generates the non-diffractive beam in wider media concentration range. Moreover, the gap of transmittance between red and green line was same and constant at every fixed concentrations.

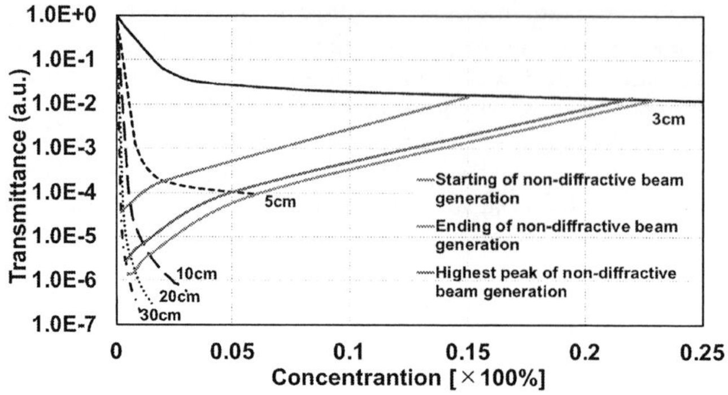

Figure 5. Relation between beam transmittance and media concentrations in generation range of non-diffractive beam.

4. Summary

In this study, we have propagated the annular beam at short distance 5 cm and 3 cm in random media, the highest peak of non-diffractive beam appeared in 5% and 22% media concentration and their intensity ratio were 2.1% and 2.4%, respectively. These media concentrations are close to densities of nimbostratus and human tissue. Then, we analyzed the variation between the center intensity ratio and media concentrations to generate the non-diffractive beam by obtained experiment results (3 cm ~ 30 cm). It was proven that the center intensity ratio of non-diffractive beam was higher in shorter distance, because the center part gets stronger forward scattering effects in the higher concentration range media. Lastly, we analyzed the relation between the transmittance of beam and media concentration. It was deduced that the concentration range which generates non-diffractive beam in the media became wider at shorter propagation distance. Moreover, the gap of transmittance between starting and red line for generation of the non-diffractive beam was same and constant at every fixed concentrations. The propagation effect of the annular beam in random media is applicable for the thick atmospheric cloud remote sensing, and human tissue pathology examination in medical field.

5. References

[1] T. L. Anderson, D. S. Covert, and R. J. Charlson, "Cloud droplet number studies with a counterflow virtual impactor," J. Geophys. Res. 99(D4), 8249-8256 (1994).

[2] M. A. Lefsky, W. B. Cohen, G. G. Parker, and D. J. Harding, "Lidar Remote Sensing for Ecosystem Studies.," Bioscience 52(1), pp.19-30 (2002).

[3] E. P. Zege, I. L. Katsev, I. N. Polonsky, "Analytical solution to LIDAR return signals from clouds with regard to multiple scattering," Applied optics B 60, 345-353 (1995).

[4] A. Deepak, U. O. Farrukh, and A. Zardecki, "Significance of higher-order multiple scattering for laser beam propagation through hazes, fogs, and clouds," Applied optics 21(3), 439-447 (1982).

[5] J. Durnin, J.J. Miceli Jr. "Diffraction-free beams," Phys. Rev. Lett. 58(15), 1499-1501 (1987).

[6] V. Garces-Chavez, D. McGloin, H. Melville, W. Sibbett, and K. Dholakia, Nature 419(6903), 145-147 (2002).

[7] A. Vasara, J. Turunen, and A. T. Friberg, J. Opt. Soc. Am. A 6(11), 1748-1754 (1989).

[8] H. T. Eyyuboglu., Y.E. Yenice, Y. K. Baykal, "Higher Order Annular Gaussian Laser Beam Propagation in Free Space," Opt. Eng. 45(3), 038002 (2006).

[9] T. Shiina, K. Yoshida, M. Ito and Y. Okamura, "In-line type micropulse lidar with an annular beam: experiment," Appl. Opt. 44(34), 7407-7413 (2005).

[10] Z. Peng and T. Shiina, "Generation control of a non-diffractive beam in random media by adjusting concentration," Optics Communications 391, 94-99 (2017).

[11] Z. Peng and T. Shiina, "Analysis of a non-diffractive beam generated from an annular beam in random media," Optics Communications 402, 349-354 (2017).

[12] D.A. Boas, "Diffuse Photon Probes of Structural and Dynamical Properties of Turbid Media," Theory and Biomedical Applications (Ph.D. Thesis), Univ. of Pennsylvania. (1996).

End-to-end simulation of a coherent Lidar in a turbulent wind field

Cariou Jean-Pierre, Thobois Ludovic, Combrexelle Sébastien
LEOSPHERE
14-16 rue Jean Rostand, 91400 Orsay, France

Abstract: For operational applications such as wind resource assessment or turbine control, to warranty the performances in velocity accuracy and data availability is required. A Lidar simulator (Simulid™) has been developed for assessing the velocity accuracy and the data availability of pulsed coherent wind Doppler Lidar system worldwide. This simulator takes into account both the Lidar parameters (instrument hardware, signal processing, beam propagation…) and the aerosol properties. Recently, Simulid™ has been coupled with a 3D wind model (Turbsim) to perform simulated wind measurements in any representative wind field, including turbulence. We present the methodology and the validation process. Velocity accuracy is compared with experimental tests performed in a frozen propagation medium and the representativeness of the simulated wind measurements is discussed.

Keywords: Coherent Laser Radar, Lidar Instrumental Simulator.

1. Introduction

Accuracy and availability of atmospheric sensors are of critical importance, especially when their measurements are the inputs of numerical models or decision-making processes. For example the idea of Lidar-assisted wind turbine control is spreading out within the wind industry as it opens the door to new possibilities in improving wind turbine design and efficiency and reducing costs. In order to develop appropriate Lidar-assisted control strategies as well as to evaluate their benefits and mitigate their limitations, the support of realistic simulation tools is needed. In this paper the methodology developed to simulate realistic Lidar wind measurement is presented.

2. Ideal Lidar measurement

It is well known that coherent Doppler Lidars cannot measure wind locally but perform instead a spatial averaging of the wind speed within the volume of atmosphere that is spanned by the pulse during acquisition [1]. More precisely, given a point of measurement $\vec{x} = [x, y, z]$, the radial wind speed seen by the Lidar can be approximated by the following convolution:

$$v_t(\vec{x}) = \int_{-\infty}^{\infty} \varphi(r) \times v_t(r, \vec{x})\, dr \quad (1)$$

where $\varphi(r)$ is the range weighting function of the Lidar derived from its pulse shape and apodization window and $v_t(r, \vec{x})$ is the local radial wind speed given by the projection of the wind field at a given timestamp t $\vec{\boldsymbol{V}}_t$ onto the unit vector of the Lidar line of sight (LOS) $\vec{\boldsymbol{n}}$**,** i.e.

$$v_t(r, \vec{x}) = < \vec{\boldsymbol{n}}, \vec{\boldsymbol{V}}_t(r\vec{\boldsymbol{n}} + \vec{x}) > \quad (2)$$

where <·,·> stands for the standard dot product (see Figure 1).

Several existing aero-elastic simulation tools use this approximation to simulate Lidar measurements. However, this approximation corresponds to an ideal Lidar measurement as it is not sensitive to the effects of atmospheric conditions, hardware and software real capabilities (e.g., detector noise, spectral resolution …). Both of these aspects are taken into account in the methodology developed in LEOSPHERE, firstly, by modeling the influence of the atmosphere/Lidar configuration on the backscattered power and, second, by reproducing the embedded signal processing on realistic Doppler spectra.

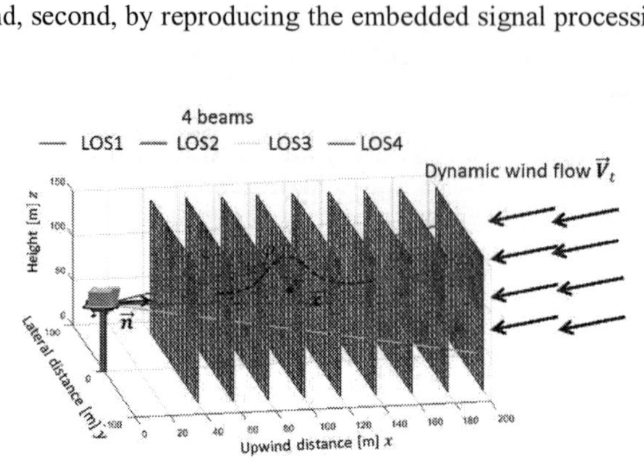

Figure 1: Four-beam Lidar in a wind-field

3. Methodology for realistic Lidar measurement simulation

3.1. Carrier-to-noise ratio simulation via Simulid™

Pulsed Doppler Lidars as any other atmospheric sensors have measurement performances that depend on local weather conditions and Lidar configurations. Both influence the so-called Carrier-to-noise ratio (CNR) which directly reflects the amplitude of the signal backscattered to the Lidar. Leosphere has developed a CNR simulation tool, Simulid™ [2], which takes into account both the instrumental and the atmospheric contributions with two distinct models (steps 1-2 of Figure 2):

- The instrumental model simulates the beam propagation based on a Gaussian model. It accounts for the instrument hardware (optics/detector efficiency, lens aperture, pulse energy and duration, noise power) and signal processing specificities (accumulation time, FFT size)
- The atmospheric model derives aerosol extinction α and backscatter β coefficients from a given atmospheric condition (either user-defined or taken from the Monitoring Atmospheric Composition & Climate (MACC model)

These two contributions are then combined to compute a CNR profile as follows

$$CNR(r) = \overbrace{F(r)}^{Instrumental\ function} \times \underbrace{\beta(r)e^{\int_0^r \alpha(r')dr'}}_{Atmospheric\ function} \quad (3)$$

where r denotes the distance between the Lidar and the point of measurement. Simulid™ also provides a CNR threshold to filter data and hence assess the availability of the measurement [2].

3.2. Doppler spectra simulation

In real Lidar systems, measured radial wind speeds are the outputs of a signal processing chain which consists in translating the received signal into a Doppler spectrum through a Fourier transform and then applying a Doppler frequency estimator. These two steps are reproduced within our simulation tool. Realistic Doppler spectra are obtained by using a generative model, called Feuilleté model [3-4]. This model takes as inputs the CNR and the radial wind speed profiles within the neighborhood of the measurement point to derive the statistics of the resulting Doppler spectrum (step 3 of Figure 2). The Doppler frequency estimator used in LEOSPHERE Lidars, based on a maximum likelihood estimator [3-4], is then applied on the simulated spectrum (step 4 of Figure 2). The overall simulation scheme is described in Figure 2.

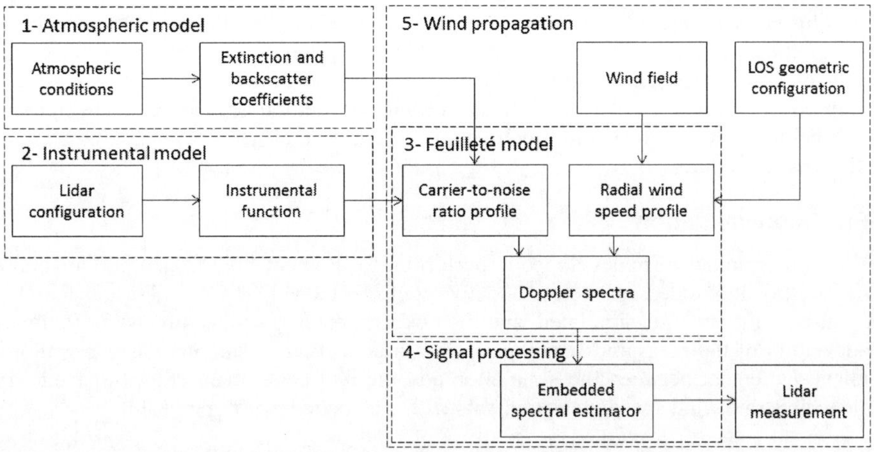

Figure 2: General flowchart for realistic Lidar measurement simulation

3.3. Temporal acquisition

The scheme described above deals with the simulation of a single Lidar measurement at a given timestamp. Time series of Lidar measurement are obtained by propagating the wind field throughout time. In this study, since 2D temporal wind fields generated by TurbSim [5] are used, wind propagation is performed via the frozen turbulence Taylor hypothesis which states that the wind field is uniformly propagated, i.e.

$$\vec{\boldsymbol{V}}_{t+dt}(x,y,z) = \vec{\boldsymbol{V}}_t\,(x-\overline{U}dt,y,z) \quad (4)$$

where $\overline{U}$ is the mean horizontal wind speed (step 5). However, it is straightforward to adapt our methodology to 3D temporal wind fields. Finally, note that the methodology developed in this study easily deals with the case of multi-beam Lidars (as in Figure 1) by taking into account several additional parameters such as the number of LOS, the LOS order sequence, switching time …

4. Results and validation

4.1. Example of simulated time series

The figure below compares for three different levels of CNR (from high to low, close to the CNR threshold) the simulated Lidar radial wind speed (green) with the true local radial wind speed (blue) and the weighted approximation (red) given in (2).

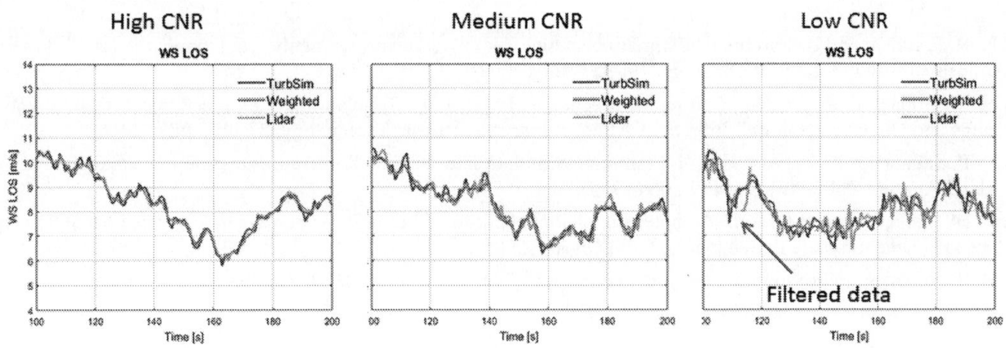

Figure 3: Example of simulated time series of radial wind speed (WS) for different levels of CNR

This figure illustrates the benefits of our methodology over (2): while the weighted approximation is simply a smoothed version of the true local radial wind which is blind to the level of CNR, our approach yields wind speed measurements that are sensitive to the level of CNR as it is observed in practice. Moreover, for a low level of CNR, the developed methodology occasionally produces data with a CNR below the CNR threshold. Such data is an outlier that will be considered as not valid by our data filtering procedure.

4.2. Precision validation

An experimental setup has been developed to perform measurements in a frozen propagation medium, i.e. with scattering motionless particles, under a controlled level of CNR. Figure 4 plots for two different pulse durations the simulated and the experimental velocity precision (defined as the standard-deviation of the measured wind speed over a collection of samples) as a function of CNR. The excellent agreement between the simulation and the real experiment shows that our simulation tool is able to generate wind speed measurements with a realistic level of precision.

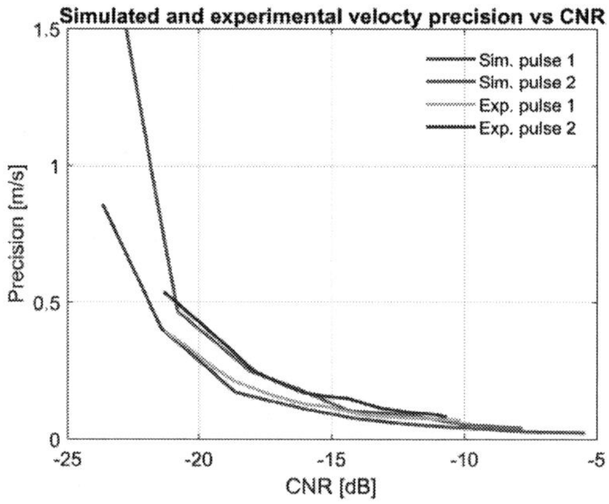

Figure 4: Simulated and experimental velocity precision as a function of CNR for two pulse durations

5. Conclusion

The Lidar simulator developed by LEOSPHERE to simulate the response of a Lidar to a dynamic wind-field that models the entire measurement chain and includes all sources of noise. It provides wind speed measurements that have been shown to be highly representative of real measurements both in terms of availability and precision.

6. References

[1] R. Frehlich, and L. Cornman, "Estimating spatial velocity statistics with coherent Doppler Lidar," Journal of Atmospheric and Oceanic Technology, 19, 355-366 (2002).

[2] M. Boquet, P. Royer, J.-P. Cariou, M. Machta, and M. Valla, "Simulation of Doppler Lidar Measurement Range and Data Availability," Journal of Atmospheric and Oceanic Technology, 33, 977-987 (2016).

[3] P. Salamitou, A. Dabas and P. H. Flamant, "Simulation in the time domain for heterodyne coherent laser radar," Applied Optics vol. 34, 499-506 (1995).

[4] M. Valla, "Etude d'un Lidar Doppler impulsionnel à laser Erbium fibré pour des mesures de champ de vent dans la couche limite de l'atmosphère, " PhD thesis (2005).

[5] B. Jonkman, "TurbSim user's guide: version 1.50," NREL/EL-500-46198 (2009).

Application of piecewise AR model estimation for approximate stationary process in decision of pulse number and range gate

Qizun Zhang (a), He Chen (a), Siying Chen (a), Pan Guo (a), and Yinchao Zhang (a)
(a) Beijing Institute of Technology

Abstract: The algorithm for wind field data inversion of CoherentWind Lidar usually uses PP, PM and ML. The premise of those algorithms is to assume that the echo signal is approximately stationary. If the pulse count is sufficiently large and the range gate is short enough, the signal is considered to be nearly stationary, but increasing the accumulated pulse number will reduce the temporal resolution and reducing the range gate length will increase the inversion error. In order to ensure the resolution of the signal without loss of accuracy, the Bayesian method based on the autoregression (AR) model is used to transform the non-stationary echo signal into a piecewise stationary signal. By using this method the optimal number of accumulated pulses and the optimal range gate division could be obtained effectively, so the resolution and accuracy of the inverse wind field can be ensured at the same time.

Vertical Velocity Statistics and Turbulence Characterization by Coherent Doppler Lidar during Typhoon MAWAR

Xiaochun Zhai(a), Songhua Wu(a, b), Xiaoquan Song(a, b)

(a)Ocean Remote Sensing Institute, College of Information Science and Engineering, Ocean University of China, Qingdao 266100, China.

(b) Laboratory for Regional Oceanography and Numerical Modeling, Qingdao National Laboratory for Marine Science and Technology, Qingdao 266100, China.

**Email: wush@ouc.edu.cn*

Abstract: This paper gives an analysis of CDL investigation of gravity wave and turbulence characteristics during typhoon MAWAR episode over the coastal zone in south China (CZSC). The FFT and wavelet analysis are used to subtract the organized wave-like structure. Radiosonde and ECMWF reanalysis data are used to demonstrate the existent of gravity wave and to explain its mechanism, respectively. The up/downdraft of vertical velocity in this case are analysed from different time periods. A wavelet decomposition technique is used to subtract the gravity wave from raw vertical velocity datasets, and higher-order moments and characteristic scales are also analysed base on gravity wave time series. As a result, comprehensive atmospheric dynamic characteristics and their relationship with gravity wave during MAWAR episode have been studied.

Keywords: Coherent Doppler lidar, gravity wave, turbulence, typhoon

1. Introduction

The coastal zone in South China (CZSC) is one of the regions with the highest level of economic development in China. It borders Nanling mountain to the north and South China Sea (SCS) to the south. Due to its special geographical location, CZSC is one of the areas that most frequently suffer from marine meteorological disasters such as typhoon, rainstorm and sea fog, and is also one of the key areas that influence the short-term climate change of China. However, due to the lack of sufficient spatial-temporal monitoring data, the understanding of the characteristics of land-ocean-atmosphere interaction and its evolution in this area is not sufficient, and the accuracy of weather prediction and forecast is not ideal as well. Therefore, there has been a pressing need for carrying out the field experiments to strengthen the knowledge of atmospheric boundary layer dynamics and thermodynamics processes and to improve the weather and short-term climate prediction. The Marine Meteorological Science Experiment Base at Bohe of Maoming (M2SE2B), located at the CZSC, is the fixed observation site for typhoon research with sophisticated and fully functional equipment. The field experiment was carried out during August - November 2017 at M2SE2B focusing on the spatial-temporal evolution of atmospheric boundary layer and air-sea interaction during typhoon landfalling. This paper presents a case study of the wind field and turbulence observations using coherent Doppler lidar (CDL) during Typhoon MAWAR episode in this experimental campaign.

2. Lidar technology and methodology

Figure 1 shows the sketch map of experimental location and the outfield the experiment at M2SE2 during August-November 2017. The spatial-temporal evolution of signal-to-noise ratio (SNR) and vertical velocity are shown in figure 2. In this case study, organized wave-like structure can be seen from SNR and vertical velocity datasets. The FFT and wavelet analysis are used to subtract the large-scale coherent signal.

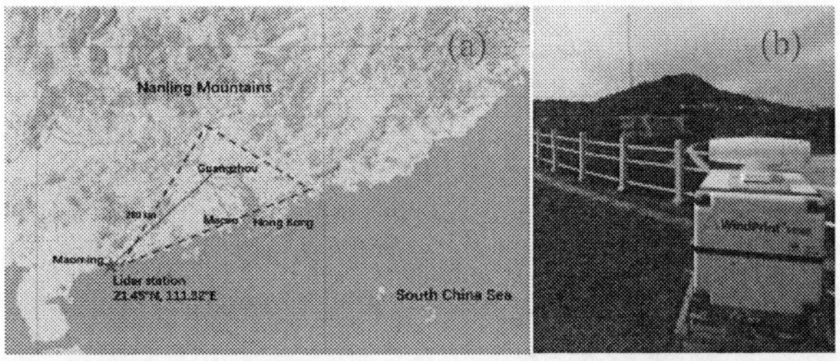

Figure 1. (a) The experimental location (marked with red star) (b) the outfield experiment at M2SE2B.

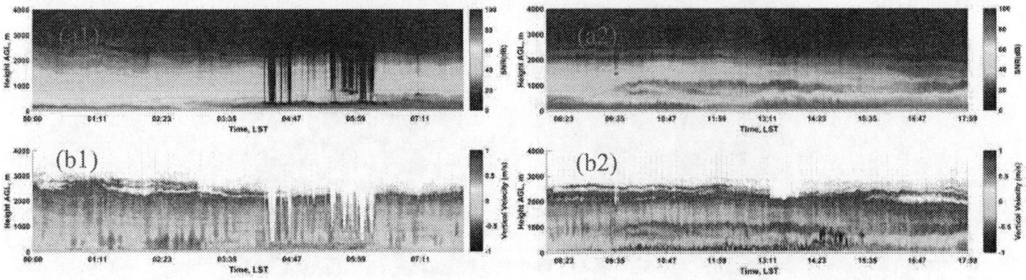

Figure 2. Time Height Intensity of (a1)(a2) SNR (dB) and (b1)(b2)vertical velocity (m/s) at 00-08 (a1,b1) and 08-18 (a2, b2) 02 Sep 2017.

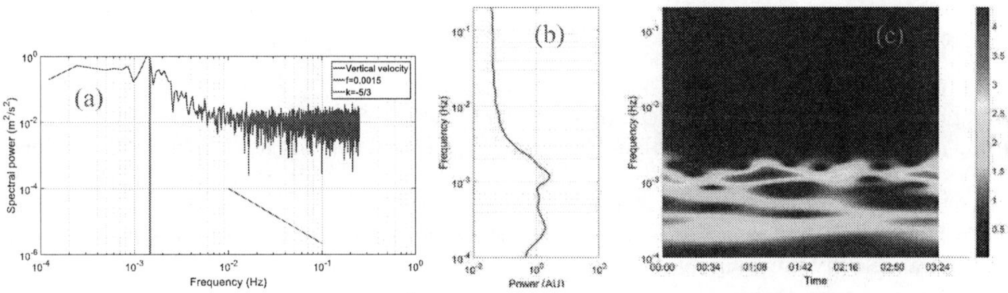

Figure 3. (a) FFT spectral power and (b)(c) wavelet analysis of vertical velocity at 02 Sep 2017: 00:00-04:00 from height 1000 m to 2000 m.

Figure 3 (a) shows that the peak frequency of FFT spectral power is about 0.0015Hz, corresponding to the time period of about 11 min, and an apparent frequency at about 0.0016 Hz using Morlet wavelet analysis can be seen in figure 3 (b)(c). According to the linear mountain wave theory [1], the waves that can propagate vertically in the atmosphere can be derived by the use of Scorer parameter $l^2 \approx N^2 / U^2$,where N is the Brunt-Vaisala frequency, and U is the cross-mountain wind speed. Based on the radiosonde data, a profile of the Scorer parameter is calculated shown in figure 4, a wave with wavelength λ and associated wave number $k = 2\pi / \lambda$ can propagate in the atmosphere if $k^2 < l^2$.

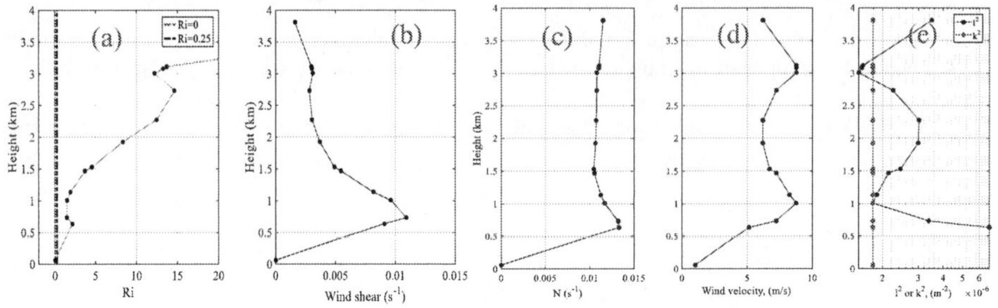

Figure 4. (a) Richardson number (b) wind shear factor (c) Brunt-Väisälä frequency (d) wind velocity (e) derived Scorer parameter (blue) and approximate wave number corresponding to observed waves.

The gravity wave production mechanism can be explained in a sense from ECMWF reanalysis dataset [2]. The cyclone structure of typhoon MAWAR can be seen obviously from figure 5 (a). The experimental location (cross point in figure 5 (b)) has a negative vorticity, which is under the control of a surface high-pressure system (anti-cyclone). The descending motion outside of typhoon can easily result in large-area inverse layer. As a result, the continuously northly wind shown in figure 5 (a) induces the gravity wave, and the relative stationary atmosphere affected by the descending motion outside of typhoon MAWAR may the main cause that the gravity wave can exist at long time period and large spatial area.

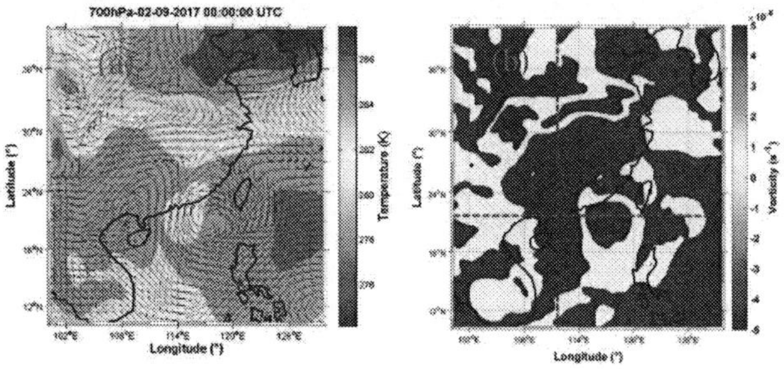

Figure 5. Synoptic conditions derived from the ECMWF ERA Interim reanalysis at 00:00 (a) Wind vectors and temperature (b) vorticity distribution.

3. Results

In order to understand this complex field of atmospheric physics, more updraft and downdraft field observations are required [3]. Figure 6 shows the up/downdraft statistics from 08-18 LST 02 Sep 2017. A lifting aerosol layer can be seen in figure 2 (b), and the vertical velocity near the ground has a distinct difference compared with the ones above. Form the updraft/downdraft statistics shown in figure 6, there also exists obvious difference below 500 m and above 500 m. The organized updraft and downdraft between 500 m and 2000 m in figure 2 (b) corresponds to relative constant occurrence, duration and mean vertical velocity shown in figure 6.

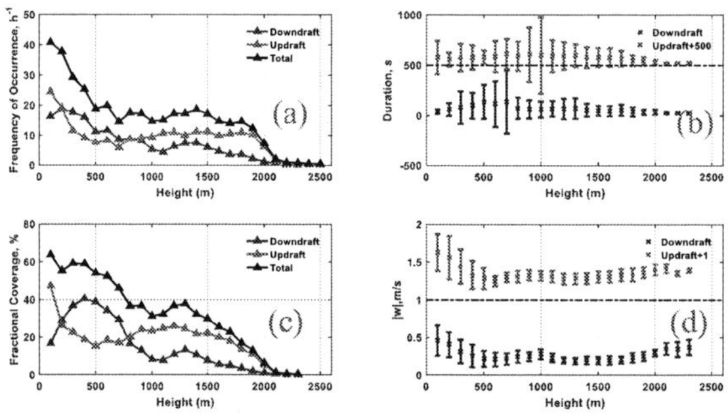

Figure 6. (a) Frequency of occurrence (b) duration (c) fractional coverage (d) mean vertical velocity of downdraft and updraft from height of 100 m to 2500 m during 08-18 LST 02 Sep 2017.

The vertical velocity can be divided into three parts: mean vertical velocity $\overline{w}$, wave structure $\overset{o}{w}$ and fine-scale turbulence w' in this study where the gravity wave is the larger-scale turbulence. A wavelet multiresolution analysis technique is used to decompose the fine-scale turbulence signal and large-scale coherent signal, that is, the gravity wave structure [4]. Figure 7 (a) shows the vertical velocity time series at different iteration process. An example of wavelet decomposition of vertical velocity is shown in figure 7 (b). It is decomposed into (middle) wave and (down) turbulence fluctuation part, respectively.

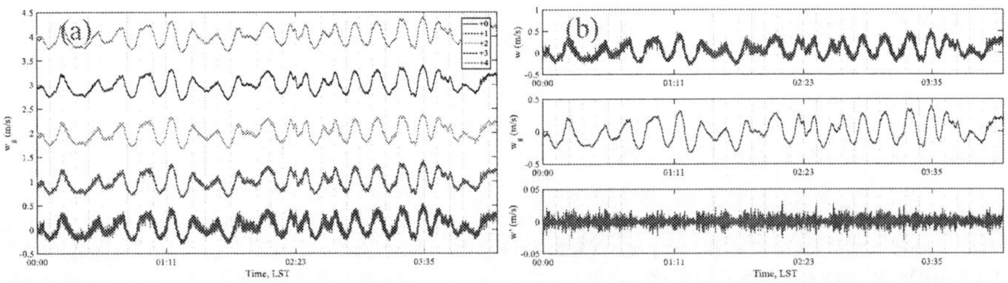

Figure 7. (a) Vertical velocity time series at different iteration process. (b) An example of wavelet decomposition of w wind component of lidar at height of 1000 m at 2017-09-02: 00:00-04:00.

The turbulence characteristic from vertical velocity can be divided into mainly two parts in the present studies. The first one is its higher-order moments, including variance, skewness and kurtosis, respectively [5]. The second is the characteristic time length scales. The contribution of $\mathring{w}$ to total turbulence is much larger than w''s. Figure 8 shows the $\mathring{w}$ turbulence statistics. Larger values of variance shown in figure 8 (a) appear at lower altitudes, which are consistent with the previous analysis. Since the time series of horizontal wind profile are unavailable in this study, the corresponding length scale cannot be obtained, but $T_{Lw} / T_{\lambda w}$ equals to L_w / λ_w at the same height, shown in figure 8 (f) can be analyzed.

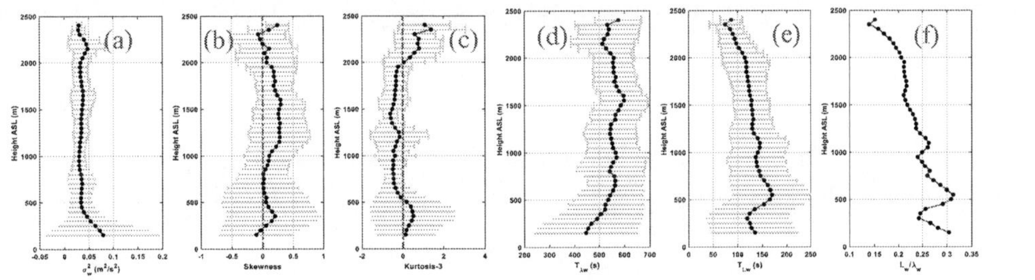

Figure 8. (a) Vertical velocity variance (b) skewness (c) kurtosis-3 (d) peak frequency time (e) integral time scale (f) the ratio of peak frequency scale and integral scale at 00-24 02 Sep 2017.

4. Conclusions

Various atmospheric dynamic processes based on CDL zenith pointing mode have been observed during this experimental campaign, including different atmospheric boundary layer type, the cloud and aerosol layer effect on vertical velocity and turbulence and so forth. This paper focus on a case study during Typhoon MAWAR episode. Organized wave-like structures can be seen from CDL detection. We analyze this case using radiosonde and ECMWF reanalysis data to demonstrate that the large-scale coherent structure is gravity wave. The FFT and Morlet wavelet analysis are used to subtract this structure, corresponding to about 11-min period. The topography effect and the resulting subsidence caused by typhoon may be the cause of long-time period existence of gravity wave structure. The up/downdraft features are specifically analyzed and found that there are significant differences below and above 500 m, and it may result from more composite effect of solar radiation at daytime and wind shear below 500 m. The wavelet decomposition technique is used to exactly separate gravity wave and fine-scale turbulence component. The gravity wave is specially analyzed based on higher-order moments and characteristic scale retrieval procedure. The relationship between gravity wave and different turbulence features will be analyzed in further study.

5. References

[1] Durran, D. R.: Mountain waves and downslope winds, in: Atmospheric Process over Complex Terrain, edited by: Blumen, W.,American Meteorological Society, Boston, 59–81, 1990

[2] Chouza Keil, Fernando, et al. "Vertical wind retrieved by airborne lidar and analysis of island induced gravity waves in combination with numerical models and in situ particle measurements." Atmospheric Chemistry and Physics (ACP)16.7 (2016): 4675-4692

[3] Ansmann, Albert, Julia Fruntke, and Ronny Engelmann. "Updraft and downdraft characterization with Doppler lidar: cloud-free versus cumuli-topped mixed layer." Atmospheric Chemistry and Physics 10.16 (2010): 7845-7858.

[4] Wang, Yansen, et al. "Investigation of nocturnal low-level jet–generated gravity waves over Oklahoma City during morning boundary layer transition period using Doppler wind lidar data." Journal of Applied Remote Sensing 7.1 (2013): 073487.

[5] Lenschow, Donald H., Volker Wulfmeyer, and Christoph Senff. "Measuring second-through fourth-order moments in noisy data." Journal of Atmospheric and Oceanic Technology 17.10 (2000): 1330-1347.

Lidar Observation of Wind Shear along Glide Path at Beijing Capital International Airport

Hongwei Zhang (a), Songhua Wu (a, b), Qichao Wang (a)
(a) Ocean Remote Sensing Institute, College of Information Science and Engineering, Ocean University of China, Qingdao 266100, China.
*(b)Laboratory for Regional Oceanography and Numerical Modeling, Qingdao National Laboratory for Marine Science and Technology，*Email: wush@ouc.edu.cn*

Abstract: The significance of wind shear to aviation lies in its effect on aircraft performance and hence its potentially adverse effect on flight safety during landing phase especially within 500m (1600 ft) at height. The lidar team in Ocean University of China (OUC) carried out the particular research of the low level wind shear along the glide path at Beijing Capital International Airport (BCIA) in 2016 by utilizing a pulse coherent Doppler lidar (PCDL). On account of the high spatial and temporal resolution, the small-scale air mass disturbances along the glide path can be captured easily. The experiment configuration, the PCDL, the observation modes, the case study of the wind shear and the statistics of wind shear intensity are proposed and experimentally demonstrated. The reliability of wind shear identification algorithms has been proved by wind shear reports from the crew and the terminal control office.

Keywords: Pulsed coherent Doppler lidar, low-level wind shear; headwind, civil aviation safety

1. Introduction

According to the definition of International Civil Aviation Organization (ICAO), the low-level wind shear is defined as a sharply change of the wind speed or wind direction in space within 2000 feet at height. Wind shear includes vertical and horizontal components which usually occurs near fronts or coast and near the ground surface [1]. Wind shears in the lowest few hundred meters of the atmosphere have been responsible for a number of aircraft accidents [2-3]. In the years 1975 to 2006, the USA National Transportation Safety Board (NTSB) attributed wind shear as the cause of 20 aircraft accidents, with a total of over 500 fatalities [4]. There were 51 accidents caused by turbulence encounter, which account for 19 percent of the total accidents between 2014 and 2017[5]. Effective wind shear alarm systems are needed to guarantee the safety of civil aviation.

The rate of wind shear related accidents was reduced as the awareness of pilot increasing and the deployment of all current wind shear protection systems (LLWAS, PWS (Predictive Wind-Shear System), TDWR, and WSP (Weather Systems Processor)) .

Due to the variable types of low-level wind shear and its uncertainty of location and quick evolution, the required high spatial and temporal resolution of the low-level wind shear detection can only be achieved by operating a PCDL with different scanning modes. In this paper, we adopt the pulse coherent Doppler lidar (PCDL) system which has been proved to be an accurate optical sensing technique to capture the small scale characteristics of wind, turbulence, wake vortices, and so forth under non-rainy weather [6-10].

The pulsed 1550 nm fiber laser source is based on the MOPA architecture with a large core fiber and excellent beam quality (M^2=1.2), whose pulsed energy is 150 μJ with a pulse repetition rate of 10 kHz and an pulse length of 200. The Fast Fourier Transform (FFT) spectral estimates are processed with FPGA (Field Programmable Gate Array) in real time. All these design of the PCDL system make it stable, reliable and high-integrated. The detailed specifications of the PCDL system are listed in the Table 1.

Table 1 Specifications of the pulsed coherent Doppler lidar system

Qualification	Specification	Qualification	Specification
Radial Spatial Resolution	30 m	Wavelength	1.5 μm
Measurement Range*	60 to 4000 m	Pulse Energy	150 μJ
Scanner Positioning Accuracy	0.1°	Pulse Duration	200 ns
Velocity Measurement Range	± 37.5 m/s	Pulse Repetition Frequency	10 kHz
Resolution of Velocity	0.1 m/s		

***Weather condition related**

PPI (Plane Position Indicator) at low elevation angles has been developed and utilized during the past low-level wind shear observation campaigns to capture the wind shear events.

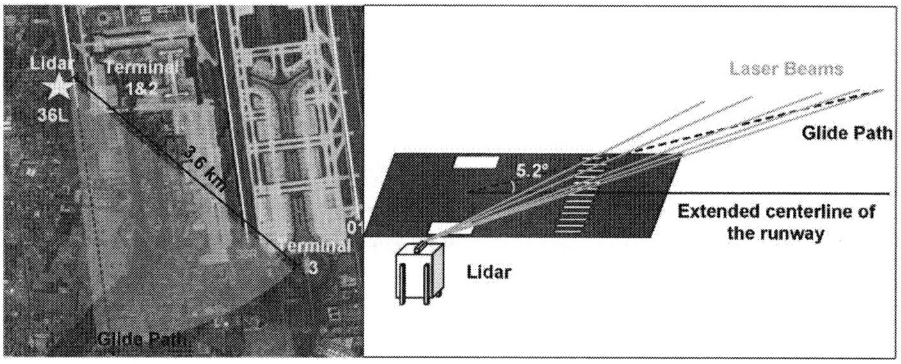

Figure 1 Location and scanning strategies of PCDL system that configured at BCIA.

Wind speed retrieved from the PPI scan data is related to the intercept angle between the laser beam pointing direction and the prevailing wind. When the scanning azimuthal angles are nearly corresponding with or nearly opposite to the prevailing wind, the LOS wind speeds are nearly equal to the real wind speed. However, when the lidar beam pointing direction is nearly perpendicular to the primary wind direction, the LOS wind speed is a project component of the wind vector. If the angle of intercept is larger than 30 degrees, the radial wind data would not be utilized to construct the headwind profiles considering the increasing underestimation of the headwind component as the across angle increases.

For the study of wind shear occurring along the glide path, the PCDL was configured at the south ending of the 36L/18R runway as Figure 1 shows. Taking the 5.2° incidence angle of aircraft landing path into consideration, a Pointing PPI scanning pattern was performed in the field wind shear observation campaign to reveal the wind that aircrafts would encounter.

2. Results

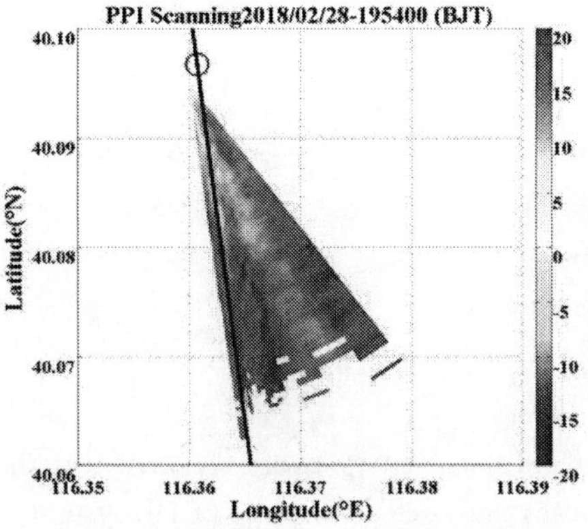

Figure 2 Glide path scanning of the 36L/18R runway at BCIA on Feb 28th, 2018.

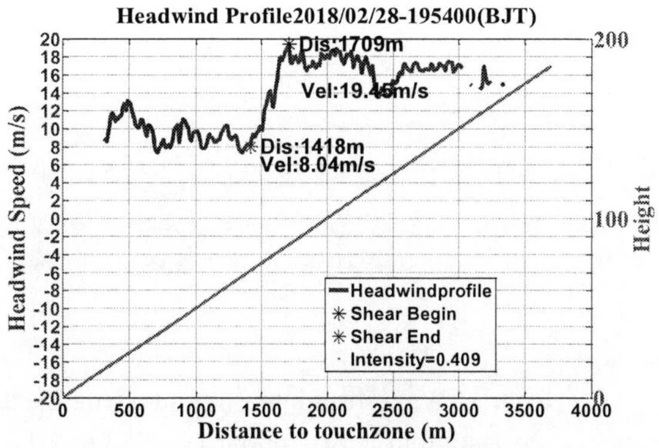

Figure 3 Wind speed profiles measured by PCDL at BCIA on Feb 28th, 2018.

We have captured the existence of small-scale airflow disturbances embedded in the prevailing wind with high spatial resolution by operating pointing PPI scan mode as Figure 2 shows. The red and blue colors indicate positive (away from lidar) and negative (towards lidar) movement, respectively, of the atmospheric particles along the laser beam. These airflow disturbances are generally small scale with serval hundred meters at horizontal dimensions and difficulty to be captured by DBS mode. The black line in Figure 2 is the extended centerlines of the runway and the width of line is about 100 m by taking the uncertainty with the flight path of the aircraft into consideration. By checking the series of PPI scans, the airflow disturbances move along glide path with the background wind flowing over the runway corridor. They are believed to arise from disruption of the northerly airflow by the artificial buildings above the ground surface.

For illustrating the transient and sporadic nature of the low level wind shear along the glide path, we construct the headwind profiles through the pointing PPI scan as Figure 3 shows. The duration of a PPI scan over the runway corridor is about 50 s (including PPI scan and data processing and data display).

Woodfield and Woods defined a severity factor I of wind shear as

$$I = (\frac{dV}{dt})(\frac{\Delta V}{V_{app}})^2 = \frac{1}{V_{app}}\left|\frac{\Delta V}{R^{1/3}}\right|^3, \tag{1}$$

where $\Delta V/R^{1/3}$ is the rate of change of wind speed, ΔV is the total change of wind, V_{app} means the normal approach speed of the aircraft (75 m/s in this study) and R represents the ramp length.

The value of the severity factor I is 0.08 with respect to the headwind speed reduced 22 kt within 300 m. The cubic root of severity factor is used to classify the turbulence alerts into two levels: "moderate turbulence (0.3~0.5)" and "severe turbulence (>0.5)". The value in this study is 0.409, a moderate turbulence.

More than 12000 PPI scans have been analyzed in this study. Wind shear severity factor calculated from headwind profiles are illustrated in Figure 4.

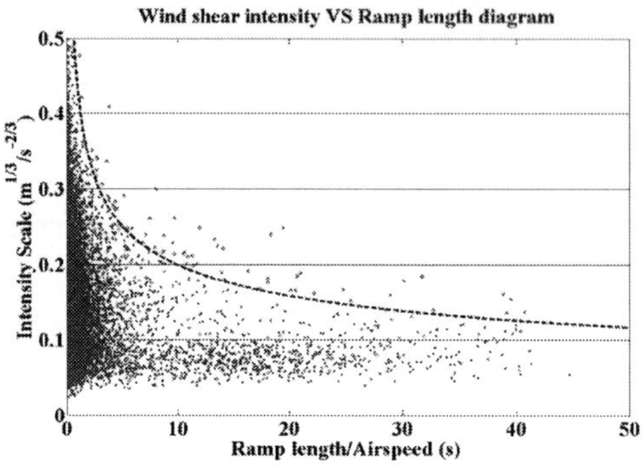

Figure 4 Wind shear intensity-ramp length diagram for wind shear reports over 36L corridor in the period from Jan 16 to Mar 16, 2018 based on wind shear ramps detected by the PCDL system.

The blue dots represent the severity factors with the maximum difference of headwind profiles less than 15 knots, while the maximum difference of red dots exceed the threshold adopted by the ICAO. The dotted line is the standard intensity scale (the speed difference is 15 knots) with different ramp length. Different pilots might have different practices in estimating the wind shear magnitude. Due to the sudden change of speed with shorter ramp length, the intensity scale may come to high value as the black circle marked. From the relationship statics of wind shear intensity scale and ramp length, we can find that the wind shear intensity scale is maintained at relative low level even under northwest prevailing wind at most part of time. It is unreliable to alert the wind shear event by identifying single factor (maximum difference of headwind profile or wind shear severity factor).

The method to identify the wind shear based on setting threshold value (15 knots) is more reliable to monitor the low level wind shear than wind shear intensity scale. The probably reason may be the shorter ramp length in this study. The wind shear intensity scale will exceed the threshold by calculating the Eq. 1. When it comes to longer ramp length (the headwind gradient will slows down with the same speed difference), the method based on setting threshold value may cause false alarms.

The reliability of algorithms have been proved by wind shear reports from the crew and the airport terminal control office.

3. Discussion and conclusions

The PCDL system has been proved to be an effective remoting tool to detect the low-level wind shear under dry conditions. It is possible to capture the wind shear events with different observation modes. From the field wind shear observation campaign at BCIA, we developed algorithms to identify the wind shear and turbulence. Unlike the traditional wind shear monitor system (such as TDWR, PWS, and WSP), the PCDL system is unable to cover the whole airport area due to the limited coverage. However, the PCDL system is able to capture the small-scale airflow disturbances with high spatial resolution and high data updating rate. The PCDL system perform well when it was used as a complementary system to radar coverage for the identification of the terrain induced wind shear.

Acknowledgements

We thank our colleagues for their kindly support during the field experiments; Yilin Qi and Jiaping Yin from Qingdao Leicent Co. Ltd for technical support. This work was jointly supported by the Boeing Company and Ocean University of China Joint Research project.

Major references

[1] International Civil Aviation Organization, CIRCULAR 186-AN/122.1987. *International Civil Aviation Organization Circular*, Montreal, Canada (1987).

[2] Fujita, T. T., *Preprints,* Downbursts and microbursts-An aviation hazard.*19th Conference on Radar Meteorology*, Miami, American Meteorological Society, 94-101 (1980).

[3] James Evans and Donald Turnbull, *Development of an automated wind shear detection system using Doppler weather radar. Proceedings of the IEEE. Vol. 77, No. 11 (1989).*

[4] Hallowell, R. G., and Cho, J. Y., Wind shear system cost-benefit analysis. *Lincoln Laboratory Journal*, 18 (2) (2010).

[5] International Civil Aviation Organization (ICAO). *ICAO Safety Report 2017 Edition.*

[6] Harris M, Young R I, Köpp F, et al, Wake vortex detection and monitoring. *Aerospace science and Technology*, 6(5): 325-331 (2002).

[7] Shun, C. M., and Chan, P. W, Applications of an infrared Doppler lidar in detection of wind shear. *Journal of Atmospheric and Oceanic Technology*, 25(5), 637-655 (2008).

[8] Hon, K. K., Chan, P. W. et al, Application of short-range LIDAR in early alerting for low-level wind shear and turbulence at Hong Kong international airport. *Advances in* Meteorology, 21.1: 74-85 (2014).

[9] Wu, S., Yin, J. et al, Wind Turbine Wake Visualization by Doppler Lidar. In *Optics and Photonics for Energy and the Environment* (pp. EM3A-4). Optical Society of America (2014).

[10] Frehlich, R., and Kelley, N., Measurements of wind and turbulence profiles with scanning Doppler lidar for wind energy applications. *IEEE Journal of Selected Topics in Applied Earth Observations and Remote Sensing*, *1*(1), 42-47 (2008).

[11] CHAN, P. W. Severe wind shear at Hong Kong International Airport: climatology and case studies. Meteorological Applications, 24.3: 397-403 (2017).

[12] CARRUTHERS, David, et al. Modelling of wind shear downwind of mountain ridges at Hong Kong International Airport. Meteorological Applications, 21.1: 94-104 (2014).

The Wind Observation over Sea Surface by UAV-borne Coherent Doppler Lidar

Qichao Wang (a), Songhua Wu (a, b)*, Jintao Liu (a), Kailin Zhang (a), Bingyi Liu (a, b)

(a) Ocean Remote Sensing Institute, College of Information Science and Engineering, Ocean University of China, Qingdao 266100, China.

(b) Laboratory for Regional Oceanography and Numerical Modeling, Qingdao National Laboratory for Marine Science and Technology, Qingdao 266100, China.

**Email: wush@ouc.edu.cn*

Abstract: A compact UAV-borne Coherent Doppler Lidar (UCDL) was developed by the Ocean University of China (OUC) for the observations of wind profiles and boundary layer structures in Marine Atmospheric Boundary Layer (MABL). The first flight campaign was conducted at Hailing Island (111.99°E, 21.62°N) from November to December 2016. The motion-correction and wind retrieval methodology of the UCDL are expounded in detail. Additionally, the backscatter signals of sea surface are used to retrieve and calibrate the wind field. During the first flight campaign, the wind observations from the UCDL are intercompared with the results from a collocated ground-based coherent Doppler lidar (GCDL). The characteristics of the wind field are also discussed in this paper.

Keywords: UAV-borne Coherent Doppler Lidar, Sea Surface Wind, Data Retrieval, Correction and Validation.

1. Introduction

Accurate and rapid wide-range observation of sea surface wind is important for the research of ocean dynamic prediction model, offshore wind resource assessment, air-sea interaction and flux. The coherent Doppler wind lidar integrated at an air-based platform provides an approach to conduct such observations [1-4]. Hence, an UAV-borne Coherent Doppler Lidar (UCDL) was developed by the Ocean University of China (OUC), it is a compact wind detection system with a light weight of 23.2 kg. The operated eye-safety laser in the UCDL generates a light pulse at the wavelength of 1.55 μm with the energy of 100 μJ. The full width at half maximum (FWHM) of the pulse width is 200 ns, which indicates a radial spatial resolution of 30 m [5].

Figure 1. Shipborne UAV ocean observation system and flight routes of the first experiment

The UCDL is carried at a shipborne UAV ocean observation system (Figure 1). The flight campaign was conducted at Hailing Island (111.99°E, 21.62°N), Guangdong Province from November to December 2016. During this campaign, more than ten flights at different heights (e.g. 800m, 1100m, 1200m, 1500m) and different measurement time (e.g. morning, noon and nightfall) were completed. The airspace and flight routes are shown in Figure 1. During several flights, for purpose of intercomparison, a ground-based coherent Doppler lidar (GCDL) was deployed simultaneously at the seashore (red star in Figure 1).

2. ALGORITHM

A modified Velocity Azimuth Display (VAD) method is applied to determine the wind speed and the direction [6]. In this method, the real radial velocities v_r obtained from the FFT spectra at different ranges are necessary. Differs from the GCDL, the mobile UCDL system measures the wind field under the condition of high speed movement over the sea surface during the observation [7], which induces the attitude and velocity of UAV into the FFT data and the recorded redial velocity v_{re}. Hence, it is necessary to correct the effect caused by the movement of UAV [8].

The attitude information of the UAV is recoded by the GPS and the IRS (25Hz, 0.05°) those are carried at the UAV. The pointing angle θ_r (azimuth and pitch angle) of each laser radial is calculated by combining the roll angle α , pitch angle β , yaw angle γ of UAV and the recorded point angle θ_{re} of each laser radial [9]. Then, the range gate R_n at which the laser reaches the sea surface in each radial is determined by using the θ_r and the height information of UAV and UCDL. Afterwards, the speed v_e caused by the movement of UAV in each radial is obtained by calculating the FFT data at range gate R_n. The real radial speed v_r caused by wind in every range gate can be obtained by subtracting the v_e from recorded radial velocities. Finally, the wind profile can be determined by taking θ_r and v_r into VAD method.

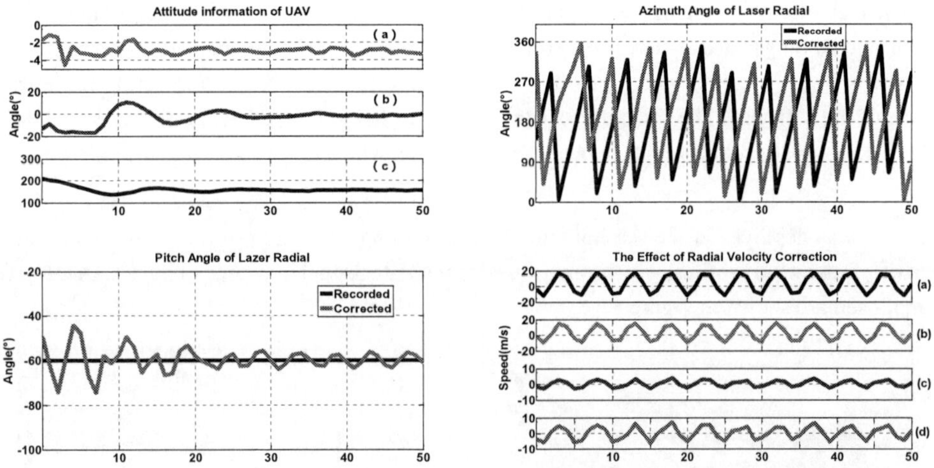

Figure 2. The attitude information of UAV and the effect of data processing

Figure 2 (1) shows the attitude information of UAV, which includes the (a) roll angle α, (b) pitch angle β and (c) yaw angle γ. Figure 2 (2, 3) shows the difference between the recorded point angle (black line) and the corrected point angle (red line) of laser radial, (2) is the azimuth angle and (3) is the pitch angle.

In Figure2 (4), (a) is the radial velocity v_{re} recorded by UCDL system, (b) is v_e, (c) is v_r, and (d) is real radial velocity v_{rh} in horizontal plane, which could be calculated by using v_r and θ_r .

3. Results

During this campaign, more than ten flights were carried out in about 30 days. The THI (Time-Height-Intensity) plot of the wind profiles over the sea surface observed at several flights (20th November, 17th ~19th December) are shown in Figure 3.

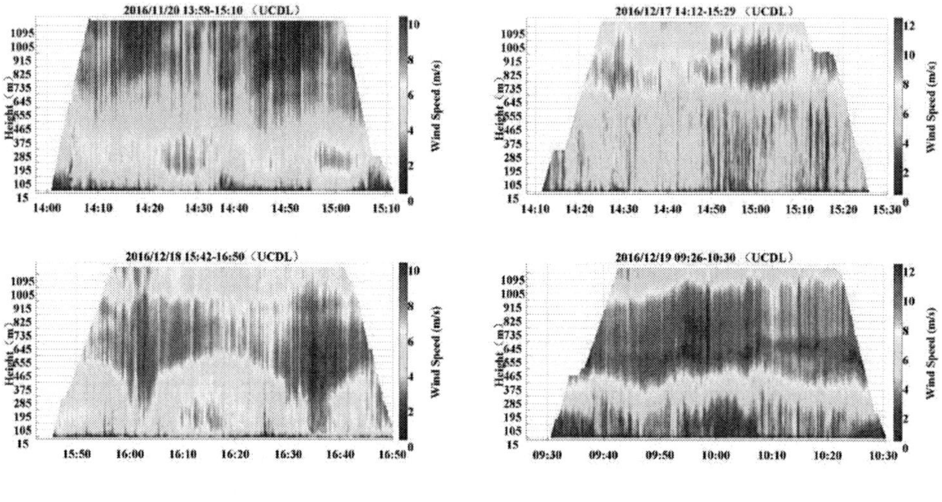

Figure 3. Wind THI diagrams of several flights observed by UCDL

During these flights, the maximum flying height of the UAV is 1200 m and the blind range of UCDL is 60 m. The lack of data on the left and right sides in each flight is limited by the height of UAV during the climb and landing phases. There is a distinct symmetry in the THI diagram of 18th December (Figure 3(3)). In this flight, the UAV choose the A-E-A route (red route), it is a reciprocating route from the seashore to the offshore. The structure of atmosphere is stable and the wind profile show the same tendency. The wind speed has slightly increased over time, which can also be concurrently observed by GCDL.

The GCDL was deployed at the seashore and kept observing the wind profile during the flights. This system offers the wind profiles, which can be used to validate the performance of the UCDL. The results of comparison are shown in Figure 4.

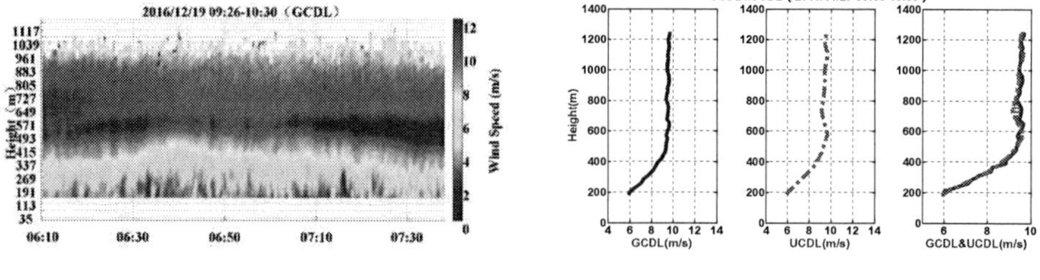

Figure 4. Wind data observed by GCDL.

Figure 4 (1) is the THI plot observed by GCDL at the same time of Figure 3 (4), which observed by

UCDL. Significant vertical stratification of wind speed is observed by both systems, and observations of two systems match well. Figure 4 (2) shows the comparison result of ten-minute average wind profile. The discrepancies between two systems are mainly caused by the difference locations and the non-homogeneous wind field. The ten-minute average wind profile of UCDL and GCDL are compared. The maximum deviation between two systems is 0.39m/s, the minimum deviation is less than 0.1m/s and the average deviation in different height is 0.14m/s.

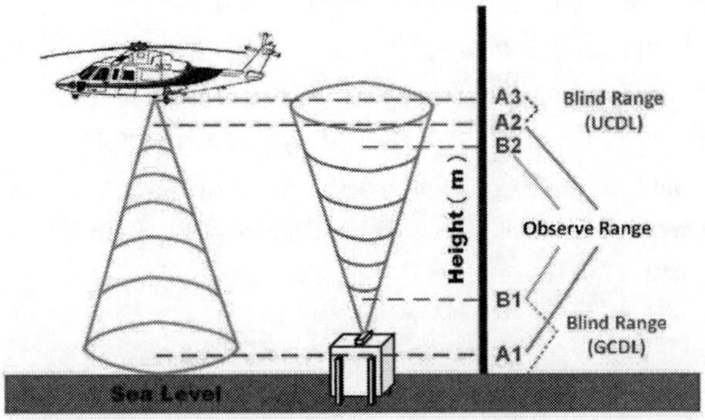

Figure 5. Observe range of UCDL and GCDL

During the flight on 19th December, The observe range of UCDL (15m~1140m above sea surface) is wider than GCDL(113m~1000m above sea surface). Because of the low concentration of aerosol and the long distance between the GCDL and particles in the upper layers, the GCDL can only provide the wind field below 1000 m (B2 in Figure 5). Comparing with the GCDL, the UCDL measured the wind profile from the sky, taking the advantage of the short distance between the UCDL and the upper layers, the UCDL performs better in high altitude wind observations. In addition, the UCDL can avoid the limitation of blind range in low altitude area, provides the wind profiles from 15m above the sea surface.

4. Conclusion

The UCDL system developed by OUC team has the ability to observe the wind over sea surface. We already completed the first flight campaign at Hailing Island, the inversion algorithm has been tested and the accuracy of UCDL has been validated by GCDL. Due to the operating position, it can lower the blind range near to the sea surface. Comparing with GCDL, UCDL performs better in high altitude wind measurement and it has the advantage of rapid wide-range observations. Further data processing and analysis will be carried out.

5. References

[1] Baker, W. E., et al, "Lidar-Measured Wind Profiles: The Missing Link in the Global Observing System", Bull. Amer. Meteor. Soc., 95, 4, 543-564 (2014).

[2] Koch, G. J., Beyon, J. Y., Cowen, L. J., Kavaya, M. J., & Grant, M. S., "Three-dimensional wind profiling of offshore wind energy areas with airborne Doppler lidar". Journal of Applied Remote Sensing, 8(1), 083662. (2014).

[3] Peña, A., Hasager, C. B., Gryning, S. E., Courtney, M., Antoniou, I., & Mikkelsen, T., "Offshore wind profiling using light detection and ranging measurements". Wind Energy, 12(2), 105-124 (2009).

[4] Hardesty, M., Tucker, S., Baidar, S., & Beubien, M., "Airborne tests of an OAWL Doppler lidar: Results and potential for space deployment". In EPJ Web of Conferences (Vol. 176, p. 02004, 2018). EDP Sciences.

[5] Wu, S., Wang, Q., Liu, B., Liu, J., Zhang, K., & Song, X., "UAV-borne coherent doppler lidar for marine atmospheric boundary layer observations". In EPJ Web of Conferences (Vol. 176, p. 02012, 2018). EDP Sciences.

[6] Browning K A., Wexler R., "The determination of kinetic properties of a wind field using Doppler radar", Journal of Applied Meteorology, (1968), 15, 1302-1306.

[7] J. Y. Beyon et al., "Airborne wind profiling algorithms for the pulsed 2-micron coherent Doppler lidar at NASA Langley Research Center," Proc. SPIE 8731, 87310K (2013).

[8] Werner, C., Leike, I., Streicher, J., Reitebuch, O., Cress, A., & Wergen, W., "Validation of doppler lidar wind measurements with the local model of the german weather service". In International Conference on Space Optics (2017, November)—ICSO 2000 (Vol. 10569, p. 105690K). International Society for Optics and Photonics.

[9] Li Z. G., Liu Z. S., Zhu J. S., Liu B. Y., Reitebuch O., "Wind Retrieval Algorithms For The Wind Products Of The Airborne Coherent Doppler Lidar", Proceeding 'Dragon 3 Mid-Term Results Symposium', Chengdu, China (ESA SP-724, November 2014).

Up to 12 km vertical wind profiling using Mitsubishi Electric's long range lidar

Ikuya Kakimoto (a), Taiji Harada (a), Shumpei Kameyama (a), Jong-Sung Ha (b), and Hong-Il Kim (b)
(a) Mitsubishi Electric Corporation
(b) Korea Aerospace Research Institute
Lead Author e-mail address: Kakimoto.Ikuya@eb.MitsubishiElectric.co.jp

Abstract: A new function of Mitsubishi Electric's long range lidar for vertical wind profiling has been developed. The observation performance evaluation was carried out by KARI, and the result shows that the measurable height of up 12 km in no precipitation condition was demonstrated. The correlation coefficient of wind speed and direction between the lidar and radio sonde are also shown. The result implies the expanding use of wind lidars for high altitude wind sensing.

Keywords: Wins sensing, High altitude

1. Introduction

Mitsubishi Electric's wind Doppler lidar "DIABREZZA A series" (Fig. 1) has been deployed for microburst / wind shear detection for several international airports such as Tokyo, Narita, Hong Kong. Further, Mitsubishi Electric has recently made a contract with Beijing New International Airport and scheduled to deliver in 2018. This lidar consists of the 1.5-μm fiber-based lidar transceiver [1] and the high-power planar waveguide laser amplifier [2]. The amplifier enables the very long measurable range of more than 30 km in the case of the slant path sensing [3]. In addition to the usual functions for the airport application, a new function for vertical wind profiling has been developed in response to the requirement from KARI (Korea Aerospace Research Institute). Here, we introduce the results for the vertical wind profiling which were performed by KARI.

Fig. 1. Appearance of the lidar.

2. Vertical wind profiling results

The lidar vertical wind profiling was performed using the digital beam switching method with 4 direction measurement. The range resolution of the lidar was 150 m, and the data accumulation time for each measurement direction was 2 minutes. This was same as the output data interval. The radio sonde went up from the ground to an height of more than 10km with the weather balloon taking for 30 minutes and output 1 minute averaged data. The wind speed and direction measured by the lidar were compared with that measured by the sonde. This inter-comparison test was carried out 25 times in the period between 10th January and 15th March in 2016.

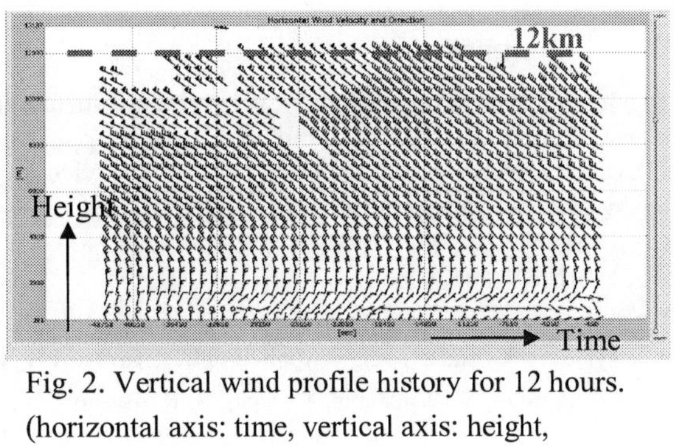

Fig. 2. Vertical wind profile history for 12 hours. (horizontal axis: time, vertical axis: height, arrows: wind vector)

Figure 2 shows the example of the wind history of vertical wind profile for 12 hours. It is shown that the measurable height reached about 12 km, which is much higher than the top of the boundary layer, in no precipitation condition. Figures 3 show examples of the above mentioned inter-comparison. The qualitative good agreements can be seen. The maximum measurable height and correlation coefficient between the lidar and sonde data are summarized in table 1. The results for the representative time slots are also shown in the table. The correlation coefficient of wind speed between the lidar and radio sonde was 0.98 on average for 25 times trials and that of wind direction was 0.87.

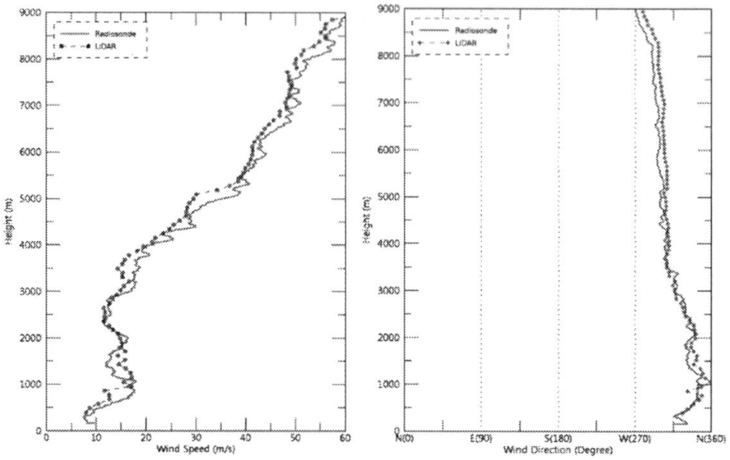

Figs. 3. Data comparison between lidar and radio sonde of wind speed (left) and direction (right).

Table 1. Maximum measurable height and correlation coefficient between lidar and sonde data.

No.	Date	Maximum measurable height	Correlation (Wind speed)	Correlation (Wind direction)
1	Jan.10 (09:40~)	7.81 km	0.99	0.97
~	~	~	~	~
14	Feb.7 (09:26~)	10.45 km	0.99	0.79
~	~	~	~	~
17	Mar.2 (10:24~)	6.88 km	0.83	0.94
~	~	~	~	~
25	Mar.15 (18:06~)	9.60 km	0.99	0.95
Average			0.98	0.87

3. Summary

In addition to the conventional low level wind shear detection function, the DIABREZZA series has had a new function for vertical wind profiling for upper height wind observation up to an altitude of 12 km. DIABREZZA has proved to be a very useful device for this application. From now on, KARI is expected to develop new applications for DIABREZZA.

4. References

[1] S. Kameyama T. Ando, K. Asaka, Y. Hirano, and S. Wadaka,"Compact all-fiber pulsed coherent Doppler lidar system for wind sensing," Applied optics, **46**, No.11, pp.1953-1962, 2007.

[2] T. Sakimura, Y. Watanabe, T. Ando, S. Kameyama, K. Asaka, H. Tanaka, T. Yanagisawa, Y. Hirano, and H. Inokuchi, "1.5-μm high-gain and high-power laser amplifier using a Er,Yb:glass planar waveguide for coherent Doppler LIDAR," Proc. of CLEO, CTu2D.7, 2012.

[3] S. Kameyama, T. Sakimura, Y. Watanabe, T. Ando, K. Asaka, H. Tanaka, T. Yanagisawa, Y. Hirano, and H. Inokuchi, "Wind sensing demonstration of more than 30 km measurable range with a 1.5 μm coherent Doppler LIDAR which has the laser amplifier using Er,Yb:glass planar waveguide," in Proceedings of SPIE, 85260E-1, 2012.

Comparison of wind measurements between virtual tower and VAD methods with different elevation angles

Xiaoying Liu[1], Songhua Wu[1, 2*], Hongwei Zhang[1], Qichao Wang[1], Xiaochun Zhai[1]
1 Ocean Remote Sensing Institute, College of Information Science and Engineering, Ocean University of China, Qingdao 266100, China.
2 Laboratory for Regional Oceanography and Numerical Modeling, Qingdao National Laboratory for Marine Science and Technology, Qingdao 266100, China.
**Email: wush@ouc.edu.cn*

Abstract: This paper compares wind retrieval methods of VT (Virtual Tower) mode and VAD (Velocity-Azimuth Display) with different scanning elevation angles. The field experiment was performed from September 23 to October 13 2017 at the campus of Ocean University of China under various weather conditions. A total of seven lidars were involved in the experiment. Three lidars carried out the staring mode to constitute the VT. The other four lidars were concurrently operating PPI scanning at different elevation angles, corresponding to different spatial average volume. The results of wind speed and direction from VT and VAD shows good correlation when lidars are well synchronized. The influence of spatial homogeneity on wind retrieval is also described in this paper.
Keywords: Lidar, Virtual Tower, VAD, Volume Average

1. Introduction

Accurate wind field information can be used to explore the subtle structure of complex wind flow, meanwhile, it can also provide the possibility to comprehend the underlying physical mechanisms and to optimize wind field models [1]. The single-lidar commonly uses the Doppler beam swinging (DBS) strategy and the velocity azimuth display (VAD) technique to obtain high accuracy [2-4]. Both of these methods are based on spatial horizontal homogeneity assumption, which is often invalid in complex wind field [5-6]. In response to this question, two or more additional lidars are used to obtain a three-dimensional wind field by spatiotemporal synchronization scanning, which can be called virtual tower (VT) [7-8]. In order to demonstrate the feasibility and accuracy of VT, the comparison between DBS mode and VT are studied in [6, 8], and the VT and sonic anemometer also show good agreement in [1, 9]. This paper introduces the results of VT and VAD mode with different scanning elevation angles.

2. Experiments

The field experiment was performed from September 23 to October 13 2017 at the campus of Ocean University of China. As shown in Figure 1, a total of seven coherent Doppler lidars were involved in the experiment. Three lidars (VT1, VT2 and VT3 in the Figure 1 (a) and (b)) constituted the virtual tower by carrying out the staring mode. They were located on the north, east and south of the playground respectively to meet the requirements that three lidar beams should be intersecting and non-coplanar [10]. The other four lidars (as shown in Figure1 (a) and (b), the VAD1, VAD2, VAD3 and VAD4) were concurrently operating PPI scanning at different elevation angles, corresponding to different spatial average volume. Four lidars on VAD mode were all placed on the west side of the playground and each of two adjacent lidars is spaced one meter apart.

During the experiment, three lidars of VT simultaneously stared at the six points (50 m to 300 m with 50 m increments) above the VAD2 (see in Figure 1(b)). The distance of these three lidars to VAD2 were all 98m, which were approximately the same. This ensured that they had the same elevation angle when measuring the same height. The three lidar beams are continuously observed for 1 min at each height at the sampling rate of 0.5Hz, which can capture the rapid atmospheric process while ensuring the high temporal resolution as much as possible. The other four lidars on VAD mode scanned at the speed of 6 degrees per second for full-cycle PPI scan. Therefore, it took 1 minute to obtain a wind profile. The experimental parameters of seven lidars are shown in Table 1.

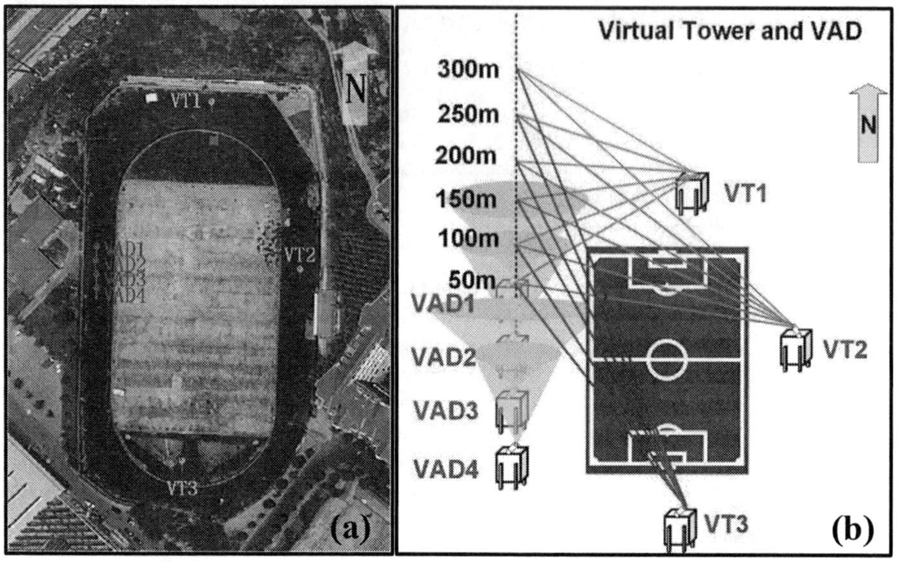

Figure 1. Schematic of experimental configuration. (a) Seven lidars' location on the playground. (b)The virtual tower stare scan and four full-cycle PPI scan with different elevation angles.

Table 1. Experimental observation parameters of seven lidars

Lidar Code	Retrieval Methods	Azimuth(°)	Elevation(°)					
			50 m	100 m	150 m	200 m	250 m	300 m
VT1	VT	226.39	26.8	45.3	56.6	63.7	68.4	71.8
VT2		270.55	26.9	45.4	56.7	63.7	68.5	71.8
VT3		338.41	26.9	45.5	56.7	63.8	68.5	71.8
VAD1	VAD	0~360	50					
VAD2			60					
VAD3			35.3					
VAD4			70					

3. Methodology

Lidar measures the movement of aerosol particles in the atmosphere to obtain a line-of-sight (LOS) velocity (or radial velocity). The LOS velocity is a projection of the wind field in the pointing direction of the laser beam, so the horizontal and vertical velocity can be retrieved from the LOS velocity by Eq.(1).

$$V_r = u\sin\theta\cos\phi + v\cos\theta\cos\phi + w\sin\phi \qquad (1)$$

where V_r is the LOS velocity, u, v, w are the zonal velocity, the meridional velocity, and the vertical velocity, respectively. And θ, ϕ are the azimuth and elevation angles, respectively.

In order to retrieve the three components of the wind field, two additional equations are needed. Therefore, the three radial velocities from three lidars of the virtual tower can form a set of wind vector equations. In Eq. (2), the terms on the left–hand side are the radial velocity measurements of the VT1, VT2 and VT3, which can be represented by **Vr**. The right-hand side includes the geometrical information in the matrix **M** and the velocity vector in meteorological coordinates **V** [10]. So the relationships among these three vector matrix can be written as Eq. (3). And the vector **V** is our concerned actual wind information.

$$\begin{bmatrix} V_{r1} \\ V_{r2} \\ V_{r3} \end{bmatrix} = \begin{bmatrix} \sin\theta_1 \cos\phi_1 & \cos\theta_1 \cos\phi_1 & \sin\phi_1 \\ \sin\theta_2 \cos\phi_2 & \cos\theta_2 \cos\phi_2 & \sin\phi_2 \\ \sin\theta_3 \cos\phi_3 & \cos\theta_3 \cos\phi_3 & \sin\phi_3 \end{bmatrix} \begin{bmatrix} U \\ V \\ W \end{bmatrix} \tag{2}$$

$$\mathbf{V} = \mathbf{M}^{-1}\mathbf{Vr} \tag{3}$$

4. Analysis and Results

The single lidar measurements need to consider the difference between scalar and vector average when compared to the point measurements. The analysis of averaged wind vector need to average these components over a period of time e.g.10 min by scalar averaging and vector averaging. The magnitude of the vector is different from the scalar, and the ratio between them and the standard deviation of the wind direction follow the law of the Bessel function as seen in Figure 2 (a). The ratio of vector to scalar average can be written in Eq. (4) [11], where V_{vector} and V_{scalar} are the vector average and the scalar average respectively, and σ is standard deviation of the wind direction in the interval time.

$$\text{Ratio} = \frac{V_{vector}}{V_{scalar}} = \frac{\sin\left(\sigma\sqrt{3}\right)}{\sigma\sqrt{3}} \tag{4}$$

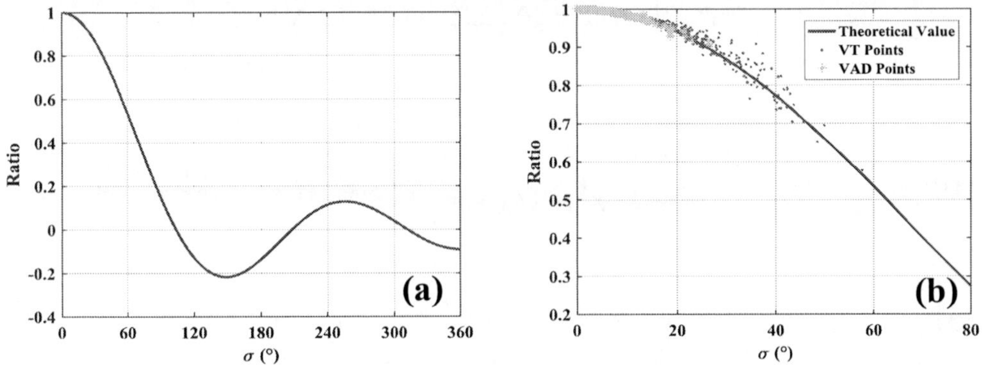

Figure 2. The Bessel function of the ratio of vector to scalar average with the standard deviation of the wind direction. (a) The results of theoretical analysis. (b) The results of experimental data with VT and VAD (The interval time is 10 min).

Analysis of VAD data (volume measurements) can be well fitted to Bessel function. VT is not point measurements in the strict sense with a certain measurement volume and is sensitive to wind direction. It is also possible to calculate the vector and scale average in the interval time. And the results of VT are also fitted to the function well. The following analysis are all based on the vector average of VT and VAD.

The comparison of horizontal velocity and direction between VT and VAD are shown in Figure 3. The data came from the observations made on September 27 and 28, 2017. The scanning elevation angle of the VAD used in the comparison is 60° . As seen from the Figure 3, the results of wind speed and direction from VT and VAD shows good correlation. The correlation coefficient (R) are the 0.93 and 0.99 in the velocity and direction, respectively. The standard deviation of 0.78 m/s and 3.8° are all in the reasonable range. Figure 4 shows the time series of horizontal velocity of VT and VAD at four elevation angles. The absence of data in the plot is due to quality control based on time synchronization criteria. The wind speed of VT and VAD show a consistent changing trend over time. When the wind speed is lower, the wind magnitude of VT and VAD are roughly equivalent. As the wind speed continues to increase, the speed of VT is significantly higher than that of VAD and the fluctuations also increase.

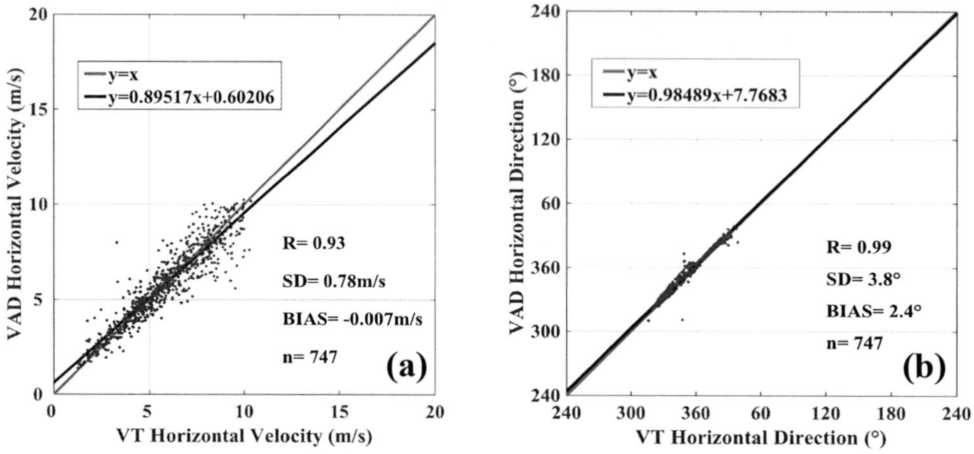

Figure 3. Comparison of the (a) horizontal velocity and (b) horizontal direction from the VT technique with the measurements made by VAD technique.

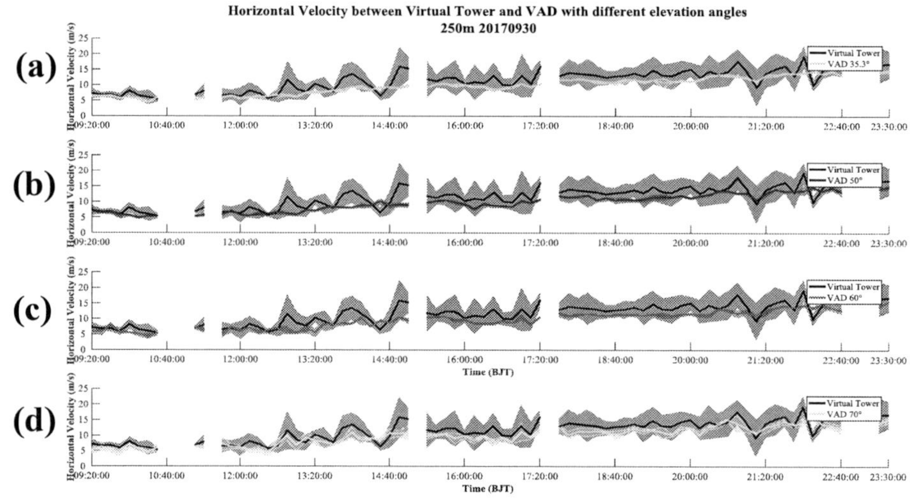

Figure 4. Time series of horizontal velocity of VT and VAD with four elevation angles, as shown (a) 35.3°, (b) 50°, (c) 60° and (d) 70° at 250 m on September 30, 2018. The shaded portion in the figure represents the fluctuation of wind speed.

Compared with the cup anemometer, lidar measures the wind with a volume average around a given height [11]. Lidar has an effective probe length corresponding to a finite height resolution and elevation angle. In this experiment, the radial resolution is 30m, and the elevation angles is different as seen in Table 1. Figure 5 shows the result of average volume, where the ratio is the wind speed of volume average around a given height to the speed of the given height. According to the figure, the larger elevation angle and the lower height have the smaller ratio, which means the more significant influence of volume average.

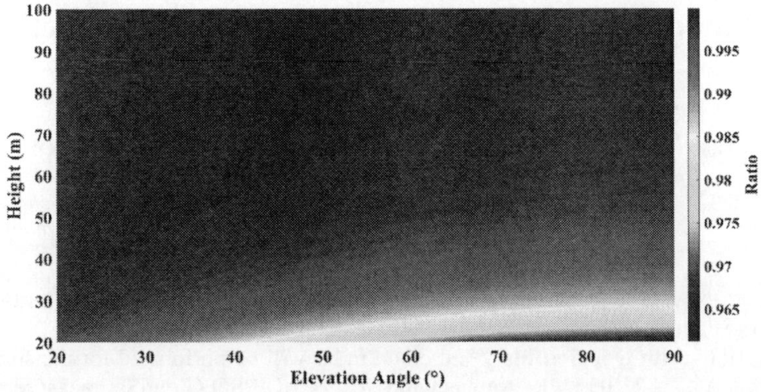

Figure 5. The ratio of volume average wind speed around a given height to the speed of the given height.

It is worth to noting that these ratios are all above 0.95. From the perspective of theoretical analysis, the effect of the volume average on wind speed is not obvious at different elevation angles with the 30 m radial resolution.

5. Conclusions

This paper introduces the synchronous observation experiments of seven lidars. The comparison between VT and VAD shows good agreement. The volume average effect is also analyzed theoretically. In the next work, it is also necessary to analyze the influence on the actual wind field with measurement data. The time series analysis indicates that VT and VAD have the same tendency.

6. Acknowledgement

This work was supported by National Natural Science Foundation of China (No.41471309) and the National Key Research and Development Program of China (016YFC1400904). The authors wish to thank Fanghan Wang for valuable discussion.

7. References

[1] Aditya Choukulkar, "Evaluation of single and multiple Doppler lidar techniques to measure complex flow during the XPIA field campaign," Atmospheric Measurement Techniques **10**, 247-264 (2017).

[2] Strauch, R. G., "The Colorado Wind-Profiling Net-work", Atmos. Ocean. Tech., **1**, 37–49 (1984).

[3] Browning, K. A. "The determination of kinematic properties of a wind field using Doppler radar", Appl. Meteorol. , **7**, 105–113 (1968).

[4] Gottschall, J, "Lidar profilers in the context of wind energy—A verification procedure for traceable measurements," Wind Energy, **15**, 147–159 (2012).

[5] Bingöl F, "Conically scanning lidar error in complex terrain," Meteorologische Zeitschrift, **18**, 189–195 (2009).

[6] Pauscher L, "An Inter-Comparison Study of Multi- and DBS Lidar Measurements in Complex Terrain," Remote Sensing, **8**, 782 (2016).

[7] Calhoun, R., "Virtual towers using coherent Doppler lidar during the Joint Urban 2003 dispersion experiment," Appl. Meteo-rol. Clim, 45, 1116–1126 (2006).

[8] Newman, Jennifer F., "Testing and validation of multi - lidar scanning strategies for wind energy applications." Wind Energy 19.12 (2016): 2239-2254.

[9] Mann, Jakob, "Comparison of 3D turbulence measurements using three staring wind lidars and a sonic anemometer." IOP Conference Series: Earth and Environmental Science.1, (2008).

[10] Fernando Carbajo Fuertes, "3D Turbulence Measurements Using Three Synchronous Wind Lidars: Validation against Sonic Anemometry", Journal of Atmospheric and Oceanic Technology 31.7, 1549-1556 (2014).

[11] Clive, P. J. M. "Compensation of bias in Lidar wind resource assessment." Wind Engineering 32.5, 415-432 (2008).

Physical model approach (PM) for the analysis of active sensor systems

Kaori Sato (a), Hajime Okamoto (a), Hiroshi Ishimoto (b), and Shoken Ishii (c)
(a) Research Institute for Applied Mechanics, Kyushu University
(b) Meteorological Research Institute, Japan Meteorological Agency
(c) National Institute of Information and Communications Technology

Abstract: New ground-based lidars are being built based on the Multiple-Field-of-view Multipe-Scattering Polarization Lidar (MFMSPL) concept (Okamoto *et al*., 2016). Joint observations with Doppler lidars and cloud radar will be performed for the simulation of EarthCARE ATLID/CPR, Aeolus wind lidar, CALIPSO/CloudSat measurements to evaluate satellite retrieval algorithms for cloud/aerosol process studies. For the analysis of the joint observation system, a physical model (PM) that simulates the polarized multiple scattered space-borne lidar signals for input vertical profiles of cloud microphysics and air molecules as well as instrument settings such as the FOV has been developed (Sato *et al*., 2018). Error analyses by Monte Carlo simulations for CALIPSO and ATLID showed stability and advantage of the PM over more statistical approach. In this paper, application and extension of the PM for the simulation/analysis of the joint ground-based lidar system and satellite active sensors is studied.

Atmospheric remote sensing by employing a blue-band continuous-wave lidar system

Hui Li (a), Zheng Kong (a), Zhi Liu (a), Lishan Zhang (a), Peng Guan (a), and Liang Mei (a)
(a) Dalian University of Technology

Abstract: A blue-band Scheimpflug lidar system is developed for atmospheric pollution monitoring. Atmospheric remote measurements were carried out for 10 days in urban area to validate the feasibility and performance of the Scheimpflug lidar system. 24-hour continuous measurement is achieved in winter on a near horizontal path with an elevation angle of 6.4°. The aerosol extinction coefficient retrieved by the Fernald-inversion algorithm shows good agreement with the variation of PM10/PM2.5 concentrations recorded by a local pollution monitoring station. The experimental result revealed that the linear ratio between the aerosol extinction coefficient and PM10 concentration under high moistures (75%-90%) is about 2-time of that in low moisture (≤75%) when PM10 concentrations are less than 100 μg/m3.

Rayleigh Lidar Observations of Tropical (13.5°N, 79.2°E) Mesospheric Inversion Layers: A Comparison between Dynamics and Chemistry

Karanam Ramesh[1,*], Sundararajan Sridharan[1], Karnam Raghunath[1]
[1] *National Atmospheric Research Laboratory (NARL), Gadanki, A.P., India.*
* *Email: karanamram@gmail.com, kramesh@narl.gov.in*

Abstract: The Indo-Japanese Rayleigh lidar system installed at National Atmospheric Research Laboratory (NARL), Gadanki (13.5°N, 79.2°E), India has been operational since 1998 with 532 nm Nd: YAG green laser of energy ~550 mJ/pulse (20 Hz PRF). The vertical temperature profiles are derived above ~25 km using molecular back scattered photons assuming the atmosphere in hydrostatic equilibrium and obeys ideal gas law. The lidar system has been upgraded with a high power laser of energy ~600 mJ/pulse (50 Hz PRF) in 2007 that significantly minimize the uncertainty in temperatures above ~70 km. The lidar temperatures with high spatial (300 m) and temporal (250 sec) resolutions are used to study the vertical propagation characteristics of short period gravity waves and their role in the formation of mesospheric inversion layers (MILs). Further the roles of dynamics and chemistry are compared for the occurrence of MILs.

Keywords: Rayleigh Lidar, Temperature, Gravity waves, Mesospheric Inversion Layer.

1. Introduction

One of the interesting and poorly understood features of the mesosphere and lower thermosphere (MLT) region is the phenomenon of Mesospheric Inversion Layers (MILs). The term *Mesospheric Inversion Layer* (MIL) refers to the inversion of temperature gradient from negative to positive superposed upon the characteristically decreasing mesospheric thermal structure with the positive lapse rate at the bottom. Though the occurrence of MILs has been reported from several sites using lidar and space borne observations, it is still unknown about their morphological characteristics. Further the causative mechanisms of MILs are quite complex and a few suggested mechanisms for their occurrence are gravity wave breaking, gravity wave-tidal interactions, chemical heating, and planetary wave critical level interaction [1, 2].

Rayleigh LIDAR (LIght Detection And Ranging) is one of the most powerful techniques for the study of mesospheric inversion layers (MILs) during nighttime with high spatial and temporal resolution. The Rayleigh lidar system installed at National Atmospheric Research Laboratory (NARL), Gadanki (13.5°N, 79.2°E) employed a Q-switched solid state (Nd:YAG) pulsed laser (PL8020) from Continuum, USA during 1998-2006. The system operates at 532 nm green laser with the pulse repetition frequency (PRF) of 20 Hz and the relatively lower energy of ~550 mJ per pulse. Later in 2007, the lidar system has been upgraded with a high power laser of energy ~600 mJ/pulse and PRF of 50 Hz and operated in monostatic biaxial mode. The initial diameter of the laser beam is 9 mm and it will be expanded to 90 mm before sending into atmosphere in order to minimize the beam divergence from ~0.45 mRad to less than 0.1 mRad so that the beam remains within the field of view (FOV) of the receiver (Newtonian telescope) at all ranges of interest. The atmospheric back scattered photons are collected by the receiver and they are allowed to fall on photo multiplier tubes (PMTs) of different gains (R-channel & U-channel). The PMT output signal will be collected and processed through the MCS-plus (Multi Channel Scalar) PC based photon count system with two data acquisition channels of the receiver (R, U channels).

2. Figures and Tables

The Indo-Japanese lidar (IJL) system at NARL consists of (i) transmitter, (ii) receiver and (iii) detector and the block diagram is as shown in the Figure 1.

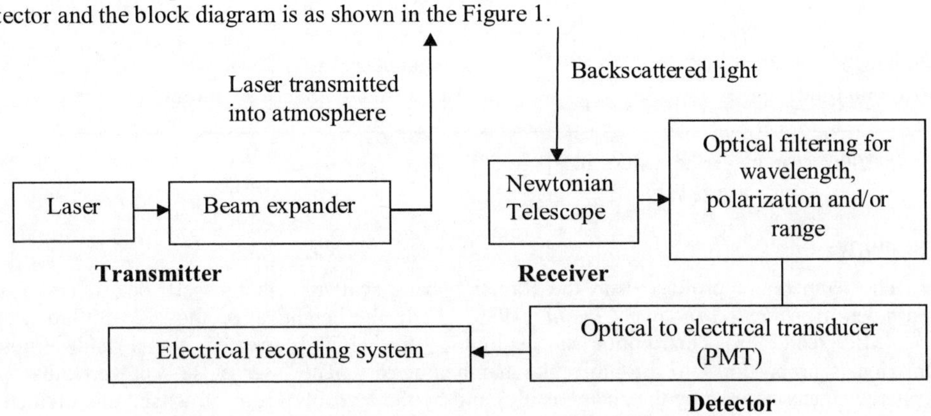

Figure 1. Block diagram of a NARL lidar system

The lidar uses laser as the source for the three special properties: low beam divergence, extreme narrow spectral width and short intense pulses. The receiver system of a lidar (normally a telescope) collects and processes the scattered laser light and then directs onto a photodetector, a device which converts the light signal to an electrical signal. The signal detection and recording of a lidar takes the light from the receiver and records the measured intensity as a function of altitude. Usually photomultiplier tubes (PMTs) are used as detectors in incoherent lidars that use visible and ultraviolet light.

The major specifications of NARL Rayleigh lidar system are given in Table 1.

Parameters	Specifications
Transmitter	
Laser Source	Nd:YAG (PL 9050 (Continum, USA) [PL 8020]*)
Operating wavelength & energy/pulse	532 nm & 600 mJ [550 mJ]*
Pulse width	7 nsec
Pulse repetition frequency (PRF)	50 Hz [20 Hz]*
Beam diameter	9 mm (Expanded to 90 mm)
Beam divergence	< 0.1 mrad (after expansion)
Receiver	
Telescope type	Newtonian with diameter of 75 cm
Field of view	1 mRad
Interference filter band width	1.07 nm
Maximum transmission	48% #
Photomultiplier tube (Head-on-type)	
Model	Hamamatsu, R3234-01
Cathode & Anode sensitivity	64.2 mA/lumen & 2170 A/lumen
Gain	34×10^7
Dark current	50 pA

Data acquisition	
Software	PC-based data acquisition system with EG&G ORTEC Multi Channel Scalar (MCS) software
Bin width	2 μsec (=300 m of height resolution)
Scan length	1024 channels
Integration time	250 sec (12500/5000* laser pulses)

* for 1998-2006; # can be varied; values in [ ] are of PL8020

3. Equations

The temperature profiles from the Rayleigh back scattered photons are determined from the method given by *Hauchecorne and Chanin*, [1980] [3]. In the height range above 25-30 km where the aerosol (Mie scattering) contribution is negligible, the range corrected signal and atmospheric transmission is proportional to the molecular number density. The laser pulse sent vertically into the atmosphere is backscattered by the air molecules and by the aerosols. If the atmosphere is divided into a number of layers of constant thickness, Δz then the backscattered signal in the ith altitude layer (z_i-$\Delta z/2$, z_i+$\Delta z/2$) can be given by the lidar equation.

The lidar equation for the backscattered photon counts can be given as follows:

$$N(z_i) = \frac{N_0 A K T^2(z_0, z_i)}{4\pi (z_i - z_0)^2} \left| n_r(z_i)\beta_r + n_m(z_i)\beta_m \right| \Delta z \qquad (1)$$

where

N_i	: Backscattered signal from i^{th} altitude layer,
N_0	: Number of emitter photons,
$n_r(z_i)$ & $n_m(z_i)$	: Air molecules and aerosols concentrations,
β_r & β_m	: Rayleigh and Mie backscattering cross sections,
z_0	: Altitude of the lidar site,
A	: Telescope area,
$T^2(z_0, z_i)$	: Atmospheric transmission,
K	: Optical efficiency of the lidar system,
Δz	: Thickness of the layer.

Using the number density taken from an appropriate model (MSIS- 90 or CIRA-86) for the height of 50km where the signal-to-noise ratio is fairly high, the constant of proportionality is evaluated and there by the density profile r(z) is derived. Taking the pressure (P) at the top of the height range (90km) from the model, the pressure profile is computed using the measured density profile, assuming the atmosphere to be in hydrostatic equilibrium. Using the ideal gas relation, the temperature at i^{th} layer $T(z_i)$ is given by

$$T(z_i) = \frac{M\, g(z_i) \Delta z}{R\, Log(1+X)} \text{ where } X = \frac{\rho(z_i) g(z_i) \Delta z}{P(z_i + \Delta z/2)} \qquad (2)$$

and the statistical standard error in the temperature is given by

$$\frac{\delta T(z_i)}{T(z_i)} = \frac{\delta Log|1+X|}{Log|1+X|} = \frac{\delta X}{(1+X)\, Log(1+X)} \qquad (3)$$

The simulated lidar signal from equation (1) is as shown in Figure 2. The back scattered photons decreases exponentially in response to decreasing density with respect to height.

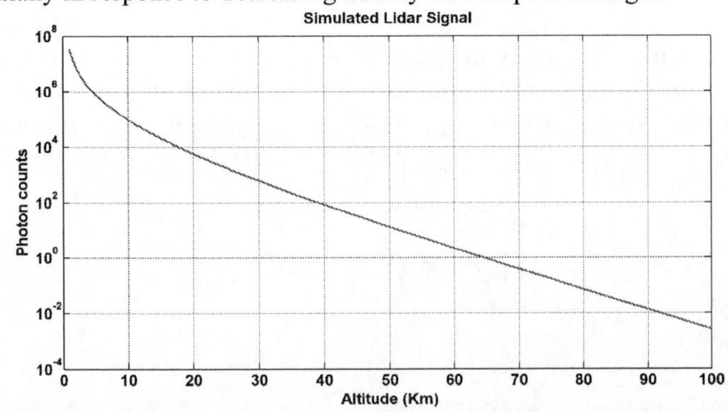

Figure 2. The simulated lidar signal for the lidar system at NARL, Gadanki.

4. Observations and Results

Figure 3a shows the nightly (19:07-23:14 LT) mean Rayleigh lidar temperature (T) over Gadanki along with TIMED-SABER (Thermosphere-Ionosphere-Mesosphere Energetics and Dynamics- Sounding of the Atmosphere using Broadband Emission) temperature profile for 12.5°N, 79.0°E at 23:58 LT. Both show large inversion layers above ~75 km with amplitude and thickness of ~92.6 K, ~83.8 K and 8.4 km, 5.2 km respectively. The MSIS-90 (Mass Spectrometer Incoherent Scatter-90) model temperature profile also included in the figure for reference. The profiles of lapse rate, Γ=-dT/dz and the Brunt-vaisala frequency square (N^2) calculated from equation (4) are shown in Figure 3b and Figure 3c respectively.

$$N^2(z) = \frac{g(z)}{T(z)}\left[\frac{dT(z)}{dz} + \frac{g(z)}{c_p}\right] \quad (4)$$

Here g(z), T(z) and c_p are acceleration due to gravity, mean temperature and specific heat at constant pressure (1004 J/K/Kg) respectively.

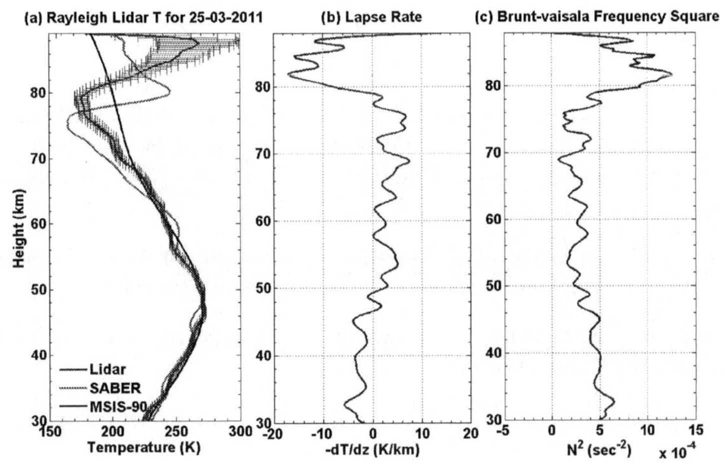

Figure 3. Vertical profiles of (a) lidar, TIMED-SABER and MSIS-90 model temperature, (b) lapse rate (-dT/dz), (c) Brunt-vaisala Frequency square (N^2) on the night of 25-03-2011 over Gadanki region.

From Figure 3b and 3c, it can be observed that the turbulent layers are not present since the lapse rate (Γ) is not attaining the adiabatic lapse rate (Γ_d~10 K/km) and the condition for convective instability ($N^2<0$) is not satisfied above ~75-80 km which indicates the negligible role of gravity wave breaking for the occurrence of inversion layer. Further the SABER chemical heating rates among H, O, O_2, O_3, OH and HO_2 (list of reactions is given in *Ramesh et al.*, [2013] [4]) for 25-03-2011 are shown in Figure 4. It is clear the heating rates increase above ~77 km and the reaction between H and O_3 dominates all the reactions with ~10 K/day. Also the total heating rates are about 20 K/day in the inversion region at ~80 km.

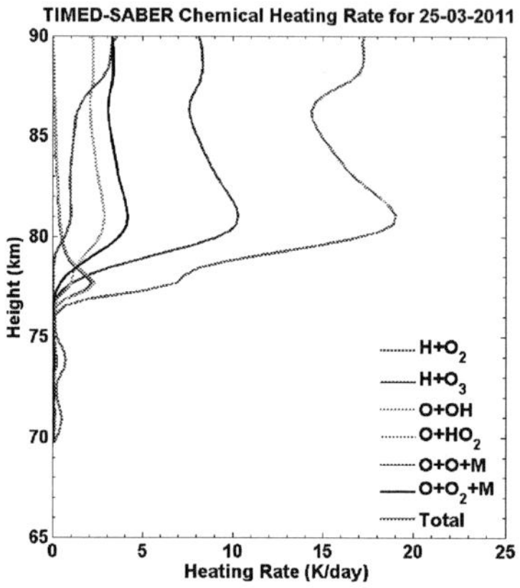

Figure 4. SABER chemical heating rates among H, O, O_2, O_3, OH, HO_2 for 25-03-2011.

5. Conclusions

In the present study, the dominant causative mechanism for the occurrence of a large MIL has been investigated mainly using the Rayleigh lidar, SABER observations for 25 March 2011 over Gadanki. It is clearly evident that the dynamics in connection with gravity wave breaking is negligible while the chemical heating due to the exothermic reactions among H, O, O_2, O_3, OH, HO_2 is more dominant for the occurrence of the MIL above ~75-80 km.

6. Acknowledgements

The authors thank the technical staff for operating and maintaining the Rayleigh lidar system at NARL, Gadanki, India. They also gratefully acknowledge the NASA's TIMED-SABER team for providing the data used in this study through the website: http://saber.gats-inc.com/.

7. References

[1] J.W. Meriwether, A. J. Gerrard, "Mesosphere inversion layers and stratosphere temperature enhancements", *Rev. of Geophys.*, 42, RG3003 http://dx.doi.org/10.1029/2003RG000133 (2004).

[2] W.J. Meriwether, C.S. Gardner, "A review of the mesospheric inversion layer phenomenon", *J. Geophys. Res.*, 105, 12405–12416 (2000).

[3] A. Hauchecorne, M.L. Chanin, "Density and temperature profiles obtained by lidar between 35 and 70 km", *Geophys. Res. Lett.*, 7, 565–568 (1980).

[4] K. Ramesh, S. Sridharan, S. Vijaya Bhaskara Rao, "Dominance of chemical heating over dynamics in causing a few large mesospheric inversion layer events during January-February 2011", *J. Geophys. Res. Space Physics*, 118, 6751–6765, doi:10.1002/jgra.50601 (2013).

AUTHOR INDEX

AUTHOR INDEX

AUTHOR INDEX

Cooperative Institute for Research in Environmental Sciences (CIRES)
University of Colorado Boulder
216 UCB
Boulder, CO 80309
USA

ISBN 978-1-5108-7033-8

CURRAN ASSOCIATES INC.
proceedings.com